STATISTICAL SIGNAL PROCESSING

STATISTICS: Textbooks and Monographs

A SERIES EDITED BY

D. B. OWEN, Coordinating Editor
Department of Statistics
Southern Methodist University
Dallas, Texas

Volume 1: The Generalized Jackknife Statistic, *H. L. Gray and W. R. Schucany*
Volume 2: Multivariate Analysis, *Anant M. Kshirsagar*
Volume 3: Statistics and Society, *Walter T. Federer*
Volume 4: Multivariate Analysis: A Selected and Abstracted Bibliography, 1957-1972, *Kocherlakota Subrahmaniam and Kathleen Subrahmaniam* (out of print)
Volume 5: Design of Experiments: A Realistic Approach, *Virgil L. Anderson and Robert A. McLean*
Volume 6: Statistical and Mathematical Aspects of Pollution Problems, *John W. Pratt*
Volume 7: Introduction to Probability and Statistics (in two parts) Part I: Probability; Part II: Statistics, *Narayan C. Giri*
Volume 8: Statistical Theory of the Analysis of Experimental Designs, *J. Ogawa*
Volume 9: Statistical Techniques in Simulation (in two parts), *Jack P. C. Kleijnen*
Volume 10: Data Quality Control and Editing, *Joseph I. Naus*
Volume 11: Cost of Living Index Numbers: Practice, Precision, and Theory, *Kali S. Banerjee*
Volume 12: Weighing Designs: For Chemistry, Medicine, Economics, Operations Research, Statistics, *Kali S. Banerjee*
Volume 13: The Search for Oil: Some Statistical Methods and Techniques, *edited by D. B. Owen*
Volume 14: Sample Size Choice: Charts for Experiments with Linear Models, *Robert E. Odeh and Martin Fox*
Volume 15: Statistical Methods for Engineers and Scientists, *Robert M. Bethea, Benjamin S. Duran, and Thomas L. Boullion*
Volume 16: Statistical Quality Control Methods, *Irving W. Burr*
Volume 17: On the History of Statistics and Probability, *edited by D. B. Owen*
Volume 18: Econometrics, *Peter Schmidt*
Volume 19: Sufficient Statistics: Selected Contributions, *Vasant S. Huzurbazar (edited by Anant M. Kshirsagar)*
Volume 20: Handbook of Statistical Distributions, *Jagdish K. Patel, C. H. Kapadia, and D. B. Owen*
Volume 21: Case Studies in Sample Design, *A. C. Rosander*

Vol. 22: Pocket Book of Statistical Tables, *compiled by R. E. Odeh, D. B. Owen, Z. W. Birnbaum, and L. Fisher*
Vol. 23: The Information in Contingency Tables, *D. V. Gokhale and Solomon Kullback*
Vol. 24: Statistical Analysis of Reliability and Life-Testing Models: Theory and Methods, *Lee J. Bain*
Vol. 25: Elementary Statistical Quality Control, *Irving W. Burr*
Vol. 26: An Introduction to Probability and Statistics Using BASIC, *Richard A. Groeneveld*
Vol. 27: Basic Applied Statistics, *B. L. Raktoe and J. J. Hubert*
Vol. 28: A Primer in Probability, *Kathleen Subrahmaniam*
Vol. 29: Random Processes: A First Look, *R. Syski*
Vol. 30: Regression Methods: A Tool for Data Analysis, *Rudolf J. Freund and Paul D. Minton*
Vol. 31: Randomization Tests, *Eugene S. Edgington*
Vol. 32: Tables for Normal Tolerance Limits, Sampling Plans, and Screening, *Robert E. Odeh and D. B. Owen*
Vol. 33: Statistical Computing, *William J. Kennedy, Jr. and James E. Gentle*
Vol. 34: Regression Analysis and Its Application: A Data-Oriented Approach, *Richard F. Gunst and Robert L. Mason*
Vol. 35: Scientific Strategies to Save Your Life, *I. D. J. Bross*
Vol. 36: Statistics in the Pharmaceutical Industry, *edited by C. Ralph Buncher and Jia-Yeong Tsay*
Vol. 37: Sampling from a Finite Population, *J. Hájek*
Vol. 38: Statistical Modeling Techniques, *S. S. Shapiro*
Vol. 39: Statistical Theory and Inference in Research, *T. A. Bancroft and C.-P. Han*
Vol. 40: Handbook of the Normal Distribution, *Jagdish K. Patel and Campbell B. Read*
Vol. 41: Recent Advances in Regression Methods, *Hrishikesh D. Vinod and Aman Ullah*
Vol. 42: Acceptance Sampling in Quality Control, *Edward G. Schilling*
Vol. 43: The Randomized Clinical Trial and Therapeutic Decisions, *edited by Niels Tygstrup, John M. Lachin, and Erik Juhl*
Vol. 44: Regression Analysis of Survival Data in Cancer Chemotherapy, *Walter H. Carter, Jr., Galen L. Wampler, and Donald M. Stablein*
Vol. 45: A Course in Linear Models, *Anant M. Kshirsagar*
Vol. 46: Clinical Trials: Issues and Approaches, *edited by Stanley H. Shapiro and Thomas H. Louis*
Vol. 47: Statistical Analysis of DNA Sequence Data, *edited by B. S. Weir*
Vol. 48: Nonlinear Regression Modelling: A Unified Practical Approach, *David A. Ratkowsky*
Vol. 49: Attribute Sampling Plans, Tables of Tests and Confidence Limits for Proportions, *Robert E. Odeh and D. B. Owen*
Vol. 50: Experimental Design, Statistical Models, and Genetic Statistics, *edited by Klaus Hinkelmann*
Vol. 51: Statistical Methods for Cancer Studies, *edited by Richard G. Cornell*
Vol. 52: Practical Statistical Sampling for Auditors, *Arthur J. Wilburn*
Vol. 53: Statistical Signal Processing, *edited by Edward J. Wegman and James G. Smith*

OTHER VOLUMES IN PREPARATION

STATISTICAL SIGNAL PROCESSING

Edited by

Edward J. Wegman
Office of Naval Reserve
Arlington, Virginia

James G. Smith
Naval Air Systems Command
Washington, D.C.

MARCEL DEKKER, INC.   New York and Basel

Statistical signal processing

(Statistics, textbooks and monographs; v. 53)
Includes indexes.
1. Signal processing—Statistical methods. I. Wegman,
Edward J., 1943-   . II. Smith, James G., 1933-
III. Series.
TK5102. 5S6925 1984 621.38'043 83-26255
ISBN 0-8247-7159-1

COPYRIGHT © 1984 by MARCEL DEKKER, INC. ALL RIGHTS RESERVED

Neither this book nor any part may be reproduced or transmitted in
any form or by any means, electronic or mechanical, including
photocopying, microfilming, and recording, or by any information
storage and retrieval system, without permission in writing from the
publisher.

MARCEL DEKKER, INC.
270 Madison Avenue, New York, New York 10016

Current printing (last digit) :
10 9 8 7 6 5 4 3 2 1

PRINTED IN THE UNITED STATES OF AMERICA

To Christopher Andrew Wegman

Preface

Analog and digital signal processing are the cornerstones of much of modern technology, particularly of the computer revolution that is swelling up around us. The relation of the computer revolution to signal processing is a dual relationship. On the one hand, the wide availability of computers implies an ever greater capability to generate and transmit information in an ever more cluttered environment. This inevitably leads to noisier communications and a requirement to process signals which are at best stochastically masked by noise and which, in fact, may be themselves stochastic in character. On the other hand, the wide availability of computing power implies that sophisticated mathematical and statistical techniques may be brought to bear on the digital processing of signals rather than only processing by the relatively simple electronic circuits of the past. The development of mathematical and statistical tools for signal processing is inextricably bound up with the availability of computational power to implement these tools. This book and the workshop from which it arose are intended to span the intellectual range of statistical signal processing methodology from the somewhat esoteric realm of abstract stochastic process theory to the exciting, fast-paced world of VLSI computing architectures for implementing these algorithms.

The book is conceptually divided into six parts roughly ordered from abstract research to applications. The first section deals with Time Series Analysis and Stochastic Processes. This section is intended to lay the foundations of statistical inference about stochastic processes, the generic archetype of signals we consider later. Emphasis in this area is on the nonstandard (i.e., non-Gaussian) finite-dimensional distributional structures that characterize so many of the present signal processing environments.

The second section focuses on Signal Estimation and Detection. Signal detection particularly is an extremely challenging statistical problem. In the stochastic process setting it is several orders of magnitude more difficult than the iid cases usually encountered in standard discussions of statistical hypothesis testing. Again, in this area, our emphasis has been on moving beyond the usual Gaussian assumptions to distributional assumptions that more realistically model actual signal and noise characteristics.

The focus of the third section shifts to Data Analysis and Modeling. The availability of computing resources has at least two profound implications on the type of data which may be collected. Data may be and are frequently taken in much larger volume and are of a higher dimensional character. This is particularly true of data such as signal data which are generated, monitored and transmitted electronically. Moreover, the volume and dimensionality of this data allow a far more incisive inference about the fine-structure of the underlying mechanism. The point to be made is that dealing with 4,000,000 observations and dealing with 200 observations are fundamentally different chores even though standard theory may say that n can equal 200 or n can equal 4,000,000. The purpose of this section is to formulate approaches to this chore as well as to document empirically some of the distribution character of signals.

The detection of a signal is usually only the first step to the useful exploitation of information. Our next section focuses on Array Processing and Target Tracking. Array processing is essentially a synonym for antenna theory or spatial processing and is based on the idea that a spatially distributed set of sensors can be used to discover directional information about signal sources. Target tracking refers to the exploitation of array processing to locate and separate distinct signal sources as they move in time. This task is obviously of prime interest to the sonar-oriented Naval community.

Our fifth section focuses on Statistical Image Processing and, of course, specializes in stochastic processes (signals) with a two-dimensional domain. Digital imaging, in fact, video in general is a topic of tremendous technical interest. The interplay of computing resources and mathematical techniques is probably tested in this area more than any other type of signal processing commonly found. Emphasis here is on innovative mathematical and statistical approaches to image processing.

Our final section deals with Architecture for Signal Processing and is potentially one of the most fascinating displays of high technology presently on the scene. Advances in physical electronics are staggering with the now available very large scale integrated-circuit (VLSI) and the possibility of one-half micron devices looming on the horizon. The sheer density of circuits on such a chip preclude the manual design of every element and therefore imply chips must contain many repetitions of relatively simple processors. Perhaps this is the area where the interplay of mathematics and computing finds its most elegant expression.

This volume is the proceedings of a Workshop on Signal Processing in the Ocean Environment held at the US Naval Academy on 11-15 May 1982. All of the work represented in this volume is work carried out with support of the Office of Naval Research (ONR) in the area of Signal Processing. The scope of this work is enormous and the logistics associated with the workshop were complex. We would particularly like to thank Dr. Douglas J. DePriest, our associate in the ONR Statistics and Probability Program, for his superb handling of local arrangements and his exceptional efforts to insure a smooth-running workshop. Dr. Tom Sanders of the Mathematics Department of the US Naval Academy was a most gracious host and we deeply appreciate his efforts on our behalf.

EDWARD J. WEGMAN
JAMES G. SMITH

Contents

Preface	v
Contributors	xi
Part I: Time Series Analysis and Stochastic Processes	1
Time Series Model Identification, Spectral Estimation, and Functional Inference Emanuel Parzen	3
A Review of Some Aspects of Robust Inference for Time Series R. Douglas Martin	19
Extremes of Nonstationary Stochastic Waveforms M.R. Leadbetter	41
Non-Gaussian Linear Processes, Phase and Deconvolution Keh-Shin Lii and Murray Rosenblatt	51
Filtering and Smoothing of Nonstationary Processes M.M. Rao	59
ARMA Time Series Modeling: A Singular Value Decomposition Approach James A. Cadzow	67
A Statistical Frequency Domain Signal Processing Method Roger F. Dwyer	79
Part II: Signal Estimation and Detection	91
Detection in a Non-Gaussian Environment Stuart C. Schwartz and John B. Thomas	93

Detection of Non-Gaussian Signals in Gaussian Noise
  C.R. Baker and A.F. Gualtierotti   *107*

Data-Adaptive Detection of a Weak, Single-Frequency, Plane-Wave
  Signal in Noise and Strong, Unidirectional Interference
  Donald W. Tufts, Ramdas Kumaresan, and Ivars Kirsteins   *117*

Optimal Detection in Linear Reverberation Noise
  Patrick L. Brockett   *133*

On Some Estimation/Detection Problems from Sampled Data
  Elias Masry   *141*

A Technique for Improving Detection and Estimation of Signals
  Contaminated by Under Ice Noise
  Roger F. Dwyer   *153*

Estimation in the Presence of Noise of a Signal Which is
  Flat Except for Jumps
  Yi-Ching Yao   *167*

Part III: Data Analysis and Modeling   *177*

Cross Validated Spline Methods for Direct and Indirect
  Sensing Experiments
  Grace Wahba   *179*

Tools for Large Data Set Analysis
  Leo Breiman and Jerome Friedman   *191*

Data Analysis and Modeling of Arctic Sea Ice Subsurface Roughness:
  A Summary
  Donald P. Gaver and Patricia A. Jacobs   *199*

Probability Density Functions of Ocean Acoustic Noise
  Processes
  Frederick W. Machell and Clark S. Penrod   *211*

Experimental and Modeled Density Estimates of Underwater
  Acoustic Returns
  Gary R. Wilson and Dennis R. Powell   *223*

Part IV: Array Processing and Target Tracking   *241*

Coherent Array Processing
  Melvin J. Hinich   *243*

On Nonlinear Filtering and Tracking
  R.R. Mohler, W.J. Kolodziej, R.S. Engelbrecht, and H.D. Brunk   *253*

Contents ix

Generalized Search Optimization
    Lawrence D. Stone   *265*

The Distribution of the Random Lighted Portion of a Curve in a
Plane Shadowed by a Poisson Random Field of Obstacles
    Shelemyahu Zacks and Micha Yadin   *273*

Some Factors Influencing Localization Accuracy
    Peter M. Schultheiss   *287*

Adaptive Range Tracking of Underwater Maneuvering Targets Using
Passive Measurements
    Richard L. Moose   *297*

Passive Sonar Delay Estimate Improvement Using *a priori* Knowledge
and Increased Number of Sensors
    R. Lynn Kirlin   *313*

Capability of Array Processing Algorithms to Resolve
Source Bearings
    Stuart R. DeGraaf and Don H. Johnson   *329*

Detection Thresholds for Multitarget Tracking in Clutter
    Thomas E. Fortmann and Yaakov Bar-Shalom   *341*

Multitarget Tracking Using Joint Probabilistic Data
Association
    Yaakov Bar-Shalom and Thomas E. Fortmann   *353*

Selection of Processing Parameters for Generating Ambiguity
Surfaces
    Joseph R. LaPointe, Jr.   *365*

Part V: Statistical Image Processing   *373*

Syntactic Approach to Signal and Image Analysis
    K.S. Fu   *375*

Application of Map Estimation Techniques to Image Segmentation
    Howard Elliot, M.F. Tenorio,
    Fred R. Hansen, and Lalita Srinivasan   *385*

A Cluster Analysis Program for Image Segmentation
    Melvin F. Janowitz   *399*

DCT Image Compression over Noisy Channels
    Jerry D. Gibson   *411*

On Segmentation of Time Series and Images in the Signal
   Detection and Remote Sensing Contexts
      Stanley L. Sclove *421*

Efficient Algorithms for Digital Processing of Remotely
   Sensed Imagery
      C.H. Chen *435*

The Sigma Filter and Its Application to Speckle Smoothing of
   Synthetic Aperture Radar Images
      Jong-Sen Lee *445*

Encoding Techniques for a Pictorial Database
   Chung-Chun Yang and Shi-Kuo Chang *461*

Part VI: Architectures for Signal Processing *471*

Impact of VLSI on Modern Signal Processing
   S.Y. Kung *473*

A VLSI SAR Processor for Ocean Surveillance
   Benjamin Friedlander *491*

Parallel Algorithms and Computational Structures for Linear
   Estimation Problems
      Gerard G.L. Meyer and Howard L. Weinert *507*

Performance of Multihops Per Bit Binary Frequency—Shift
   Keying Frequency Hopping (BFSK-FH) Spread Spectrum Syst
   in the Partial Band Jamming Environments
      J.S. Lee and Y.K. Hong *517*

Index *531*

Contributors

C. R. BAKER, Department of Statistics, University of North Carolina, Chapel Hill, North Carolina

YAAKOV BAR-SHALOM, Department of Electrical Engineering and Computer Science, University of Connecticut, Storrs, Connecticut

LEO BREIMAN, Department of Statistics, University of California, Berkeley, California

PATRICK L. BROCKETT, Department of Finance and Applied Research Laboratories, University of Texas, Austin, Texas

H.D. BRUNK, Departments of Statistics and Mathematics, Oregon State University, Corvallis, Oregon

JAMES A. CADZOW, Department of Electrical and Computer Engineering, Arizona State University, Tempe, Arizona

SHI-KUO CHANG, Information Systems Research Laboratory, University of Illinois at Chicago, Chicago, Illinois

C.H. CHEN, Department of Electrical and Computer Engineering, Southeastern Massachusetts University, North Dartmouth, Massachusetts

STUART R. DeGRAAF, Department of Electrical Engineering, Rice University, Houston, Texas

ROGER F. DWYER, Surface Ship Sonar Department, Naval Underwater Systems Center, New London, Connecticut

HOWARD ELLIOTT, Department of Electrical and Computer Engineering, University of Massachusetts, Amherst, Massachusetts

R.S. ENGELBRECHT, Department of Electrical and Computer Engineering, Oregon State University, Corvallis, Oregon

THOMAS E. FORTMANN, Automated Systems Department, Bolt, Beranek and Newman, Inc., Cambridge, Massachusetts

BENJAMIN FRIEDLANDER, Systems Control Technology, Inc., Palo Alto, California

JEROME FRIEDMAN, Stanford Linear Accelerator Center, University of California, Berkeley, California

K.S. FU, School of Electrical Engineering, Purdue University, West Lafayette, Indiana

DONALD P. GAVER, Department of Operations Research, Naval Postgraduate School, Monterey, California

JERRY D. GIBSON, Department of Electrical Engineering, Texas A&M University, College Station, Texas

A.F. GUALTIEROTTI,* Department of Mathematics, Federal Institute of Technology (EPFL), Lausanne, Switzerland

FRED R. HANSEN, System Studies Department, Sandia National Laboratories, Livermore, California

MELVIN J. HINICH, Departments of Government and Economics, University of Texas, Austin, Texas

Y.K. HONG, J.S. Lee Associates, Inc., Arlington, Virginia

PATRICIA JACOBS, Department of Operations Research, Naval Postgraduate School, Monterey, California

MELVIN F. JANOWITZ, Department of Mathematics and Statistics, University of Massachusetts, Amherst, Massachusetts

DON H. JOHNSON, Department of Electrical Engineering, Rice University, Houston, Texas

R. LYNN KIRLIN, Department of Electrical Engineering, University of Wyoming, Laramie, Wyoming

---

Current Affiliation:
*IDHEAD, Universite de Lausanne, Lausanne, Switzerland

IVARS KIRSTEINS, Department of Electrical Engineering, University of Rhode Island, Kingston, Rhode Island

W.J. KOLODZIEJ, Department of Electrical and Computer Engineering, Oregon State University, Corvallis, Oregon

RAMDAS KUMARESAN, Department of Electrical Engineering, University of Rhode Island, Kingston, Rhode Island

S.Y. KUNG, Department of Electrical Engineering, University of Southern California, Los Angeles, California

JOSEPH LaPOINTE, JR., Analytical Technology Applications Corporation, Mountain View, California

M.R. LEADBETTER, Department of Statistics, University of North Carolina, Chapel Hill, North Carolina

J.S. LEE, J.S. Lee Associates, Inc., Arlington, Virginia

JONG-SEN LEE, Aerospace Systems Division, U.S. Naval Research Laboratory, Washington, DC

KEH-SHIN LII, Department of Statistics, University of California, Riverside, California

FREDERICK W. MACHELL, Applied Research Laboratories, University of Texas, Austin Texas

R. DOUGLAS MARTIN, Department of Statistics, University of Washington, Seattle, Washington

ELIAS MASRY, Electrical Engineering and Computer Sciences Department, University of California, San Diego, La Jolla, California

GERARD G.L. MEYER, Electrical Engineering and Computer Science Department, The Johns Hopkins University, Baltimore, Maryland

R. R. MOHLER, Department of Electrical and Computer Engineering, Oregon State University, Corvallis, Oregon

RICHARD L. MOOSE, Department of Electrical Engineering, Virginia Polytechnic Institute and State University, Blacksburg, Virginia

EMANUEL PARZEN, Statistics Department, Texas A&M University, College Station, Texas

CLARK S. PENROD, Applied Research Laboratories, University of Texas, Austin, Texas

DENNIS R. POWELL, Applied Research Laboratories, University of Texas, Austin, Texas

M.M. RAO, Department of Mathematics, University of California, Riverside, California

MURRAY ROSENBLATT, Department of Mathematics, University of California, La Jolla, California

PETER M. SCHULTHEISS, Department of Electrical Engineering, Yale University, New Haven, Connecticut

STUART C. SCHWARTZ, Department of Electrical Engineering and Computer Science, Princeton University, Princeton, New Jersey

STANLEY L. SCLOVE, Departments of Mathematics and Quantitative Methods, University of Illinois at Chicago Circle, Chicago, Illinois

LALITA SRINIVASAN, Department of Electrical Engineering, Colorado State University, Fort Collins, Colorado

LAWRENCE D. STONE, Daniel H. Wagner Associates, Sunnyvale, California

M.F. TENORIO, Department of Electrical and Computer Engineering, University of Massachusetts, Amherst, Massachusetts

JOHN B. THOMAS, Department of Electrical Engineering and Computer Science, Princeton University, Princeton, New Jersey

DONALD W. TUFTS, Department of Electrical Engineering, University of Rhode Island, Kingston, Rhode Island

GRACE WAHBA, Department of Statistics, University of Wisconsin, Madison, Wisconsin

HOWARD L. WEINERT, Electrical Engineering and Computer Science Department, The Johns Hopkins University, Baltimore, Maryland

GARY R. WILSON, Applied Research Laboratories, University of Texas, Austin, Texas

MICHA YADIN, Faculty of Industrial Engineering and Management, Technion, Israel Institute of Technology, Technion City, Haifa, Israel

CHUNG-CHUN YANG, Systems Research Branch, Aerospace Systems Division, U.S. Naval Research Laboratory, Washington, DC

YI-CHING YAO,* Statistics Center, Massachusetts Institute of Technology, Cambridge, Massachusetts

---

Current Affiliation:
*Department of Statistics, Colorado State University, Fort Collins, Colorado

SHELEMYAHU ZACKS, Department of Mathematical Sciences, State University of New York, Binghamton, New York

STATISTICAL SIGNAL PROCESSING

Part I: Time Series Analysis and Stochastic Processes

Time Series Model Identification, Spectral Estimation, and Functional Inference*

Emanuel Parzen

Statistics Department
Texas A&M University
College Station, TX

1 FUNCTIONAL INFERENCE FORMULATION OF PARAMETER ESTIMATION

A general formulation of statistical theory as methods of analysis of statistical data assumes that a statistical question starts with a probability model for the observed data set, or sample, which is a function of a parameter to be estimated;

$$f(\text{sample}|\text{parameter})$$

represents the probability density function of the sample as a function of the parameter.

Classical statistical inference assumes that the parameter is a finite dimensional vector $\theta = (\theta_1, \ldots, \theta_k)$. Functional inference assumes that the parameter is a function, such as $f(\omega)$, $0 \leq \omega \leq 1$.

Parameter estimators are best determined by one of two general statistical principles: *Bayes theorem or minimum divergence.*

Bayes theorem assumes a prior distribution for the parameter, and computes the posterior distribution of the parameter given sample, denoted

$$f(\text{parameter}|\text{sample}).$$

A minimum divergence method introduces a function $I(f; f_\theta)$ which measures

---

*Research supported in part by the Office of Naval Research under contract ONR N00014-82-MP-2001.

the divergence between the true density f and a proposed density f_θ. If one can form a raw estimator $\tilde{f}$ of the probability density of the sample, then an estimator $\hat{\theta}$ is determined by minimizing $I(\tilde{f}; f_\theta)$ with respect to θ. We call $\hat{\theta}$ a minimum divergence estimator. When θ is a finite dimensional vector, $\hat{\theta}$ often exists; we call $f_{\hat{\theta}}$ a parametric-exact estimator of the true f. When θ is a function we have two main approaches to determine estimators $\hat{\theta}$ which we call *nonparametric penalty* and *parametric-select*.

A nonparametric penalty method determines $\hat{\theta}$ by minimizing

$$I(\tilde{f}; f_\theta) + \lambda\, J(\theta)$$

where $J(\theta)$ is a measure of the smoothness of the function (tending to ∞ as θ becomes less smooth) and λ is a penalty parameter which is chosen by the researcher to balance the fidelity measure $I(\tilde{f}; f_\theta)$, assumed to tend to 0 as θ becomes less smooth, with the smoothness measure $J(\theta)$.

A parametric-select method determines $\hat{\theta}$ by approximating θ by a vector θ_m of suitable dimension m, called the order, and setting $\hat{\theta} = \hat{\theta}_m$ where $\hat{\theta}_m$ minimizes over θ_m the divergence $I(\tilde{f}; f_{\theta_m})$. The problem of selecting the "best" order m is a problem of model identification; the problem of then estimating $\hat{\theta}_m$ can often be treated as a problem of classical statistical estimation of a finite dimensional vector.

When an estimator, denoted $\hat{\theta}$, is used as an estimator of θ, one has to take into account two kinds of errors, called respectively *bias* and *variance*. Bias is a measure of the deterministic difference between θ_m and θ, while variance is a measure of the stochastic distance between $\hat{\theta}_m$ and θ_m. As m increases bias decreases while variance increases. This is an example of the fundamental problem of empirical spectral analysis which is how to achieve an optimal balance between bias and variance. [When one uses autoregressive spectral estimation, this problem reduces to a question of determining the order m of the approximating autoregressive scheme.]

2 PARAMETER ESTIMATION AND INFORMATION THEORY

A general approach to determining divergence measures is provided by information theory. Let $f(y)$ and $g(y)$ be two probability densities on the real line, $-\infty < y < \infty$. The information divergence of index α of a (model) g from (a true density) f is defined for $\alpha = 1$ (index 1) by

$$I_1(f; g) = \int_{-\infty}^{\infty} \{-\log \frac{g(y)}{f(y)}\} f(y)\, dy$$

and for $\alpha > 0$ (but $\alpha \neq 1$) by [Renyi (1961)]

$$I_\alpha(f; g) = \frac{-1}{1-\alpha} \log \int_{-\infty}^{\infty} \{\frac{g(x)}{f(x)}\}^{1-\alpha} f(x)\, dx.$$

Information divergence of index 1 has a preferred role because it has an important decomposition

$$I_1(f; g) = H(f; g) - H(f)$$

Model Identification

defining

$$H(f; g) = \int_{-\infty}^{\infty} \{-\log g(y)\} f(y) \, dy,$$
$$H(f) = H(f; f) = \int_{-\infty}^{\infty} \{-\log f(y)\} f(y) \, dy.$$

We call $H(f; g)$ the cross-entropy of f and g, and $H(f)$ the entropy of f.

Information divergence of index 1 is usually referred to just as information divergence.

Another fundamental decomposition concerns the information $I(Y|X)$ about a continuous random variable Y in a continuous random variable X, defined by

$$I(Y|X) = I_1(f_{Y|X}; f_Y)$$
$$= E_X I_1(f_{Y|X=x}; f_Y).$$

The entropy of Y and conditional entropy of Y given X are defined by

$$H(Y) = H(f_Y)$$
$$H(Y|X) = H(f_{Y|X}) = E_X H(f_{Y|X=x}).$$

One can show that

$$I(Y|X) = H(Y) - H(Y|X).$$

To apply these concepts to the problem of identifying a probability model for a sequence of random variables $Y(1), Y(2), \ldots$, we define the information divergence between the probability densities for the infinite sequence by

$$I_1(f; g) = \lim_{T \to \infty} \frac{1}{T} E_f \left[-\log \frac{g(Y(1), \ldots, Y(T))}{f(Y(1), \ldots, Y(T))} \right].$$

3 INFORMATION DIVERGENCE OF SPECTRAL DENSITY FUNCTIONS

We next consider a time series $Y(t)$, $t = 0, \pm 1, \ldots$ which is a zero mean Gaussian stationary time series with covariance function

$$R(v) = E[Y(t) Y(t + v)],$$

and correlation function

$$\rho(v) = R(v)/R(0) = \text{Corr}\, [Y(t), Y(t + v)].$$

Despite the possible confusion with a probability density, we use f to denote the spectral density function

$$f(\omega) = \sum_{v=-\infty}^{\infty} e^{-2\pi i v \omega} \rho(v), \quad 0 \leqslant \omega \leqslant 1.$$

assuming $\sum_{v=-\infty}^{\infty} |\rho(v)| < \infty$. The frequency variable ω is usually assumed to vary in the interval $-0.5 \leqslant \omega \leqslant 0.5$. But only the interval $0 \leqslant \omega \leqslant 0.5$ has physical significance. We prefer the interval $0 \leq \omega \leq 1$ for mathematical reasons.

A theorem of Pinsker (1963) can be interpreted as providing a formula for the information divergence between the probability density of a zero mean Gaussian stationary time series (normalized by its variance) with (true) spectral density $f(\omega)$, and the probability density of a zero mean Gaussian stationary times series (normalized by its variance) with (proposed model) spectral density $f_\theta(\omega)$:

$$I_1(f(\omega); f_\theta(\omega)) = \frac{1}{2} \int_0^1 \left\{ \frac{f(\omega)}{f_\theta(\omega)} - \log \frac{f(\omega)}{f_\theta(\omega)} - 1 \right\} d\omega.$$

It has an information decomposition:

$$I_1(f(\omega); f_\theta(\omega)) = H(f(\omega); f_\theta(\omega)) - H(f(\omega)),$$

defining cross entropy

$$H(f(\omega); f_\theta(\omega)) = \frac{1}{2} \int_0^1 \left\{ \log f_\theta(\omega) + \frac{f(\omega)}{f_\theta(\omega)} \right\} d\omega$$

and entropy

$$H(f(\omega)) = \frac{1}{2} \int_0^1 \{\log f(\omega) + 1\}$$

From a time series sample $Y(t)$, $t = 1, 2, \ldots, T$ one can form a raw estimator $\tilde{f}(\omega)$ of the true spectral density $f(\omega)$ by

$$\tilde{f}(\omega) = \left| \sum_{t=1}^T Y(t) e^{-2\pi i \omega t} \right|^2 \div \sum_{t=1}^T Y^2(t).$$

The sample correlation function

$$\hat{\rho}(v) = \sum_{t=1}^{T-v} Y(t) Y(t+v) \div \sum_{t=1}^T Y^2(t)$$

is computed by (for $0 \leq v < Q - T = M$)

$$\hat{\rho}(v) = \frac{1}{Q} \sum_{k=0}^{Q-1} \exp\left(2\pi i \frac{k}{Q} v\right) \tilde{f}\left(\frac{k}{Q}\right).$$

Estimating the parameters θ [of a parametric model $f_\theta(\omega)$] by minimizing the information divergence

$$I(\tilde{f}(\omega); f_\theta(\omega))$$

or equivalently the cross-entropy

$$H(\tilde{f}(\omega); f_\theta(\omega))$$

is asymptotically equivalent to the method of maximum likelihood.

An important example of the foregoing general considerations is the autoregressive model (of order m) for a spectral density. It has parameters $\theta = (\sigma_m^2, \alpha_m(1), \ldots, \alpha_m(m))$ and is defined by

$$f_m(\omega) = \sigma_m^2 |g_m(e^{2\pi i \omega})|^{-2},$$

where
$$g_m(z) = 1 + \alpha_m(1)z + \ldots + \alpha_m(m)z^m$$
has all its roots in $|z| \geq 1$, the exterior of the unit circle in the complex z-domain. Then
$$2H(\tilde{f}; f_m) = \sigma_m^2 + \frac{1}{\sigma_m^2} \int_0^1 \tilde{f}(\omega)|g_m(e^{2\pi i\omega})|^2 d\omega$$
is minimized by $\hat{\sigma}_m^2, \hat{\alpha}_m(1), \ldots, \hat{\alpha}_m(m)$ satisfying the sample Yule-Walker equations. The autoregressive spectral estimator
$$\hat{f}_m(\omega) = \hat{\sigma}_m^2 |\hat{g}_m(e^{2\pi i\omega})|^{-2}$$
is a parametric-exact estimator when the time series $Y(t)$ obeys an autoregressive scheme of order m, and is a parametric-select estimator when the autoregressive scheme is adopted as an approximating model.

Maximum entropy characterization of AR spectra. The spectral density $f(\omega)$ that maximizes entropy $H(f(\omega))$ among all $f(\omega)$ satisfying the constraints
$$\int_0^1 e^{2\pi i\omega j} f(\omega) \, d\omega = \rho(j), \quad j = 0, \pm 1, \ldots, \pm m$$
for specified correlations $\rho(j)$ is the autoregressive spectral density $f_m(\omega)$ with coefficients determined by the Yule-Walker equations. A "one-line" proof of this fundamental fact, originally stated by Burg, is as follows: from
$$H(f; f_m) = \frac{1}{2}\{\log \sigma_m^2 + 1\}$$
it follows that
$$0 \leq I_1(f; f_m) = H(f; f_m) - H(f)$$
$$= H(f_m) - H(f)$$
and
$$H(f) \leq H(f_m).$$

The foregoing simple proof of the maximum entropy character of autoregressive spectral densities is analogous to a proof of the maximum entropy character of exponential model probability densities. [Parzen (1982)]

4 MODEL IDENTIFICATION, PREDICTION THEORY, AND MEMORY

Discussions of general statistical principles are usually concerned with the principles of parameter estimation. The more important problem of model identification does not yet receive the systematic attention and emphasis merited by its crucial importance. I believe we are in a position to describe qualitatively the types of models that are usually fitted to "Gaussian" time series that are analyzed by methods related to spectral analysis. We distinguish 4 model types which we call [Parzen (1981)]

1. No memory or white noise
2. Short memory or stationary

3. Long memory (or nonstationary)
3a. Long memory: transform to short memory
3b. Long memory: long memory plus short memory.

In the definition and identification of these models, we use the ideas of prediction theory. The information about a times series $Y(t)$ at time t in the m most recent values $Y(t-1), \ldots, Y(t-m)$ is denoted

$$I_m = I(Y|Y_{-1}, \ldots, Y_{-m})$$
$$= I(Y(t)|Y(t-1), \ldots, Y(t-m))$$

For a Gaussian stationary time series

$$I_m = -\frac{1}{2} \log \sigma_m^2$$

where

$$Y^{\mu,m}(t) = E[Y(t)|Y(t-1), \ldots, Y(t-m)]$$
$$Y^{\nu,m}(t) = Y(t) - Y^{\mu,m}(t)$$
$$\sigma_m^2 = E[|Y^{\nu,m}(t)|^2] \div E[|Y(t)|^2].$$

As m tends to ∞, I_m tends to

$$I_\infty = I(Y|Y^-) = I(Y(t)|Y(t-1), \ldots)$$
$$= -\frac{1}{2} \log \sigma_\infty^2 = -\frac{1}{2} \int_0^1 \log f(\omega) \, d\omega = -(H(f(\omega))) + \frac{1}{2}.$$

Further, if $H(f) > -\infty$, then $H(f_m) \to H(f)$ and $I(f; f_m) \to 0$.
We then define a time series $Y(t)$ to be

no memory: $\sigma_\infty^2 = 1$, $I_\infty = 0$
short memory: $0 < \sigma_\infty^2 <$, $0 < I_\infty < \infty$
long memory: $\sigma_\infty^2 = 0$, $I_\infty = \infty$.

In terms of the dynamic range of the spectral density $f(\omega)$,

$$DR(f) = \max_{0 \leq \omega \leq 1} f(\omega) \div \min_{0 \leq \omega \leq 1} f(\omega)$$

we define intuitively

no memory: $DR(f) = 1$
short memory: $1 < DR(f) < \infty$
long memory: $DR(f) = \infty$.

The models we build for a time series depend on its memory type. A model corresponds to a transformation of the time series to a white noise series. Therefore a no memory or white noise time series requires no further modeling, although one may be interested in determining such statistical characteristics as the mean, variance, and probability distribution.

5 ARMA MODEL IDENTIFICATION FOR SHORT MEMORY TIME SERIES

A short memory time series $Y(t)$ is modeled by an invertible filter which transforms it to white noise:

$$Y(t) \rightarrow \boxed{\text{innovations filter } g_\infty} \rightarrow \epsilon(t) = Y^\nu(t)$$

The infinite memory prediction errors are denoted

$$Y^\nu(t) = Y(t) - Y^\mu(t) = g_\infty(L) \, Y(t),$$

$$g_\infty(z) = 1 + \alpha_\infty(1) \, z + \ldots + \alpha_\infty(m) z^m + \ldots ,$$

$$L \, Y(t) = Y(t-1).$$

We call $Y^\nu(t)$ the innovation series and it is a white noise time series with variance $\sigma_\infty^2 R(0)$. In general a short memory time series is modelled by representing it, or approximating it, by an ARMA (p, q) scheme:

$$Y(t) + \alpha_p(1) \, Y(t-1) + \ldots + \alpha_p(p) \, Y(t-p)$$
$$= \epsilon(t) + \beta_q(1) \, \epsilon(t-1) + \ldots + \beta_q(q) \epsilon(t-q)$$

where the polynomials

$$g_p(z) = 1 + \alpha_p(1)z + \ldots + \alpha_p(p) \, z^p$$
$$h_q(z) = 1 + \beta_q(1)z + \ldots + \beta_q(q) \, z^q$$

are chosen so that all their roots in the complex z-plane are in the region $\{z : |z| > 1\}$ outside the unit circle. Then $g_p(z)$ and $h_q(z)$ are the transfer functions of invertible filters. $\epsilon(t)$ is assumed to be a white noise time series which we identify with the innovations $\epsilon(t) = Y^\nu(t)$;

$$\sigma_{p,q}^2 = E[\epsilon^2(t)] \div E[Y^2(t)]$$

is an estimator of σ_∞^2. The spectral density of an ARMA (p,q) scheme is

$$f_{p,q}(\omega) = \sigma_{p,q}^2 \frac{|h_q(e^{2\pi i \omega})|^2}{|g_p(e^{2\pi i \omega})|^2}.$$

The process of identifying ARMA (p,q) schemes which are adequate (and parsimonious) approximating models for a time series can be studied rigorously, and various at least semiautomatic methods are available which are based on order determining schemes.

The conditions under which the exact (or true) models is an AR(p) or ARMA (p,q) can be stated in terms of information numbers. Define the information about Y in X_2 conditional on X_1 by

$$I(Y|X_1; X_1, X_2) = H(f_{Y|X_1}) - H(f_{Y|X_1,X_2})$$
$$= H(Y|X_1) - H(Y|X_1, X_2).$$

Then Y is $AR(p)$ is equivalent to (where Y^- denotes the infinite past $Y_{-1}, Y_{-2}, \ldots$)

$$0 = I(Y|Y_{-1}, \ldots, Y_{-p}; Y^-) = I_\infty - I_p;$$

Y is ARMA(p,q) is equivalent to

$$0 = I(Y|Y_{-1}, \ldots, Y_{-p}, Y_{-1}^\nu, \ldots, Y_{-q}^\nu; Y^-)$$

Given a time series sample, $Y(t)$, $t = 1, 2, \ldots T$, of length T, one can calculate successively (using Fast Algorithms such as the Yule-Walker equations) estimators

$$\hat{I}_p = -\frac{1}{2} \log \hat{\sigma}_p^2, \quad p = 1, 2, \ldots,$$

which can be regarded as test statistics for testing white noise, or more precisely AR(0) against AR(p). The work of Akaike (1974, 1977) and Hanan and Quinn (1949) leads one to conjecture that a universal test for white noise (whose theory needs further study) is of the form (for a suitable choice of constant $c \geq 0$, say $c = 1$)

$$-\frac{1}{2} \log \hat{\sigma}_p^2 - \frac{p}{T} \log \log T \leq \frac{c}{T} \text{ for } p = 1, 2, \ldots.$$

A related conjecture is that optimal orders p of approximating autoregressive schemes can be identified by first determining the orders at which are attained the absolute and relative minima of order determining criteria which determine orders p for which $\hat{I}_\infty - \hat{I}_p$ is not significantly different from zero.

Akaike's order determining criterion AIC is defined by

$$\text{AIC}(m) = \log \hat{\sigma}_m^2 + \frac{2m}{T};$$

it seeks to determine the order of an exact autoregressive scheme which the time series is assumed to obey. One can raise the objection against it that it does not consistently estimate the order, which is done by a criterion due to Hannan and Quinn (1979):

$$\text{AICHQ}(m) = \log \hat{\sigma}_m^2 + \frac{m}{T} \log \log T$$

Parzen (1974), (1977) introduced an approximating autoregressive order criterion called CAT (criterion autoregressive transfer function), defined by

$$\text{CAT}(m) = \frac{1}{T} \sum_{j=1}^{m} \left(1 - \frac{j}{T}\right) \hat{\sigma}_j^{-2} - \left(1 - \frac{m}{T}\right) \hat{\sigma}_m^{-2}.$$

One chooses the value of CAT(0), such as

$$\text{CAT}(0) = -\left(1 + \frac{1}{T}\right),$$

in order to accept the hypothesis of white noise when it is true a specified percentage (say 90%) of the time. In practice CAT and AIC lead in many examples to exactly the same orders. It appears reassuring that quite different conceptual foundations can lead to similar conclusions in practice.

Model Identification

An important application of fitting an approximating autoregressive scheme $AR(\hat{p})$ to a time series is the estimation of information numbers which are used to determine the goodness of fit of $ARMA(p,q)$ schemes. It should be emphasized that the ultimate decision on the adequacy of a model should be based on a definition of "parsimony of a model" which requires that the spectral distribution function of the residuals $(Y|\text{variables in model})^\nu(t)$ be "parsimoniously" not significantly different from white noise.

The spectral density of the memory p prediction errors $Y^{\nu,p}(\cdot)$ can be expressed in terms of $g_p(z)$, the autoregressive transfer function of order p, by

$$f_{Y^\nu,p}(\omega) = \frac{1}{\sigma_p^2} |g_p(e^{2\pi i\omega})|^2 f(\omega).$$

If the time series $Y(\cdot)$ is in fact $AR(p)$, then its spectral density equals the *approximating autoregressive spectral density* $f_p(\omega) = \sigma_p^2 |g_p(e^{2\pi i\omega})|^{-2}$
A time series $Y(\cdot)$ can be regarded as approximated by an $AR(p)$ if

$$\bar{f}_p(\omega) = \frac{f(\omega)}{f_p(\omega)}$$

can be regarded as not "nonsignificantly" different from a constant. In this way a test of the hypothesis that a time series $Y(\cdot)$ is $AR(p)$ can be converted to a test of the hypothesis that the prediction error time series is white noise.

Extremely useful diagnostics concerning model identification and fit are provided by spectral distribution functions

$$F(\omega) = 2 \int_0^\omega f(\omega') \, d\omega', \quad 0 \leq \omega \leq 0.5$$

6 MODEL IDENTIFICATION OF LONG MEMORY TIME SERIES

A time series is diagnosed as being long memory when it is not no memory or short memory. There are two important models for a long memory time series $Y(t)$.

(a) *Transformable to a short memory time series by a non-invertible filter.*

$Y(t)$ —[Noninvertible Filter]→ $\tilde{Y}(t)$ —[Innovations Filter]→ $\tilde{Y}^\nu(t) = Y^\nu(t)$

Long Memory   Short Memory   No Memory

The noninvertible filter is chosen to be a difference operator of the form

$$\tilde{Y}(t) = G(L) Y(t) = Y(t) + A(1) Y(t-1) + \ldots + A(M) Y(t-M).$$

Then $\tilde{Y}(t)$ has sample spectral density

$$\tilde{f}_{\tilde{Y}}(\omega) = |G(e^{2\pi i\omega})|^2 \tilde{f}_Y(\omega).$$

Finding the noninvertible filter can be regarded as an additive decomposition of the log spectral densities: choose G so that

$$\log \tilde{f}_{\tilde{Y}}(\omega) = \log \tilde{f}_Y(\omega) + \log |G(e^{2\pi i\omega})|^2$$

looks like the sample spectral density of a short memory time series. ARARMA schemes [Parzen (1982)] choose G by best lag nonstationary autoregression.

(b) *Representable as the sum $Y(t) = S(t) + N(t)$ of a long memory signal plus a short memory noise.*

$$Y(t) \begin{cases} S(t) \text{ long memory} \\ N(t) \text{ short memory} \end{cases}$$

A usual approach to finding $S(t)$ is to model it as a sum of pure harmonics:

$$S(t) = \sum_{j=1}^{k} \{A_j \cos 2\pi\omega_j t + B_j \sin 2\pi\omega_j t\}.$$

It is difficult to identify how many terms, and what frequencies ω_j, to include in $S(t)$. A new approach could be based on regarding this model as an additive decomposition of the sample spectral density:

$$\tilde{f}_Y(\omega) = \tilde{f}_S(\omega) + \tilde{f}_N(\omega).$$

One defines $S(t)$ by first forming $\tilde{f}_S(\omega)$ which is chosen so that

$$\tilde{f}_N(\omega) = \tilde{f}_Y(\omega) - \tilde{f}_S(\omega)$$

looks like the sample spectral density of a short memory time series. One determines a threshold value C which is to be subtracted from $\tilde{f}_Y(\omega)$ to form $\tilde{f}_S(\omega)$ by

$$\tilde{f}_S(\omega) = \{\tilde{f}_Y(\omega) - C\}_+ .$$

One determines the threshold value by treating $\tilde{f}(k/Q)$ as a data batch to be studied by nonparametric data modeling methods using quantile functions [see Parzen (1979)].

7 THE ARRAY OF SPECTRAL ESTIMATORS

Given a time series sample $\{Y(t), t = 1, 2, \ldots, T\}$ a bewildering array of estimated spectral densities $f(\omega)$ can be formed. [Beamish and Priestley (1981), Priestley (1981).]

(a) *Preprocessing.* To analyze a time series sample $Y(t)$, $t = 1, \ldots, T$, one will proceed in stages which often involve the subtraction of or elimination of strong effects in order to see more clearly weaker patterns in the time series structure.

The aim of preprocessing is to transform $Y(\cdot)$ to a new time series $\tilde{Y}(\cdot)$ which is short memory. Some basic preprocessing operations are memoryless transformation (such as square root and logarithm), detrending, "high pass" filtering, and differencing. One usually subtracts out the sample mean $\bar{Y} = \frac{1}{T}\sum_{t=1}^{T} Y(t)$; then the time series actually processed is $Y(t) - \bar{Y}$. If the mean $\bar{Y}$ is a large number, it should be subtracted; the variations in $Y(t)$ are then the variations of $Y(t)$ about its mean. The sample mean $\bar{Y}$ and sample variance $\hat{R}(0)$ should always be recorded.

Model Identification

(b) *Sample Fourier Transform by Data Windowing, extending with Zeroes, and Fast Fourier Transform.* Let $Y(t)$ denote a preprocessed time series. The first step in the analysis could be to compute successive autoregressive schemes using operations only in the time domain. [Davis, Newton, Pagano (1982)]. An alternative first step is the computation of the sample Fourier transform

$$\tilde{\psi}(\omega) = \sum_{t=1}^{T} Y(t) \exp(-2\pi i \omega t)$$

at an equispaced grid of frequencies in $0 \leqslant \omega \leqslant 1$, of the form $\omega = \dfrac{k}{Q}$, $k = 0, \ldots, Q-1$. We call Q the spectral computation number. One should chose $Q \geqslant T$, and we recommend $Q \geqslant 2T$.

Prior to computing $\psi(\omega)$, one should extend the length of the time series by adding zeroes to it. Then $\tilde{\psi}(\omega)$, $\omega = \dfrac{k}{Q}$, can be computed using the Fast Fourier transform.

If the time series may be long memory one should compute in addition a sample "data windowed" Fourier transform

$$\tilde{\psi}_W(\omega) = \sum_{t=1}^{T} Y(t) W\left(\frac{t}{T}\right) \exp(-2\pi i \omega t).$$

To understand the effect of the window, one replaces $Y(t)$ by a spectral representation $Y(t) = \int_0^1 \exp(2\pi i \lambda t) \, d\psi(\lambda)$; then

$$\tilde{\psi}_W(\omega) = \int_0^1 w_T(\omega - \lambda) \, d\psi(\lambda) \text{ where } w_T(\lambda) = \sum_{t=1}^{T} W\left(\frac{t}{T}\right) \exp(-2\pi i \lambda t).$$

Considerations involved in the choice of data windows are discussed in Harris (1978).

(c) *Sample Spectral Density.* The sample spectral density $\tilde{f}(\omega)$ is obtained essentially by squaring and normalizing the sample Fourier transform;

$$\tilde{f}(\omega) = \frac{|\tilde{\psi}(\omega)|^2}{\dfrac{1}{Q}\sum_{k=0}^{Q-1} \left|\tilde{\psi}\left(\dfrac{k}{Q}\right)\right|^2}, \quad \omega = \frac{k}{Q}, \, k = 0, 1, \ldots, Q-1.$$

(d) *Nonparametric kernel spectral density estimator.* An estimator $\hat{f}(\omega)$ of the spectral density is called: *parametric* when it corresponds to a parametric model for the time series (such as an AR or ARMA model); *nonparametric* otherwise. A general form of non-parametric estimator is the kernel estimator.

$$\hat{f}(\omega) = \sum_{v=-\infty}^{\infty} k\left(\frac{v}{M}\right) \hat{\rho}(v) \, e^{-2\pi i \omega v}, \; 0 \leqslant \omega \leqslant 1.$$

Two popular choices of kernel are the Parzen window [Parzen (1961)]

$$\begin{aligned} k(t) &= 1 - 6t^2 + 6t^3, & |t| &\leqslant 0.5, \\ &= 2(1 - |t|)^3, & 0.5 &\leqslant |t| \leqslant 1, \\ &= 0, & 1 &\leqslant |t|. \end{aligned}$$

and the spline-equivalent window [Parzen (1958), Cogburn and Davis (1974), Wahba (1980)]

$$k(t) = \frac{1}{1 + t^{2r}}$$

where $r \geqslant 2$ is usually chosen to equal 2 or 4. The problem of determining optimum truncations points M has no general solution; one approach is to choose a large value of M to obtain a preliminary smoothing of the sample spectral density.

(e) *Autoregressive spectral density estimators.* The Yule-Walker equations are solved to estimate innovation variances $\hat{\sigma}_m^2$, to which are applied order determining criteria (AIC, CAT) to determine optimal orders m and also to test for white noise. The value of $\hat{\sigma}_m^2$ and the dynamic range of the autoregressive spectral estimator $\hat{f}_m(\omega)$ are used to determine the memory type of the time series [Parzen (1982)]. One should determine a best and second best AR order.

(f) *ARMA spectral density estimations.* When a time series is classified as short memory the approximating AR scheme is used to form the MA(∞) coefficients which are used to form a subset regression procedure for determining the best fitting ARMA scheme, and the corresponding ARMA spectral density estimator.

We do not believe that spectral estimation is a nonparametric procedure to be conducted independently of model identification. The final form of spectral estimator should be based on an identification of the type (AR, MA, or ARMA) of the whitening filter of a short memory time series.

(g) *Nonstationary autoregression.* When a time series is classified as long memory, more accurate estimators of autoregressive coefficients are provided by minimizing a "forward and backward" least squares criterion

$$\sum_{t=m+1}^{T} (Y(t) + \alpha_m(1) \, Y(t-1) + \ldots \alpha_m(m) \, Y(t-m))^2$$
$$+ \, (Y(t-m) + \alpha_m(1) \, Y(t-m+1) + \ldots + \alpha_m(m) \, Y(t))^2,$$

or by Burg estimators [for references to descriptions of Burg's algorithm, see Kay and Marple (1981)].

When several harmonics are present in the data, whose frequencies are close together, least squares autoregressive coefficient estimators are more effective than Yule-Walker autoregressive coefficient estimators in providing autoregressive spectral estimators which exhibit the split peaks one would like to see in the estimated spectral density.

(h) *Spectral density estimators based on inverse correlations and cepstral correlations.* Additional insight into the peaks and troughs to be given significance in the final estimator of the spectrum of a short memory time series can be provided by forming nonparametric kernel estimators of $f^{-1}(\omega)$ and $\log f(\omega)$.

For a spectral density $f(\omega)$ obeying suitable conditions, one can define the inverse-correlation function [see Cleveland (1972), Parzen (1974), Chatfield (1979)]

$$\rho i(v) = \int_0^1 e^{2\pi i v \omega} f^{-1}(\omega) \, d\omega \div \int_0^1 f^{-1}(\omega) \, d\omega$$

and the *cepstral-correlation function* [see Wahba (1980) for an application]

$$\gamma(v) = \int_0^1 e^{2\pi i v \omega} \log f(\omega) \, d\omega.$$

It should be noted that the inverse-correlation function is nonnegative definite. However the cepstral-correlation function is not. These new types of correlation functions are introduced because they may provide more parsimonious parametrizations in the sense that they decay to 0 faster than does the correlation function. Statistical inference (from a sample) of the probability law of a time series often achieves greatest statistical efficiency by using the most parsimonious parametrizations. Thus to form estimators $\hat{f}(\omega)$ of the spectral density $f(\omega)$ from a raw estimator $\tilde{f}(\omega)$, greater precision may be attained by first forming estimators $\{f^{-1}(\omega)\}$ and $\{\log f(\omega)\} \cdot$ of the inverse of logarithm or the spectral density. Autoregressive spectral estimation may be regarded as an approach to estimating $f(\omega)$ by first estimating $f^{-1}(\omega)$ [Durrani & Arslanian (1982)].

REFERENCES

[1] Akaike, H. (1970). A fundamental relation between predictor identification and power spectrum estimation. *Ann. Inst. Statist. Math.* 22, 219-223.

[2] _____ (1974). A new look at the statistical model identification, *IEEE Trans. Autom. Contr.*, AC-19, 716-723.

[3] _____ (1977). On entropy maximization principle, *Applications of Statistics*, P.R. Krishnaiah, ed., North-Holland, Amsterdam, 27-41.

[4] Beamish, M. and M.B. Priestley, (1981). A study of autoregressive and window spectral estimation. *Appl. Statist.*, 30, 41-58.

[5] Carmichael, J.P. (1976). *The Autoregressive Method*, Ph.D. thesis, Statistical Science Division, State University of New York at Buffalo.

[6] _____ (1978). Consistency of the Autoregressive Method of Density Estimation, Technical Report, Statistical Science Division, State University of New York at Buffalo.

[7] Chatfield, C. (1979). Inverse autocorrelations, *J. R. Statistic. Soc.* A 142, 363-77.

[8] Childers, D.G. (1978). *Modern Spectrum Analysis*, New York: IEEE Press.

[9] Cleveland, W.S. (1972). The inverse autocorrelations of a time series and their applications. *Technometrics* 14, 277-93.

[10] Cogburn, R. and H.T. Davis. (1974). Periodic splines and spectral estimation. *Annals of Statistics*, 2, 1108-1126.

[11] Davis, H.T., H.J. Newton, and J. Pagano (1982). "A Toeplitz Gram-Schmidt Algorithm for Autoregressive Modeling," Technical Report N-33, Institute of Statistics, Texas A&M University.

[12] Durrani, T.S. and A.S. Arslanian (1982). Windows associated with high resolution spectral estimators. Submitted for publication.

[13] Grenander, U. (1981). *Abstract Inference* New York: Wiley.

[14] Hannan, E.J. and B.G. Quinn (1979). The determination of the order of an autoregression, *Journal of the Royal Statistical Society*, 41, 190-195.

[15] Harris, F. (1978). On the use of windows for harmonic analysis with the discrete Fourier Transform, *Proc. IEEE*, 66, 51-83.

[16] Kay, S.M. and Stanley L. Marple, Jr. (1981). Spectrum Analysis—A Modern Perspective. *Proceedings of the IEEE*, 69, 1380-1419.

[17] Kromer, R.E. (1969). Asymptotic properties of the autoregressive spectral estimator. Ph.D. Thesis. Statistics Dept., Standford University.

[18] Pagano, M. (1980). Some recent advances in autoregressive processes. *Directions in Time Series*, ed. D.R. Brillinger and G.C. Tiao, Institute of Mathematical Statistics, 280-302 (Comments by H.T. Davis).

[19] Parzen, E. (1957). "On consistent estimates of the spectrum of a stationary times series" *Ann. Math. Statist.*, 28, 329-348.

[20] _____ (1958). "On asymptotically efficient consistent estimates of the spectral density function of a stationary time series" *J. Roy. Statist. Soc., B.*, 20, 303-332.

[21] _____ (1961). "Mathematical considerations in the estimation of spectra" *Technometrics*, 3, 167-190.

[22] _____ (1964). "An approach to empirical time series analysis" *J. Res. Nat. Bur. Standards, Sec. D*, 68D, 937-951.

[23] _____ (1967). "Empirical multiple time series analysis" *Proc. of the Fifth Berkeley Symposium on Mathematical Statistics and Probability*, ed. L. Le Cam and J. Neyman, University of California Press, Vol. I, pp. 305-340.

[24] _____ (1967). "The role of spectral analysis in time series analysis" *Review of the International Statistical Institute*, 35, 125-141.

[25] _____ (1968). "Statistical spectral analysis (single channel case) in 1968)" *Proceedings of NATO Advanced Study Institute on Signal Processing*, Enchede, Netherlands.

[26] _____ (1969). "Multiple time series modeling" *Multivariate Analysis-II*, ed. P. Krishnaiah, Academic Press: New York, 389-409.

[27] _____ (1970). "Statistical inference on time series by RKHS methods, II" *Proceedings 12th Seminar Canadian Mathematical Congress*, ed. R. Pyke, Canadian Mathematical Congress, Montreal 1-37.

[28] _____ (1974). "Some Recent Advances in Time Series Modeling" *IEEE Transactions on Automatic Control*, AC-19, 723-730.

[29] _____ (1977). "Multiple Time Series: Determining the Order of Approximating Autoregressive Schemes" *Multivariate Analysis-IV*, ed. P. Krishnaiah, North Holland: Amsterdam, 283-295.

[30] _____ (1979). "Nonparametric Statistical Data Modeling" *Journal of the American Statistical Association*, (with discussion), 74, 105-131.

[31] _____ (1979). "A Density-Quantile Function Perspective on Robust Estimation" *Robustness in Statistics*, ed. R. Launer and G. Wilkinson, New York: Academic Press, 237-258.

[32] _____ (1980). "Time Series Modeling, Spectral Analysis, and Forecasting" *Directions in Time Series Analysis*, ed. D.R. Brillinger and G.C. Tiao, Institute of Mathematical Statistics.

[33] _____ (1981). Autoregressive Spectral Estimation, Log Spectral Smoothing, and Entropy," *IEEE Spectral Estimation Workshop* (Hamilton, Canada).

[34] _____ (1981). "Time Series Model Identification and Prediction Variance Horizon," *Applied Time Series Analysis II*, ed. D. Findley, New York: Academic Press, 415-447.

[35] _____ (1982). "Maximum entropy interpretation of autoregressive spectral densities," *Statistics and Probability Letters*, 1, 7-11.

[36] _____ (1983). Autoregressive Spectral Estimation, *Handbook of Statistics III*, ed. D. Brillinger and P. Krishnaiah, Amsterdam: North Holland.

[37] Pinsker, M.S. (1963). *Information and Information Stability of Random Variables and Processes*, San Francisco: Holden Day.

[38] Priestley, M.B. (1981). Spectral Analysis and Time Series, London: Academic Press. Two volumes.

[39] Renyi, A. (1970). *Probability Theory* (Appendix: Introduction to Information Theory) Amsterdam: North Holland.

[40] Wahba, Grace (1980). Automatic smoothing of the log periodogram, *Journal of the American Statistical Assn.*, 75, 122-132.

A Review of Some Aspects of Robust Inference for Time Series*

R. Douglas Martin

Department of Statistics
University of Washington
Seattle, WA

1 INTRODUCTION

The body of theoretical work on time series utilizes primarily one of two mathematically convenient fictions, namely either (i) a second-order description, or (ii) a Gaussian assumption, in which a case second-order description is a complete description. The second-order formulation is at the base of many important concepts and structures in time series, including Wold's decomposition, the spectral representation, and prediction theory. In all of these one has the convenience of utilizing Hilbert space methods (for details see the appropriate sections of the recent book by Grenander, 1981). On the other hand the Gaussian assumption allows one to utilize the parametric method of maximum likelihood for time series models, early work in this area being due to Whittle (1953, 1962). *The* nonparametric method for time series consists of estimating the *spectrum*, a second-order description in the frequency domain, by a variety of methods based on the periodogram.

Unfortunately, many time series encountered in practice are quite decidedly non-Gaussian, as many practitioners know, and correspondingly, second-order descriptions are far from adequate. Series often contain anomalies of numerous kinds, including local bumps or bursts, shifts in level, nonstationarities of various kinds, and isolated outliers. Least-squares and other Gaussian maximum-likelihood procedures are quite non-robust toward such phenomena. Here we shall be primarily concerned with methods which are geared to deal well with a not-too-large fraction of local bumps or bursts, and isolated outliers.

It cannot be stressed too strongly that: (i) second-order descriptions are woefully inadequate for representing such phenomena, and (ii) a Gaussian mar-

---
*This research was supported by the Office of Naval Research under Contract N00014-82-K-0062.

ginal distribution for a series hardly insures that potent versions of such phenomena do not exist. For a striking and graphic portrayal of these two facts see the example displayed in Figures 4 through 11 of Martin and Thomson (1982). The essence of the example in these figures is that a time series often has a moderate to large amount of low frequency energy, with corresponding sample paths having broad peaks and valleys, so that outliers and bumps can be modest to small on the scale of the process (e.g., as measured by the range of the data), while being quite large on a local scale and clearly visible to the eye.

This last observation leads us to give the following loose definition of an *outlier* in a time series. An outlier y_t is a data value which lies well outside of the central mass (say 95% of the mass) of conditional density $f(y_t | Y^{t-1})$ where the conditioning variables Y^{t-1} consist of all the past observations $Y^{t-1} = (y_1, \ldots, y_{t-1})$. This density is often called the *observation prediction density*. Since we seldom get our hands on such a conditional density, it is convenient and natural to cast the definition somewhat differently. Let $\hat{y}_t^{t-1}$ denote a "good" predictor of the y_t given the past Y^{t-1}. In particular $\hat{y}_t^{t-1}$ should have the kind of resistance/robustness properties discussed in the next section, so that this predictor is not unduly affected by outliers in Y^{t-1} (such a predictor appears in Section 8). Then y_t is an outlier if the prediction residual $r_t = y_t - \hat{y}_t^{t-1}$ has magnitude large compared with a good scale measure s_r for all of the residuals r_t, $t = 1, \ldots, n$. For example one might well take s_r to be the suitably scaled interquartile distance of the r_t. These definitions can be generalized in a more or less obvious way to cover the case of a "patch" or "bump" of outliers, $y_t, \ldots, y_{t+k}$.

The above comments should make the following point clear. One cannot hope to have a good method for dealing with outliers in time series by using only an instantaneous nonlinear transformation of the data, i.e., treatment of the form $\tilde{y}_t = g(y_t)$. True, some time series will contain outliers which are large on the scale of the process, and in those cases such a procedure may prevent the worst consequences. Note, however, that $\tilde{y}_t$ will in general still be an outlier in the sense given above, for this value is specified without regard to the neighboring values y_{t-1}, y_{t+1}, etc. of the series. More sophisticated procedures are called for and these will be discussed in Sections 5, 7 and 8. Sections 2 and 3 review robustness concepts for independent observations and for time series, respectively. Some time series outlier models are mentioned in Section 4. Some robust alternatives to least-squares and Gaussian maximum-likelihood procedures are introduced in Section 5. Section 6 comments on fully-adaptive estimates. Section 7 deals with some aspects of robustness toward dependency, both with and without outliers simultaneously present. Finally Section 8 briefly describes robust data smoother-cleaner algorithms, and gives an application to radar glint noise.

2 ROBUSTNESS CONCEPTS FOR INDEPENDENT OBSERVATIONS

The following comprise four *robustness* concepts in moderately wide use today: *(1) Resistance; (2) Efficiency Robustness; (3) Min-Max Robustness; (4) Qualitative Robustness*. These concepts have been applied mainly to situations involving only independent observations until quite recently.

Resistance, a term due to J. W. Tukey (1976), is in fact a term distinct from *robustness*. It is the data-oriented version of the probability based word robust. As such it is the basic primitive form of robustness which captures the essential goals of robust estimation, namely *large changes* in a *smallish fraction* of the data, e.g., gross outliers, should have only a *small effect* on the estimate. *Small changes* in *all* the data, e.g., rounding (or fine quantization), should have only a *small effect* on the estimate. As is well known, least-squares and other Gaussian maximum-likelihood procedures lack resistance, and hence resistant/robust procedures have been invented.

Of the three bonafide robustness terms, the notion of efficiency robustness (Tukey, 1960; Mosteller and Tukey, 1977) is the oldest and least mathematical concept, and hence the one most accessible to applied statisticians. Let $V_S(F)$ denote a variance standard of reference of data distribution F, and for the moment assume we are in one of those special situations where unbiased estimates exist. $V_S(F)$ might be the Cramer-Rao bound for either asymptotic or finite-sample cases. It would preferably be the Pitman bound in the latter case, when dealing with problems such as location and scale where the Pitman bound can by some means be evaluated (Pregibon and Tukey, 1981). Alternatively, $V_S(F)$ may be simply the variance of the best known estimate at distribution F. With $V_T(F)$ the variance of estimate T at distribution F, the efficiency of T at F is

$$EFF(T,F) = \frac{V_S(F)}{V_T(F)}. \qquad (1)$$

An *efficiency-robust estimate* T is one whose efficiency is high at the nominal distribution F_0 (often Gaussian), and also at strategically chosen alternative distributions $F_1, F_2, \ldots, F_K$ (usually heavy-tailed outlier-generating distributions). Often efficiencies, $REFF(T,T_{LS};F)$ relative to least-squares or other Gaussian maximum-likelihood estimates, are used with the variance V_{LS} or V_{GMLE} replacing V_S in (1). For problems where bias is unavoidable, and this is the case for almost all truly realistic robustness problem formulations, one will use mean-squared errors in place of variances in (1), and also compare biases as well.

Huber (1964) introduced *min-max* robust estimates in his by-now classic paper on robust estimates of location. Here the asymptotic variance $V(T,F)$ of estimator T at distribution F is the loss and the statistician wishes to minimize, over a family $\mathscr{T}$ of estimates, the maximum of $V(T,F)$ over a family $\mathscr{F}$ of distributions. Huber showed that such min-max estimates exist in the class of location M-estimates $\hat{\mu} = T$ obtained by solving

$$\min_\mu \Sigma_1^n \rho\left(\frac{y_i - \mu}{c \cdot \hat{s}}\right) \qquad (2)$$

with ρ symmetric and convex, the y_i independent and identically distributed (i.i.d.), and $y_i \sim F(\cdot - \mu)$. Here $\hat{s}$ is a robust scale estimate and c is a tuning constant adjusted to obtain high efficiency robustness. Equivalently $\hat{\mu}$ is a solution of

$$\Sigma_1^n \psi\left(\frac{y_i - \hat{\mu}}{c \cdot \hat{s}}\right) = 0 \qquad (3)$$

with *psi function* $\psi = \rho'$. We henceforth choose $\hat{s} = 1$ and absorb c into the definition of ψ for notational convenience. Huber's (1964) famous min-max solution is based on an ϵ-contaminated family with standard Gaussian central distribution, and the *saddle-point pair* (T_0, F_0) has $T_0 = \hat{\mu}_0$ obtained from (3) with $\psi = \psi_0$ given by

$$\psi_0(t) = \begin{cases} t & |t| < K \\ K \operatorname{sgn}(t) & |t| \geq K \end{cases} \quad (4)$$

with $K = K(\epsilon)$ determined by the contamination fraction ϵ. Other families yield other saddle-point ψ-functions (see for example, Huber, 1981).

Qualitative robustness was introduced by Hampel (1968, 1971), and this is a fundamental continuity property which is the probabilistic counterpart of Tukey's data-oriented term *resistance*. Let $Y_1, \ldots, Y_n$ be i.i.d. with values in R^k and common distribution F, and let $T_n = T_n(Y_1, \ldots, Y_n)$ define a sequence of estimates with values in R^p for sample sizes $n = 1, 2, \ldots$. This sequence induces the sequence of maps

$$T_n : F \to L_{T_n}(F) \quad (5)$$

where $L_{T_n}(F)$ is the *law* of T_n at F. Then T_n is said to be qualitatively robust at F (or in a neighborhood of F, or everywhere) if the sequence of maps (5) is equicontinuous at F (or in a neighborhood of F, or everywhere), using the Prohorov distance on the metric spaces where F and $L_{T_n}(F)$ are elements. The Prohorov metric incorporates the possibility of both gross outliers and rounding errors in ϵ-neighborhoods in a natural manner, and thus is extremely attractive for use in a robustness definition.

When $\{T_n\}$ is obtained from a functional $T = T(F)$ defined on a subset $\mathscr{F}_s$ of the family of all distributions by evaluation of T at the empirical distribution function (e.d.f.) F_n, $T_n = T(F_n)$, one set of sufficient conditions for $\{T_n\}$ to be robust at F is: (i) $T_n = T(F_n)$ is a continuous function on R^n for each $n = 1, 2, \ldots$, and (ii) T is continuous at F. For Huber's class of location M-estimations (3) T is defined implicitly by

$$\int \psi(y - T(F)) dF(y) = 0. \quad (6)$$

In essence robustness is achieved by choosing ψ to be bounded and monotone (in addition, uniqueness of the solution $T_0(F)$ at F is needed—see Huber, 1981).

Of the above concepts I regard resistance and qualitative robustness as fundamental, with efficiency robustness a close companion. Qualitative robustness is a principle which should be regarded on a par with other principles of statistics such as sufficiency, unbiasedness, etc. Whenever possible a statistic should be selected to have the property of qualitative robustness, all other things being relatively equal. Thus from now on the term robust, without other qualifiers, will be taken to mean qualitatively robust.

Since some rather ridiculous estimates such as $T \equiv c$, with c a constant are robust, one needs to combine the principle with some other measure, and efficiency robustness is a nautral candidate (see Beran, 1977a, 1977b, for notable efforts to obtain full efficiency and robustness simultaneously).

Min-max robustness is more or less frosting on the cake: it is nice to have, but one shouldn't lose any sleep over not obtaining it. Also one should not, as has been done in some of the recent engineering literature, take min-max robustness as the *guiding* concept, at least not without some circumspection. The main justification for concentrating on min-max robustness would be that one already has a basic continuity property in hand, but that the modulus of continuity is so bad that something like a good min-max solution would be appealing. Note, however, that one must demonstrate that the modulus of continuity is indeed bad, and this is a somewhat subjective matter.

There are two important concepts affiliated with the core ideas of robustness which are also due to Hampel. The first is the *breakdown point* (Hampel, 1968, 1971), a global (asymptotic) measure which is essentially the largest fraction of contamination which an estimator can stand without breaking down completely by virtue of being taken to the boundary of the parameter space. The second concept, the *influence curve* (Hampel, 1974) is an asymptotic infinitesimal (or local) measure which gives the effect of a vanishingly small fraction of contamination of specific value on an estimate as the sample size tends to infinity.

Influence curve considerations lead one to use psi-functions (e.g., ψ in Equation 5) that are continuous. In the sequel we take boundedness and continuity of ψ to be the essential features needed for robustness. Nonmonotone ψ can be used by computing 1-step Newton solutions to equations like (5), starting with a near-solution obtained with a monotone ψ.

Both the above concepts have finite-sample versions. Tukey's sensitivity curves or stylized sensitivity curves (see Andrews, et al., 1972), and Mallows' empirical influence curves (Mallows, 1976) are finite sample versions of the influence curve. Hodges (1967) introduced the precursor of the breakdown point, and recently Donoho (1982) has stressed the relative importance of finite-sample breakdown points.

Bounded-influence regression is an approach to regression which was stimulated by the notion that an estimator's influence curve should be bounded. This problem area has seen vigorous attention by a small group of researchers (Hampel, 1975, 1978; Mallows, 1976; Krasker and Welsch, 1982; Maronna, Bustos and Yohai, 1979). This topic deserves a brief introduction, both for its own sake, and also because the approach may be adapted for robust estimation of certain times series models. Consider the regression model

$$y_i = \mathbf{x}_i^T \beta + \epsilon_i, \; i = i, \ldots, n \quad (7)$$

where the ϵ_i are i.i.d. with common *symmetric* distribution F_ϵ, and $\beta^T = (\beta_1, \ldots, \beta_p)$. M-estimates $\hat{\beta}_M$ for regression are solutions of the estimating equation

$$\sum_{i=1}^{n} \mathbf{x}_i \psi(y_i - \mathbf{x}_i^T \hat{\beta}_M) = 0 \quad (8)$$

obtained by minimizing the regression analogue of (2). It is assumed that ψ is bounded, continuous, and monotonic.

First suppose that the x_i are known *exactly* (i.e., are observed without error) and the specification (7) with regard to the $\mathbf{x}_i^T \beta$ is *correct*. Then the only source of distributional difficulty is the ϵ_i which may contain outliers due to F_ϵ being heavy-tailed. In this formulation $\hat{\beta}_M$ is robust according to Hampel's asymptotic definition. There may, however, still be some finite sample problems caused by so-called *X-leverage points* (see Huber, 1981, Chapter 7; Belsley, Kuh and Welsch, 1980).

On the other hand, suppose the $\mathbf{x}_i$ are occasionally observed with large errors (say keypunch errors for example), and/or the specification (7) is incorrect in any one of a variety of ways (e.g., a mixture model for β with $P(\beta = \beta_0) = 1 - \gamma$ and $P(\beta = \beta_1) = \gamma$ with γ small). Then M-estimates $\hat{\beta}_M$ are not at all robust. In order to obtain regression estimates which are robust against such possibilities, it is desirable to use a bounded-influence (BI) regression estimate $\hat{\beta}$ which is the solution of an equation of the form

$$\sum_{i=1}^{n} \phi(\mathbf{x}_i, y_i - \mathbf{x}_i^T \hat{\beta}) = 0 \qquad (9)$$

where $\phi(\cdot,\cdot)$ is a bounded and continuous function on $R^p \times R^1$. This will guard against outliers/model uncertainty in both the *independent variables*, or *carriers* $\mathbf{x}_i$ and the residuals ϵ_i. It would be quite dangerous to rely on the M-estimate $\hat{\beta}_M$ if one were not quite sure about the purity of the $\mathbf{x}_i$.

The reasons for pointing out the above features of ordinary regression M-estimates and BI regression alternatives are twofold. First of all there are certain problems in communications theory (and practice) where exact knowledge of the $\mathbf{x}_i$ is virtually assured. This is the case for example where $\mathbf{x}_i^T \beta$ represents a signal of known structure, such as a constant signal (i.e., a location problem) or a sinusoidal signal with unknown amplitude (where $p = 1$), or with unknown amplitude and phase (where $p = 2$). We discuss such problems in Section 7. On the other hand when one is fitting autoregressive (AR) or autoregressive-moving-average (ARMA) models, and one has an additive outliers (AO) model, as discussed in Section 4, the *carriers* are quite definitely contaminated and observed with error. For this situation autoregression M-estimates are hopelessly bad, and some form of bounded-influence regression is called for.

Among the topics which deserve mention, but are otherwise beyond the scope of this paper, I would mention: (i) *quantitative robustness* (see Huber, 1981, Chapter 1), (ii) a *decision theoretic framework for robustness* (Millar, 1981); (iii) *asymptotically shrinking $\sqrt{n}$-neighborhood formulations* (Bickel, 1982); (iv) *finite-sample min-max results for testing and confidence intervals* (Huber, 1981, Chapter 10); (v) *Hampel's extremal problem* (see Huber, 1981, Chapter 11).

3 ROBUSTNESS CONCEPTS FOR TIME SERIES

Although the fundamental continuity idea behind robustness has a simple and immediate appeal, both the definition and the proofs of sufficient conditions are highly technical (even the need for the *equi*continuity part of the definition requires a little explanation). This is unfortunate because it makes all levels of detail quite inaccessible to the practitioner or engineer. *Resistance* is a much more palatable concept in this regard, but even this concept may require careful verification for complex estimates. Things get even more complicated when one tries to provide an adequate definition of qualitative robustness for time series problems.

On the other hand, it is quite important to have a solid theory as a cornerstone from which to build. If the theory is complex, as is now the case, then the theoretician has a responsibility to communicate the central concepts and results as clearly and simply as possible to potential users of proposed robust procedures.

Parameter Estimation

In recognition of the need for a suitable version of qualitative robustness for time series parameter estimates the following researchers have made contributions to the problem: Papantoni-Kazakos and Gray (1979), Cox (1981), Bustos (1981) and Boente, Fraiman and Yohai (1982).

An issue arising in the time series case is that of specifying the metric, and hence the topology, for the space of sample paths. There are a variety of ways to do this, as is reflected in the above references, and what is required is a reasonable balance so that the topology is neither too weak (in which case no estimates are robust) nor too strong (in which case all estimates are robust).

Papantoni-Kazakos and Gray (1979) work with the so-called $\bar{\rho}$ (rho-bar) metric. Their definition has a defect in the arbitrariness of the per-letter metric ρ_0 used to arrive at a final $\bar{\rho}$ metric. In order to deal with arbitrarily heavy-tailed processes for example, it is necessary to choose ρ_0 bounded. Cox's (1981) definition circumvents this difficulty, but only applies to estimates whose functional versions (analogous to $T(F)$ in (6)) depend on only a finite-dimensional marginal distribution for the process.

The Boente, Fraiman and Yohai (1982) work, initiated by Yohai, seems to be the most attractive. A major feature of their definition is that the metric d_γ^n they use for sample paths of length n is extremely natural and transparent:

$$d_\gamma^n = \inf\left\{\gamma: \frac{\#\{i: |y_i - y_i'| \geq \gamma\}}{n} \leq \gamma\right\} \qquad (10)$$

where $\#\{i: |y_i - y_i'| \geq \gamma\}$ is the number of coordinates in the two observed sample paths $y = (y_1, \ldots, y_n)$ and $y' = (y_1', \ldots, y_n')$ which differ by at least γ. Thus d_γ^n is the smallest γ such that the fraction of coordinates whose difference exceeds γ is no greater than γ. This is a data-based distance which allows for both rounding up to an amount γ, and a fraction γ of gross errors in a γ-neighborhood. Of course the final definition of robustness involves some additional structure, and also letting $n \to \infty$.

Consider an estimate T_n obtained by solving the estimating equation of rather general form

$$\Sigma_{i=1}^n \psi_i(y_1', \ldots, y_n'; T_n) = 0 \qquad (11)$$

where $y_1', \ldots, y_n'$ is the observed segment of a time series. The essential requirement which needs to be met to insure robustness is that the psi-functions ψ be bounded and continuous. Specific examples are given in Section 5.

Filtering and Smoothing Problems

In filtering and smoothing problems we have as many estimates, call them $\hat{x}_t$, $t = 1, \ldots, n$, as there are observed data values $y_1, \ldots, y_n$. Thus a filter or a smoother is a mapping S_n from R^n to R^n. It is not clear exactly what constitutes an appropriate definition of qualitative robustness for problems of this type. We surely want some form of continuity for the sequence of maps $\tilde{S}_n: \mu \to \mu_{S_n}(\mu)$ where μ is the measure for the stationary process y_t and $\mu_{S_n}(\mu)$ is the measure for $\hat{x}_1, \ldots, \hat{x}_n$. Consistency is not a possibility in filtering and smoothing problems, and evidently equicontinuity may not be as crucial here. However, this remains to be determined.

At the very least, we would require a resistance version of robustness for the $\hat{x}_t$, $t = 1, \ldots, n$. This amounts to requiring that the map S_n defines a bounded and continuous functional of μ_n, the measure for $y_1, \ldots, y_n$. Boundedness insures that no single y_t can spoil the $\hat{x}_t$, and continuity insures that small rounding errors cannot have a large effect. Thus we would require that $S_n = S_n(\mu_n)$ be a *weakly continuous function* on the space $\mathscr{F}_n$ of measures μ_n for $y^T = (y_1, \ldots, y_n)$ (Compare this with Huber, 1981, Chapter 1). Linear filters and smoothers lack resistance—appropriate bounded and continuous nonlinearity is required to achieve robust/resistance filters and smoothers. The smoother-cleaners of Section 8 have this property.

4 TIME SERIES MODELS FOR OUTLIERS

In some previous work I have concentrated on the robust estimation of AR and ARMA model parameters, and robust spectral density estimation, utilizing the following two distinct outlier generating models for observed time series y_t (see Martin, 1981, and Martin and Thomson, 1982, and the reference therein):

The Innovations Outliers (IO) Model

$$x_t = \mu + \Sigma_{l=0}^{\infty} h_l \epsilon_{t-1} \tag{12}$$

where the ϵ_t are i.i.d. with common distribution F which is symmetric and possibly heavy-tailed, $\Sigma h_l^2 < \infty$ and μ is the *location* parameter for x_t. Then let

$$y_t = x_t \tag{12'}$$

be perfect observations of the x_t process.

The Additive Outliers (AO) Model

$$x_t = \mu + \Sigma_{l=0}^{\infty} h_l \epsilon_{t-l} \tag{13}$$

with ϵ_t i.i.d. Gaussian, $\Sigma h_l^2 < \infty$ and

$$y_t = x_t + v_t \tag{14}$$

where $P(v_t = 0) = 1 - \gamma$ with γ small. The AR and ARMA models are special cases of the general linear processes (12) and (13).

For the AR case the IO model corresponds roughly to a finite parameter linear regression model with heavy-tailed error distribution. However, some quirks of the model exist, and they will be mentioned in the next section. The v_t in the AO model represent outliers, either in patches or in isolation, and in the AR case we have the analogue of a linear regression model with Gaussian residuals, but with errors in the variables (EV).

The AO model is a special case of a more general kind of x_i perturbation model

$$y_t = (1 - z_t)x_t + z_t w_t \qquad (15)$$

with z_t a binary series with $P(z_t = 1) = \gamma$ (see Yohai and Bustos, 1982). We shall also refer to this as an AO model, even though the term replacement model might equally well be used.

ARCH Autoregressions

Recently we have also been studying the properties of the following type of ARCH autoregressions, and associated parameter estimation problems (Nemec and Martin, 1983). Let

$$y_t = \gamma + \phi_1 y_{t-1} + \ldots + \phi_p y_{t-p} + \epsilon_t \qquad (16)$$

with ϵ_t an ARCH process as defined by Engle (1981):

$$\epsilon_t | \mathscr{E}^{t-1} \sim N(0, h(\mathscr{E}^{t-1})) \qquad (17)$$

where $\mathscr{E}^{t-1}$ is the past history of the ϵ_t. The intercept γ accounts for a non-zero mean for y_t. The ϵ_t are uncorrelated, but *not* independent. The functions h which we have concentrated on are of the same form which Engle (1981) emphasizes in the regression context:

$$h(\mathscr{E}^{t-1}) = \alpha_0 + \alpha_1 \epsilon_{t-1}^2 + \ldots + \alpha_p \epsilon_{t-p}^2. \qquad (18)$$

The parameters α_i must satisfy certain minimal constraints to insure wide-sense stationarity, and more severe constraints to insure existence of higher order moments (see Engle, 1981). The usual Gaussian autoregression is a special case of (18) obtained by $\alpha_1 = \cdots = \alpha_p = 0$, and $\alpha_0 = \sigma^2$.

The marginal density for ϵ_t is more or less heavy-tailed, depending on the values of the α_i. This statement may be inferred by checking that certain higher order moments do not exist, depending on the values of the α_i, and by empirical checks based on the (easily) simulated ARCH type ϵ_t. None-the-less, an open problem concerning the ϵ_t process itself is that of determining an analytic form for the stationary distribution of the ϵ_t, even in the simplest case where

$$h(E^{t-1}) = \alpha_0 + \alpha_1 \epsilon_{t-1}^2.$$

ARCH autoregressions are potentially much more useful than IO autoregressions mainly because their sample paths seem more realistic representations of many time series sample paths arising in practice.

Regression with Non-Gaussian AR Residuals

In Section 7 we discuss robust point estimation of β in the following model:

$$y_t = \mathbf{x}_t^T \beta + u_t \qquad (19)$$

with the very special assumptions that the x_t are known exactly, and

$$u_t = \phi_1 u_{t-1} + \ldots + \phi_p u_{t-p} + \epsilon_t \qquad (19')$$

where ϵ_t is a possibly heavy-tailed outlier producing mechanism. The ϵ_t could be i.i.d., or an ARCH process. This setup includes the special case of estimating location with non-Gaussian AR errors. Except for the location case where some work has been done (Portnoy, 1977; Wegman and Carroll, 1977), this problem has not been studied at all in the previous literature.

5 LEAST-SQUARES AND ROBUST ESTIMATES OF AUTOREGRESSIONS

Let's focus *solely on the autoregression* versions of IO and AO models, and the AR ARCH models described in the previous section. Discussion of moving average models is omitted here for the sake of brevity. A perfectly observed Gaussian autoregression is regarded as the nominal model, with IO, AO and AR ARCH models particular types of non-Gaussian deviations from this nominal model.

Consider the pth-order autoregression version of the regression M-estimate (8) for a $\mu = 0$ version of (12) and (13):

$$\Sigma_{t=p+1}^{n} z_t \psi(y_t - z_t^T \hat{\phi}_M) = 0 \qquad (20)$$

where $z_t^T = (y_{t-1}, \ldots, y_{t-p})$. This includes the least-squares estimate $\hat{\phi}_{LS}$ as a special case. Now $\hat{\phi}_{LS}$ has a rather notable property at finite variance IO models: its asymptotic covariance matrix depends only upon ϕ, and not upon the distribution of the ϵ_t (Whittle, 1962; Martin, 1982a). This was cited as a robustness property by Whittle.

However, several points are in order. First of all, unlike $\hat{\phi}_M$, $\hat{\phi}_{LS}$ lacks efficiency robustness at IO models (Martin, 1982). Secondly, $\hat{\phi}_{LS}$ is disastrously nonrobust toward AR ARCH models (Nemec and Martin, 1983). We conjecture that $\hat{\phi}_M$ is robust toward AR ARCH models, but this remains to be established. More importantly, neither $\hat{\phi}_{LS}$ or $\hat{\phi}_M$ are robust toward AO models of either the specific type (14) or the general type (15); both type of estimates suffer from severe biases as well as inflated variances (Denby and Martin, 1979).

Since AO models are included in arbitrarily small Prohorov neighborhoods of a Gaussian autoregression (see, for example, Cox, 1981) both $\hat{\phi}_{LS}$ and $\hat{\phi}_M$ lack qualitative robustness! Following the comments made in conjunction with (11), we require estimating equations whose summands are bounded and continuous functions of the data, and this is not the case with the M-estimate defined by (20). The point is that AO models give rise to errors in the z_t which can have quite potent effects.

Three classes of robust estimates have been proposed for this setup: *(i) Bounded-Influence Autoregression (BIFAR); (ii) RA-Estimates; (iii) Robust Data Cleaning followed by Least-Squares*. The first class utilizes *bounded-influence regression* type estimates, or generalized M-estimates (GM-estimates) applied to autoregressions. The two main variants are the Hampel-Krasker-Welsch version and the Mallows version (see Martin, 1981, and the references therein).

The class of RA-estimates, due to Yohai and Bustos (1982) are obtained as follows. First, one computes *robust* covariances $\tilde{\gamma}_k = \tilde{\gamma}_k(\phi)$ of lag-k residuals:

$$\tilde{\gamma}_k = \frac{1}{n}\Sigma_{t=1}^{n-k}\tilde{\psi}(r_t, r_{t+k}) \tag{21}$$

where $r_t = r_t(\hat{\phi}) = y_t - (\hat{\phi}_1 y_{t-1} + \ldots + \hat{\phi}_p y_{t-p})$ are the residuals. Then the $\tilde{\gamma}_k$ are substituted for the conventional covariance estimates $\hat{\gamma}_k$, obtained when $\tilde{\psi}(r_t, r_{t+k}) = r_t \cdot r_{t+k}$, in the usual least-squares equations expressed in terms of $\hat{\gamma}_k$ (see Yohai and Bustos, 1982, for details).

Robustness is achieved by choosing $\tilde{\psi}$ to be a bounded and continuous function on R^2. One choice for $\tilde{\psi}$ is $\tilde{\psi}(u, v) = \psi(u)\psi(v)$ for some bounded, continuous ψ on R^1. The essential idea is that the estimates yield zero values for robust lag-k correlation estimates of the residuals, for $k = 1, \ldots, p$, incorporated in a manner which results in high efficiency. Hence the name *RA-estimates* stands for (robust) residual-autocorrelation-based estimates.

The third class of estimates is obtained by iterative application of a *robust smoother-cleaner* to remove outliers, followed by application of the usual least-squares estimate (Kleiner, Martin and Thomson, 1979; Martin, Samarov and Vandaele, 1982; Martin and Thomson, 1982). The smoother-cleaner has the property that at a gross-outlier position (in the sense described in Section 1), the outlier is replaced by an interpolate based on all the other cleaned data. An algorithm for smoother-cleaners is given in Section 8. Robustness is obtained for this method by virtue of the smoother-cleaner being a bounded and continuous function of the data.

All three of the above classes of estimates may be modified to cover the case of nominally Gaussian ARMA models with varying degrees of elegance, and success yet to be fully determined.

A careful comparative study of the three approaches is not yet available. Yohai and Bustos (1982) should have good comparative results on classes (1) and (2) for AR(1) and MA(1) models in the very near future. Both BIFAR and RA estimates are consistent and highly efficient at the nominal Gaussian AR model (Fisher consistency), while being robust for well chosen psi-functions. They are typcially asymptotically normal as well, and have small biases at AO models (one might well call the latter feature *bias robustness*). I believe that the RA-estimates will be generally preferred to BIFAR estimates for at least two good reasons aside from their efficiency and bias robustness. Assuming the latter are on at least a roughly even par with BIFAR estimates, the RA-estimates are (i) quite natural for time series models, and can be applied in principle to models of considerable complexity, and (ii) they can be designed with just one efficiency tuning constant whose values are relatively easy to determine (compare this with the difficulty involved in choosing tuning constants for BIFAR estimates implied by Peters, Samarov and Welsch's (1982) discussion in the general regression context).

The method of robust data-cleaning, followed by least squares in an iterative manner, is a quite natural and attractive one. Note, however that it requires the use of a BIFAR or RA-estimate to provide a reasonably good starting point for iteration, as the overall procedure is highly nonlinear. It is even some kind of approximation to a non-Gaussian M.L.E. if an appropriate filter-smoother is used (Martin, 1981), and it fits in nicely with a robust prewhitening approach to spectral density estimation (Kleiner, Martin and Thomson, 1979; Martin and Thomson, 1982). The method has a drawback whose importance is somewhat debatable, namely the method is not Fisher consistent. This is certainly quite objectionable from a theoretical point of view, and there unfortunately seems to be no easy way to get around the problem other than through some form of adaption. This we intend to pursue in the near future. On the other hand certain calculations show that the asymptotic bias at the nominal Gaussian model will be so small as to have little practical consequence (Martin and Thomson, 1982, Section 6).

6 FULL ADAPTION VERSUS ROBUSTNESS

During the course of the workshop for which this talk was prepared, the following extemporaneous remarks were made.

Some attention was given by several speakers to density estimation and score function approximation, where the *(efficient) score function* is $\Psi = -f'/f$, f being a density for presumably i.i.d. data. Such attention is presumably motivated by a desire to use blatantly adaptive methods. This prompted recollection of Stone's (1975) Monte Carlo results presented at the end of his asymptotic treatment of adaptive, asymptotically efficient, location estimates $\hat{\mu}$. These estimates are obtained by solving

$$\Sigma_1^n \hat{\Psi}_n\left(\frac{y_i - \hat{\mu}}{\hat{s}}\right) = 0 \qquad (22)$$

where $\hat{\Psi}_n$ is an estimate of Ψ and $\hat{s}$ is a robust scale estimate. Stone used $\hat{\Psi}_n(r) = [-\hat{f}'_n(r)/\hat{f}_n(r)] \cdot d_n(r)$ where $\hat{f}_n, \hat{f}'_n$ are kernel density estimates using a Gaussian density type kernel, and $d_n(r)$ truncates $[-\hat{f}'_n(r)/\hat{f}_n(r)]$ to zero outside a symmetric interval $[-a_n, a_n]$ with $a_n \to \infty$ as $n \to \infty$.

A question frequently raised about such fully adaptive estimates is "how large must n be in order for the asymptotics to set in?" Somewhat surprisingly, n needn't be so large, as Stone's Monte Carlo for sample size $n = 40$ showed. His results give $EFF(\hat{\mu}, f) \geq .86$ for f ranging over the Gaussian, Laplace, Contaminted Normal (contamination fraction = .1, contamination variance = 9) and Cauchy distributions.

While Stone's Monte Carlo results are quite encouraging, his results need to be contrasted with the fact that: (i) comparable results are achieved with a robust location M-estimate of the type (3) using a good $\hat{s}$, an appropriate value for c, and a good redescending psi-function ψ—for example Tukey's bisquare psi-function (see Mosteller and Tukey, 1977), and (ii) such an M-estimate is computationally *much* simpler than the fully adaptive estimate (22).

It is doubtful that there are many applications where going the additional 10% or so, from around 90% to full efficiency, is worth the computational effort and complexity of the fully adaptive estimate. A counter argument is that if staying as close as possible to full efficiency is really cheap, then why not? Of course we should really check to determine at what (small) sample size full adaption becomes untenable.

7 ROBUSTNESS AND DEPENDENCY

In this section we wish to make two main points. The first is that relatively small amounts of serial correlation can seriously affect the level (or false alarm) of a test, or equivalently the error rate of a confidence interval. This is true even in the completely Gaussian case, where it is a surprisingly unadvertised fact that tests and confidence intervals are very nonrobust toward dependency. Here we use the word robust very loosely and intuitively—the definitions of qualitative robustness for time series given in Section 3 may need to be modified for this kind of problem.

The second point is made in connection with the *very special* model assumptions made in connection with equation (19)-(19'). Namely, ordinary location *M*-estimates are not adequate for estimation of location with non-Gaussian autoregressive errors, unless the dependency is quite weak. They can be quite inefficient compared with *proper M-estimates*, i.e., true M.L.E. type estimates for the actual model. Similar comments apply to problems of linear regression with non-Gaussian autoregressive errors.

We hasten to add that proper *M*-estimates for the model (19)-(19') are not at all qualitatively robust with regard to any full neighborhood of the Gaussian case of (19)-(19'). Other estimates are needed to achieve qualitative robustness.

The Student's t Confidence Interval with Dependency

Consider the usual Student's t 95% confidence interval which has error rate of 5%: $CI = (\bar{y} - t_{.025, n-1} S/\sqrt{n},\ \bar{y} + t_{.025, n-1} S/\sqrt{n})$, where $\bar{y}$ is the sample mean of $y_1, y_2, \ldots, y_n$, and S^2 is the usual sample variance estimate. Suppose that in fact the y_t are given by the special case of (19)-(19') where $\mathbf{x}_i^T \boldsymbol{\beta} = \mu$, a location parameter, and that u_t in (19') is a zero mean Gaussian AR(1) process with transition parameter ϕ. If in fact $\phi = 0$, then CI has the stated error rate of .05. However when $\phi \neq 0$, and the sample size is large the results are as follows:

ϕ	Error Rate
.25	.13
.5	.27
.7	.42
.9	.66

The results are dramatic. For $\phi = .25$ the error rate has more than doubled, and things get rapidly worse with increasing ϕ. The problem is that as $n \to \infty$

$$S^2 \to \mathrm{VAR}\, y_1 = \frac{\sigma_\epsilon^2}{1 - \phi^2} \neq \mathrm{VAR}_\infty \sqrt{n}\, \bar{y} = \frac{\sigma_\epsilon^2}{(1-\phi)^2} = S_u(0) \quad (23)$$

where VAR_∞ denotes the asymptotic variance, and $S_u(f)$ is the spectral density for the error process u_t. It should be noted that the right hand equalities hold quite generally; we needn't restrict ourselves to AR or even ARMA processes (Grenander, 1981). What we need in order to studentize $\bar{y}$ with dependency present is an estimate of $S_u(0)$, the spectral density of the error process at the origin. The same is true with regard to setting the threshold for tests.

Heidelberger and Welch (1980) have studied nonparametric methods for doing this. The author and a student have checked the behavior of autoregressive type estimates of $S_u(0)$ with Akaike's (1977) order selection rule AIC, in a casual way via Monte Carlo. This also seems to work, with the proviso that jackknifing must be done to remove the $0(n^{-1})$ bias in the autoregressive coefficient estimates if the sample size is not large enough relative to the amount of correlation (this remains to be determined with care, but for an AR(1) process, $\phi = .8$ and $n = 50$ definitely requires such bias removal).

Robust Estimation of Location

P. Huber's (1964) M-estimates $\hat{\mu}_{OM}$ of location, obtained by solving (3), were introduced in the context of independent and identically distributed observations y_t. The new subscript notation "OM" stands for *ordinary* location M-estimate, for reasons which will become obvious shortly. The behavior of $\hat{\mu}_{OM}$ when the y_t are both dependent and non-Gaussian has received relatively little attention. However, some relatively recent work includes that of Portnoy (1977) and Wegman and Carroll (1977). The main conclusions of Portnoy's work are: (i) if the y_t have only weak correlation structure then $\hat{u}_{OM}$ has high absolute efficiency for heavy-tailed distributions associated with moving-average type errors; (ii) weak dependency and heavy-tailedness seems to motivate the use of a redescending psi-function.

Unfortunately, ordinary location M-estimates cannot compete with *proper* location M-estimates with non-Gaussian ARMA model errors when the correlation structure is moderate to strong. By *proper* M-estimate we mean true maximum-likelihood type estimates appropriate for the model. These are obtained as follows.

Let y_t be given by the location model special case of (19)

$$y_t = \mu + u_t \qquad (24)$$

where the u_t are now an ARMA(p,q) generalization of (19') process

$$u_t + \phi_1 u_{t-1} + \ldots + \psi_p u_{t-p} = \epsilon_t + \theta_1 \epsilon_{t-1} + \ldots + \theta_q \epsilon_{t-q}. \qquad (24')$$

Heavy-tailed F's give rise to outliers in the ϵ_t, and hence in the u_t and y_t. This model may be written in the equivalent form

$$y_t + \phi_1 y_{t-1} + \ldots + \phi_p y_{t-p} = \gamma + \epsilon_t + \theta_1 \epsilon_{t-1} + \ldots + \theta_q \epsilon_{t-q} \qquad (25)$$

where the expression for the intercept γ is

$$\gamma = \mu(1 + \Sigma \phi_i). \qquad (26)$$

Let $\alpha = (\gamma, \phi, \theta)$ denote the true parameter vector for (24)-(24') or (25)-(26), and let α' denote an arbitrary value in the region where the process y_t is stationary and invertible. For a given α' one can generate residuals $r_t(\alpha')$ from the recursion, using appropriate initial conditions, in the usual way (see for example, Box and Jenkins, 1976). An M-estimate $\hat{\alpha}$ of α is a solution of the minimization problem

$$\min_{\alpha'} \Sigma_{t=1}^n \rho \left\{ \frac{r_t(\alpha')}{c \cdot \hat{s}} \right\}. \qquad (27)$$

For $\rho(t) = -\log f(t)$, this yields a conditional maximum likelihood estimate (conditioned on $y_1, \ldots, y_p$ and the initial conditions for the ϵ_t) which is asymptotically efficient. Consistency and asymptotic normality of "one-step" M-estimates are established in Lee and Martin (1983).

Now given the M-estimates $\hat{\alpha} = (\hat{\gamma}, \hat{\phi}, \hat{\theta})$, the relation (26) leads to the *proper* location M-estimate

$$\hat{\mu} = \frac{\hat{\gamma}}{1 + \Sigma \hat{\phi}_i}. \tag{28}$$

In the special case where $\rho(t) = -\log f(t)$ this yields the conditional M.L.E. of μ. The above estimate is the one which is really the appropriate M-estimate of μ for the model (24)-(24').

Detailed comparisons of the asymptotic and finite-sample behaviors of $\hat{\mu}_{OM}$ and $\hat{\mu}_M$ are given for AR(1) and MA(1) models by Lee and Martin (1983). It is shown that the efficiency of $\hat{\mu}_{OM}$ can be quite small relative to that of $\hat{\mu}$.

Robust Estimation of Signal Parameters

The regression model (19) contains as special cases some of the classical models of communication theory, where one is estimating signal parameters. For example, estimation of signal amplitude deals with the case $\mathbf{x}_t^T = \beta \cos 2\pi f_0 t$, while estimation of signal amplitude and phase is based on the case where $\mathbf{x}_t^T = \beta_1 \cos 2\pi f_0 t + \beta_2 \sin 2\pi f_0 t$. For these models it turns out that the ordinary least-squares estimate is asymptotically efficient when the ϵ_t in (19') are Gaussian, and even under much more general assumptions for Gaussian u_t (Grenander and Rosenblatt, 1957, Grenander, 1981).

However, when the ϵ_t are non-Gaussian and heavy-tailed, the situation is much the same as in the location problem just discussed. An alternative to least squares is required, but *ordinary* M-estimates lack efficiency robustness. One requires a *proper* M-estimate geared to the model (19)-(19'), and such estimates are unfortunately a bit more complicated than in the simple case of estimating location. One possibility for computing proper M-estimates for regression models with non-Gaussian AR errors, is via a straightforward robustification of Durbin's (1960) two-stage least-squares procedure. Details may be found in Martin (1982b).

8 ROBUST DATA SMOOTHER CLEANERS

As was mentioned in Section 5, so-called *smoother-cleaners* form a building block for robust parameter estimation. They also form a basis for robust spectrum estimation via a robust prewhitening approach. Since details are provided in the references cited in Section 5, only the briefest of descriptions and an example are provided here.

Consider the AO model (14), with x_t an AR(p) process having a state-variable representation $\mathbf{X}_t = \Phi \mathbf{X}_{t-1} + \mathbf{U}_t$, with $x_t = (\mathbf{X}_t)_1$ being the first component of the p-vector $\mathbf{X}_t$, and similarly $\epsilon_t = (\mathbf{U}_t)_1$. In the first pass the data y_t is processed in forward time with the *filter-cleaner* algorithm

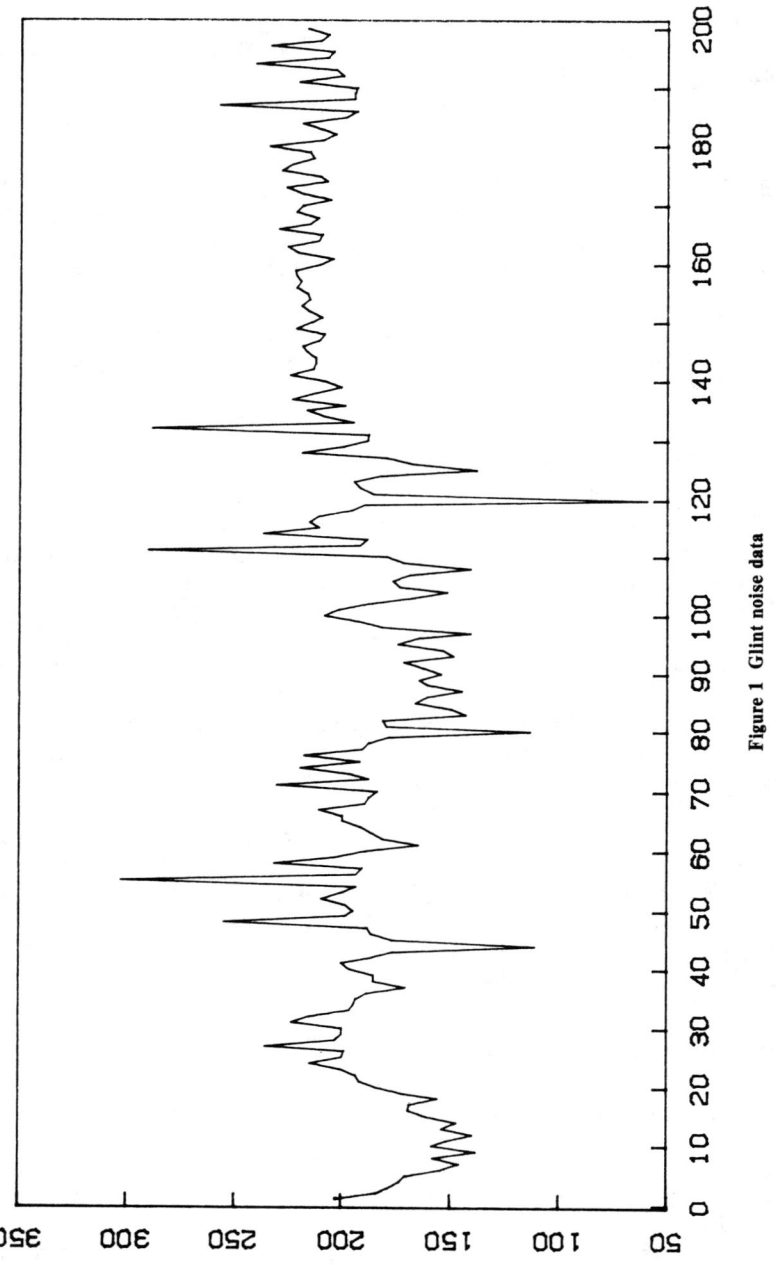

Figure 1 Glint noise data

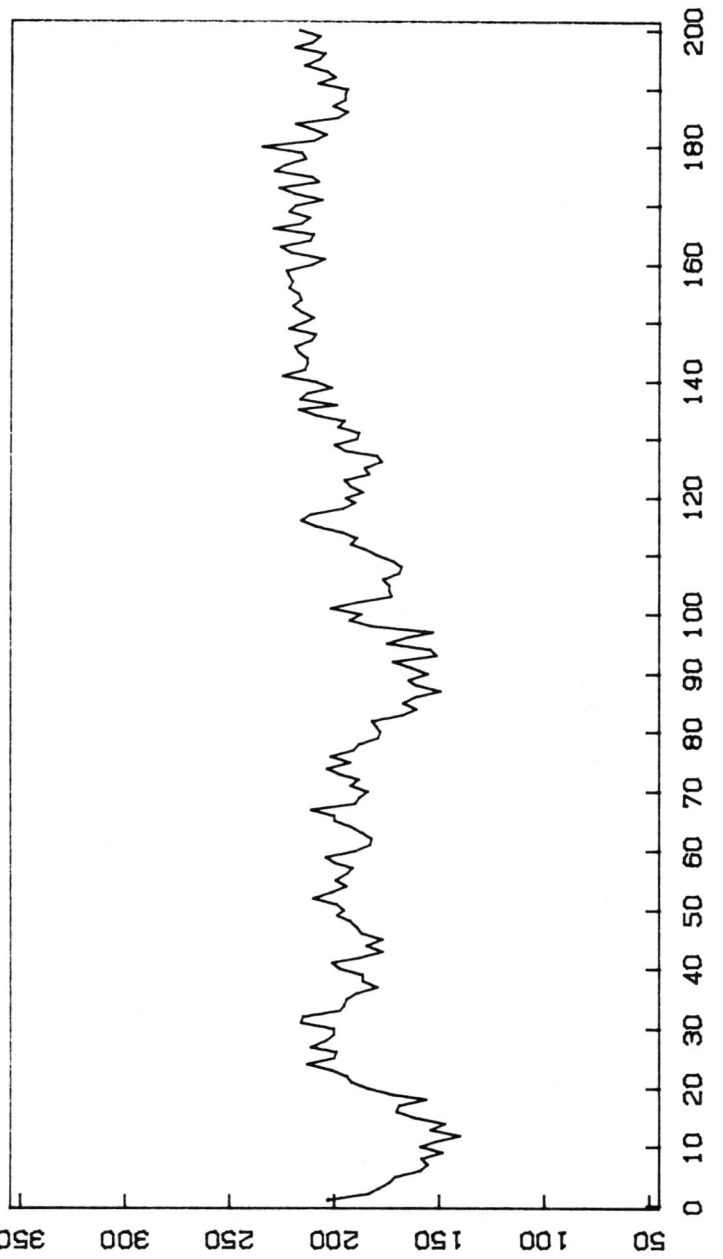

Figure 2 Smoother-clean of glint noise data

$$\hat{\mathbf{X}}_t = \Phi\hat{\mathbf{X}}_{t-1} + \mathbf{m}_t s_t \psi\left(\frac{y_t - \hat{y}_t^{t-1}}{s_t}\right) \qquad (29)$$

where

$$\hat{y}_t^{t-1} = (\Phi\hat{\mathbf{X}}_{t-1})_1 \qquad (30)$$

is a robust one-step-ahead predictor, as was mentioned in Section 1; here ψ is bounded and continuous, and the "gain" $\mathbf{m}_t$ and time-varying scale s_t are computed from auxiliary recursions. In essence (29)-(30) is a robustified Kalman filter with data-dependent gain and scale sequences.

The smoother-cleaner output is then obtained by the reverse-time pass

$$\hat{\mathbf{X}}_t^n = \hat{\mathbf{X}}_t + A_t(\hat{\mathbf{X}}_{t+1}^n - \Phi\hat{\mathbf{X}}_t), \quad t = n-1, n-2, \ldots, 1 \qquad (31)$$

with initial condition $\hat{\mathbf{X}}_n^n = \hat{\mathbf{X}}_n$. Here the $\hat{\mathbf{X}}_t$ come from (29), and the A_t are computed from quantities appearing in the auxiliary recursions for (29). This algorithm is a robustified form of the optimal linear smoothing algorithm due to Meditch (1969).

As an example of the efficacy obtainable through use of the smoother-cleaner (29)-(30), consider the *glint noise* sample path in Figure 1. This highly spiky non-Gaussian data is obtained from radar measurements of position of an aircraft target. The composite, reverberation-like nature of the radar return is the cause of the glint spikes, which result in an unnecessarily high observation-noise variance at the input of a target tracking loop. These spikes can be nicely eliminated, and the observation noise level thereby tremendously reduced by use of a smoother-cleaner, as shown in Figure 2, where a 3rd-order autoregressive approximation for the data was used. For details concerning the application of smoother-cleaners to glint noise data, see Section VII of Martin and Thomson (1982).

REFERENCES

Akaike, H. (1977), "On entropy maximization principle," *Proceedings of the Symposium on Applications of Statistics*, Wright State Univ., Dayton, OH.

Andrews, D.F., Bickel, P.J., Hampel, F.R., Huber, P.J., Rogers, W.H. and Tukey, J.W. (1972), *Robust Estimates of Location—Survey and Advances*, Princeton University Press, Princeton, NJ.

Belsley, D.D., Kuh, E. and Welsch, R.E. (1980), *Regression Diagnostics: Identifying Influential Data and Sources of Collinearity*, Wiley, New York.

Beran, R. (1977a), "Robust location estimates," *Annals Statist.*, 5, 431-444.

Beran, R. (1977b), "Minimum Hellinger distance estimates for parametric models," *Annals Statist.*, 5, 445-463.

Bickel, P.J. (1982), "Robust regression based on infinitesimal neighborhoods," Tech. Rep. No. 16, Dept. of Statistics, Univ. of California, Berkeley.

Boente, G., Fraiman, R. and Yohai, V. (1982), "Qualitative robustness for general stochastic processes," Tech. Rep. No. 26, Dept. of Statistics, Univ. of Washington, Seattle.

Box, G.E.P. and Jenkins, G.M. (1976), *Time Series Analysis: Forecasting and Control*, Holden-Day, San Francisco.

Bustos, O. (1980), "Qualitative robustness for general processes," unpublished manuscript, Instit. de Matematica Purae Aplicada, Rio de Janeiro, Brazil.

Cox, D. (1981), "Metrics on stochastic processes and qualitative robustness," Tech. Rep. 3, Dept. of Statistics, Univ. of Washington, Seattle.

Denby, L. and Martin, R.D. (1979), "Robust estimation of the first order autoregressive parameter," *Jour. Amer. Stat. Assoc.*, 74, No. 365, 140-146.

Donoho, D. (1982), "Breakdown properties of multivariate location estimators," Ph.D. Qualifying Paper, Statistics Dept., Harvard Univ., Cambridge, MA.

Durbin, J. (1960), "Estimation of parameters in time series regression models," *Jour. Royal Stat. Soc.*, 22, 139-153.

Engle, R.F. (1981), "Autoregressive conditional heteroscedasticity with estimates of the variance of United Kingdom inflation," discussion paper (second revision), Dept. of Economics, Univ. of California, San Diego.

Grenander, U. and Rosenblatt, M. (1957), *Statistical Analysis of Stationary Time Series*, Wiley, New York.

Grenander, U. (1981), *Abstract Inference*, Wiley, New York.

Hampel, F.R. (1968), "Contributions to the theory of robust estimation," Ph.D. Dissertation, Dept. of Statistics, Univ. of California, Berkeley.

Hampel, F.R. (1971), "A general qualitative definition of robustness," *Annals Math. Stat.*, 42, 1887-1895.

Hampel, F.R. (1974), "The influence curve and its role in robust estimation," *Jour. Amer. Stat. Assoc.*, 69, 383-393.

Hampel, F.R. (1975), "Beyond location parameters: robust concepts and methods," *Proceedings of I.S.I. Meeting, 40th Session*, Warsaw.

Hampel, F.R. (1978), "Optimally bounding the gross-error-sensitivity and the influence of position in factor space," *1978 Proceedings of the A.S.A. Statistical Computing Section*, Amer. Stat. Assoc., Washington, DC.

Heidelberger, P. and Welch, P.D. (1980), "A spectral method for simulation confidence interval generation and run control length," Research Report RC 8264 (#35526), IBM Thomas J. Watson Research Center, Yorktown Heights, NY.

Hodges, J.L. (1967), "Efficiency in normal samples and tolerance of extreme values for some estimates of location," *Proceedings Fifth Berkeley Symposium on Math. Statistics and Prob.*, Vol. 1, 163-186.

Huber, P. (1964), "Robust estimation of a location parameter," *Annals Math. Stat.*, 35, 73-101.

Huber, P. (1981), *Robust Statistics*, Wiley, New York.

Kleiner, B., Martin, R.D. and Thomson, D.J. (1979), "Robust estimation of power spectra" (with discussion), *Jour. Royal Stat. Soc. B*, 41, No. 3, 313-351.

Krasker, W.S. and Welsch, R.E. (1982), "Efficient bounded-influence regression estimation," *Jour. Amer. Stat. Assoc.*, 77, No. 379, 595-604.

Lee, C.H. and Martin, R.D. (1982), "M-estimates for ARMA processes," Tech. Rep. No. 23, Dept. of Statistics, Univ. of Washington, Seattle.

Lee, C.H. and Martin, R.D. (1983), "M-estimates of location for dependent data," Tech. Rep. No. 29, Dept. of Statistics, Univ. of Washington, Seattle.

Mallows, C.L. (1976), "On some topics in robustness," Bell Labs. Tech. Memo, Murray Hill, New Jersey. (Talks given at NBER Workshop on Robust Regression, Cambridge, MA, May 1973, and at ASA-IMS Regional Meeting, Rochester, NY, May 21-23, 1975.)

Maronna, R.A., Bustos, O. and Yohai, V. (1979), "Bias and efficiency robustness of general M-estimates with random carriers," *Smoothing Techniques for Curve Estimation*, (Proceedings, Heidelberg, 1979), edited by T. Gasser and M. Rosenblatt, Springer-Verlag, New York.

Martin, R.D. (1981), "Robust methods for time series," in *Applied Time Series II*, edited by D.F. Findley, Academic Press, New York.

Martin, R.D. (1982a), "The Cramer-Rao bound and robust M-estimates for autoregressions," *Biometrika*, 69.

Martin, R.D. (1982b), "Robust estimation of signal parameters with dependent data," *Proceedings 21st IEEE Conference on Decision and Control*, 433-436.

Martin, R.D. Samarov, A. and Vandaele, W. (1982), "Robust methods for ARIMA models," to appear in proceedings of ASA-CENSUS-NBER Conference on Applied Time Series Analysis of Economic Time Series, Oct. 13-15, 1981.

Martin, R.D. and Thomson, D.J. (1982), "Robust resistant spectrum estimation," *IEEE Proceedings*, Vol. 70, No. 9, 1097-1115.

Meditch, J.S. (1969), *Stochastic Optimal Linear Estimation and Control*, McGraw-Hill, New York.

Millar, P.W. (1981), "Robust estimation by minimum distance methods," *Z. Wahrschein.*, 55, 73-89.

Mosteller, F. and Tukey, J.W. (1977), *Data Analysis and Regression*, Addison-Wesley, Reading, MA.

Nemec, A. and Martin, R.D. (1983), "Some properties of some ARCH type autoregressions and least square estimates," in preparation.

Papantoni-Kazakos, P. and Gray, R.M. (1979), "Robustness of estimators on stationary observations," *Annals Probability*, 7, No. 6, 989-1002.

Portnoy, S.L. (1977), "Robust estimation in dependent situations," *Annals Statist.*, 2, 1127-1137.

Pregibon, D. and Tukey, J.W. (1981), "Assessing the behavior of robust estimates of location in small samples: introduction to configural polysampling," Tech. Rep. No. 185, Series 2, Dept. of Statistics, Princeton Univ., Princeton, NJ.

Peters, S.C., Samarov, A.M. and Welsch, R.E. (1982), "Computational procedures for bounded-influence and robust regression," MIT/CCREMS Tech. Report No. TR-30.

Stone, C.J. (1975), "Adaptive maximum-likelihood estimates of a location parameter," *Annals Statist.*, 3, 267-284.

Tukey, J.W. (1960), "A survey on sampling from contaminated distributions," in *Contributions to Probability and Statistics*, edited by I. Olkin, Stanford Univ. Press, Stanford, CA.

Tukey, J.W. (1976), "Useable resistant/robust techniques of analysis," *Proceedings of First ERDA Statistical Symposium*, Los Alamos, New Mexico, Nov. 3-5, 1975, edited by W.L. Nicholson and J.L. Harris, Battelle Pacific N.W. Labs.

Wegman, E.J. and Carroll, R.J. (1977), "A Monte Carlo study of robust estimators of location," *Comm. Statist.-Theor. Meth.*, A6, 9, 795-812.

Whittle, P. (1953), "Estimation and information in stationary time series," *Arch. Math.*, I2, 423-434.

Whittle, P. (1962), "Gaussian estimation in stationary time series," *Bull. Int. Stat.*, *39*, 105-129.

Yohai, V.J. and Bustos, O. (1982), "Robust RA-type estimates for ARMA models," Tech. Rep., Dept. of Mathematics, Ciudad Universitaria, Buenos Aires, Argentina.

Extremes of Nonstationary Stochastic Waveforms*

M. R. Leadbetter

Department of Statistics
University of North Carolina
Chapel Hill, NC

1 INTRODUCTION

The statistical properties of long term maximum values recorded by a random waveform have obvious importance in many areas of application. This is particularly apparent for direct applications such as the design of a dyke to withstand all likely floods or enforcement of an environmental standard and estimation of violation probabilities. But it is also the case in less direct situations—ranging from the work of D. G. Kendall [5] in determining whether the ancient creators of stone arrays used a definite unit of measurement, all the way to designing a signal detection system in which a large maximum value indicates the presence of a target.

By way of simple illustration consider a long term long range acoustic surveillance of a large area. The waveform being received will be random in nature. The received waveform is processed in some way and a target is declared present if the maximum of this processed waveform exceeds some threshold level. Essential parameters of such a system are

(a) the false alarm probability (i.e., the probability of "seeing a target" when none is really present) and

(b) the detection probability (i.e., that of detecting a target which is actually present).

The evaluation of (a) and (b) require a discussion of the statistical properties of the waveform maximum in a long time period. One of the features of the received waveform is that—in spite of a possibly erratic random component—it is nevertheless significantly correlated at nearby points of time.

---

*Work supported under Contract N00014-75-C-0809, Task NR 042-214 with the Office of Naval Research.

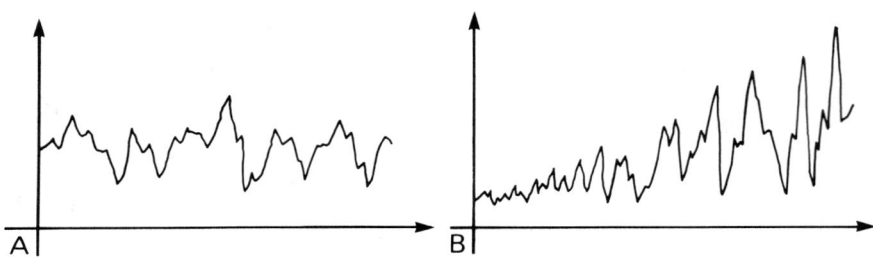

Figure 1 (a) Stationary, (b) Non-stationary waveforms

That is the waveform exhibits a "dependence structure." Another feature is that while the general characteristics (mean level, mean size of fluctuation, etc.) may stay constant for periods of time (which we call "stationarity") they may, and usually will, eventually change (leading to "nonstationary" behavior). Further the presence of a target for a period of time will itself introduce nonstationarity.

It is thus important to be able to discuss the probabilistic properties of the maximum of a waveform exhibiting both dependence structure and nonstationary. (See Figure 1.)

The classical setting for discussion of maxima of random quantities assumes a discrete time situation, stationarity and no dependence, i.e., being concerned with

$$M_n = \max(X_1, X_2 \ldots X_n) \quad (1.1)$$

where $X_1, X_2 \ldots$ are independent, identically distributed random variables. We discuss the classical results—forming so-called "Extreme value theory"—briefly in the next section. In Section 3 it will be shown how dependence may be introduced into the theory, followed by recent results for non-stationarity in Sec. 4. The discussion throughout will use the classical discrete time framework. Analogous results may be obtained in continuous time cf. [8] or one may regard a continuous waveform as having been "time-sampled," a common practice under current processing procedures. Finally a numerical illustration is provided in Section 5.

2 CLASSICAL THEORY OF EXTREMES

As noted above the classical theory is concerned with the maxmum M_n of n independent and identically distributed (i.i.d.) random variables (r. v.) $X_1 \ldots X_n$ (cf. (1.1) as n becomes large. If the common distribution function d.f. of each X_i is F, the d.f. of M_n is simply $F^n(x)$. This is an exact result, but may not be too useful when n is large, if F is not known exactly. However, it often happens that M_n has a limiting distribution under a linear normalization in the sense that

$$P\{a_n(M_n - b_n) \leq x\} \xrightarrow{d} G(x) \text{ as } n \to \infty \quad (2.1)$$

for some constants $a_n > 0$, b_n, and some nondegenerate d.f. G. This provides a useful approximation for the distribution of M_n since the class of limiting d.f.'s G which may occur is restricted to only three general "types" as the following central result of the classical theory shows.

Theorem 2.1. (Extremal Types Theorem). Let $X_1, X_2 \ldots$ be i.i.d. random variables. If $M_n = \max(X_1 \ldots X_n)$ satisfies (2.1) for some $\{a_n > 0\}, \{b_n\}$ and some nondegenerate d.f. G, then $G(x)$ is one of the following "types:"

$$\text{Type I} \quad G(x) = \exp\{-e^{-x}\} \quad -\infty < x < \infty$$

$$\text{Type II} \quad G(x) = \exp\{-x^{-\alpha}\} \quad x > 0, \ (\alpha > 0)$$

$$\text{Type III} \quad G(x) = \exp\{-(-x)^\alpha\} \quad x < 0, \ (\alpha > 0)$$

(In these replacement of x by $ax + b$ for some $a > 0$, b is permitted).

The usefulness of this result arises from the fact that it is not necessary to know the detailed form of the d.f. F of each X_i in order to determine which (if any) G occurs as a limit in (2.1). Rather only the general behavior of the tail $[1 - F(x)]$ as x becomes large needs be known. The precise conditions for making this determination are contained in the classical "domain of attraction" criteria (cf. [2], [7]). One important special case is that for which each X_i is a standard normal r.v. for which $F(x) = \Phi(x)$, the standard normal d.f. In this case (2.1) holds with the constants

$$a_n = (2 \log n)^{1/2}$$
$$b_n = (2 \log n)^{1/2} - \tfrac{1}{2}(2 \log n)^{-1/2}(\log \log n + \log 4\pi) \tag{2.2}$$

and

$$G(x) = \exp(-e^{-x}).$$

There are numerous results in the classical theory other than the central distributional Theorem 2.1. However we mention just one which is especially relevant here and is useful in obtaining results regarding domains of attraction.

Theorem 2.2. Let $X_1, X_2 \ldots$ be i.i.d. with common d.f. F and let $\{u_n\}$ be a sequence of constants. Let $0 \leqslant \tau \leqslant \infty$. Then

$$P\{M_n \leqslant u_n\} \to e^{-\tau} \text{ as } n \to \infty \tag{2.3}$$

if and only if

$$n[1 - F(u_n)] \to \tau \text{ as } n \to \infty. \tag{2.4}$$

This result is trivially proved when $0 < \tau < \infty$ and easily extended to the cases $\tau = 0$ or ∞. It might be noted that this refers to a fixed sequence u_n whereas (2.1) involves a family, having the special (linear) form $u_n = u_n(x) = x/a_n + b_n$.

3 DEPENDENT SEQUENCES

In most cases of practical interest it is clear that the independence assumptions required by the classical theory cannot be strictly correct. If temperature and other weather related quantities were statistically independent from day to day, forecasting would be impossible. Even in physical situations when observed values are independent, filtering and other processing techniques quickly introduce dependence. This happens even for the simplest classical examples—such as coin tossing or the changing of failed light bulbs—where sums of the i.i.d. random variables quickly replace the variables themselves as objects of study.

Interest in developing extremal theory for dependent situations arose as early as 1954 with a small but significant work by G.S. Watson [12] concerning so-called "m-dependent" stationary sequences and subsequent work by R. M. Loynes [9] under "strong mixing" assumptions. Since that time there has been considerable interest in the literature, in a variety of related areas involving dependent situations.

The main thrust of these results has been to give conditions under which the dependent sequences behave as if they were independent (or approximately so) from the standpoint of extremal behavior. In particular it can be shown that unless the dependence structure is "almost totally overpowering," then Theorem 2.1 still applies and, under a further restriction, so does Theorem 2.2. Further, under these conditions, the same criteria may be used for domains of attraction as in the classical case.

The specific dependence conditions used are described in [6] and [7]. The main one—used for the Extremal Types Theorem—is a very weak condition of "mixing type" involving a distributional restriction on the dependence between the two sets $(X_i \ldots X_p), (X_{p+m} \ldots X_n)$ when the separation m between the groups increases. We do not pursue this in detail here but note that for a stationary Gaussian sequence with autocorrelations $\{r_n\}$, the required conditions hold under the very weak condition of S. M. Berman [1], viz.

$$r_n \log n \to 0 \text{ as } n \to \infty. \qquad (3.1)$$

Specifically if (3.1) holds for the stationary Gaussian sequence, then so does the limit (2.1) with a a_n, b_n given by (2.2) and $G(x) = \exp\{-e^{-x}\}$.

From a practical viewpoint the principal consequence of these results is that the classical theory of extreme values may be widely used in cases where the terms are still identically distributed (in fact stationary) but where the independence assumptions no longer hold. This holds for a wide range of dependence structures. In cases of extremely high dependence other than the classical limits may occur (cf. [10]). But in a substantial number (and probably almost all) practical cases, the classical theory applies.

4 NONSTATIONARY SEQUENCES

Thus far we have been discussing random sequences which are time-correlated, but still identically distributed—and indeed stationary. This is certainly a case which is useful in practice. However it would obviously be more useful still if commonly occurring forms of nonstationarity can be included. For example one often wants to consider the addition to a stationary sequence of a deterministic component—such as a trend, seasonal or periodic effect.

The familiar time series technique of subtracting a deterministic component, analyzing the remainder, and adding it back, will not work for maxima since obviously a maximum will be more likely to occur where the added component is high. However it does prove possible to obtain results for this situation. To be definite we focus on Gaussian sequences X_1, $X_2 \ldots$ which need not be stationary. (Non-Gaussian cases may also be dealt with (cf. [4]) but involve much more complication of conditions and notation.)

It is convenient to approach the discussion of the maximum through the more general question "If u_{ni} are constants for $1 \leq i \leq n$, $n \geq 1$, under what conditions does

$$P\{\bigcap_{i=1}^{n} (X_i \leq u_{ni})\} \tag{4.1}$$

converge to a limit as $n \to \infty$? Even though the X_i are nonstationary we may simply change the u_{ni} to $v_{ni} = (u_{ni} - EX_i)/(\text{var } X_i)^{1/2}$ and thereby assume that they have zero means and unit variances. (Of course the correlation structure will still exhibit nonstationarity in general).

The problem may now be reduced to the consideration of *independent* standard normal X_i by simply obtaining conditions under which

$$P\{\bigcap_{i=1}^{n} (X_i \leq u_{ni})\} - \prod_{i=1}^{n} \Phi(u_{ni}) \to 0 \text{ as } n \to \infty. \tag{4.2}$$

If we can show this, then clearly if (4.1) has a limit for independent standard normal random variables, it has the same limit for the nonstationary X_i considered. The relevant theorem giving (4.2) is the following:

Theorem 4.1. Let the Gaussian sequence $\{X_n\}$ have zero means, unit variances, and correlations $r_{ij} = \text{corrn } (X_i, X_j)$ where $|r_{ij}| \leq \rho_{|i-j|}$ for $i \neq j$, $\{\rho_n\}$ being constants such that $\rho_n < 1$, and $\rho_n \log n \to 0$ as $n \to \infty$. Let $\{u_{ni}\}$ be constants such that

$\sum_{i=1}^{n}[1 - \Phi(u_{ni})]$ is bounded and $\lambda_n = \min_{1 \leq i \leq n} = u_{ni} \to \infty$. Then (4.2) holds. □

While we do not give the detailed proof of this result it is of interest to note that it depends on what is often called "Slepian's Lemma" which bounds the absolute value of the difference in (4.2) by

$$K \sum_{1 \leq i \leq j \leq n} |r_{ij}| \exp - \frac{1}{2}(u_{ni}^2 + u_{nj}^2)/(1 + |r_{ij}|) \tag{4.3}$$

(K being a constant). The expression (4.3) may be shown to tend to zero, though in this degree of generality it requires some delicate maneuvering, which may be achieved using a technique due to Hüsler. The derivation of Slepian's lemma appears in various forms in the literature (e.g., [7]).

Thus under the conditions stated, the quantity (4.1) has a limit if it does so under i.i.d. assumptions, i.e., if $\prod_{i=1}^{n} \Phi(u_{ni})$ converges. But as long as $\lambda_n = \min_{1 \leq i \leq n} u_{ni}$ tends to infinity it is readily shown that this latter quantity converges if and only if $\sum_{i=1}^{n} [1 - \Phi(u_{ni})]$ converges, leading at once to the main result.

Theorem 4.2. Let the Gaussian sequence $\{X_n\}$ have zero means, unit variances and correlation r_{ij} where $|r_{ij}| \leq \rho_{|i-j|}$ for $i \neq j, \rho_n^* < 1$, and $\rho_n \log n \to 0$ as $n \to \infty$. If $\{u_{ni}\}$ are such that

$$\sum_{i=1}^{n} [1 - \Phi(u_{ni})] \to \tau \text{ as } n \to \infty,$$

then

$$P\{\bigcap_{i=1}^{n} (X_i \leq u_{ni})\} \to e^{-\tau}. \qquad \square$$

This result is a very general form of the elementary Theorem 2.2. By specializing it, we may find the asymptotic distribution of the maximum of a Gaussian sequence of the type considered in the theorem together with an added deterministic component that is

$$M_n = \max(Y_1, Y_2 \ldots Y_n)$$

with $Y_i = X_i + m_i$ ($EX_i = 0$, var $X_i = 1$, cov $(X_i, X_j) = r_{ij}$) where m_i are constants. Then it may be shown that the limiting law (2.1) holds with $G(x) = \exp(-e^{-x})$ provided that the added constants m_i do no increase altogether too rapidly—specifically we require that

$$\max_{1 \leq i \leq n} m_i = o[(\log n)^{\frac{1}{2}}/\log \log n] \text{ as } n \to \infty. \qquad (4.4)$$

The normalizing constant a_n is still as in (2.2), viz. $a_n = (2 \log n)^{\frac{1}{2}}$ but the constant b_n needs modification. Specifically we have

$$P\{a_n(M_n - b_n - m_n^*) \leq x\} \to \exp(-e^{-x})$$

where a_n, b_n are as in (2.2) and m_n^* is chosen to satisfy

$$\frac{1}{n} \sum_{i=1}^{n} e^{(m_i - m_n^*)a_n - \frac{1}{2}(m_i - m_n^*)^2} \to 1 \text{ as } n \to \infty. \qquad (4.5)$$

As is clear, this result is simply obtained from Theorem 4.2 by writing

$$u_{ni} = \frac{x}{a_n} + b_n + m_n^* - m_i$$

and checking that $\sum_{i=1}^{n} [1 - \Phi(u_{ni})] \to e^{-x}$.

The relation (4.5) defining m_n^* (e.g., by putting the left hand side equal to 1 for all n) may, of course, be difficult to solve. However in some cases there is a simplification. For example if the m_i are bounded, it is possible to obtain an explicit expression for the constant m_n^* in terms of the m_i. Such a case was considered by J. Horowitz ([3]). However it has been pointed out by H. Rootzén that the constants given there are correct only in special cases—e.g. where most of the m_i are equal. (The error in [3] arose from the lack of certain required uniformity of convergence in the derivation). It has been further pointed out by H. Rootzén that for unbounded m_n not growing too fast with n, various iterations are possible to obtain m_n^*. It is thus clear that by suitable alteration of constants, a wide variety of trends and seasonal effects can be added to at least a Gaussian sequence and still preserve the double exponential limit for the asymptotic distribution of the maximum M_n.

5 AN ILLUSTRATION

The following example involves an environmental application in which a periodic seasonal effect is present in the data. There are, of course, obvious analogies with a signal processing situation in which a periodic signal is buried in noise. This data (taken from Roberts [1979]) given in Table 1 below consists of 19 years of maximum 1-hour averages of sulfur dioxide concentrations at Long Beach, California.

Table 1 Maximum monthly 1 hr. SO_2 averages parts per hundred million

Year	Jan	Feb	Mar	Apr	May	Jun	Jul	Aug	Sep	Oct	Nov	Dec	Yearly max
1956	47	31	44	12	13	3	14	21	33	26	40	32	47
1957	22	19	20	32	20	23	18	16	13	14	41	25	41
1958	15	13	20	12	24	13	37	20	32	27	27	68	68
1959	20	32	20	15	3	6	8	15	17	15	20	20	32
1960	22	18	23	20	8	13	14	9	13	16	27	20	27
1961	25	20	20	16	10	10	8	10	12	16	14	43	43
1962	20	13	15	18	10	12	10	10	11	11	14	7	20
1963	12	18	27	21	2	7	4	4	15	10	18	18	27
1964	16	10	3	3	19	9	16	25	4	14	18	21	25
1965	16	18	9	14	8	10	18	18	14	12	17	14	18
1966	27	33	25	10	17	30	13	18	22	15	25	23	33
1967	30	40	32	10	8	7	8	26	10	40	18	17	40
1968	51	30	18	22	10	19	22	25	26	29	50	40	51
1969	37	13	55	14	9	10	13	17	33	13	15	44	55
1970	23	19	10	11	15	12	25	40	25	20	12	8	40
1971	22	36	20	28	10	15	20	55	38	41	26	25	55
1972	30	32	18	27	37	13	23	19	21	31	25	13	37
1973	10	8	8	12	11	16	25	16	11	28	10	23	28
1974	8	9	9	13	8	14	9	9	25	11	19	15	34
Ave.	23.8	21.6	20.8	16.3	12.7	13.7	16.0	19.6	19.7	20.4	22.9	25.0	

An inspection of the averages given in the last row clearly indicates a seasonal fluctuation. As is often the case with small, nonnegative measurements, a lognormal distribution appears to give a reasonable fit. It further appears reasonable to model the data, after logarithmic transformation, as a Gaussian sequence with added seasonal effect. If the model does apply, the above results would suggest that the annual maxima of 1 hour averages should have a distribution approximately of double exponential form. In Fig. 2 the 19 available values are shown on a double exponential probability plot and the evident linearity certainly suggests a reasonable fit.

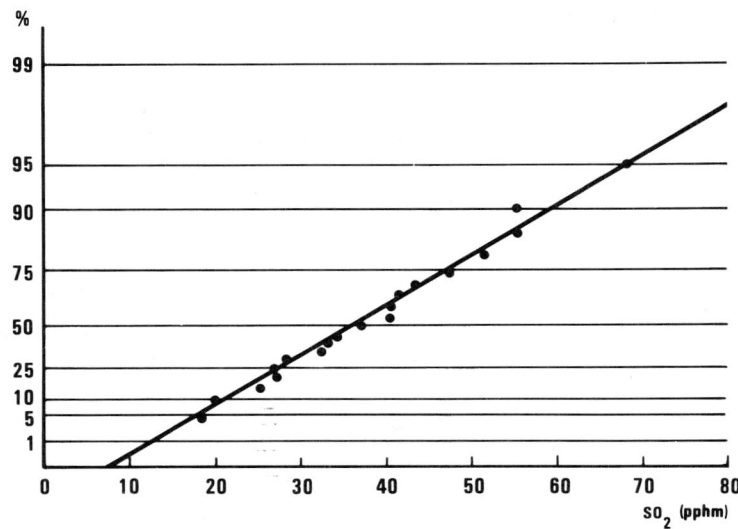

Figure 2 Maximum annual 1 hour averages of sulfur dioxide concentration on a double exponential probability plot

The normalizing (location and scale) constants describing the distribution of yearly maxima are of course dependent on the precise nature of the seasonal component—in accordance with the theory given above. These constants, therefore, provide a potential means of making inferences about the seasonal component (or about the nature of a signal buried in noise, for signal processing applications). We have not made a detailed numerical study in this illustration but note that a modified least squares procedure given by Roberts [11] leads to the estimated d.f. for the yearly maxima given by

$$P\{M(\text{year}) \leqslant u\} = \exp\{-e^{-0.81}(u - 31.5)\}$$

as shown in Fig. 2.

REFERENCES

[1] Berman, S.M. (1962). Limiting distribution of the maximum term in a sequence of dependent random variables. Ann. Math. Statist. 33, 894-908.

[2] Haan, L. de (1970). On regular variation and its application to the weak convergence of sample extremes. Amsterdam Math. Centre Tracts 32.

[3] Horowitz, J. (1980). Extreme values from a nonstationary stochastic process: An application to air quality analysis. Technometrics 22, 409-478.

[4] Hüsler, J. (1981). Asymptotic approximation of crossing probabilities of stationary random sequences. Tech. Report, Dept. of Math. Statist. Univ. of Bern.

[5] Kendall, D.G. (1974). Hunting quanta. Phil. Trans. Roy. Soc. A 276, 231-266.

[6] Leadbetter, M.R. (1974). On extreme values in stationary sequences. Z. Wahrsch. verw. Gebiete 28, 289-303.

[7] Leadbetter, M.R., Lindgren, G. and Rootzén, H. (1983). Extremes and related properties of random sequences and processes. Springer Statistics Series.

[8] Leadbetter, M.R. and Rootzén, H. (1982). Extreme value theory for continuous parameter stationary processes. Z. Wahrsch. verw. Gebiete. 1-20.

[9] Loynes, R.M. (1965). Extreme values in uniformly mixing stationary stochastic processes. Ann. Math. Statist. 36, 993-999.

[10] Mittal, Y. and Ylvisaker, D. (1975). Limit distributions for the maxima of stationary Gaussian processes. Stochastic Process. Appl. 3, 1-18.

[11] Roberts, E.M. (1979). Review of statistics of extreme values with applications to air quality data, Part II, applications. J. Air Pollution Control Assoc. 29, 733-740.

[12] Watson, G.S. (1954). Extreme values in samples from m-dependent stationary processes. Ann. Math. Statist. 25, 798-800.

Non-Gaussian Linear Processes, Phase and Deconvolution

Keh-Shin Lii

Department of Statistics
University of California
Riverside, CA

Murray Rosenblatt*

Department of Mathematics
University of California
La Jolla, CA

1 INTRODUCTION

The time series literature on finite parameter models is typically dominated by a Gaussian assumption or by insights based on what occurs in the case of a Gaussian model. A class of stationary non-Gaussian linear processes is considered in this paper. Under broad conditions it is shown that structural characteristics that cannot be estimated (are not identifiable) in the Gaussian case can be resolved in the non-Gaussian case. In particular this is the case with phase information.

The basic model is now introduced. Let v_t, $t = \ldots, -1, 0, 1, \ldots$ be independent identically distributed random variables with mean zero and variance one. $\{\alpha_j\}$ is a sequence of real constants with

$$\sum_{j=-\infty}^{\infty} \alpha_j^2 < \infty.$$

The linear process generated by $\{\alpha_j\}$ and $\{v_t\}$ is

$$x_t = \sum_{j=-\infty}^{\infty} \alpha_j v_{t-j}. \tag{1}$$

---

*Research supported in part by Office of Naval Research Contract N00014-81-K-0003.

The function $\alpha(z) = \sum_j \alpha_j z^j$ is the z-transform of the process $\{x_t\}$. $\alpha(e^{-i\lambda})$ (λ real) is the frequency response function or transfer function of the process $\{x_t\}$. A primary object is the estimation of $\alpha(e^{-i\lambda})$ in terms of observations on the process $\{x_t\}$ alone (the v_t's are not observed). The linear non-Gaussian process (1) has been proposed as a model for types of seismic exploration (see Donoho (1981), Godfrey and Rocca (1981), and Wiggins (1978)). The coefficients α_j can be thought of as weights (or a wavelet) due to a disturbance passing through a medium and the random terms v_j as the reflectivity of slabs in the layered medium. In a number of geophysical contexts the data (the x_t's) are noticeably non-Gaussian. A basic concern is to deconvolve and estimate the α_j's and v_j's.

If the process $\{x_t\}$ is normal (Gaussian) the probability structure is completely determined by the spectral density

$$f(\lambda) = \frac{1}{2\pi} |\alpha(e^{-i\lambda})|^2$$

or equivalently by $|\alpha(e^{-i\lambda})|$. The phase information, $\arg\{\alpha(e^{-i\lambda})\}$, is not identifiable in the case of normal $\{x_t\}$. Consider the case of a rational function $\alpha(z)$

$$\alpha(z) = A(z)/B(z)$$

with $A(z)$, $B(z)$ polynomials

$$A(z) = \sum_{k=0}^{q} a_k z^k, \quad a_0 \neq 0$$

$$B(z) = \sum_{k=0}^{p} b_k z^k, \quad b_0 = 1.$$

The process $\{x_t\}$ is then a finite parameter autoregressive moving average process so that

$$\sum_{j=0}^{p} b_j x_{t-j} = \sum_{k=0}^{q} a_k v_{t-k}. \tag{2}$$

In the case of a normal process, one cannot distinguish between a real root $z \neq 0$ of $A(z)$ or $B(z)$ and its inverse, or between pairs of nonzero conjugate roots and their paired conjugated inverses. The probability structure (except possibly for variance whch can be adjusted by scaling) is the same. For this reason, it is conventional in the Gaussian case to assume that all roots of $A(z)$ and $B(z)$ have modulus greater than one. However, in the case of a non-Gaussian stationary process (ARMA process) satisfying (2) the different specification of roots, which correspond to the same probability structure in the Gaussian case, actually correspond to different probability structures. The usual procedures based on ideas stemming from Gaussian models cannot distinguish between roots inside the unit disc and roots outside the unit disc. The procedures we present allow one to distinguish between nonzero roots inside $|z| < 1$ and roots outside $|z| < 1$. The discussion follows that given in Rosenblatt (1980). A detailed derivation of the results given together with a more extensive discussion of related issues can be found in Lii and Rosenblatt (1982).

2 HIGHER ORDER SPECTRA

The following proposition is basic in the development of the technique proposed and follows the argument given in Rosenblatt (1980).

Proposition. Assume $\{x_t\}$ a non-Gaussian linear process and for convenience let all moments of the independent random variables $\{v_t\}$ be finite. Further let

$$\sum |j| |\alpha_j| < \infty$$

with $\alpha(e^{-i\lambda}) \neq 0$ for all λ. Then on the basis of observations on $\{x_t\}$ alone (say $x_1, \ldots, x_n$) the function $\alpha(e^{-i\lambda})$ is identifiable (as $n \to \infty$) up to an undetermined integer a in a factor $e^{ia\lambda}$ and the sign of $\alpha(1) = \sum \alpha_k$. For the result it is sufficient actually to have moments of order $k > 2$ finite with cumulant $\gamma_k \neq 0$.

The assumption that the v_t's are non-Gaussian with all moments finite implies that there is a cumulant (semi-invariant) of v_t, $\gamma_k \neq 0$, with smallest subscript $k > 2$. The spectral density of kth order cumulants of $\{x_t\}$ is

$$b_k(\lambda_1, \ldots, \lambda_{k-1}) = \frac{1}{(2\pi)^{k-1}} \sum_{j_1, \ldots, j_{k-1}} \text{cum}(x_t, x_{t+j_1}, \ldots, x_{t+j_{k-1}})$$

$$\exp\left(-\sum_{s=1}^{k-1} ij_s \lambda_s\right) \quad (3)$$

$$= \frac{\gamma_k}{(2\pi)^{k-1}} \alpha(e^{-i\lambda_1}) \ldots \alpha(e^{-i\lambda_{k-1}}) \alpha(e^{i(\lambda_1 + \ldots + \lambda_{k-1})}).$$

If we set

$$h(\lambda) = \arg\left\{\alpha(e^{-i\lambda}) \frac{\alpha(1)}{|\alpha(1)|}\right\}$$

it follows that

$$\left\{\frac{\alpha(1)}{|\alpha(1)|}\right\}^k \gamma_k = (2\pi)^{\frac{k}{2}-1} b_k(0, \ldots, 0)\{f(0)\}^{-k/2}$$

and

$$h(\lambda_1) + \ldots + h(\lambda_{k-1}) - h(\lambda_1 + \ldots + \lambda_{k-1})$$

$$= \arg\left[\left\{\frac{\alpha(1)}{|\alpha(1)|}\right\}^k \gamma_k^{-1} b_k(\lambda_1, \ldots, \lambda_{k-1})\right] \quad (4)$$

because $h(-\lambda) = \overline{h(\lambda)}$. Notice that

$$h'(0) - h'(\lambda) = \lim_{\Delta \to 0} \frac{1}{(k-2)\Delta}\{h(\lambda) + (k-2)h(\Delta) - h(\lambda + (k-2)\Delta)\}.$$

We can rewrite $h(\lambda)$ as

$$h(\lambda) = \int_0^\lambda \{h'(u) - h'(0)\} du + c\lambda = h_1(\lambda) + c\lambda$$

with $c = h'(0)$. The α_j's are real and so one must have

$$h(\pi) = a\pi$$

for some integer a. Now

$$h(\pi) = h_1(\pi) + c\pi.$$

If we set

$$h_1(\pi)/\pi = \delta$$

it follows that $c = k - \delta$. The integer k can't be specified without additional information because it corresponds to the indexing of the v_t's. Also the sign of $\alpha(1)$ can't be determined since one can change the sign of all the α_j's and all the v_t's without altering the observed process $\{x_t\}$. Notice that $h_1(\lambda)$ can be estimated consistently by making use of kth order cumulant spectral estimates (see Brillinger and Rosenblatt (1967)). Since

$$\alpha(e^{-i\lambda}) = |2\pi f(\lambda)|^{\frac{1}{2}} \exp\{ih(\lambda)\}$$

the proposition follows from the observations made above.

Estimation of the Phase. Treatments of the estimation of second order spectral densities can be found in many places (see Anderson (1971) and Box and Jenkins (1976)). We will consider the estimation of $h(\lambda)$ when the third order cumulant γ_3 of v_t is nonzero. Corresponding analyses of the case in which γ_4 is nonzero are currently being developed and will be presented elsewhere. As already suggested above, it is actually $h_1(\lambda)$ that will be estimated. Equation (4) is

$$h'(0) - h'(\lambda) = \lim_{\Delta \to 0} \frac{1}{\Delta} \{h(\Delta) + h(\lambda) - h(\lambda + \Delta)\}$$

where $k = 3$ and up to a sign

$$h(\lambda) + h(\Delta) - h(\lambda + \Delta) = \arg\{b_3(\lambda, \Delta)\}.$$

From now on the subscript is deleted and it is to be understood that we deal with the bispectral density $b(\lambda, \mu)$. $_n b(\lambda, \mu)$ is an estimate of $b(\lambda, \mu)$ based on a sample of size n. We take

$$\theta_n(\lambda, \mu) = \arctan(\mathrm{Im}\,_n b(\lambda, \mu) \mathrm{Re}\,_n b(\lambda, \mu))$$

as an estimate of

$$\phi(\lambda, \mu) = \arg b(\lambda, \mu).$$

Now consider estimating $h_1(\lambda)$. Let $\Delta = \Delta(n)$, $k\Delta = \lambda$ and $\Delta = \Delta(n) \to 0$ as $n \to \infty$. Assume $b(0,0)$ positive. Later we show how the case in which $b(0,0)$ is negative is handled. Notice that

$$\begin{aligned}
h_1(\lambda) &= h(\lambda) - h'(0)\lambda \\
&\cong h(k\Delta) - \frac{h(\Delta)}{\Delta} k\Delta \\
&= -\sum_{j=1}^{k-1} \{h(j\Delta) + h(\Delta) - h((j+1)\Delta)\} \\
&= -\sum_{j=1}^{k-1} \arg b(j\Delta, \Delta).
\end{aligned}$$

Because of this we take

$$H_n(\lambda) = -\sum_{j=1}^{k-1} \arg{_nb(j\Delta,\Delta)}$$

as an estimate of $h_1(\lambda)$. An estimate of δ is given by $H_n(\pi)/\pi$. The sign of $b(0,0)$ is estimated by using the real part of $_nb(\Delta,\Delta)$. If it is negative, all $_nb(j\Delta,\Delta)$ are multiplied by a minus sign. The estimate $H_n(\lambda)$ is then taken as

$$H_n(\lambda) = -\sum_{j=1}^{k-1} \arg\{-_nb(j\Delta,\Delta)\}.$$

The detailed conditions under which our estimate of $h_1(\lambda)$ is consistent are given in Lii and Rosenblatt (1982). Expressions are given there for asymptotic estimates of bias and variance.

3   COMPUTATIONS

In this section we make a few remarks on computations. Consider a sample (of x observations) of size $n = mN$. Center and normalize the observations so that they have sample mean zero and variance one. The sample is decomposed into m disjoint sections on length N with N chosen so that the variance of the bispectral estimate from each section is not too large (see Lii and Helland (1981) for a discussion of an algorithm for bispectral estimates). Choose a grid of points $\lambda_j = j\Delta$ in $(0, 2\pi)$, $j = 1, \ldots, M$, $\Delta = 2\pi L/N$. Compute a bispectral estimate $_Nb(j\Delta, \Delta)$ with a weight function of bandwidth Δ from each subsection. Then compute $\theta_n(j) = \arg\{_nb(j\Delta, \Delta)\} + 2k\pi$ with the integer k chosen to ensure continuity (neighboring values as close to each other as possible) of $H_n(l\Delta) = H_n(\lambda_l) = -\sum_{j=1}^{l-1} \theta_n(j)$, $l = 2, \ldots, M+1$. One starts with $l = 2$. Given that $h(0) = 0$ set $H_n(0) = 0$ and estimate $H_n(\Delta) = H_n(\lambda_1)$ by interpolating between 0 and $H_n(\lambda_2)$, $\lambda_2 = 2\Delta$. Since

$$\alpha_k = \frac{1}{2\pi}\int_0^{2\pi} \alpha(e^{-i\lambda})e^{ik\lambda}d\lambda$$

a suggested estimate $\hat{\alpha}_k$ of α_k would be given by

$$\begin{aligned}\hat{\alpha}_k &= \frac{1}{2\pi}\int_0^{2\pi} \hat{\alpha}(e^{-i\lambda})e^{ik\lambda}d\lambda \\ &\cong \frac{1}{(M+2)}\sum_{j=0}^{M+1}\sqrt{2\pi f_n(\lambda_j)}\exp\left\{i\left[H_n(\lambda_j) - \frac{H_n(\pi)}{\pi}\lambda_j + k\lambda_j\right]\right\}.\end{aligned} \quad (5)$$

Of course, these are estimates up to an undetermined shift of subscript.

A few examples were generated by Monte Carlo simulation. These were moving average schemes $x_t = v_t + \alpha_1 v_{t-1} + \alpha_2 v_{t-2}, t = 1, \ldots, 640$ where $v_t = v'_t - Ev'_t$ and the v_t's were independent Pareto random deviates with density $f(v) = 4v^{-5}$ for $v \geq 1$ and $f(v) = 0$ for $v < 1$. These were generated by using the GGUW subroutine in IMSL and transforming the uniform deviates. $\{x_t\}_1^{640}$ was partitioned into five equal disjoint sections of 128 points. The bispectral estimate $_{128}b^{(i)}(j\Delta,\Delta)$, $j = 1, \ldots, 13$, $i = 1, \ldots, 5$ was computed by using the algorithm given in Lii and Helland (1981). Here

$\Delta = \dfrac{18}{128} = .442$. The final bispectral estimate is

$$_{640}\hat{b}(j\Delta, \Delta) = \frac{1}{5}\sum_{k=1}^{5} {}_{128}b^{(k)}(j\Delta, \Delta).$$

Further one obtains

$$\theta_n(j) = \arg\{{}_n\hat{b}(j\Delta, \Delta)\}$$
$$= \arctan(\operatorname{Im}{}_n\hat{b}(j\Delta, \Delta)/\operatorname{Re}{}_n\hat{b}(j\Delta, \Delta))$$

by taking the principal value and

$$-H_n(j\Delta) = -H_n(\lambda_j) = \sum_{i=1}^{j-1}\theta_n(i), \quad j = 2, \ldots, 14.$$

Estimates of the α_k's are computed by using formula (5). A standard estimate of the spectral density with bandwidth Δ was used. The four cases specified in Table 1 were considered.

Table 1 Coefficients, roots and estimated coefficients for four cases

Case	Coefficients			Roots		Estimated coefficients		
	α_0	α_1	α_2	r_1	r_2	$\hat{\alpha}_0$	$\hat{\alpha}_1$	$\hat{\alpha}_2$
1	1.0	−.833	0.167	2.0	3.0	.9274	−.7443	.0847
2	1.0	−2.333	0.667	.5	3.0	.9317	−2.236	.7812
3	1.0	−3.50	1.50	2.0	0.333	1.220	−3.322	1.607
4	1.0	−5.0	0.0	0.5	0.333	1.298	−4.046	5.529

The second case was used to illustrate our deconvolution procedure. The estimated model is

$$x_t = .9317v_t - 2.236v_{t-1} + .7812v_{t-2}.$$

Given $\hat{\alpha}(B) = .9317 - 2.236B - .7812B^2$ (B the backward shift operator) we set

$$\hat{v}_t = \sum_{j=-q}^{q} \hat{a}_j x_{t-j} \approx \frac{1}{\hat{\alpha}(B)} x_t. \tag{6}$$

The formal expansion of the right hand side of (6) has been truncated to obtain $\hat{v}_t$. The first line of the graph (Figure 1) is case 2 generated from $t = 10$ to $t = 234$. The independent Pareto sequence generating x_t is given in the second line. The third line gives the estimated $\hat{v}_t$ deconvolved by our procedure. The difference $v_t - \hat{v}_t$ is graphed in the fourth line. The standard deviation of the signal x_t is .435 while the square root of the estimated mean square error of our deconvolution is 5.75×10^{-2}. The last line gives the minimum phase deconvolution of x_t. Our deconvolution has retained the general shape of the orginal v_t sequence very well. The minimum phase deconvolution has no similarity to the original v_t sequence. It is interesting to notice that the deconvolution given here in the case of Pareto noise is as effective as that given in Lii and Rosenblatt (1982) for exponential noise even though the moment properties of the Pareto distribution are somewhat worse than those of the exponential distribution.

Non-Gaussian Linear Processes

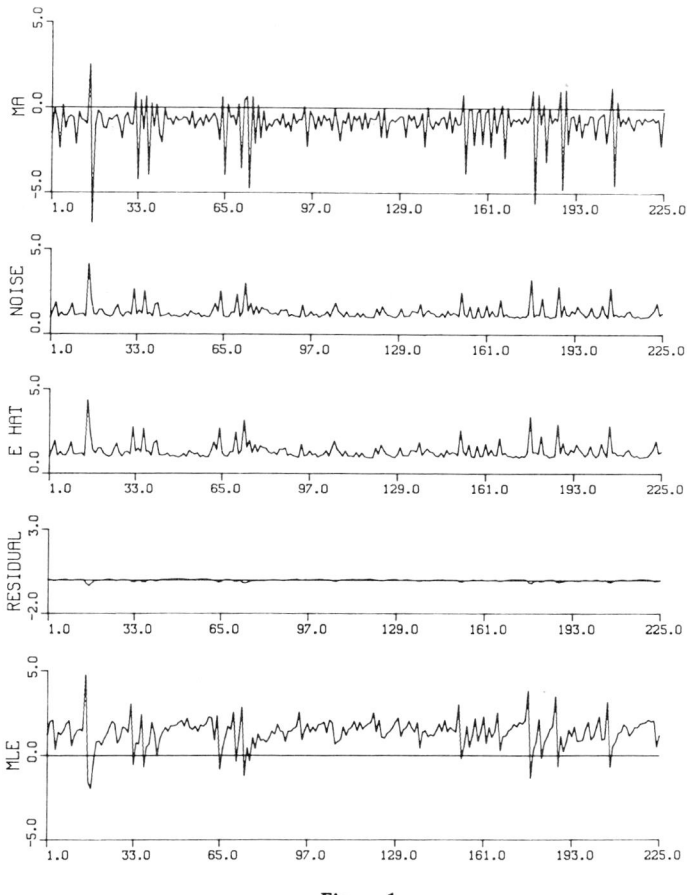

Figure 1

REFERENCES

Anderson, T.W. (1971) *The Statistical Analysis of Time Series,* John Wiley.

Box, G.E.P. and Jenkins, G.M. (1976) *Time Series Analysis, Forecasting and Control,* Holden-Day.

Brillinger and Rosenblatt (1967) "Asymptotic theory of estimates of kth order spectra", in *Spectral Analysis of Time Series* (ed. B. Harris) 153-188, John Wiley.

Donoho, D. (1981) "On minimum entropy deconvolution" in *Applied Time Series Analysis* II (ed. D.F. Findley) 565-608.

Godfrey, R. and Rocca, F. (1981) "Zero memory nonlinear deconvolution" Geophysical Prospecting, 29, 189-228.

Lii, K.S. and Helland, K.N. (1981) "Cross-bispectrum computation and variance estimation" *ACM Trans. Math. Software* 7, 3.

Lii, K.S. and Rosenblatt, M. (1982) "Deconvolution and estimation of transfer function phase and coefficients for non-Gaussian linear processes," Ann. Statist. 10, 1195-1208.

Rosenblatt, M. (1980) "Linear processes and bispectra" J. Appl. Prob. 17, 265-270.

Wiggins, R.A. (1978) "Minimum entropy deconvolution" Geoexploration 17.

Filtering and Smoothing of Nonstationary Processes*

M. M. Rao

Department of Mathematics
University of California
Riverside, CA

1 INTRODUCTION

Let $\{X(t), t \in \mathbb{R}\}$ be a mean zero, second order (complex valued) stochastic process on a probability space (Ω, Σ, P). Suppose the process is expressible as:

$$X(t) = Y(t) + Z(t), t \in \mathbb{R}, \qquad (1)$$

where $Y(t)$ and $Z(t)$ are processes of the same type. Here $X(t)$ is the output, $Y(t)$—the signal and $Z(t)$— the noise process. It is desired to "filter" the signal from noise in an optimal manner in that one likes to get a best estimator of $Y(t)$ using the mean-square-error criterion as the loss function. Also the Y- and Z-processes can be correlated. In order to solve the problem, one assumes only the covariance functions of these processes as known. The following are the desirable structures of these covariance functions.

Let $r_x\colon (s, t) \mapsto E(X_s \bar{X}_t) = \int_\Omega X_s \bar{X}_t dP$, be the covariance function. Then the process is said to be of *Cramér class* if r_x can be expressed as:

$$r_x(s, t) = \int_\mathbb{R} \int_\mathbb{R} g_x(s, \lambda) \overline{g_x(t, \lambda')} F_x(d\lambda, d\lambda'), \ s, \ t \in \mathbb{R}, \qquad (2)$$

relative to a positive definite (or covariance) function F_x of bounded variation on each finite rectangle of $\mathbb{R}^2$, and $\{g_x(s, \cdot), s \in \mathbb{R}\}$ is a suitable class of known functions. If F_x of (1) concentrates on the diagonal $\lambda = \lambda'$, then the resulting class is called a *Karhunen process*. If on the other hand $g_x(s, \lambda) = e^{is\lambda}$ in (2) so that F_x is of bounded variation on $\mathbb{R}$ itself, then the process is *strongly* (or Loève) *harmonizable*. These classes were introduced respectively in [2], [7] and [9]. In the event that F_x concentrates on the diagonal, then a strongly har-

---
*This research is supported under the ONR Contract No. N00014-79-C-0754.

monizable process becomes the familiar (weakly) *stationary* process. If in (1) F_x is not necessarily of bounded variation, but only satisfies a weaker condition, called Fréchet variation, on $\mathbb{R}^2$ and $g_x(s, \lambda) = e^{is\lambda}$, the resulting class is *weakly harmonizable*. The function F_x of (1) is called the *spectral function* of the process in all cases. The last class was introduced independently in [1] and [18] under different names (*V*-boundedness in [1], 'harmonizability' in [18]). Using the concept of Fréchet variation in (2) it is possible to define a *weak Cramér* class, and then the following inclusions hold:

strongly harmonizable $\subset$ weakly harmonizable

Stationary

∩ ∩

⊂ Karhunen class $\subset$ Cramér class $\subset$ weak Cramér class.

The following observation shows why one needs to consider classes which are more general than stationary ones. Let $\{X_t, t \in \mathbb{R}\}$ be a stationary process and $I \subset \mathbb{R}$ be a subset. If $\{V_t, t \in \mathbb{R}\}$ is the truncation of the first one on I, in that $V_t = X_t, t \in I; = 0, t \in \mathbb{R} - I$, then $\{V_t, t \in \mathbb{R}\}$ is *not* stationary if the X_t's have positive variance, and it is strongly harmonizable if I is finite but it is always weakly harmonizable whatever $I \subset \mathbb{R}$ is. Similar considerations lead to a study of other general classes in applications. A first serious analysis of the weakly harmonizable case was given in [13], and [10]. A comprehensive account completing and complementing these studies, with all the technical details, is given in [16].

In what follows the problems of filtering and smoothing on some of the above noted nonstationary processes will be discussed and certain general results described. This gives an overview of the work being done for classes of nonstationary processes.

2 FILTERING PROBLEMS

A general concept of a filter on stationary processes has been given by Hannan ([5], [6]). In the present context it would be of the form: if $\{W_t, t \in \mathbb{R}\}$ is one of the classes introduced above, then a linear mapping T on the space $L^2(P)$ is a filter and $\{X_t, t \in \mathbb{R}\}$ is a filtered process, if $X_t = TW_t, t \in \mathbb{R}$, and T also commutes with translations, i.e., if $U_t W_s = W_{s+t}$ and $\{U_t, t \in \mathbb{R}\}$ defines a family of unitary operations on $L^2(P)$, then one must have $U_t(TW_s) = T(U_t W_s), s, t \in \mathbb{R}$. Here the family $\{U_t, t \in \mathbb{R}\}$ is called the *shifts* on $\{W_s, s \in \mathbb{R}\}$. While a shift always exists for the stationary classes, it need not exist for harmonizable or Cramér classes. In [3], Getoor has modified the Karhunen class so that in the representation of the covariance r_x for (2), the functions $g_x: \mathbb{R} \times \mathbb{C} \to \mathbb{C}$ are assumed to exist and then F_x concentrates on the diagonal of $\mathbb{C} \times \mathbb{C}$ where "$\mathbb{C}$" is the complex numbers. Thus a *modified Karhunen class* is one whose covariance function admits a representation:

$$r_x(s, t) \int_{\mathbb{C}} g_x(s, \lambda) \overline{g_x(t, \lambda)} F_x(d\lambda), \quad s, t \in \mathbb{R}. \qquad (3)$$

This is called a "normal process" if $g_x(s, \lambda) = e^{s\lambda}, \lambda \in \mathbb{C}, s \in \mathbb{R}$. For this class a shift operator exists; it is a normal operator on $L^2(P)$. If $\mathbb{C}$ is replaced by the reals $\mathbb{R}$, then the "normal process" becomes stationary. In this sense, for our classes of nonstationary processes shifts may not exist. So it is necessary to consider and study special forms of the filters T that are interesting in applications.

For an analysis of these problems, it is also essential that the condition (2) (or (3)) be translated to a representation of the process itself. Thus if the $X(t)$ is such that its covariance function is representable as (2), then it can be shown that (for a proof, see [2] and for the general case [16])

$$X(t) = \int_{\mathbb{R}} g_x(t, \lambda) \tilde{Z}_x(d\lambda) \tag{4}$$

where $\tilde{Z}_x(\cdot)$ is a stochastic measure on the Borel sets of $\mathbb{R}$. It has orthogonal increments (= "orthogonally scattered") if $X(t)$ is either a Karhunen or a stationary process and not so otherwise. One has

$$E(\tilde{Z}_x(A) \overline{\tilde{Z}_x(B)}) = \int_A \int_B F_x(d\lambda, d\lambda'), \tag{5}$$

where F_x is as in (2). The integral in (4) is a stochastic integral but can equally be interpreted as a Dunford-Schwartz integral. It should be noted that for the weakly harmonizable case a similar representation holds, but there are technical problems (cf. [14]), and for (2) one has to employ a Morse-Transue integration technique [11]. In the modified Karhunen case Z_x is defined on the Borel sets of $\mathbb{C}$ with similar properties.

With this preamble, it is possible to present a solution for the optimal filter, for the model (1) in reasonable generality with the mean-square-error criterion. The ideas for the solution are analogous to those of the Bayesian estimation in the theory of statistical inference.

THEOREM 1. *Let X, Y, Z be of Cramér class and satisfy*

$$X(t) = Y(t) + Z(t), \; t \in \mathbb{R} \tag{6}$$

so that $X(t)$ is the output, $Y(t)$ the signal and $Z(t)$ the noise. Let $a \in \mathbb{R}$ be a fixed point. If F_x, F_y, F_z are the spectral functions of the respective processes, $E(Y(s)\overline{Z(t)}) = 0$, s, $t \in \mathbb{R}$, then relative to the mean-square-error criterion, there is an optimal filter $\hat{Y}(a)$ at a of $Y(t)$ given by

$$\hat{Y}(a) = \int_{\mathbb{R}} k_a(\lambda) \tilde{Z}_x(d\lambda) \tag{7}$$

where $\tilde{Z}_x(\cdot)$ is the representing stochastic measure of the output process X, and k_a satisfies (8) below, and is a unique solution of the integral Equation (9):

$$\int_{\mathbb{R}} \int_{\mathbb{R}} k_a(\lambda) \overline{k_a(\lambda')} F_x(d\lambda, d\lambda') < \infty, \tag{8}$$

$$\int_{\mathbb{R}} \int_{\mathbb{R}} k_a(\lambda) \left(F_y(d\lambda), d\lambda' \right) + F_z(d\lambda, d\lambda')) = \int_{\mathbb{R}} \int_{\mathbb{R}} g_y(a, \lambda) F_y(d\lambda, d\lambda'). \tag{9}$$

If X, Y, Z are Karhunen processes with the $g(\cdot)$'s-bounded, then the solution becomes simpler. In this case, if one also assumes that F_x, F_y, F_z have densities, f_x, f_y, f_z, then the k_a satisfying (8) and (9) is given by:

$$k_a(\lambda) = g_y(a, \lambda) f_y(\lambda) [f_y(\lambda) + f_z(\lambda)]^{-1}. \tag{10}$$

[Since $|k_a(\lambda)| \leq |g_y(a, \lambda)|$, and g_y is bounded, k_a is f_x-integrable.] In particular if the processes are stationary, then $g_x(a, \lambda) = g_y(a, \lambda) = g_z(a, \lambda) = e^{ia\lambda}$, and then (10) reduces to the familiar solution, first obtained by Grenander in ([4], p. 273). The Cramér class extension was discussed in [17]. If Y and Z

processes are also correlated and their covariance r_{yz} is representable as (1), then the solution is slightly more involved, but is similarly obtainable. The details of this computation are given in [8]. To save space, it will not be described here.

The function k_a of (8), (9) or (10) is called the *response function*. An algorithm for its calculation is desirable in the general case.

3 POLYNOMIAL FILTERS AND SMOOTHING

Let us turn to more specialized filters than those of the preceding section. Considering the discrete parameter processes, let $TY(n) = X(n)$, be the difference operator. Thus more explicitly written:

$$TY(n) = \sum_{k=0}^{m} a_k Y_{n-k} = X_n, \ a_k \in \mathbb{C}, \ n = 0, \pm 1, \pm 2, \ldots \quad (11)$$

Consider the characteristic polynomial $p(\cdot)$ associated with T (hence the name polynomial filter):

$$p(z) = \sum_{k=0}^{m} a_k z^k, \ Q = \{\lambda : p(e^{-i\lambda}) = 0\}.$$

The problem here is this: If X is the output, Y the input, T is specified by (11), under what conditions on the coefficients a_k can one *recover* the strongly harmonizable input (or signal) if the output is a known strongly harmonizable process? Here is a solution of the problem.

THEOREM 2. *Let the filter T be defined by (11). Then $TY = X$, admits a strongly harmonizable solution Y for a given strongly harmonizable output X iff*

(i) $|F_x|(Q \times Q) = 0$, *and* (ii) $\int_{Q^c} \int_{Q^c} \dfrac{|F_x|(d\lambda, d\lambda')}{|p(e^{-i\lambda})\overline{p}(e^{-i\lambda'})|} < \infty,$

where F_x is the spectral function of X and $|F_x|(\cdot, \cdot)$ is its variation. The solution is unique iff $Q = \phi$. If $Q \neq \phi$, then there is exactly one Y-series solution satisfying $|F_y|(Q \times Q) = 0$. When these conditions hold, the solution is given by:

$$Y_n = \int_{Q^c} e^{in\lambda}(p(e^{-i\lambda}))^{-1} \tilde{Z}_x(d\lambda), \ n = 0, \pm 1, \ldots, \quad (12)$$

where $\tilde{Z}_x(\cdot)$ is the stochastic measure associated with X on the Borel sets of $(-\pi, \pi]$. (Thus $Q^c = (-\pi, \pi] - Q$, the complement in $(-\pi, \pi]$.)

If the process is stationary, then F_x concentrates on the diagonal, and the work simplifies. The result then agrees with the work of [12]. In the harmonizable case, this result is essentially given in Kelsh [8]. Since the form of our T may be regarded as smoothing of the input data, it was called a smoothing operation in [12], and the method of solution was termed *inversion*. The present terminology is more descriptive of the situation.

With the existence and uniqueness given in the preceding result, it is natural to ask the following: When is the filter physically realizable, or in detail, when does the solution depend only on the past and present (but *not* the future)? This question can be answered as follows.

THEOREM 3. *Let the existence conditions of Theorem 2 be satisfied for the filter defined by (11). If all the roots of the characteristic polynomial $p(z) = 0$ lie outside the unit circle of the complex plane, then one has*

$$Y_n = \sum_{k \geq 0} b_k X_{n-k},$$

and the filter is physically realizable. Here the coefficients $\{b_k, k \geq 0\}$ are given by the power series expansion of $(p(z))^{-1}$ for $|z| < \rho, \rho > 1$. If there are roots $\{z_j\}_{j \geq 1}$ of $p(z) = 0$ on the unit circle and if they are simple, let $z_j = e^{-i\lambda_j}$. If the $\{\lambda_j\}_{j \geq 1}$ form a set of μ-measure zero, where $\mu(A) = |F_x|(A \times (-\pi, \pi])$ so that μ is the marginal of $|F_x|$, then again physical realizability obtains. If, however, there are some roots of $p(z) = 0$, inside the unit disc, then the filter is not physically realizable.

These results give a reasonable extension of the stationary theory to the strongly harmonizable case. The ideas of the proofs are (nontrivial) extensions of those of [12], and can be obtained from [8].

Two immediate questions to ask at this point are:

(i) What are the continuous parameter analogs of these results and are there multivariate extensions?

(ii) What about these studies for the weakly harmonizable processes, and for the other classes of Section I?

Some remarks on these questions will be given in the next section.

4 FURTHER DEVELOPMENTS

The preceding results, given for the discrete parameter case, have analogs in the continuous parameter. Here

$$(TY)(t) = \int_{\mathbb{R}} f(s) Y(t - s) ds = X(t), \quad t \in \mathbb{R},$$

and $f \in L^1(\mathbb{R})$ such that its Fourier transform $\hat{f}$ is $|F_x|$ integrable. The set $Q = \{t \in \mathbb{R} : \hat{f}(t) = 0\}$ and p is replaced by $\hat{f}$. With this format, the results extend. Details are in [8]. The integral above is a simple stochastic integral discussed in [7] and [4], but it is a specialization of the Bochner integral of vector functions relative to a scaler measure. However, the multivariate extensions have some real technical problems including rank etc. This is being investigated.

If the processes X, Y, Z are weakly harmonizable, then most of the computations for the preceding theorems fail since the integral for (1) is not absolutely convergent. As noted before, one has to use the theory of bimeasures of [11], (cf. also [19] and [16]). However, with the appropriate tools and different proofs, it is possible to extend several of the considerations of [8]. Though similar techniques appear to work for the Cramér (both weak and the usual) and Karhunen cases, the details have not been set down rigorously. As noted in the introduction, the weakly harmonizable processes are perhaps the most useful from an applicational point of view (even if one starts with the stationary data) since they have nice closure properties. The vector case and the rest of the problems should be developed primarily to this class, and that is what is being done at present.

REFERENCES

[1] Bochner, S., "Stationarity, boundedness, almost periodicity of random valued functions." *Proc. Third Berkeley Symp. Math Statist. and Prob. 2* (1956), 7-27.

[2] Cramér, H., "A contribution to the theory of stochastic processes," *Proc. Second Berkeley Symp. Math. Statist. and Prob.* (1951), 329-339.

[3] Getoor, R.K., "The shift operator for nonstationary stochastic processes," *Duke Math. J. 23* (1956), 175-187.

[4] Grenander, U., "Stochastic processes and statistical inference," *Ark Mat. 1* (1950), 195-227.

[5] Hannan, E.J., "The concept of a filter," *Proc. Cambr. Phil. Soc. 63* (1967), 221-227.

[6] Hannan, E.J., *Multiple Time Series*, John Wiley, New York, 1970.

[7] Karhunen, K., "Über lineare Methoden in der Wahrscheinlichkeitsrechnung," *Ann. Acad. Scient. Fennicae* Ser. AI. No. 37 (1947), 1-79.

[8] Kelsh, J.P., "Linear analysis of harmonizable time series," unpublished Ph.D. thesis, University of California, Riverside, 1978.

[9] Loève, M., "Fonctions aléatoires du second ordre" Note in P. Lévy's *'Processus Stochastiques et Mouvement Brownien'* Gauthier-Villars, Paris (1948), 299-352.

[10] Miamee, A.G. and H. Salehi, "Harmonizability, V-boundedness, and stationary dilations of stochastic processes," *Indiana Univ. Math. J. 27* (1978), 37-50.

[11] Morse, M. and W. Transue, "₵-bimeasures and their integral extensions," *Ann. Math. 64* (1956), 480-504.

[12] Nagabhushanam, K., "The primary process of a smoothing relation," *Ark. Mat. 1* (1950), 421-488.

[13] Niemi, H., "Stochastic processes as Fourier transforms of stochastic measures," *Ann. Acad. Scient. Fennicae Ser. AI No. 591* (1975), 1-47.

[14] Rao, M.M., "Representation of weakly harmonizable processes," *Proc. Nat. Acad. Sci. 78* (1981), 5288-89.

[15] Rao, M.M., "Covariance analysis of nonstationary time series" in *Developments in Statistics*, Vol. 1, Academic Press, New York (1978), 171-225.

[16] Rao, M.M., "Harmonizable processes: Structure theory," *L'Enseignement Mathématique*, 28 (1982), 295-351.

[17] Rao, M.M., "Inference in stochastic processes-III," *Z. Wahrschein.* 8 (1967), 49-72.

[18] Rozanov, Yu. A., "Spectral theory of abstract functions," *Theor. Prob. Appl.,* 4 (1959), 271-287.

[19] Ylinen, K., "On vector bimeasures," *Ann. Mat. Pura. Appl. 117* (1978), 115-138.

ARMA Time Series Modeling: A Singular Value Decomposition Approach*

James A. Cadzow

Department of Electrical and Computer Engineering
Arizona State University
Tempe, AZ

1 ALGEBRAIC CHARACTERIZATION OF ARMA PROCESSES

The time series $\{x(n)\}$ is said to be an ARMA random process of order (p,q) if its elements arise as the response of the linear recursive operator

$$\sum_{k=0}^{p} a_k x(n-k) = \sum_{k=0}^{q} b_k \epsilon(n-k) \tag{1}$$

to the excitation time series $\{\epsilon(n)\}$ which is composed of a sequence of zero mean, unit variance, uncorrelated random variables (i.e., white noise). This model's a_k autoregressive and b_k moving average parameters may be directly obtained through exact knowledge of the time series $\{x(n)\}$ autocorrelation sequence,

$$r_x(n) = E\{x(n+k)\bar{x}(k)\} \tag{2}$$

in which E and $-$ denote the operations of expectation and complex conjugation, respectively. As we will now show, this ability is a direct consequence of the Yule-Walker equations which **govern recursive relationship (1), that is**

$$\sum_{k=0}^{p} a_k r_x(n-k) = \sum_{k=0}^{q} b_k \bar{h}(k-n) \tag{3}$$

where $\{h(n)\}$ denotes the unit-sample (Kronecker delta) response of linear

---

*Work supported under Contract N00014-82-K-0257, Task NR 277-306 with the Office of Naval Research.

operator (1). Once the a_k and b_k parameters have been determined, the time series' associated power spectral density function (i.e., the Fourier transform of $r_x(n)$) is specified by

$$S_x(e^{j\omega}) = \left| \frac{\sum_{n=0}^{q} b_n e^{-j\omega n}}{\sum_{n=0}^{p} a_n e^{-j\omega n}} \right|^2 = \left| \frac{B_q(e^{j\omega})}{A_p(e^{j\omega})} \right|^2. \qquad (4)$$

A brief description of the procedure for determining an ARMA model's parameters from exact autocorrelation lag information is now given.

Autoregressive Parameter Determination

The autoregressive parameters are readily obtained through use of the so-called extended Yule-Walker equations. Namely, under the typical restriction that linear operator (1) is causal, the Yule-Walker equations (3) for indices $n > q$ simplify to the homogeneous relationships

$$\sum_{k=0}^{p} a_k r_x(n-k) = 0 \qquad \text{for all } n > q. \qquad (5)$$

These extended Yule-Walker equations are seen to involve the a_k parameters in a linear fashion. With this in mind, the determination of the autoregressive parameters are readily obtained upon evaluating expression (5) over any of p (or more) distinct index values satisfying $n > q$, and, then solving the resultant system of consistent linear equations for the a_k parameters subject to the normalizing constraint that $a_0 = 1$. In particular, this evaluation over the contiguous set of t indices as specified by $q + m < n \leq q + m + t$ where $m \geq 1$ and $t \geq p$ results in

$$R_m \underline{a} = \underline{0} \qquad m \geq 1 \qquad (6)$$

where $\underline{0}$ is the $t \times 1$ zero vector, $\underline{a}$ is the $(p+1) \times 1$ autoregressive parameter vector

$$\underline{a} = [1, a_1, a_2, \ldots, a_p]' \qquad (7)$$

where the prime symbol (') denotes the operation of vector transposition, and, R_m is the $t \times (p+1)$ autocorrelation matrix with elements

$$R_m(i,j) = r_x(q + m + i - j) \quad \begin{array}{l} 1 \leq i \leq t \\ 1 \leq j \leq p+1 \\ m \geq 1. \end{array} \qquad (8)$$

If the correlation elements $r_x(n)$ used in forming the autocorrelation matrices R_m correspond to an ARMA (p,q) process, it follows that the rank of the R_m will be p so long as $t \geq p$. It then follows that the required autoregressive parameter vector (7) will correspond to any properly normalized eigenvector (i.e., its first component is one) of the $(p+1) \times (p+1)$ matrix $R_m^* R_m$ associated with a zero eigenvalue, that is

$$R_m^* R_m \underline{a} = \underline{0} \quad m \geqslant 1 \tag{9}$$

where the symbol * denotes the operation of complex conjugate transposition. It is important to note that the required autoregressive parameter vector $\underline{a}$ will satisfy this relationship for all positive integer values of m.

Moving Average Determination

Once the ARMA model's autoregressive parameters have been determined using relationship (9), the moving average parameters may be obtained by filtering the time series $\{x(n)\}$ with the nonrecursive filter

$$A_p(z) = \sum_{n=0}^{p} a_n z^{-n}.$$

It is readily shown that the resultant filtered response time series will be a moving average process of order q whose autocorrelation elements are given by

$$\sum_{k=n}^{q} b_k \bar{b}_{k-n} = \sum_{k=0}^{p} \sum_{i=0}^{p} a_k \bar{a}_i r_x(n + i - k)$$
$$-q \leqslant n \leqslant q. \tag{10}$$

Since the elements a_k and $r_x(n)$ are given, the right side of this expression may be evaluated. Using the method of spectral factorization, one may then solve for the required b_k coefficients.

2 GENERALIZED HIGH PERFORMANCE METHOD

To implement the modeling procedure as described in Section 1, it is necessary to have exact autocorrelation lag information. In the more commonly encountered situation, however, it is necessary to generate an ARMA model based only upon a set of raw time series observations as represented by

$$x(1), x(2), \ldots, x(N). \tag{11}$$

It is possible to adapt the spirit of the previous section in order to obtain such a model. This will entail carrying out the following two step procedure.

(i) generate autocorrelation lag estimates from the observation set (11),

(ii) substitute these estimates into the structure (8) so as to generate the $t \times (p + 1)$ autocorrelation matrix estimates $\hat{R}_m$ for $1 \leqslant m \leqslant s$ where s is some fixed integer.

There exists a variety of procedures for effecting the autocorrelation estimates required in step (i) (e.g., see Reference [4]). The standard biased autocorrelation estimate as given by

$$\hat{r}_x(n) = \frac{1}{N} \sum_{k=1}^{N-n} x(n+k) \bar{x}(k) \quad \text{for } 0 \leqslant n \leqslant N - 1 \tag{12}$$

has been found to provide particularly satisfactory modeling performance although other choices may be more appropriate in a given application.

Once the autocorrelation matrix estimates $\hat{R}_m$ have been formulated, it will be generally found that they will each have full rank (i.e., min $(p + 1, t)$) even when the time series is an ARMA (p,q) process. This is consequence of the errors inherent in the autocorrelation lag estimate procedure. Due to this factor, it will not be possible to find a nontrivial autoregressive parameter vector $\underline{a}$ for which $R_m \underline{a} = \underline{0}$. Nonetheless, it is still desirable that this relationship be reasonably satisfied over a suitable range of values on m. A functional which measures the degree to which these homogeneous relationships are being approximated is given by

$$f(\underline{a}) = \sum_{k=1}^{s} w_k \underline{a}^* \hat{R}_k^* \hat{R}_k \underline{a} \qquad (13)$$

where the w_k are nonnegative weights and s is a positive integer. The integers s and t are normally chosen so that the largest autocorrelation lag estimate argument (i.e., the variable n of $r_x(n)$) appearing in $f(\underline{a})$ is less than N. This is found to result in the inequality

$$s + t \leq N - q. \qquad (14)$$

The selection of s and t play an essential role in the performance capability of the method to be now described. Generally, improved performance is achieved upon selecting s and t reasonably large subject to the inequality constraint (14). Moreover, it has been found that letting $s = t$ yields a satisfactory selection for the s and t variables.

Our objective is to then find an autoregressive parameter vector with first component equal to one which will minimize quadratic functional (13). This minimization will have the effect of causing each of the vectors $R_k \underline{a}$ to be nearly equal to the zero vector as required in the theoretial expression (6). It is readily shown that this optimum autoregressive parameter vector estimate is obtained by solving the following consistent system of linear equations

$$S\underline{a} = \alpha \underline{e}_1 \qquad (15)$$

where $\underline{e}_1$ is the $(p + 1) \times 1$ vector whose components are all zero except for its first which is equal to one, and, α is a normalizing constant selected so that the first component of $\underline{a}$ is one as required. The $(p + 1) \times (p + 1)$ matrix S has as its components

$$S(i,j) = \sum_{k=1}^{s} w_k \sum_{m=1}^{t} \overline{\hat{R}}_k(m,i) \hat{R}_k(m,j)$$

$$1 \leq i,j \leq p + 1 \qquad (16)$$

where it is recalled that the element $\hat{R}_k(m,i)$ is an estimate of the autocorrelation lag term $r_x(q + k + m - i)$. The autoregressive parameter vector selection procedure (15) is a generalization of the high performance method of ARMA spectral estimation (e.g., see References [1]-[4]) in which all the weights w_k are zero except for $w_1 = 1$. The high performance method has been found to often provide a better spectral estimate than such contemporary procedures as the Burg method, periodogram, and the Box-Jenkins method [1]-[5].

3 SINGULAR VALUE DECOMPOSITION

It is possible to achieve a significant improvement in the modeling performance of the generalized method (15) by making use of the algebraic characterization of ARMA processes as outlined in Section 1. Namely, let us consider the extended $t \times (s + p)$ autocorrelation matrix R whose general (i,j) element is given by

$$R(i,j) = r_x(s + q + i - j) \quad \begin{array}{c} 1 \leq i \leq t \\ 1 \leq j \leq s + p \end{array} \quad (17)$$

where s is a positive integer variable corresponding to the upper index of the quadratic functional (13). We are particularly interested in this matrix since it contains within its makeup the $t \times (p + 1)$ matrices, $R_1, R_2, \ldots R_s$ whose estimates form the matrix S used in expression (15). Specifically the right most $p + 1$ column of R form R_1, while R_2 is identified as the $p + 1$ columns immediately to the left of the rightmost column of R and so forth.

If the autocorrelation lags used in formulating R by means of (17) correspond to an ARMA process of order (p,q), it follows that the rank of R would be p for all values of s and t that exceed or equal p. When autocorrelation lag estimates are substituted into (17) so as to obtain the extended autocorrelation matrix estimate $\hat{R}$, however, it is generally found that the rank of $\hat{R}$ is full (i.e., min (s,t)). In order to remove this rank incompatibility, it would be desirable to find a matrix of rank p, denoted by $\hat{R}^{(p)}$, which is closest to $\hat{R}$ in the mean square sense. This best rank p approximation is readily found by first making a singular value decomposition (SVD) of $\hat{R}$ (e.g, see Reference [6]), that is

$$\hat{R} = U \Sigma V^* \quad (18)$$

where U and V are $t \times t$ and $s \times s$ unitary matrices (i.e., $U^{-1} = U^*$ and $V^{-1} = V^*$), respectively, and, the $t \times s$ matrix Σ has all zero components except for nonnegative singular values λ_{nn} which appear along its diagonal which are ordered in a nonincreasing fashion (i.e., $\lambda_{nn} \geq \lambda_{n+1,n+1}$). The best rank p approximation to the extended autocorrelation matrix estimate is then readily

$$\hat{R}^{(p)} = U \Sigma_p V^* \quad (19)$$

where U and V are as in Equation (18) while Σ_p is obtained from Σ by setting to zero all but its p largest singular values. The general (i,j)th element of this best rank p approximation is thereby found to be

$$\hat{R}^{(p)}(i,j) = \sum_{n=1}^{p} \lambda_{nn} u_n(i) \bar{v}_n(j) \quad (20)$$

where $\underline{u}_n$ and $\underline{v}_n$ denote the nth column vectors of the unitary matrices U and V respectively.

Using this best rank p matrix approximation of $\hat{R}$, we then extract the best rank p approximations of the R_k for $k = 1, 2, \ldots, s$ along the lines suggested below Equation (17). Next, these matrix elements are substituted into relationship (15) to generate the best rank p adapted matrix $S(p)$. It is readily shown that this matrix has elements

$$S^{(p)}(i,j) = \sum_{k=1}^{s} w_k \sum_{m=1}^{t} \bar{R}^{(p)}(m,s+i-k) \hat{R}^{(p)}(m,s+j-k)$$

$$1 \leqslant i, j \leqslant p + 1 \tag{21}$$

where expression (20) is incorporated in evaluating these matrix elements. The required autoregressive parameter vector is then obtained by solving the following system of $(p + 1)$ linear equations

$$S^{(p)}\underline{a} = \alpha \underline{e}_1 \tag{22}$$

in which the normalizing scalar α is selected so that the first component of a is one.

4 NUMERICAL EXAMPLE

In this section, the effectiveness of the generalized ARMA modeling procedure herein developed will be demonstrated for the specific time series composed of two sinusoids in additive white noise as represented by

$$x(n) = \sin(2\pi f_1 n) + \sin(2\pi f_2 n) + w(n) \tag{23}$$

in which the white noise process $\{w(n)\}$ has variance one and the sinusoidal frequencies are specified by

$$f_1 = 0.2 \quad \text{and} \quad f_2 = 0.215.$$

This time series serves as an excellent vehicle for testing the resolution capabilities of spectral estimators because of the closeness of the sinusoidal frequencies (i.e., $\Delta f = 0.015$) and the prevailing low individual sinusoidal signal-to-noise ratios of zero dB. Using the time series description (23), ten different sample sequences (i.e., different white noise samples) each of length 128 were generated. These observation sets were then used to test the effectiveness of this paper's generalized spectral estimation method as well as the Burg method and the Periodogram. The spectral estimates which arose from the ten different observation sets using these three methods are shown in Figures 1, 2, and 3.

The ten ARMA spectral estimates of order (p,p) which arose when using this paper's method are shown superimposed in Figure 1 in a plot of $|A_p(e^{j\omega})|^{-2}$. Three separate ARMA spectral models are here shown. Namely, the standard high performance method (15) for order $p = 8$ and $s = 1$, $t = 70$ with $w_k = 1$ yielded the ten spectral estimates depicted in Figure 1a. Although spectral energy was detected in the region about $f = 0.2$, a resolution of the two sinusoids was not achieved in any of the estimates. Clearly, the ARMA order was insufficient for the task at hand. In recognition of this, the high performance method for order $p = 12$ and $s = 1$, $t = 70$ with $w_k = 1$ was next tried and resulted in the ten plots shown in Figure 1b. In this case, the two sinusoids are resolved in all of the estimates.

To illustrate the improvement provided by a singular value decomposition in ARMA modeling, the SVD generalized high performance method (22) of order $p = 4$ and $s = t = 50$ with $w_k = 1$ was used and gave rise to the plots shown in Figure 1c. In this case, all estimates produced a nearly identical resolution plot. It is noteworthy that this resolution was obtained with the lowest ARMA model order consistent with the given time series (i.e., $p = 4$). To

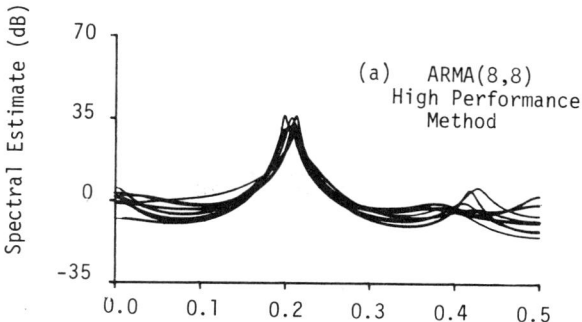

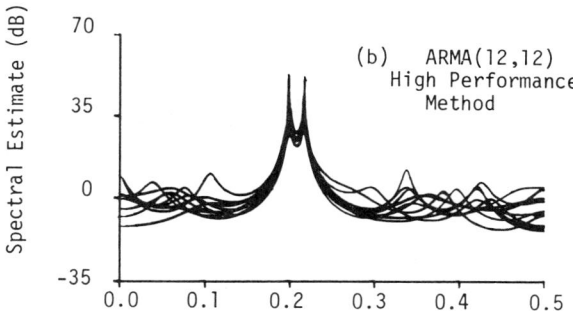

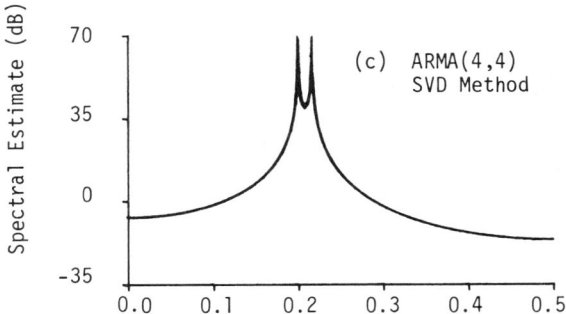

Figure 1 Ten superimposed ARMA (p. 8) spectral models each using 128 time series observations,

(a) high performance method t = 70, p = 2-8,

(b) high performance method t = 70, p = q = 12,

(c) SVD generalized method t = S = 50, p = q = 4

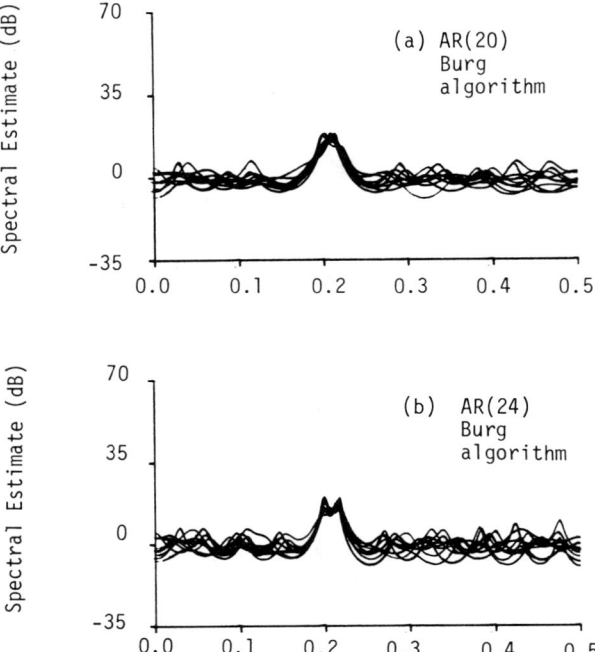

Figure 2 Ten superimposed Burg CAR(p) spectral estimates each using 128 time series observations, (a) p = 20, (b) p = 24

illustrate the usefulness of the singular value in determining model order, the fifty singular values associated with a typical estimate were found to be

$$\lambda_1 = 25, \lambda_2 = 24.7, \lambda_3 = 15.3, \lambda_4 = 15.1,$$
$$\lambda_5 = 1.2, \lambda_6 = 1.1, \ldots, \lambda_{50} = 0.03.$$

From these results, the four largest singular values are seen to be obviously dominant thereby dictating the correct $p = 4$ model order choice.

The quality of the frequency and associated pole magnitude (theoretically equal to one) estimates of the SVD ARMA (4,4) model were also very good as indicated in Table 1. In this table, the terms $\bar{f}_k$ and $\bar{p}_k$ denote the sample mean of the frequency and pole magnitude estimates, and, σ_{f_k} and σ_{p_k} the associated sample standard deviations as obtained for the ten spectral estimates. Clearly, both the frequency and pole magnitude estimates are of high quality.

Table 1 Statistics of ARMA Estimates

k	f_k	$\bar{f}_k$	σ_{f_k}	$\bar{p}_k$	σ_{p_k}
1	0.20	0.1996	0.0007	0.9991	0.0018
2	0.215	0.2160	0.0005	0.9984	0.0032

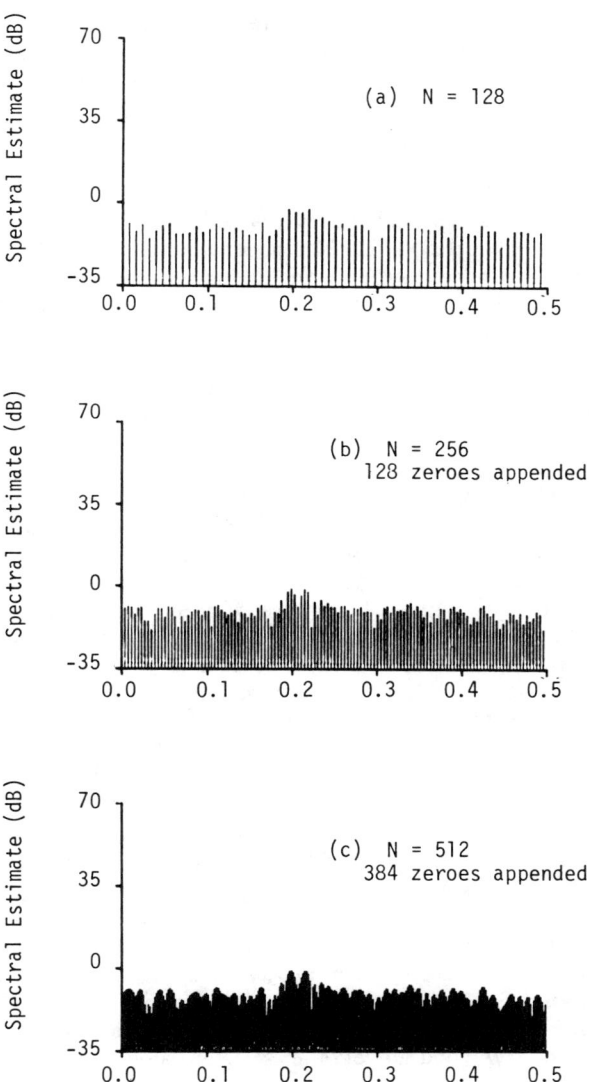

Figure 3 MA spectral estimates using 128 times series observations employing an FFT implementation of the periodogram (a) N = 128, (b) N = 256 with 128 point zero padding, (c) N = 512 with 384 point zero padding

The ten spectral estimates which arose from the Burg method on the same set of ten observations for different AR model orders are shown in Figure 2. The ten Burg AR spectral estimates of order 20 are shown superimposed in Figure 2a. As in the ARMA (8,8) estimates, the AR (20, 20) Burg estimates indicate the presence of spectral energy in a neighborhood of $f = 0.2$ but a resolution of the two sinusoids was not obtained. Upon increasing the AR order to 24, however, 9 out of the 10 estimates provided the desired spectral resolution. It is noteworthy that the *twelfth* order high performance ARMA model and the *fourth* order SVD ARMA generalized high performance model outperformed the *twenty-fourth* order Burg AR model for this sinusoidal class of time series. This is a direct consequence of the fact that an ARMA model is more compatable with a time series composed of sinusoids in white noise than is an AR model.

When the periodogram method as implemented by the fast Fourier transform (FFT) was applied to the same ten different observation sets, the resultant ten spectral estimates were remarkably similar. A typical 128 point FFT periodogram from one of these runs is shown in Figurere 3a. From this plot (and in the nine other plots), it is possible to detect only one of the two sinusoids with a number of other ambiguous detections (one true and others false) being feasible. In order to ease this ambiguity, the 128 time series observations were next padded with 128 zeroes. The resultant 256 point padded FFT periodogram is shown in Figure 3b. A clear improvement is evident through the process of padding, whereby the required resolution is achieved. Finally, the given 128 time series observations were padded with 384 zeroes and the resultant 512 point padded FFT periodogram is shown in Figure 3c. As in the 256 padded case, both sinusoids are detected.

5 CONCLUSIONS

The generalized high performance ARMA modeling methods as represented by expressions (15) and (22) are based on the algebraic properties possessed by ARMA random processes. The particular implementation involving the SVD as given by relationship (22) has been found to provide exceptional spectral estimates on the limited number of examples treated to date. An illustration of such was given in the numerical example section. A by-product of the SVD approach is that of model order selection. It has been found that the singular value behavior of the extended autocorrelation matrix estimate R provides an exceptionally good model order selection procedure.

REFERENCES

1. J.A. Cadzow, ARMA Time Series Modeling: An Effective Method, IEEE Trans. on Aerospace and Electronic Systems, Vol. AES-19, No. 1, pp. 49-58, Jan. 1983.

2. J.A. Cadzow, ARMA Modeling of Time Series, Special Issue on Digital Signal and Waveform Analysis—IEEE Trans. on Patt. Analy. and Mach. Intell., Vol. PAMI-4, No. 2, pp. 124-128, Mar. 1982.

3. J.A. Cadzow, High Performance Spectral Estimation—A new ARMA Method, IEEE Trans. on Acoust., Speech, Sig. Proc., Vol. ASSP-28, pp. 524-529, Oct., 1980.

4. J.A. Cadzow, Spectral Estimation: An Overdetermined Rational Model Equation Approach, Special Issue of the IEEE proceedings on Spectral Estimation, Vol. 9, No. 9, pp. 907-939, Sept., 1982.

5. Y.H. Pao and D.T. Lee, Additional Results on the Cadzow ARMA Method for Spectral Estimation, Proc. of the 1981 ICASSP, Atlanta, GA, pp. 480-487, March, 1981.

6. V.C. Klema and A.J. Laub, The Singular Value Decomposition: Its Computation and Some Applications, IEEE Trans. on Auto. Control, Vol. 25, No. 2, pp. 164-176, April, 1980.

A Statistical Frequency Domain Signal Processing Method*

Roger F. Dwyer

*Surface Ship Sonar Department
Naval Underwater Systems Center
New London, CT*

1 INTRODUCTION

In many important signal processing applications, including underwater acoustics, an estimate of the power spectral density (PSD) of the received data is often employed for signal detection. The data are first transformed into the frequency domain by utilizing the discrete Fourier transform (DFT) which can be efficiently executed by an algorithm called the fast Fourier transform. At this point, the data are considered to be in the frequency domain and an estimate of the PSD can be easily obtained.

The PSD is essentially a sum of the estimates of the second order moments for both the real and imaginary parts for each frequency component in the frequency domain. If the frequency domain signals are not Gaussian distributed, then higher order moments of the complex frequency components would contain additional information that may be utilized in signal processing. The object of this paper is to compare the PSD technique for signal processing with a new method which computes the frequency domain kurtosis (FDK) (1) for the real and imaginary parts of the complex frequency components.

2 FREQUENCY DOMAIN KURTOSIS

Let $x(i,q) = x((i + (q-1)M)h)$, $i = 0, 1, \ldots, M-1$, $q = 1, 2, \ldots, n$ represent the real discrete data where h is the interval between successive observations of the process. The DFT is defined as

$$X(q,F_p) = \sqrt{h/M} \sum_{i=0}^{M-1} W_i x(i,q) \exp(-jF_p i) \qquad (1)$$

---

*Work supported under Work Request N0001483WR30054, Task NR 042-499 with the Office of Naval Research.

where $j = \sqrt{-1}$, $F_p = 2\pi f_p h$ is the p-th radian frequency component, $p = 0, 1, \ldots, M - 1$ and $f_p = p/Mh$ Hz.. For simplicity, we shall assume the window weights $W_i = 1$ for all i and $h = 1$.

The input $x(i,q)$ will be a zero mean process which is composed of an additive mixture of signal and noise of the form

$$x(i,q) = N(i,q) + m(i,q)s(i,q) \qquad (2)$$

where $m(i,q)$ will either represent the effects of the propagation medium or reflect a physical characteristic of the transmitted or radiated signal $s(i,q)$. The components $N(i,q)$ and $s(i,q)$ are zero mean stationary processes and $N(i,q)$, $m(i,q)$ and $s(i,q)$ will be assumed to be mutually independent from each other.

Our model for the fading received signal $m(i,q)s(i,q)$ assumes that the total effects of the amplitude fluctuations due to multipath interference or to nonstationarities of the source, receiver, or of the medium can be simply included in the multiplicative function $m(i,q)$. On the other hand the phase fluctuations of the signal will be contained in the function $s(i,q)$ itself. This approach applies to sound propagating in the ocean (2) and to electromagnetic communication systems (3).

The power spectrum estimate is defined as

$$\hat{P}(F_p) = (1/n) \sum_{q=1}^{n} X(q,F_p) X^*(q,F_p) \qquad (3)$$

where the asterisk represents complex conjugate. In the PSD estimate considered here, n non-overlapped DFT segments are averaged to insure that each frequency component represents a consistent PSD estimate (4). Thus, Equation (3) is an asymptotically unbiased estimate of the power spectral density (4). It should also be pointed out that overlapped segments have also been studied to reduce the variance in the PSD estimate for another application (5), but this technique will not be treated here.

Substituting Equation (2) into Equation (3) and taking the expected value, we obtain the result

$$E(P(F_p)) = (1/M) \sum\sum (R_N(i_1 - i_2) \exp(-jF_p(i_1 - i_2)))$$
$$+ (1/M) \sum\sum (E(m(i_1,q)m(i_2,q)) R_s(i_1 - i_2) \exp(-jF_p(i_1 - i_2))) \qquad (4)$$

where $R_n(i_1 - i_2) = E(N(i_1,q)N(i_2,q))$
$R_s(i_1 - i_2) = E(s(i_1,q)s(i_2,q))$

and the sums are from $(i_t = 0$ to $i_t = M - 1)$ $t = 1, 2$. When it is obvious, in succeeding equations, we shall not indicate the index for the sum symbols.

The FDK is defined separately for the real and imaginary parts of Equation (1). Only the results for the real part will be discussed here. The imaginary part derivation is identical. In practice, both parts are computed, since each can contribute information. Reference (1) discusses an analysis using the FDK method for arctic underice ambient noise data. Generally, the kurtosis measures the peakedness of the data.

For a Gaussian distribution, the kurtosis is 3 within a confidence bound determined by the number of samples used in the estimate (6). Techniques for

Frequency Domain Signal Processing

optimumly processing signals contaminated by underice ambient noise as well as other noise environments are presented in References (7) and (8).

The real FDK estimate is defined as

$$K_r(F_p) = (1/n) \sum_{q=1}^{n} (X^r(q,F_p))^4 / \left\{ (1/n) \sum_{q=1}^{n} (X^r(q,F_p))^2 \right\}^2 \tag{5}$$

where r signifies the real part of Equation (1). Since we are only considering the real part, we shall refrain from indicating r in subsequent equations in order to simplify the notation.

The FDK is defined by taking the expected value of the 4th order central moment and the square of the expected value of the 2-nd order central moment separately, and then forming the ratio. The result of this operation is

$$K(F_p) = E((X(q,F_p))^4)/(E((X(q,F_p))^2))^2 \tag{6}$$

and after some manipulations, the terms of this expression are given by

$$E((X(q,F_p))^4) = (1/M^2) \text{OP}_4 \{ E(N(i_1,q)N(i_2,q)N(i_3,q)N(i_4,q) \}$$
$$+ 6(1/M) \text{OP}_2 \{ R_N(i_1 - i_2) \} (1/M) \text{OP}_2 \{ E(m(i_1,q)m(i_2,q)) R_s(i_1 - i_2) \}$$
$$+ (1/M^2) \text{OP}_4 \{ E(m(i_1,q)m(i_2,q)m(i_3,q)m(i_4,q)) \tag{6a}$$
$$E(s(i_1,q)s(i_2,q)s(i_3,q)s(i_4,q)) \},$$

and,

$$E((X(q,F_p))^2) = (1/M) \text{OP}_2 \{ R_N(i_1 - i_2) \}$$
$$+ (1/M) \text{OP}_2 \{ E(m(i_1,q)m(i_2,q)) R_s(i_1 - i_2) \} \tag{6b}$$

where we have defined the operators

$$\text{OP}_4 \{ \ \} = \sum\sum\sum\sum \{ \ \} \cos(F_p i_1) \cos(F_p i_2) \cos(F_p i_3) \cos(F_p i_4)$$

$$\text{OP}_2 \{ \ \} = \sum\sum \{ \ \} \cos(F_p i_1) \cos(F_p i_2).$$

It was proven in Reference (6) that Equation (5) is an asymptotically (as $n \to \infty$) unbiased estimate of the kurtosis if the distribution is Gaussian. In practice the true value of the FDK can only be approached asymptotically. We will assume here that n is sufficiently large and only be concerned with examining the properties of the FDK as defined in Equation (6).

CASE NO. 1. Independent Noise

Let $s(i,q) = 0$ and assume that the noise samples are identically distributed and statistically independent. Then, as $M \to \infty$ $K(F_p) \to 3$ for each p and for any distribution function of $N(i,q)$. The proof of this result can be obtained by expanding the first term of Equation (6a), defining $R_N(i_1 - i_2) = R_n \delta_{i_1 i_2}$ and letting M become large. This result suggests that as $M \to \infty$ for independent samples at the input, the output distribution function of the DFT will approach a Gaussian distribution independent of the input distribution as measured by the FDK.

Hereafter, we shall assume that the input noise is independent and Gaussian with PSD R_N. The real part of the noise PSD is the $R_N/2$.

3 SLOWLY FADING SIGNAL

Signal fluctuations on the order of seconds have been measured in underwater acoustic propagation studies (9, 10, 11, 12, 13). In some cases the received signal, over a period of seconds, has been measured to fluctuate as much as 50 dB (13). The actual amplitude distribution function for a purely sinusoidal transmitted signal, however, couldn't be exactly specified, although it was close to Rayleigh or Gaussian for a particular experimental scenario (10, 11, 13). Reference (14) presents data which shows that the amplitude fluctuations can follow a Rayleigh or log-normal distribution depending upon the propagation path. In addition there is a much slower fluctuation on the order of 10 minutes or more superimposed on those "faster" fluctuations (9, 12).

For the rest of the paper we will only consider cases where the fluctuations are slow with respect to the data interval Mh, but may be considered changing rapidly over the total detection interval nMh. These cases are supported by the measured fluctuations, reported in the references cited above, and from practical limitations of implementing large DFT's.

To summarize, if $m(i,q)$ is slowly changing with respect to Mh, or equivalently to the length of the DFT, but rapidly changing over the detection interval, then we may approximate $m(i,q) = m(q)$ and therefore, we can rewrite Equation (6) as follows

$$K(F_p) = (3 + 6A_1 + FA_2)/(1 + A_1)^2 \qquad (7)$$

where

$$A_1 = E(m^2)(1/M)\text{OP}_2\{R_s(i_1 - i_2)\}/R_N/2$$
$$A_2 = (E(m^2))^2(1/M^2)\text{OP}_4\{E(s(i_1,q)s(i_2,q)s(i_3,q)s(i_4,q))\}/(R_N/2)^2$$

and we have defined the parameter

$$F = E(m^4)//(E(m^2))^2.$$

CASE NO. 2. Sinusoidal Signal

Let the signal be given by $s(i,q) = \cos(F_k i + \phi)$ where we have neglected the index q for convenience and ϕ represents a random phase distributed uniformly between $0 - 2\pi$.

After substituting into Equation (7) the sinusoidal signal, we find, after some simplification, that

$$K(F_p) = \{3 + 3A_1 + 3/8FA_1^2\}/\{1 + 1/2A_1\}^2 \qquad (8)$$

where $.5 \cos(F_k(i_1 - i_2))$ replaces $R_s(i_1 - i_2)$ in Equation (7). If M is sufficiently large, then $(1/M)\text{OP}_2\{\cos(F_k(i_1 - i_2))\} \to M/2$ if $F_p = F_k$ and is zero otherwise.

The signal-to-noise ratio, as measured by the PSD, is given by the parameter A_1. The Appendix provides a proof showing that the signal-to-noise ratio for the PSD and the FDK estimates are equal.

If the signal-to-noise ratio $A_1 \to \infty$ then $K(F_p) \to 1.5F$. This result could be used to advantage to determine the propagation conditions. For example, suppose the PSD shows a large frequency component but the FDK is 1.5. This would indicate that $F = 1$ and no fading or fluctuation exists in the propagation

Table 1

DENSITY NAME	F	DENSITY FUNCTION
HALF-NORMAL	1.5	$f(x) = (2/\pi\sigma^2)^{1/2} \exp(-[1/2\sigma^2]x^2) \quad x \geq 0$
UNIFORM	1.8	$f(x) = 1/a, \quad 0 \leq x \leq a$
RAYLEIGH	2.0	$f(x) = (x/\sigma^2) \exp(-[1/2\sigma^2]x^2) \quad x \geq 0$
GAUSSIAN	3.0	$f(x) = (1/2\pi)^{1/2} \exp(-[1/2\sigma^2]x^2) \quad -\infty \leq x \leq \infty$
GAMMA	$\Gamma(\eta) \dfrac{(\eta+3)!}{((\eta+1)!)^2}$ $\eta \to 1 \; F \to 6$ $\eta \to \infty \; F \to 1$	$f(x) = \dfrac{\lambda^\eta}{\Gamma(\eta)} x^{\eta-1} \exp(-\lambda x) \quad x \geq 0, \; \lambda > 0$
LOG-NORMAL	$\exp(4\sigma^2)$ $\sigma^2 = .25, \; F = 2.72$ $\sigma^2 = .5, \; F = 7.4$ $\sigma^2 = .75, \; F = 20$ $\sigma^2 = 1.0, \; F = 55$	$f(x) = (\sigma x \sqrt{2\pi})^{-1}$ $\exp(-1/2\sigma^2 (\log x - u)^2) \quad x \geq 0$

path of the signal. In sonar, this result could be indicative of a short range signal propagating without multipath. On the other hand, if $K(F_p)$ is, say, greater than 3 and at the same time the PSD indicates a large frequency component, then this would suggest that $F > 1$ and fading or multipath propagation existed. For a Rayleigh fading environment $F = 2$, and for a log-normal environment F can be very large depending on the variance of the distribution. However, the measured data currently available (14) suggests that F is between 2 and 3 depending upon frequency. We included a larger spread in F, which is theoretically justified, to show its effect on the FDK estimate. Nevertheless there may be cases of practical interest where F is larger than the current data suggests. For example the effect of receiver and source motion and also high sea states on F needs to be experimentally measured.

Table I lists various values of F for some specific distribution functions which have been associated with fading and fluctuating signals in underwater acoustics, communication, and radar applications.

4 RANDOM NARROWBAND SIGNAL

In many applications, the radiated or transmitted signal is not a pure sinusoid but rather, a periodic random signal. For example, radiated ship noise is composed of random narrowband noise (15, 16). We shall now examine the FDK for a narrowband Gaussian signal. The results of the following discussion are limited to the multiplicative model given in Equation (2).

CASE NO. 3. Narrowband Gaussian Signal

Let the signal be Gaussian so that

$$\begin{aligned} E(s(i_1,q)s(i_2,q)s(i_3,q)s(i_4,q)) &= E(s(i_1,q)s(i_2,q))E(s(i_3,q)s(i_4,q)) \\ &+ E(s(i_1,q)s(i_3,q))E(s(i_2,q)s(i_4,q)) \\ &+ E(s(i_1,q)s(i_4,q))E(s(i_2,q)s(i_3,q)). \end{aligned} \quad (9)$$

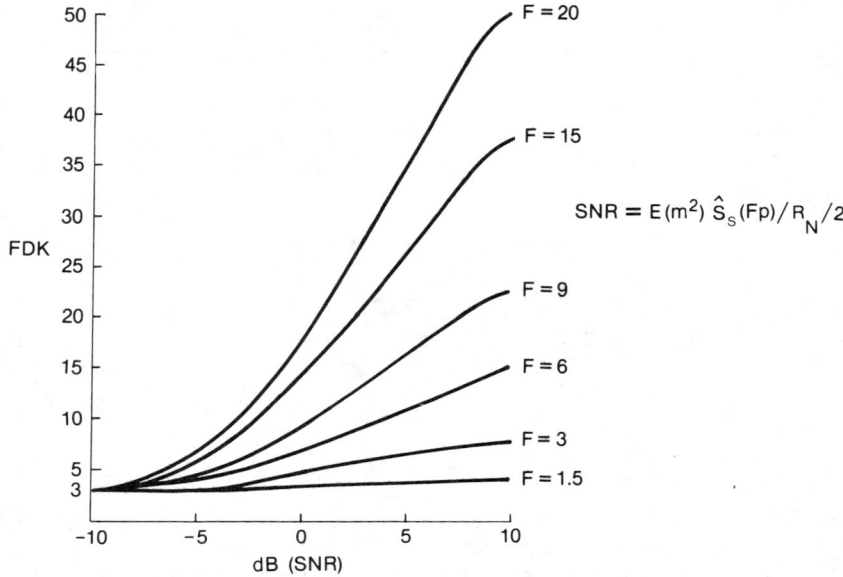

Figure 1 FDK versus SNR for a narrowband Gaussian signal in a fluctuating environment indicated by the values of F.

After substituting Equation (9) into Equation (6) and employing the slow fading condition as well as assuming that the noise is independent and Gaussian, we obtain for the FDK the expression

$$K(F_p) = 3(1 + 2A_1 + FA_1^2)/(1 + A_1)^2 \qquad (10)$$

where A_1, as defined in Equation (7), represents a signal-to-noise ratio (SNR).

For $F = 1$, then $K(F_p) = 3$. However, for all other values of $F > 1$ the FDK is greater than 3.

Figure 1 represents a plot of FDK as a function of F and SNR. We mention again that the SNR, given by the parameter A_1, represents the output PSD estimate SNR. From Figure 1 we can see that if F is greater than, say, 3 the FDK gives high values for the kurtosis even if the SNR is small. This theory suggests the possibility of detecting signals via the FDK at lower SNR than the PSD method. The data presently available to support this theory is the arctic underice ambient noise data (1) which show that the FDK can detect, in some case, non-Gaussian signals whereas the PSD method cannot. The non-Gaussian signals in the underice data were due to ice movement which suggests a different statistical model for $m(i,q)$. Here we are representing a physical characteristic of the signal by the multiplicative function $m(i,q)$ for convenience. Therefore, in order to include the underice data in the FDK formula, we will assume that $m(q) = am_1(q)$, where a is a random variable which takes on two values, $a = 0$, $a = 1$, with the probabilities $P(a = 0) = 1 - L$, $P(a = 1) = L$. The other component $m_1(i,q)$ will represent the propagation

Frequency Domain Signal Processing

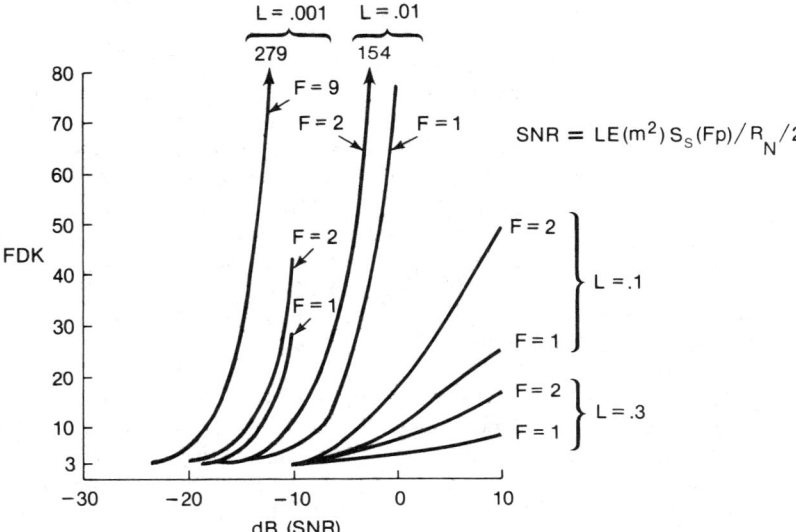

Figure 2 FDK versus SNR for a randomly occurring narrowband Gaussian signal in a fluctuating environment indicted by the values of L and F, respectively.

conditions as discussed above. The parameter allows us to consider transient, as well as frequency modulated signals. In the latter case, we simply assume that the frequency modulation causes the signal to appear in a particular frequency location randomly and, therefore, over the total interval will have a certain probability L of occurring. The actual modulation is assumed to be unknown. Therefore, taking the random variable a into account, we obtain the relationship for the FDK as

$$K(F_p) = 3(1 + L(2A_1 + FA_1^2))/(1 + LA_1)^2 \qquad (11)$$

where A_1 is defined above. If $L = 1$, then Equation (11) reduces to the results given in Figure 1. The parameters L and F are actually functions of frequency. However, we have not explicitly shown this dependence for notational simplicity.

Figure 2 represents a plot of specific values for $K(F_p)$ under different combinations of parameters L and A_1. We have also defined the parameter eff SNR which represents the effective SNR as measured by the PSD estimate and is equal to LA_1. The figure is actually plotted in terms of this parameter in dB. Also plotted for comparison is Equation (11) for values of $K(F_p)$ when F is greater than 1. These results suggest that for a transient or a frequency modulated signal, the FDK estimate may be a better detection statistic than the PSD method. If the medium is also a fading environment, which occurs often in underwater acoustics, then the FDK estimate is significantly enhanced as shown in the figure.

5 REAL DATA EXAMPLES

In order to verify the results of this paper we have included real data examples, shown in Figures 3 and 4, which represent signals generated from ice movement. A complete description of this data can be found in Reference 1. Essentially, ice movement produces transient and frequency modulated signals.

The data were processed by a 1024 point fast Fourier transform (FFT). In Figures 3 and 4 the top curve represents the PSD estimate in dB vs normalized frequency (Horizontal scale). The data in Fgure 3 was first filtered through a lowpass filter with a bandwidth of 2500 Hz and then sampled at a 10 kHz rate, giving a resolution of 10 Hz. The PSD estimate was obtained by appropriately averaging the output of 1000 consecutive FFT's which gave an overall time interval of 1.7 minutes.

The bottom curve of Figures 3 and 4 is the corresponding FDK estimate for the real part of the FFT output. Similarly, the FDK estimate for Figure 3 was also obtained by appropriately averaging over the 1000 consecutive FFT outputs. Many of the FDK estimates in Figure 3 deviate significantly from the Gaussian assumption over a wide bandwidth and thereby indicate the location of non-Gaussian signals. This conclusion cannot be obtained from the PSD estimate alone. In fact, it would be difficult to make a detection given the PSD estimate in Figure 3, since one would also have to know the PSD noise level.

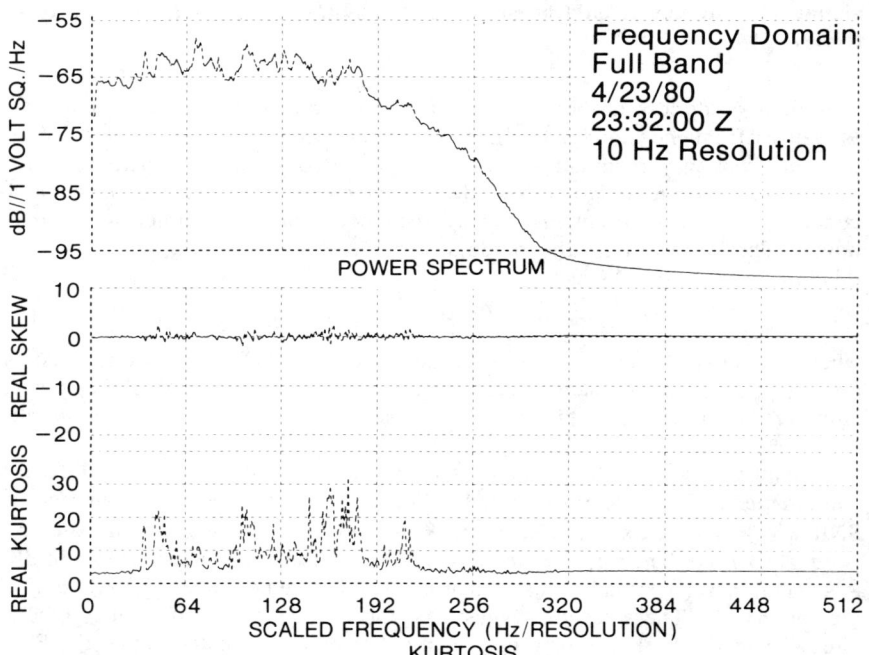

Figure 3 Real arctic data at 10 Hz resolution. **Top curve: power spectrum density estimate. Bottom curve: corresponding frequency domain Kurtosis estimate.**

Frequency Domain Signal Processing 87

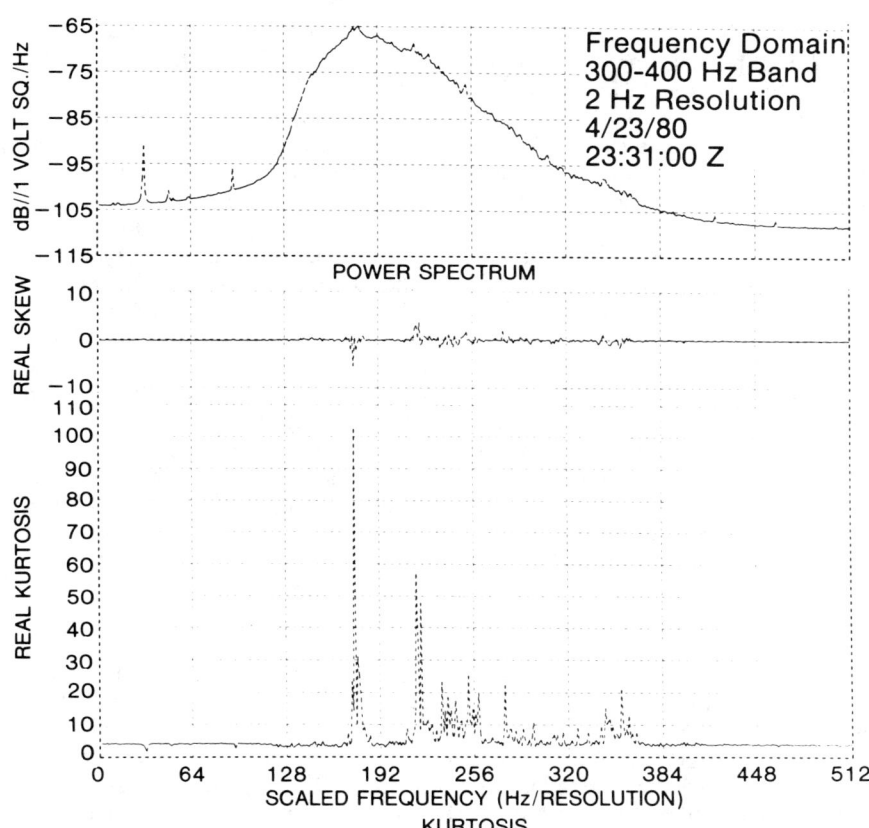

Figure 4 Real arctic data at 2 Hz resolution. Top curve: power spectrum density estimate. Bottom curve: corresponding frequency domain Kurtosis estimate.

Therefore, the advantage of the FDK estimate is that its actual value is significant, since we are looking for deviations from the Gaussian value of 3, whereas, the PSD estimate represents a relative value.

In Figures 3 and 4 we also show an estimate of the skew, in the middle curves, for comparison purposes only. A discussion of the properties of the skew will not be given in this paper, however, Reference 1 addresses the experimental significance of this measurement.

Figure 4 represents a different data set than Figure 3. The data were first passed through a 100 Hz bandpass filter centered at 350 Hz and then sampled at 2 kHz rate which gave a resolution of 2 Hz. The estimates were obtained by averaging 750 consecutive FFT outputs giving an overall time interval of 6.25 minutes. In the top curve, on the left, in Figure 4 is a 60 Hz tonal due to an electrical ground loop. The corresponding FDK estimate, indicated in the bottom curve by a "dip,"! has a value of 1.8. The next "dips" at higher frequencies are due to harmonics of the 60 Hz ground loop. These results were predicted by Equation (8) in the text since as SNR $\to \infty$, $K(F_p) \to 1.5$ F. For a ground loop $F = 1$ because the signal is not being propagated through the medium. However due to system noise the FDK estimate is prevented from reaching 1.5.

Within the passband of the filter, which is clearly seen in the top curve in Figure 4, many frequency locations deviate significantly from Gaussian indicating the presence of non-Gaussian signals. We also notice some small indication of a signal present in the corresponding PSD estimate at some of the frequency locations. However, the PSD estimate does not contain information on the non-Gaussian nature of the signals and, therefore, in order to detect these signals the noise only PSD estimate must be known. These results again show the advantage of the FDK estimate over the PSD in detecting non-Gaussian signals in some cases of practical importance.

In closing, we shall make one additional observation concerning the bandwidth of the signal. If $s(i,q)$ was a broadband signal Equation (11) would still be valid. The FDK estimate, however, would be constant over the bandwidth of the signal. Therefore, non-Gaussian broadband signals would still register kurtosis estimates but across the entire band. This is an advantage over the PSD method for detecting non-Gaussian signals, since the actual FDK estimate is significant and not its relative value as is true for the PSD method when the noise level is not known.

6 SUMMARY

We have considered using the frequency domain kurtosis estimate to obtain additional information about the frequency components of a received non-Gaussian signal. The FDK measure was defined in the frequency domain for the real and imaginary parts of each frequency component as the ratio of the averaged value of the fourth order central moment, and, the square of the second order central moment, of a DFT output. Both sinusoidal and narrowband Gaussian signals were investigated which were being propagated through a fading or multipath environment. In addition, transient and frequency modulated signals were modeled by introducing a mixing parameter L. A theoretical derivation shows that under some practical conditions which are known to exist in underwater acoustics as well as other environments, the FDK estimate gives an additional measure for non-Gaussian signals which may, under certain conditions outlined above, be more significant than the PSD estimate. In other cases, the FDK estimate reflects the propagation condition and, therefore, enhances the information gained by the PSD estimate.

APPENDIX

The signal-to-noise ratio of the PSD and the FDK estimates can be shown to be equal by defining the relationship (9)

$$R_s(i_1 - i_2) = \int_{-1/2}^{1/2} S_s(f) \exp(j\omega(i_1 - i_2)) df \qquad (12)$$

where $S_s(f)$ is the two-sided spectrum of the real symmetric correlation function $R_s(i_1 - i_2)$. Substituting Equation (12) into the expression $(1/M)OP_2\{R_s(i_1 - i_2)\}$ which represents the estimate of $S_s(F_p)$ for the real part of the PSD, we obtain for $F_p \neq 0$

$$\hat{S}_s(F_p) = 1/2 \int_{-1/2}^{1/2} S_s(f)(W_M(\omega - F_p))^2 df \qquad (13)$$

where we have neglected some terms which converge to zero as $M \to \infty$ and

$$(W_M(x))^2 = (1/M)(\sin(M/2x)/\sin(1/2x))^2$$

is called the Bartlett spectrum window (17, 18). The PSD estimate of $S_s(F_p)$ is equal to twice the value obtained in Equation (13). However, since the real part of the noise PSD estimate is also reduced by 2, the signal-to-noise ratio is the same for both the PSD and the FDK estimates.

If $F_p = 0$ then the SNR of the FDK estimate and the PSD estimate are also equal which can be easily verified by comparing Equations (4) and (6b).

ACKNOWLEDGMENT

The author wishes to thank Dr. D. Tufts of the University of Rhode Island and Dr. S. Schwartz of Princeton University for their support and helpful comments.

REFERENCES

1. R. Dwyer, "FRAM II Single Channel Ambient Noise Statistics," NUSC Tech. Document 6583, 25 Nov. 1981. Also presented at the 101st meeting of the Acoustical Society of America, 19 May 1981.

2. I. Dyer, "Statistics of Sound Propagation in the Ocean," J. Acoust. Soc. Am. 48, pp. 337-345 (1970).

3. P.A. Bellow and B.D. Nelin, "The Influence of Fading Spectrum on the Binary Error Probabilities of Incoherent and Differentially Coherent Matched Filter Receivers," IRE Trans. CS-10, pp. 160-168, June 1962.

4. M.J. Hinich and C.S. Clay, "The Application of the Discrete Fourier Transform in the Estimation of Power Spectra, Coherence, and Bispectra of Geophysical Data," Reviews of Geophysics, Vol. 6, No. 3, Aug. 1968.

5. G.C. Carter, C.H. Knapp and A.H. Nuttall, "Estimation of the Magnitude-Squared Coherence Function via Overlapped Fast Fourier Transform Processing," IEEE Trans. on Audio Electroacoustics, Vol. AU-21, pp. 337-344.

6. E.S. Pearson, "A Further Development of Tests for Normality," Biometrika 16, pp. 237-249, 1930.

7. Y. Ching and L. Kurz, "Nonparametric Detectors Based on M-Interval Partitioning," IEEE Trans. Inf. Theory, IT-18, pp. 241-250 (1972).

8. R. Dwyer, "Detection of Partitioned Signals by Discrete Cross-Spectrum Analysis," IEEE ICASSP Convention Record, 1980.

9. R.J. Urick, "Amplitude Fluctuations of the Sound from a Low-Frequency Moving Source in the Deep Sea," NOLTR 74-43, 26 Feb. 1974.

10. R.J. Urick, "Models for the Amplitude Fluctuations of Narrow-band Signals and Noise in the Sea," J. Acoust. Soc. Am., Vol. 62, No. 4, Oct. 1977.

11. G.E. Stanford, "Low-Frequency Fluctuations of a CW Signal in the Ocean," J. Acoust. Soc. Am., Vol. 55, No. 5, May 1974.

12. R.H. Nichols and H.J. Young, "Fluctuations in Low-Frequency Acoustic Propagation in the Ocean," J. Acoust. Soc. Am., Vol. 43, No. 4, 1968.

13. K.V. MacKenzie, "Long-Range Shallow-Water Signal-Level Fluctuations and Frequency Spreading," J. Acoust. Soc. Am., Vol. 34, No. 1, 1962.

14. P.F. Worcester, "Reciprocal Acoustic Transmission in a Midocean Environment: Fluctuations," J. Acoust. Soc. Am. Vol. 66, No. 4, Oct. 1979.

15. R.J. Urick, "Principles of Underwater Sound for Engineers," McGraw-Hill, 1967, New York, Chap. 10.3, pp. 266-285.

16. I. Vigness, "Random Vibration," Vol. 2, MIT Press, 1963, Cambridge, Mass., Chap. 8.

17. G.M. Jenkins and D.G. Watts, "Spectral Analysis and its Applications," 1968, Holden-Day, San Francisco, Chap. 6.

18. J. Capon, R. Greenfield and P. Kolker, "Multidimensional Maximum-likelihood Processing of a Large Aperture Seismic Array," Proceeding of the IEEE, Vol. 55, No. 2, Feb. 1967.

Part II: Signal Estimation and Detection

Detection in a Non-Gaussian Environment[*]

Stuart C. Schwartz and John B. Thomas

*Department of Electrical Engineering
and Computer Science
Princeton University
Princeton, NJ*

1 INTRODUCTION

In most engineering studies, there is usually a tradeoff between model complexity and analytical tractability; the more complicated (and realistic) the model is assumed to be, the more difficult the subsequent analysis becomes. This balance is especially delicate in the area of non-Gaussian signal processing. If departures from Gaussian statistics are investigated, it is usually assumed that the sequence of observations is independent and, often, identically distributed. With these basic assumptions, analytical results are often available. When non-Gaussian dependencies are taken into account, results are often only obtainable by means of Monte Carlo simulation since multivariate distributions are usually not available in analytical form.

It is within this framework that we summarize some of our preliminary results for detection in a non-Gaussian environment. We first consider the detection of a signal in nearly Gaussian skewed noise. Surprisingly, a small degree of skewness can lead to a significant degradation in performance of the linear detector as measured by false-alarm rate. A detector which exhibits the desired robustness is introduced. Some of the potential difficulties in using adaptive procedures are illustrated in the context of under-ice ambient noise data.

A general adaptive procedure is then outlined which uses the skewness-kurtosis plane to measure departures from Gaussian statistics. Overlays, which specify the probability density family of the observations, are introduced and used to determine the form of the nonlinear processor.

---

[*]Work supported under Contract N00014-81-K-0146, Task SRO 103 with the Office of Naval Research.

Non-Gaussian statistics are then characterized by the derivative of the logarithm of the probability density. This expression, $f'(x)/f(x)$, is estimated using the observations and is then used to form an optimum detector. It is shown by example that, under reasonable conditions, assuming a particular family of probability densities does not significantly degrade detector performance when the observations actually come from a density outside the family. This general conclusion is also arrived at using another approach, in which quantiles are used to measure the tail behavior of heavy-tailed probability densities.

In the last section, a class of multivariate non-Gaussian probability densities is defined. For this class, the locally optimum detector is seen to separate into a zero-memory nonlinear part and a part with memory. The results of a simulation to evaluate a number of detectors are presented. It is observed that simplified versions of the optimal processor also lead to improved performance when compared to conventional detectors.

2 DETECTION IN SKEWED NOISE

In robust and general nonparametric studies (Refs. 1 and 2), a symmetry assumption on the underlying noise density is usually made. (Another frequent nonparametric assumption is that the noise density is unimodal or that the median is 1/2.) In this section, we report on an investigation which assumes nonzero skewness and which studies the sensitivity of the linear and sign detectors to this lack of symmetry. Surprisingly, a small amount of skewness can lead to marked deterioration in system performance for the linear detector as measured by false-alarm rate. A modified detector is also introduced which exhibits a desired robustness to skewness in the observations.

Consider the detection problem

$$H_0: X_i = N_i, \quad i = 1, 2, \ldots k$$
$$H_1: X_i = \Theta + N_i$$

where the constant signal $\Theta > 0$. The noise is independent, identically distributed with first three moments

$$E(X_i) = 0, \quad E(X_i^2) = \sigma^2, \quad E(X_i^3) = \mu_3 > 0.$$

The skewness is defined as

$$\zeta = \mu_3/\sigma^3$$

and is assumed to be small.

If skewness were zero and the noise Gaussian, the optimum receiver is the sample mean or linear detector

$$T_1(x) = \frac{1}{k} \sum_{i=1}^{k} X_i.$$

With a small amount of skewness and nearly Gaussian noise, the test statistic T can be represented by the first few terms of the Cornish-Fisher expansion (Refs. 3 and 4). Under hypothesis H_0

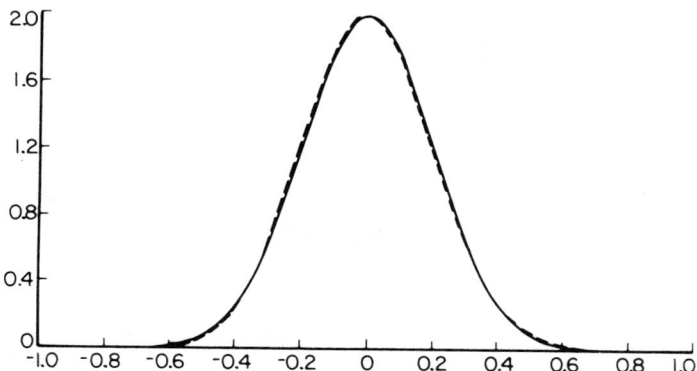

Figure 1 Probability density functions ———Gaussian, ————skewed, $\xi_n = 0.5$

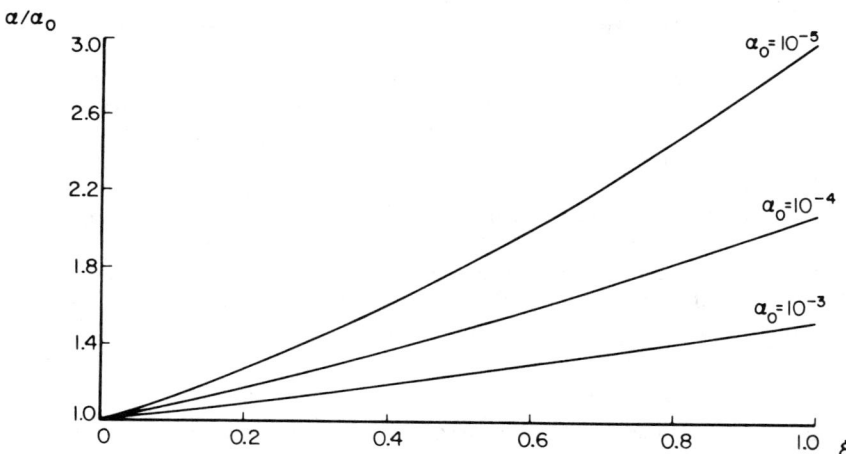

Figure 2 False alarm (normalized) versus skewness (number of observation for a decision = 100)

$$T_2(x) = \frac{\sigma}{\sqrt{k}} Z + \frac{\mu_3}{6\sigma^2 k}(Z^2 - 1) + 0(k^{-3/2})$$

where Z is the standard normal variable. (Under hypothesis H_1 the above expression includes a shift by the signal.)

Utilizing the first two terms of this expansion, one can generate the probability density function for T_2, along with analytic expressions for the false-alarm rate and detectability. Details can be found in Refs. 3 and 4.

Figure 1 compares the Gaussian density to the density of T_2 with skewness $\zeta = 0.5$. Figure 2 gives the normalized false-alarm rate as a function of skewness. The number of observations for the test is 100. The constant α_0 is the false-alarm rate under the strict Gaussian assumption ($\zeta = 0$) and provides

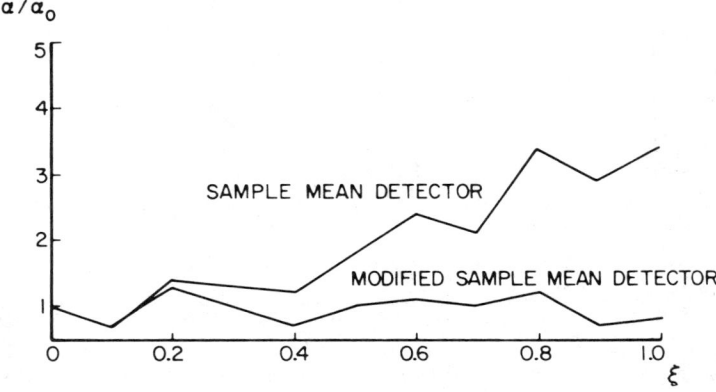

Figure 3 False alarm rate via simulation $\alpha_0 = 10^{-4}$, sample size = 100, 10^4 trials per point

the threshold setting for T_2. A slight departure from symmetry is hardly discernible (Fig. 1). Yet, for $\zeta = 0.5$, there is an increase in false-alarm rate of over 80% for a nominal $\alpha_0 = 10^{-5}$. With smaller α_0, the degradation is even more severe (Ref. 4).

Clearly, the linear detector can be modified so as to account for skewness. One approach is to use a nonparametric test such as the sign detector, which keeps the false alarm rate relatively constant (see Ref. 3 for details). A second approach is to directly modify the linear detector. Here, a natural way to proceed would be to "subtract off" that part of the observation due to skewness. The following test statistic is the simplest version of this approach:

$$T_3(x) = T_1(x) - \frac{\zeta}{6\sigma} (T_1(x))^2$$

where T_1 is the sample mean detector given above. Analytic expressions can again be developed for false-alarm rate and power (Ref. 4). Here, we choose to present the results of a Monte Carlo simulation which also serves to verify the accuracy of the analytic approximations. The results for false-alarm rates in Fig. 3 indicates that the modified sample mean detector has the desired robustness.

Implicit in the improved performance is the ability to measure the skewness accurately. This is clearly illustrated with a simple experiment performed with under-ice ambient noise. Fram II data (Ref. 5 and also discussed elsewhere in these proceedings) was used to compute empirically the false-alarm rate for the linear detector and the modified version discussed above. Required estimates of the variance and skewness were obtained by straightforward sample moment methods over (assumed stationary) blocks of data. Figure 4 summarizes the results of our first experiment. Performance of both detectors is essentially the same, in sharp contrast to the results obtained from computer-generated data discussed in the above paragraph. The primary reason for this is, we suspect, related to the nature of the nonstationary of the data (see Ref. 5, p. 8 and Fig. 5 below) and the need to use more accurate estimates of skew-

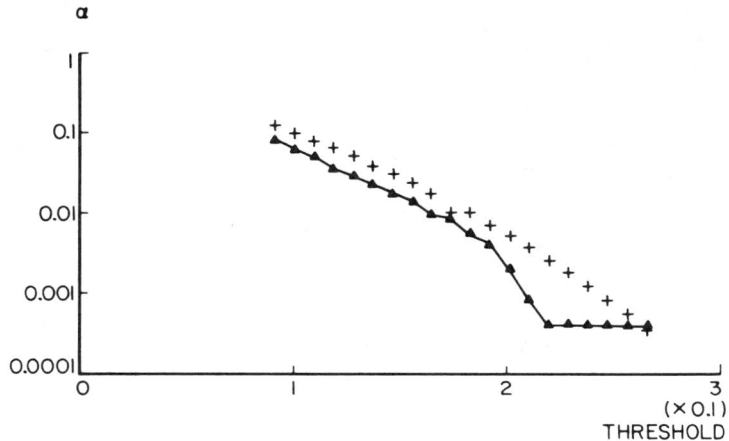

Figure 4 False alarm rate vs threshold

+ + + sample mean detection in Gaussian noise

⎯⎯⎯ sample mean detection (under ice noise)

△ △ △ △ modified sample mean detection (under ice noise)

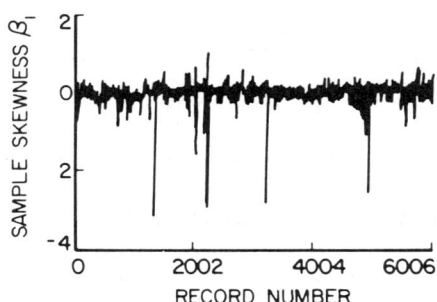

Figure 5 Sample skewness versus record number

ness with better tracking properties. This aspect of data-adaptive estimation and detection is an area of current research activity.

To conclude this section, we will outline one possible adaptive system which uses sample moments. Computed skewness and kurtosis are shown vs time for a representative Fram II data set in Figs. 5 and 6 (see Ref. 7 for further details). Kurtosis is defined as the normalized fourth central moment:

$$\beta_2 = E(X - m)^4/\sigma^4.$$

Figure 7 presents the same skewness (β_1) vs kurtosis (β_2), with time an implicit parameter. Observe the cluster of points around $\beta_1 = 0$, $\beta_2 = 3$, which are the values for a Gaussian density. The overlaid lines are different regions of

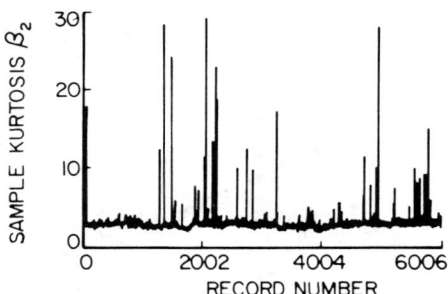

Figure 6 Sample Kurtosis versus record number

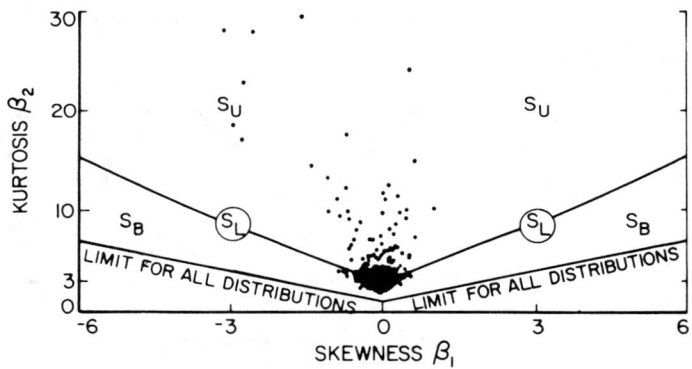

Figure 7 Scatter-plot of (β_1, β_2) for Johnson family

the Johnson family of densities. (The Johnson family can be defined as a nonlinear transformation of Gaussian variates.) The Symbol S_u, for example, represents the sinh transformation, while S_l is a logarithmic transformation, leading to a lognormal random variable (see Ref. 6, Section 2.2 for further details).

Each point in the skewness-kurtosis plane defines a unique member of the Johnson family. This one-to-one relationship can be utilized in a data-adaptive detector in the following manner. A region around the Gaussian point ($\beta_1 = 0$, $\beta_2 = 3$) can be defined. For points lying in this region, it is assumed that the observations are governed by Gaussian statistics and the optimum processor is the linear detector. When the computed moments fall outside the region, one declares that the observations are non-Gaussian and another detector is switched in to process the data. The point in the β_1-β_2 plane would determine the appropriate density which then specifies the likelihood processor (locally optimum detector in the weak signal case.) In practice, the β_1-β_2 plane would be quantized into, say, rectangular regions. Then, either due to nonstationary statistics or because of sampling variances of the moment or other estimators, the point wanders around in the region. When it leaves one region, another likelihood (nonlinear) processor can be switched in. Clearly, there are two key steps. The first is to obtain good tracking estimates. The second is to specify which family of probability densities to overlay on the β_1-β_2 plane, e.g., the Johnson, Pearson, or mixture model. This point will be discussed in more detail in the next section.

3 DENSITY FAMILIES AND ADAPTIVE DETECTION

Rather than focus on a particular moment measure such as skewness or kurtosis to characterize non-Gaussian statistics, one can attempt to generalize, but still parameterize, the problem in the following manner.

The score function of the density $f(x)$ is related to $f'(x)/f(x)$ and plays an important role in estimation theory [8]. In addition, $-f'(x)/f(x)$ determines the form of the processor for the locally most powerful test for the detection of a weak signal in noise. (This expression is also related to the test which maximizes efficacy in the small signal case.)

These observations lead us to focus on using f'/f as a characterization of non-Gaussian statistics. Observe that for the Gaussian density f'/f is linear, so departures from Gaussian statistics can be conveniently measured by departures from linearity. A straightforward way to proceed is to specify f'/f as a ratio of two polynomials. Specializing to first-order numerator and second-order denominator polynomials gives the classical Pearson family of probability densities:

$$\frac{f'(x)}{f(x)} = \frac{a + x}{b_0 + b_1 x + b_2 x^2}$$

which includes the Gaussian, gamma and beta among its members.

The procedure would now be to use the observations to estimate the above coefficients. This determines the particular density of the Pearson family and the corresponding detector. The two main approaches for estimating the coefficients are maximum likelihood estimates of the parameters directly, and indirect estimates using sample moments.

Of immediate interest is the question: what happens when there is a mismatch, i.e., when we use data generated by a non-Pearson density to fit a Pearson-type detector as described above? We proceed, using the non-Pearson mixture model ([7]):

$$f(x) = (1 - \epsilon) \frac{1}{\sqrt{2\pi}\sigma} e^{-x^2/2\sigma^2} + \frac{\epsilon\alpha}{2} e^{-\alpha|x|}.$$

The ratio of the component variances is defined to be

$$\gamma = (\alpha\sigma)^2/2.$$

This mixture density has been used to model heavy-tailed impulsive noise (see Ref. 9 and the references cited therein).

For the results that follow, the first four moments of the mixture model are used to obtain analytically the parameters a, b_0, b_1, and b_2. This then determines the processor, in this case the locally optimum detector. Figure 8 shows the curve $g(x) = -f'(x)/f(x)$ which is the zero-memory nonlinear part of the locally optimum detector. This is for a symmetric density ($a = b_1 = 0$),

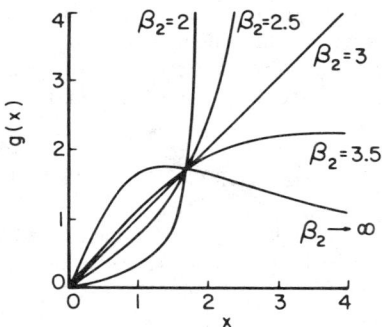

Figure 8 Pearson locally optimal ZNL's

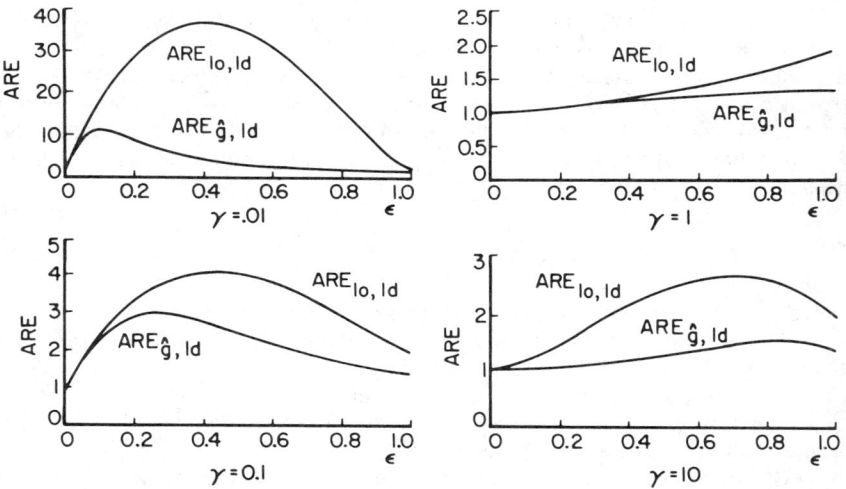

Figure 9 $ARE_{lo, ld}$ and $ARE_{\hat{g}, ld}$ for Gauss-Laplace mixture

with kurtosis shown as a parameter. Figures 9a through 9d present a comparison of ARE, as a function of the mixing constant ϵ, with γ as a parameter. (The ARE, asymptotic relative efficiency, is the ratio of efficacies. Efficacies, in turn, are incremental signal-to-noise ratio measures in the small signal case.) $ARE(l_0, ld)$ is the ratio of the locally optimum to the linear detector, while $ARE(\hat{g}, ld)$ compares the Pearson fit as discussed above to the linear detector. It is seen that for small ϵ and γ, the Pearson-fit detector compares quite favorably to the locally optimum, even though $\hat{g}(x)$ was determined from a non-Pearson density. (Addition details and further examples can be found in Ref. 7.)

Based on these preliminary computations, it would appear that, for nearly Gaussian noise (a mixing parameter of $\epsilon < 0.1$), the assumption of operating

Non-Gaussian Environment Detection

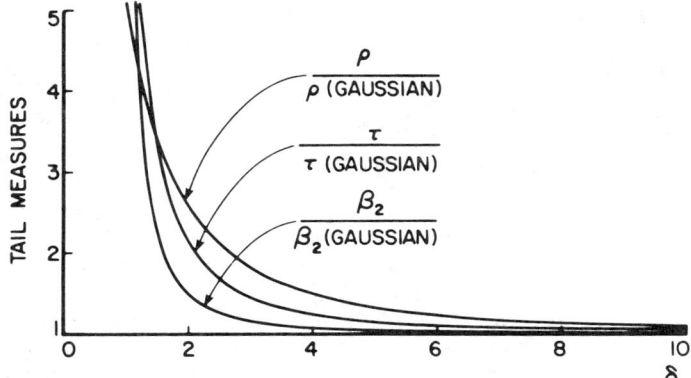

Figure 10 Johnson's SU system: β, τ, ρ vs δ

within a Pearson family does not significantly degrade detector performance when the true density turns out to be a Gaussian mixture with a Laplace contaminant.

One of the difficulties with moment estimators is the potentially large sampling variance for the higher-order moments. Oftentimes, it is more appropriate to use quantile measures to characterize the underlying statistics. We now outline a study using these measures (see Refs. 10 and 11 for details).

Let $F(x)$ be the cumulative distribution function for the noise. Let p_1 represent a quantile on the tail of the density, e.g., $p_1 = 0.999$, and let $p_2 < p_1$ be a lower quantile. Then, two other measures which can be used to characterize the density (and its tails) are:

$$\tau = F^{-1}(p_1)/F^{-1}(p_2)$$

and

$$\rho = \frac{d}{dp_1} \ln F^{-1}(p_1).$$

Figure 10 illustrates the one-to-one relationship between these measures (and kurtosis) and the Johnson S_u family as one of the two parameters δ of this density varies. The other parameter is determined from the variance (measured or given) and δ (see Ref. 10 or 11, Chap. 4).

Using these measures, one can parallel the adaptive procedure outlined above. That is, the data is used to estimate δ. This, in combination with an estimate of σ, uniquely determines the Johnson probability density and, hence, the nonlinear detector. This has been simulated when the noise does not actually come from a Johnson density but, rather, from other heavy-tailed densities (Gaussian-Gaussian mixtures, Laplacian noise). The results of a preliminary computer simulation are encouraging: the detector based on Johnson statistics performs close to the optimum detector (which is based on exact knowledge of the noise statistics) and always better than the linear detector ([11]). Again we see that detector performance does not appear to be critically dependent on

4 DETECTION IN MULTIVARIATE NOISE

specifying the correct family of densities. More important, apparently, are accurate estimates of the parameters which characterize the family and the corresponding detector.

The previous discussion assumed the observations were independent; in order to account for the dependencies, multivariate probability densities need to be considered. In this section, we study nonlinear transformations of a known multivariate density. This type of noise, called transformation noise in Refs. 7 and 12, is a multivariate generalization of the Johnson family discussed earlier. Any output marginal can be obtained by prescribing a zero-memory nonlinear transformation. It is difficult, however, to obtain analytically the output dependencies. Figure 11 is a schematic of the transformation noise generation. The output multivariate density is denoted by $f(\underline{n})$ and the input density by $\phi(\underline{v})$. With $\underline{v}$ multivariate Gaussian, there is a further decomposition, since one can easily define the linear transformation $\underline{v} = \underline{L}\underline{z}$, where $\underline{z}$ is a vector of independent Gaussian variates, and $\underline{L}$ defines the covariance structure.

For the case of a weak signal in noise, the appropriate receiver is the locally optimum detector. It is shown in Fig. 12, where it is assumed that the nonlinearity g is twice differentiable. The symbol $\otimes$ denotes element-by-element vector multiplication and $\odot$ is the vector dot product. Observe that the detector consists of a zero-memory nonlinear part (g, g', etc...) and the locally optimal nonlinearity with memory, $\nabla \phi / \phi$, for a signal in noise with a density ϕ. With $\underline{v}$ multivariate Gaussian, $\nabla \phi / \phi$ reduces to the usual linear matched filter, $R^{-1}\underline{x}$.

Figure 13 summarizes the results of a Monte Carlo simulation to evaluate ARE (see Refs. 7 and 12 for details). The output marginal was specified as Laplacian:

$$f_1(n) = \frac{\alpha}{2} \exp(-\alpha|n|)$$

which was generated by a suitable transformation of multivariate Gaussian noise. The Gaussian vector was assumed to be m-dependent, with the correlation function taken as triangular:

$$\rho_i = 1 - |i/m|, \quad |i| \leq m$$
$$\rho_i = 0, \quad |i| > m.$$

The nonlinearities required in the detector are determined from $f_1(n)$, and the signal was taken as a constant.

Four detectors were simulated. The first was the locally optimum detector, and the second assumed independent noise. The third was a simplification of the first; the indicated vector multiplication was removed. The fourth detec-

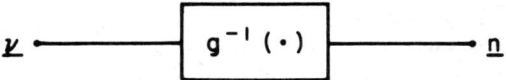

Figure 11 Transformation noise

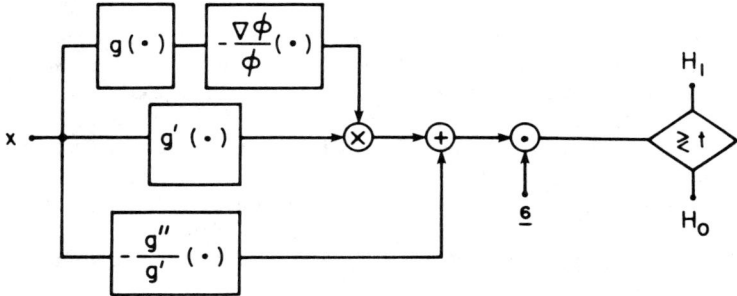

Figure 12 Locally optimum detector

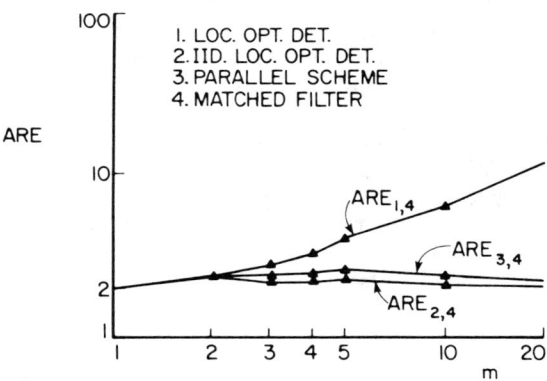

Figure 13 ARE comparison

tor was the linear matched filter. As indicated above, values for the ARE were computed via simulation; consequently, it is difficult to make general statements. Nevertheless, some observations seem appropriate.

It is clear that for small correlation time, $m \leq 3$ or 4, an independence assumption does not significantly degrade detector performance. Second, with some sort of nonlinear processing, i.e., taking into account non-Gaussian statistics, a reasonable improvement can usually be obtained: for any of the nonlinear detectors compared to the linear one, the ARE for this simulation does not fall below 2. Note that detectors 2 and 3 give similar ARE values. This is interesting since one detector assumes independent noise, while the other simplifies the optimal detector, but keeps the dependency assumption. Clearly, one would like to learn what are the essential common features of the various detectors that lead to improved performance. Then, only these features need to be incorporated into a practical receiver.

5 SUMMARY

Techniques for detection of weak signals in non-Gaussian noise have been considered. The importance of both learning and robust procedures was illustrated by means of an example where a modest deviation from a Gaussian noise assumption (in skewness) led to a substantial increase in the false-alarm rate for a linear detector. In contrast, a modified (nonlinear) detector was robust and maintained a relatively constant false-alarm rate for a wide range of skewness. These analytical results were verified by computer simulation.

Under-ice ambient noise was used in an experiment to illustrate some of the practical difficulties with adaptive procedures; in this case, because of the nonstationary of the noise, there was a need to incorporate into the adaptive detector parameter estimates with enhanced tracking properties.

The skewness-kurtosis plane was presented as a convenient graphical display to place in evidence the time-varying nature of the non-Gaussian statistics. The notion of overlays was also introduced to define density families and the corresponding likelihood processors. Densities were also characterized in terms of a function related to the score function and tail measures using quantiles. Our preliminary results seem to indicate that detector performance does not appear to be critically dependent on specifying the correct family of densities. More important to performance are accurate estimates of the moments, tail measures, or other parameters which are used to specify the detector.

A particular class of multivariate non-Gaussian densities was defined and the canonical form of the locally optimum detector derived. The results of a simulation to evaluate ARE for the optimum detector and simplified versions were presented. These results indicate that some sort of nonlinear processing which takes into account deviations from Gaussian statistics leads to a large portion of the improved performance.

ACKNOWLEDGMENT

Much of the work covered in this paper is an explication of the doctoral dissertations of Y.F. Huang, A.B. Martinez, E.J. Modugno, and P.F. Swaszek done under the supervision of J.B. Thomas. It is always a pleasure to interact with and acknowledge bright, serious-minded graduate students as they begin their professional careers.

REFERENCES

1. J.B. Thomas, "Nonparametric Detection," *Proceedings of the IEEE,* Vol. 58, pp. 623-631, 1970.

2. R.D. Martin and S.C. Schwartz, "Robust Detection of a Known Signal in Nearly Gaussian Noise," *IEEE Trans. Information Theory,* Vol. IT-17, pp. 50-56, 1971.

3. Y.F. Huang, "Alternatives to Optimal Detection, Ph.D. Thesis, Dept. of Electrical Engineering and Computer Science, Princeton University, October 1982.

4. Y.F. Huang and J.B. Thomas, "Detection of Constant Signals in Skewed Noise," *Proceedings, Nineteenth Annual Allerton Conference on Communications, Control and Computing,* pp. 607-616, 1981.

5. R.F. Dwyer, Fram II Single Channel Ambient Noise Statistics, Naval Underwater Systems Center T.D. 6583, November 1981.

6. R.A. Tapia and J.R. Thompson, Nonparametric Probability Density Estimation, The Johns Hopkins University Press, Baltimore, 1978.

7. A.B. Martinez, Non-Gaussian and Multivariate Noise Models for Signal Detection, Ph.D. Thesis, Dept. of Electrical Eng. and Computer Science, August 1982.

8. S.S. Wilks, Mathematical Statistics, John Wiley, N.Y., 1982.

9. J.H. Miller and J.B. Thomas, "The Detection of Signals in Impulsive Noise Modeled as a Mixture Process," *IEEE Trans. on Comm.,* pp. 559-563, May 1976.

10. E.J. Modugno and J.B. Thomas, "Measures of Tail Behavior and Their Relationship to Signal Detection," *Proceedings, Fifteenth Annual Johns Hopkins Conference on Information Sciences and Systems,* March 25-27, pp. 26-31, 1981.

11. E.J. Modugno, The Detection of Signals in Impulsive Noise, Ph.D. Thesis, Dept. of Electrical Engineering and Computer Science, February 1982.

12. A.B. Martinez, P.F. Swaszek, and J.B. Thomas, "Locally Optimal Detection in Multivariate Non-Gaussian Noise," Submitted for publication to IEEE Trans. on Information Theory.

Detection of Non-Gaussian Signals in Gaussian Noise*

C. R. Baker

*Department of Statistics
University of North Carolina
Chapel Hill, NC*

A.F. Gualtierotti**

*Department of Mathematics
Federal Institute of Technology (EPFL)
Lausanne, Switzerland*

1 INTRODUCTION

Detection of a non-Gaussian signal in additive and dependent Gaussian noise can be viewed as the canonical detection problem for active sonar in a reverberation-limited environment (when reverberation is the predominant source of noise).

Previous work on this problem has included partially-parametric methods, and a parametric approach under the assumption that signal and noise are independent. Here we present a new expression for the likelihood ratio under a minimum set of conditions. Additional assumptions must be introduced in order to provide methods for calculations. With these assumptions, the parameters in the expression for the likelihood ratio can be estimated from data ensembles that are representative of the detection problem. Thus, implementation requires two ensembles of training data, one having statistical properties similar to those of the noise and the other having properties similar to those of the signal-plus-noise for which detection is desired.

---
*Research supported by ONR Contracts N00014-75-C-0491 and N00014-81-K-0373.
**Current Affiliation: IDHEAP, Universite de Lausanne, Lausanne, Switzerland

The mathematical rationale for the model described in this paper is as follows. During the past 15 years, there has been a great deal of work done on determining the likelihood ratio when the noise is a Wiener process. This is not a realistic model for sonar; the paths of the Wiener process are much too irregular to represent typical received waveforms in sonar. The success with the Wiener process rests largely on the fact that it is a martingale [1]. Thus, the basic idea of the research summarized here is to use the results already obtained on likelihood ratio for the Wiener process and other Gaussian martingales, but to apply them to realistic (nonmartingale) stochastic processes. This is done by representing the noise as a sum of filtered martingales. The results obtained in this manner differ from those obtained in previous detection analyses involving Gaussian noise in two major respects:

(1) The noise can be any mean-square continuous Gaussian process;
(2) No assumptions are made on the probability distributions of the signal process.

This paper is largely expository. We first sketch the background to the detection problem, and then outline our solution. No proofs are included; they will appear elsewhere. A discussion is given of the approximations and parameter estimates that must be made to implement the likelihood ratio.

2 DATA MODELS

We have mentioned that the detection of a non-Gaussian signal in additive dependent Gaussian noise can be viewed as the basic detection problem for reverberation-limited active sonar. The physical justification for this is straightforward. In the noise-only situation, the principal noise component will be due to the superposition of reflections from a large number of small inhomogeneities (marine organisms, air bubbles, surface irregularities, etc.). Each of these inhomogeneities scatters a small amount of energy incident from the transmitted waveform back to the receiver, and each can be modeled as a random scatterer. These scatterers are typically assumed to have the same statistical properties, and to have a Poisson spatial distribution. The central limit theorem can then be applied to obtain a Gaussian reverberation process at the receiver. For discussion, see [2]-[4].

When a target such as a submarine is present, it will have few scatterers compared to those producing the reverberation. Moreover, in many situations (depending on aspect angle, etc.) a small subset of the target scatterers will predominate. Each scatterer can again be modeled as producing a random scattered waveform from the incident transmitted waveform. However, due to the paucity of scatterers, the central limit theorem cannot be applied, and so the target return (received signal) can be a sample function from a highly non-Gaussian stochastic process. Since the waveform incident on the target is first scattered by the inhomogeneities producing the reverberation, as is the composite target reflection, the (received) signal and the noise will be statistically dependent.

So far as smoothness properties of the data are concerned, this is affected by various inertias and time delays in the sensing equipment, as well as by the smoothing properties of the ocean. One can thus assume that the received

Non-Gaussian Signals

waveform will be a sample function from a mean-square continuous stochastic process. Finally, the reverberation can be assumed to be zero at the beginning of the observation period. This is always true if the observation interval is assumed to begin while the signal is being transmitted. If the more realistic assumption is made that the observation interval begins after the completion of transmitting the signal, then there will still be no reverberation at the instant the receiving system is switched on, due to unavoidable time delays and inertias in the equipment.

Thus, from consideration of only the physical aspects, we have the following detection problem:

Hypothesis H_0: Received waveform is a sample function from a mean-square continuous Gaussian noise process (N_t), $0 \leqslant t \leqslant T$. Also, $N_t = 0$ at $t = 0$ (with probability one).

Hypothesis H_1: Received waveform is a sample function from a mean-square continuous stochastic process $(S_t + N_t)$, $0 \leqslant t \leqslant T$. The signal process (S_t) can be non-Gaussian, and will not be statistically independent of (N_t).

Moreover, it is no restriction to assume that the noise process (N_t) has zero mean function $(EN_t = 0$ for $0 \leqslant t \leqslant T)$. Noise processes in sonar typically have this property.

The detection problem summarized above is the problem to be considered here.

3 SINGULAR OR NONSINGULAR DETECTION

Probability of detection (P_D) is the probability of a correct detection decision when signal is present. Probability of false alarm (P_{FA}) is the probability of an incorrect decision (false alarm) when signal is absent. In a nonsingular detection problem $P_{FA} = 0$ implies $P_D = 0$, and $P_D = 1$ implies $P_{FA} = 1$. Detection problems that do not have one or both of these properties are said to be singular. It is our feeling, which seems to be true for the sonar community, that any realistic sonar detection problem is nonsingular. When this is true, then the likelihood ratio (Radon-Nikodym derivative) dP_{S+N}/dP_N will exist. P_{S+N} is the probability measure induced on the space of sample functions by the process $(S_t + N_t)$, while P_N is the probability induced on the space of sample functions by the noise process (N_t). For the detection problem as outlined in the preceding section, the space of sample functions can be taken as $L_2[0,T]$ (Lebesgue-square-integrable functions on $[0,T]$).

4 DETECTION ALGORITHMS

When the likelihood ratio exists, it is the preferred detection algorithm under the usual reasonable criteria: Bayes, minimax, Neyman-Pearson [5]. However, the probability distributions of the $S + N$ process must usually be known in order to determine dP_{S+N}/dP_N. This will frequently not be the case.

If one cannot determine the likelihood ratio, then detection algorithms must be formulated that do not require a full statistical description of the data. Covariance and mean functions can frequently be estimated with reasonable accuracy. Since the noise is Gaussian, this will fully characterize (N_t). These considerations have led to the use of the deflection criterion to determine an algorithm.

To apply the deflection criterion, one first selects an appropriate family of possible test statistics (operations on the data), say τ. If x is the received waveform, then the deflection criterion instructs one to choose the element Λ in τ to maximize the deflection $D(\Lambda)$,

$$D(\Lambda) \equiv \frac{(E_{S+N}\Lambda(x) - E_N\Lambda(x))^2}{E_N\Lambda^2(x) - [E_N\Lambda(x)]^2}.$$

E_{S+N} denotes expectation with respect to the probability P_{S+N}, while E_N denotes expectation with respect to P_N. Heuristically, the deflection criterion seeks a test statistic whose distribution has widely-separated means under the two hypotheses, and with small variance under the noise-only hypotheses.

Analyses of the detection problem using the deflection criterion have been given for the case of a quadratic operation [6] and a quadratic-linear operation [7]. Implementation of the algorithm requires knowledge of only the covariance and mean functions of the $S + N$ and N processes. The performance of this algorithm has been examined using ensembles of experimental sonar data [8]. In the case of discrete-parameter data, the optimum quadratic operation is given by

$$\Lambda(x) = x^* \mathbf{R}_N^{-1} \mathbf{K}_S \mathbf{R}_N^{-1} x$$

where $\mathbf{R}_N$ is the noise covariance matrix, $\mathbf{R}_{S+N}$ is the signal-plus-noise correlation matrix, and $\mathbf{K}_S = \mathbf{R}_{S+N} - \mathbf{R}_N$.

In some cases it will be feasible to determine probability distributions of (S_t) and (N_t) separately. One could then implement a likelihood ratio by assuming that (S_t) and (N_t) are statistically independent. An analysis of the detection of a non-Gaussian signal in independent Gaussian noise is given in [9]. Under additional assumptions, those results can be used to write the likelihood ratio as

$$[dP_{S+N}/dP_N](x) = \int_Y [dP_{y+N}/dP_N](x) \, dP_S(y)$$

where Y is the space of signal sample functions, and P_{y+N} is the probability measure induced by $(N_t + y(t))$, $0 \leq t \leq T$, with y being a fixed signal sample path. Since N is Gaussian, dP_{y+N}/dP_N can be calculated [10]. However, even if this expression for dP_{S+N}/dP_N can be determined, the assumption of independent signal and noise will not be valid for the model being considered here (although it will frequently be valid for detection problems in passive sonar).

The work reported in this paper is directed toward obtaining the likelihood ratio for the detection problem as previously stated. As will be seen, its actual calculation involves estimation of parameters from ensembles of real data representative of the detection problem (unless one has the required analytical models—which is not likely). Thus, any evaluation of performance will be influenced by the skill with which these parameters are estimated. Moreover, the computation procedure requires one to restrict the class of admissible noise processes. Thus, it will be of interest to compare the performance of this approximation to the optimum detection algorithm, whose actual calculation involves restrictive assumptions and several parameter estimates, with the performance of sub-optimum algorithms, such as that obtained from the deflection criterion.

5 REPRODUCING KERNEL HILBERT SPACES AND MULTIPLICITY THEORY

Suppose that R is a real-valued covariance function on $[0,T] \times [0,T]$; i.e., it is symmetric and nonnegative definite. Then there exists a unique set H_R of real-valued functions on $[0,T]$ and a unique inner product $<\cdot,\cdot>_R$ on H_R such that:

(1) H_R is a Hilbert space with respect to the norm obtained from $<\cdot,\cdot>_R$;
(2) $R_s \equiv R(\cdot,s)$ is in H_R for all fixed s in $[0,T]$;
(3) $<R_s,f>_R = f(s)$ for all f in H_R.

H_R (with the inner product $<\cdot,\cdot>_R$) is called the reproducing kernel Hilbert space (RKHS) of the covariance function R [11]. It plays an important role in studying Gaussian processes. As an example of such a space, suppose that $R(t,s) = e^{-\alpha|t-s|}$ for some $\alpha > 0$. The RKHS H_R then consists of the set of all functions f on $[0,T]$ that are absolutely continuous and have derivative in $L_2[0,T]$.

In cases of practical interest, one will have $\int_0^T \int_0^T R^2(t,s)\,dsdt < \infty$. H_R and $<\cdot,\cdot>_R$ can then be found as follows. H_R will consist of real-valued functions f in $L_2[0,T]$ such that

$$\sum_{n \geq 1} [\int_0^T f(t)e_n(t)\,dt]^2/\mu_n < \infty, \qquad (1)$$

where $\{\mu_n, n \geq 1\}$ and $\{e_n, n \geq 1\}$ are the nonzero eigenvalues and associated orthonormal eigenvectors of the integral operator $\mathbf{R}$ in $L_2[0,T]$ which has R as its kernel: $[\mathbf{R}g](t) = \int_0^T R(t,s)g(s)\,ds$. Moreover, every such function f that satisfies (1) is equal almost everywhere on $[0,T]$ to an element of H_R. The inner product is given by

$$<f,g>_R = \sum_{n \geq 1} \frac{\left[\int_0^T f(t)e_n(t)\,dt\right]\left[\int_0^T g(s)e_n(s)\,ds\right]}{\mu_n}.$$

The RKHS plays an important role in the Cramér-Hida multiplicity theory for stochastic processes. Suppose that $\{X_t\}$, t in $[0,T]$ is a real-valued stochastic process on a probability space (Ω,β,P). Suppose that $\{X_t\}$ is mean-square continuous, has zero mean function, and $X_t = 0$ for $t = 0$ (with probability one).

Then (X_t) has the Cramér-Hida representation

$$X_t = \sum_{i=1}^{M} \int_0^t F_i(t,s)\,dB_i(s) \tag{2}$$

where $M \leq \infty$, the functions (F_i) are deterministic, and each B_i is a stochastic process. Moreover, each B_i is a zero-mean, mean-square continuous process with orthogonal increments, and B_i and B_j are mutually orthogonal for $i \neq j$ ($EB_i(t)B_j(s) = 0$ for all s,t in $[0,T]$). The integral $\int_0^t F_i(t,s)\,dB_i(s)$ is the mean-square limit of Stieltjes sums, of the form

$$\sum_{k=1}^{K} F_i(t,s_k)[B_i(s_{k+1}) - B_i(s_k)].$$

The equality in (2) is in the mean-square sense. M is said to be the multiplicity of the process (X_t). For further details on the Cramér-Hida representation, see [12], [13].

Suppose that (X_t) is as above, with covariance function R. The RKHS H_R can be characterized in terms of the Cramér-Hida representation. First, for each process B_i in the representation (2), the variance $EB_i^2(s)$ is a nondecreasing function on $[0,T]$. Thus, one can define a measure on the Borel sets of $[0,T]$ in the usual way, with the measure of the interval $[a,b]$ being given by $EB_i^2(b) - EB_i^2(a)$. Let this measure be denoted by λ_i. The RKHS H_R is then the set of all functions g of the form

$$g(t) = \sum_{i=1}^{M} \int_0^t F_i(t,s) y_i(s)\,d\lambda_i(s) \tag{3}$$

where each y_i is a deterministic function on $[0,T]$ satisfying $\int_0^T y_i^2(s)\,d\lambda_i(s) < \infty$ (thus y_i belongs to $L_2([0,T];\lambda_i)$). For $M = \infty$, the equality in (3) holds in the sense of convergence in H_R.

6 NOISE AND SIGNAL-PLUS-NOISE REPRESENTATION

With the assumptions already given on the noise process, and using the Cramér-Hida theory, one can write the noise process (N_t) as

$$N_t = \sum_{1}^{M} \int_0^t F_i(t,s)\,dB_i(s) \tag{4}$$

where the functions (F_i) and processes (B_i) have the properties described in the preceding section. Since (N_t) is Gaussian, $\{B_i,\ i \geq 1\}$ will be a Gaussian family. This implies that each B_i has independent increments and that B_i and B_j are mutually independent for $i \neq j$.

The selection of a representation for the signal process is made with the objective of taking advantage of the Cramér-Hida representation for the noise. Thus, it is assumed that the sample paths of the signal process (S_t) belong to the RKHS H_R with probability one; hence

$$S_t = \sum_{1}^{M} \int_0^t F_i(t,s) y_i(s)\,d\lambda_i(s) \tag{5}$$

where the functions (F_i) and measures (λ_i) are as previously described, but the

(y_i) are now stochastic processes, with the sample paths of y_i belonging (probability one) to $L_2([0,T]; \lambda_i)$. This yields the following representation for the signal-plus-noise process:

$$S_t + N_t = \sum_1^M \int_0^t F_i(t,s) y_i(s) d\lambda_i(s) + \sum_1^M \int_0^t F_i(t,s) dB_i(s). \quad (6)$$

One should note at this point that there is reason to believe [14] that a representation such as (6) is necessary in order that the detection problem be nonsingular. Thus, our representation for the noise has no additional assumptions beyond those made from physical considerations, and the assumption on the form of the signal process appears justified in order to have a realistic detection problem.

7 DEVELOPMENT OF THE LIKELIHOOD RATIO

The development of the likelihood ratio to be described here is based on the representations of (N_t) and $(S_t + N_t)$. Let

$$Z_i(t) \equiv \int_0^t y_i(s) d\lambda_i(s) + B_i(t). \quad (7)$$

In mean-square sense, one then has

$$N_t = \sum_{i=1}^M \int_0^t F_i(t,s) dB_i(s)$$

$$S_t + N_t = \sum_{i=1}^M \int_0^t F_i(t,s) dZ_i(s) \quad (8)$$

$(\mathbf{B}(t))$ and $(\mathbf{Z}(t))$ are vector stochastic processes, with each component having sample paths almost surely in $C[0,T]$, the space of continuous functions on $[0,T]$. $(\mathbf{B}(t))$ and $(\mathbf{Z}(t))$ induce probability measures $P_\mathbf{B}$ and $P_\mathbf{Z}$ on the M-fold product of $C[0,T]$. The basic idea is to develop the likelihood ratio dP_{S+N}/dP_N in terms of the likelihood ratio $dP_\mathbf{Z}/dP_\mathbf{B}$, when the latter exists. The first step is to show the existence of a measurable map T from E^M into $L_2[0,T]$, where $E = C[0,T]$. This can be done under the assumption that $dP_\mathbf{Z}/dP_\mathbf{B}$ exists. With some additional work, one then obtains a very general form of the likelihood ratio,

$$[dP_{S+N}/dP_N](x) = \int_{E^M} [dP_\mathbf{Z}/dP_\mathbf{B}](y) dP_{\mathbf{B}|N=x}(y) \quad (9)$$

where $P_{\mathbf{B}|N=x}$ is the probability measure for the $\mathbf{B}$ vector when x is observed and one assumes that x represents noise only.

In the above form, there are essentially no meaningful restrictions on the detection problem. However, from (9) one sees that to calculate dP_{S+N}/dP_N it is necessary to find $dP_\mathbf{Z}/dP_\mathbf{B}$ and $P_{\mathbf{B}|N}$. This requires further assumptions. An intermediate set of assumptions can be made which will provide an explicit formula for dP_{S+N}/dP_N; however, here we shall omit this intermediate expression and proceed directly to the form for which calculations can be made. The assumptions are then as follows:

(1) $M = 1$;
(2) B_1 is the Wiener process (W_t);

(3) (Z_t) is the solution of a stochastic differential equation, and is a diffusion:

$$Z_t = \int_0^t \alpha[Z(s)]\,ds + W_t. \tag{10}$$

These three assumptions are not unreasonable in terms of engineering practice. Assumptions (1) and (2) imply that (N_t) has the representation

$$N_t = \int_0^t F(t,s)\,dW_s. \tag{11}$$

This is recognized to be a causal linear operation on (formal) white Gaussian noise. White Gaussian noise is frequently used to model practical engineering problems; moreover, in the finite-dimensional discrete-time case, any Gaussian process can be obtained as a causal linear operation on white Gaussian noise. Assumption (3) implies that (Z_t) is a diffusion. Since (W_t) is also a diffusion, assumptions 1) and 2) imply that (N_t) is obtained by a causal linear operation on a diffusion. By (8), assumption 3) implies that $(S_t + N_t)$ is obtained by this same linear operation on another diffusion. This seems to be plausible from consideration of the physical aspects of the problem (same ocean medium acting on the scattered waveforms, etc.) and from physical-mathematical considerations ((W_t) and (Z_t) should have similar properties if signal detection is a difficult problem, the case of most interest).

In order to simplify the presentation here, a final assumption is made: the functions $\{F_t,\ t \in [0,T]:\ F_t(s) = F(t,s)\}$ span $L_2[0,T]$. This assumption is made *only* to simplify our presentation. In this case, one obtains

$$\frac{dP_{S+N}}{dP_N}(x) = \lim_n \exp\left\{\sum_{i=1}^{n-1} \alpha[m(x_{t_i})]\,[m(x_{t_{i+1}}) - m(x_{t_i})]\right. \tag{12}$$

$$\left. - \frac{1}{2}\sum_{i=1}^{n-1} \alpha^2[m(x_{t_i})][t_{i+1} - t_i]\right\}.$$

If $E\sum_1^M \int_0^T y_i^2(s)\,d\lambda_i(s) < \infty$, then the function m is defined by

$$m[x_t] = \lim_{k \to \infty} \sum_{n=1}^k \frac{<x,e_n><f_t,e_n>}{\mu_n}$$

where $<\cdot,\cdot>$ is the $L_2[0,T]$ inner product, and $\{\mu_n,\ n \geq 1\}$, $\{e_n,\ n \geq 1\}$ are the nonzero eigenvalues and associated orthonormal eigenvectors of the noise covariance operator **R**. The function f_t is defined by $f_t(s) = \int_0^t F(s,u)\,du$. Other expressions for m can be given.

8 COMPUTATION OF THE LIKELIHOOD RATIO

Calculation of the likelihood ratio requires that one have knowledge of the following parameters:
 (1) The functions α, m, F and $\{e_n,\ n \geq 1\}$;
 (2) the eigenvalues $\{\lambda_n,\ n \geq 1\}$.

Then parameters can be estimated if one has an ensemble of noise sample vectors representative of the detection problem, and one or more representative signal-plus-noise sample vectors. Suppose then that one has sample vectors $X_1, X_2, \ldots, X_K$, each X_i being a vector in L-dimensional Euclidean space obtained by sampling noise data, with $X_{i_j} = X_i(t_j)$. Suppose also that one or more vectors X' are given, representing signal-plus-noise. The procedure is then as follows:

(1) Estimate the covariance matrix R of the noise.

(2) The noise has the discrete-time approximation $N = F(\Delta W)$, where $F(i,j) = F(t_i, t_j)$, and $(\Delta W)_i = W_{t_{i+1}} - W_{t_i}$. Moreover, $R = kFF^*$ when uniform sample spacing k is employed. The second step is to calculate F from the estimated R.

(3) Signal-plus-noise has the discrete-time approximation $S + N = F(\Delta Z)$, where $(\Delta Z)_i = Z_{t_{i+1}} - Z_{t_i}$. Using a given $S + N$ sample vector, one obtains a Z sample vector by $\Delta Z = F^{-1}(S + N)$, and using the fact that $Z_{t_0} = Z_0 = 0$.

(4) Each Z vector is assumed to be a discretized solution to a stochastic differential equation

$$Z(t) = \int_0^t \alpha[Z(s)]ds + W_t.$$

Using one of various methods ([15],[16]), estimate the function α. If several $S + N$ waveforms are available, average the corresponding α's.

(5) Calculate the eigenvalues and eigenvectors of the noise covariance matrix R. Use these quantities to calculate the function m as already defined.

(6) Given m and α, calculate an approximate likelihood ratio by

$$\frac{dP_{S+N}}{dP_N}(x) = \exp\left\{\sum_{i=1}^{L-1} \alpha[m(x_i)][m(x_{i+1}) - m(x_i)] \right. \tag{13}$$
$$\left. - \frac{k}{2}\sum_{i=1}^{L-1} \alpha^2[m(x_i)]\right\}$$

where k is the width of the sampling interval.

The sequence of calculations outlined here represents a straightforward approach to the computation of the approximate likelihood ratio. Improvements on this sequence may well result when more experience is gained on the computational aspects.

REFERENCES

1. R.S. Lipster and A.N. Shiryayev, *Statistics of Random Processes: General Theory*, Springer, New York (1977).

2. P. Faure, Theoretical Model of Reverberation Noise, *J. Acoustical Society of America*, 36, 259-266 (1964).

3. H.L. van Trees, Optimum Signal Design and Processing for Reverberation-Limited Environments, *IEEE Trans. on Military Electronics*, 9, 212-229 (1965).

4. D. Middleton, A Statistical Theory of Reverberation and Similar First-Order Scattered Fields, *IEEE Trans. on Information Theory*, 13, 372-414 (1967).

5. C.W. Helstrom, *Statistical Theory of Signal Detection*, Pergammon, New York (1960).

6. C.R. Baker, Optimum Quadratic Detection of a Random Vector in Gaussian Noise, *IEEE Trans. on Communication Technology*, 14, 802-805 (1966).

7. C.R. Baker, On the Deflection of a Quadratic-Linear Test Statistic, *IEEE Trans. on Information Theory*, 15, 16-21 (1969).

8. R.E. Cunningham and C.R. Baker, A Monte Carlo Evaluation of Sonar Receivers based on Experimental Data, *J. Underwater Acoustics (USN)*, 19, 343-369 (1969).

9. C.R. Baker, On Equivalence of Probability Measures, *Annals of Probability*, 1, 690-698 (1973).

10. C.R. Rao and V.S. Varadarajan, Discrimination of Gaussian Processes, *Sankhyā*, 25A, 303-330 (1963).

11. N. Aronsajn, Theory of Reproducing Kernels, *Trans. American Mathematical Society*, 68, 337-404 (1950).

12. H. Cramér, On some classes of non-stationary stochastic processes, Proc. Fourth Berkeley Symp. on Math. Stat. and Probability, 2, 57-77 (1961).

13. T. Hida, Canonical Representation of Gaussian Processes and their applications, *Mem. Coll. Science, Univ. Kyoto, 33A*, 109-155 (1960).

14. G.A. Mel'ničenko, The structure of processes that are absolutely continuous with respect to Gaussian processes, *Uspehi Mat. Nauk 32*, 197-198 (1977).

15. S. Geman, An Application of the Method of Sieves: Functional Estimator for the Drift of a Diffusion, *Reports in Pattern Analysis*, 92, Div. Applied Math., Brown Univ. (1980).

16. U. Grenander, *Abstract Inference*, John Wiley & Sons, New York (1981).

Data-Adaptive Detection of a Weak, Single-Frequency, Plane-Wave Signal in Noise and Strong, Unidirectional Interference*

Donald W. Tufts, Ramdas Kumaresan, and Ivars Kirsteins

Department of Electrical Engineering
University of Rhode Island
Kingston, RI

1 INTRODUCTION

In a companion paper we have described a new method for data-adaptive signal detection (2), and we have applied it to one-dimensional (e.g., time or space) detection problem. Here, we consider a two-dimensional example. This method is based upon a technique for estimating "low-rank" or coherent components from noisy data (1). The method of detection is motivated by a form of the likelihood-ratio-test statistic for the case of a low-rank gaussian signal in the presence of low-rank gaussian interference and high-rank (or nearly white) gaussian noise (3). In this form of the likelihood-ratio test, an estimate of the interference is formed by projecting the observed data vector onto the principal eigenvectors of the known covariance matrix of the interference.

As opposed to the case of zero-mean gaussian interference with known covariance matrix (3), we have been working on examples in which the covariance matrix of the interference is not known, and the interference and signal are non-gaussian (2). However, we do make the assumption that the interference has enough spatial or temporal coherence to be of low-rank. This will be explained in the next section.

---

*Work supported under Contract N00014-81-K-0144, Task NR SRO 104 with the Office of Naval Research.

The details of the application of our detection method to the array problem of the paper are presented in Sections 2 and 3, and in section 4 we present some experimental results from a simulation of the method.

2 BROADBAND UNIDIRECTIONAL INTERFERENCE AS A LOW-RANK SPATIALLY COHERENT, TWO-DIMENSIONAL WAVEFORM

Let us suppose that gaussian, broadband, plane-wave interference arrives at a line array of receivers from the broadside direction. The receivers associated with each receiving element are identical and filter the output of the receiving element to a band of frequencies of width B Hz. Each filtered waveform is complex-translated to be centered at zero-frequency and represented by a sequence of mutually independent, complex-valued, gaussian samples.

The sequence of samples from any receiver is the same as the sequences from the other receivers. This is because any piece of the transmitted interference waveform arrives simultaneously at each receiver. In practice this can be approximately arranged through delay steering or mechanical turning of the array, for the important case in which the interference is sufficiently strong relative to the receiver noise and the signal.

To be more specific let us consider an example in which the array consists of ten receiving elements which are uniformly spaced over a segment of a line in space. And the processing interval is such that ten uniformly-spaced complex-valued samples are taken from each receiver over the given time interval. The resulting ten-by-ten matrix of complex-valued interference samples can be written down in the following way:

$$D_1 = \begin{bmatrix} x(1) & x(1) & \cdots & x(1) \\ x(2) & x(2) & & x(2) \\ \cdot & \cdot & & \cdot \\ \cdot & \cdot & & \cdot \\ \cdot & \cdot & & \cdot \\ x(10) & x(10) & \cdots & x(10) \end{bmatrix} \quad (1)$$

That is, $d(k, m)$, the k'th sample the m'th receiver, is given by the formula

$$d(k, m) = x(k) \quad (2)$$

for $k = 1, 2, \ldots 10$ and $m = 1, 2, \ldots 10$ in which $(x(1), x(2), \ldots, x(10))$ are ten, independent, complex, gaussian random variables. Clearly the interference data matrix D_1 has rank one, because the columns of D_1 are identical. Also, each row of D_1 can be written as a complex scale factor times the row vector $(1, 1, 1, \ldots 1)$.

Let us use an asterisk to denote the complex conjugate transpose of a matrix or the complex conjugate of a scalar. Then an estimate of the spatial covariance matrix of the interference, except for a scale factor, is $D_1^* D_1$. This matrix has rank one because each of its elements is the same real constant. Similarly, an estimate of the temporal covariance matrix of the interference, again, except for a scale factor, is $D_1 D_1^*$. This matrix has rank one, because each of its rows is a complex scale factor times the row vector $(x(1), x(2), \ldots, x(10))$.

3 SIGNAL DETECTION IN THE PRESENCE OF INTERFERENCE AND RECEIVER NOISE

Let us reconsider the procedure for gathering data in space and time which was described in the previous section. That is, ten snapshots of complex-valued samples are taken from a line array of ten receiving elements in a frequency band of width B Hz. We assume, in this section, that, in addition to the interference which was discussed in Section 2, independent white noise is generated locally at each receiving element. And a monochromatic, plane-wave signal may also arrive at the array over the time interval of the snapshots.

Imagine that a preliminary analysis of the ten snapshots, by beamforming and spectrum analysis or by use of a two-dimensional Fourier transform, has revealed that a strong, broadband interference is arriving from a given direction. We would like to reliably decide whether or not, in addition to the interference, there is a monochromatic plane wave signal arrival. The interesting case to consider is that in which the signal arrives from almost the same direction as the interference. Otherwise, conventional beamforming and spectrum analysis can be used.

Our method for signal detection consists of the following four steps:

(1) By rotation or delay adjustment, steer the array so that to a good approximation, the same interference waveform is processed by each receiver. Because the interference is assumed to be broadband and stronger than the other waveforms in this frequency band, the necessary time delays can be accurately estimated.

(2) Estimate the two-dimensional interference waveform by the method of Reference (1), temporarily treating the interference as a desired waveform which is to be enhanced.

(3) Subtract the interference estimate from the two-dimensional data array.

(4) Test for the presence of the signal by matched filtering in two dimensions, treating the detection as a problem in detecting the presence of a monochromatic, plane-wave of inexactly known frequency and wavenumber in approximately white noise. The likelihood-ratio rationale for this approach is described in References 2 and 3.

With regard to Step (2) of the above procedure, we recall from Section 2 above that, if the observed data matrix consisted only of interference, the rank of the matrix would be one. In order to estimate the interference, we form a best rank-one approximation to the observed data matrix (1). If the interference is strong, then it will be well estimated and subtracted. However, if the arrival angle of the signal is very close to that of the interference, then some of the signal energy is lost in the subtraction process, and the apparent arrival angle of the residual signal differs from the true arrival angle.

In practice, the major difficulty in using conventional beamforming and spectrum analysis in this case is the proper setting of the decision threshold, because the threshold setting must be derived directly from the data, which is dominated by the strong interference. We shall see later that, if one knew where to set the threshold, the conventional system can have very good performance.

4 EXPERIMENTAL RESULTS

The example described was simulated using a computer. The interference matrix D_1 is as described by formula [2]. The signal matrix is denoted by S and is given by the formula

$$S(k, m) = A_S e^{j2\pi f_1 k} e^{j2\pi f_2 m} \quad \text{for} \quad \begin{matrix} k = 1, 2, \ldots 10 \\ n = 1, 2, \ldots 10 \end{matrix} \quad (3)$$

where f_1 and f_2 are the temporal and spatial frequencies.

The independent receiver noise matrix is denoted by W and is given by the formula

$$W(k, m) = b(k, m) \quad \text{for} \quad \begin{matrix} k = 1, 2, \ldots 10 \\ m = 1, 2, \ldots 10 \end{matrix} \quad (4)$$

where $b(k, m)$ is a set of complex-valued, mutually independent, zero-mean, gaussian random variables. The variance of each real and imaginary part is σ_1^2.

The interference-to-noise power ratio (INR), is

$$\text{INR} = 10 \log_{10}\left(\frac{2\sigma_2^2}{2\sigma_1^2}\right) \quad (5)$$

where σ_2^2 is the variance of the real or imaginary part of each gaussian interference random variable $x(k)$ of formula (2).

The number of dB the weak signal power is below the interference power is given by

$$\Delta \text{dB} = 10 \log_{10}\left(\frac{2\sigma_2^2}{|A_S|^2}\right). \quad (6)$$

In our example we let INR = 20 dB, $f_2 = .06$ cycles per sampling period and with f_1 being varied from .01 to .1 cycles per sampling period in .01 increments.

The following cases were looked at:
(1) interference plus noise (Hypothesis H_0)

$$D_3 = D_1 + W \quad (7)$$

where D_1 is the interference matrix of formula (1) and W is the independent receiver noise matrix of formula (4).

(2) interference, weak signal, and noise (Hypothesis H_1)

$$D_3 = D_1 + S + W \quad (8)$$

where S is the weak signal matrix of formula (3).

The method used for detecting the weak signal in the presence of the interference is as described in the previous section.

The singular value decomposition of D_3 is given by $U \Sigma V^*$ where U contains the left singular vectors of D, Σ is a diagonal matrix whose elements are the singular values of D_3 and V contains the right singular vectors of D_3.

This method consists of the following two steps:
(1) Compute the residual data matrix

$$X = D_3 - D_1' \qquad (9)$$

where D_1' is the estimate of the interference matrix which is obtained from $U\Sigma^+ V^*$ where Σ^+ is obtained from Σ by setting all except the first principal singular value to zero.

(2) The two-dimensional Fourier transform of the residual data matrix X is computed at eleven points where $f_1 = .00, .01, .02 \ldots .11$ cycles per sample period and with f_2 remaining fixed at 0.06 cycles per sample period.

To measure the effective loss in signal-to-noise ratio (SNR) the previous two cases are repeated except that the interference is removed. The data matrix is then directly Fourier transformed.

These two methods are repeated over 50 independent trials using the same set of 50 interference and independent receiver noise matrices. Using this data we estimated the mean and standard deviation of the squared magnitude of the two-dimensional Fourier transform at the specified frequency points.

The criterion that was used for judging how well the methods performed is the separation of the sets of values of the squared magnitude of the Fourier transform which correspond to H_0 and H_1. The boundaries of the two regions are $\pm$ one standard deviation about the mean of the squared magnitude of the Fourier transform at the given frequency points. If these two regions did not intersect and remained apart the signal-detectability performance was judged to be satisfactory.

The experimental results are presented in Table 1, Figure 2a-b, Figure 3a-c, and Figure 4a-c. Figure 2 has a plot and table of the increase in signal power which is required to achieve the signal-detectability performance of the interference-free comparison case.

To verify our experimental results we make the following argument: The spatial covariance matrix of the interference alone, $D_1 D_1^*$, will have as its principal eigenvector the column vector

$$u_1' = A(1, 1, \ldots 1)^T \qquad (10)$$

where A is a scale factor. Now form the signal-plus-interference matrix

$$D_2 = D_1 + S \qquad (11)$$

where D_1 and S are the same as before. The signal-plus-interference covariance matrix $D_2 D_2^*$ will have rank 2. Since in our examples the interference power is considerably larger than the signal power, the principal eigenvector u_1 will only be slightly perturbed and will essentially consist of the interference eigenvector of formula (10), u_1'.

The second eigenvector u_2 of $D_2 D_2^*$ must contain the residual signal because $D_2 D_2^*$ is of rank 2. Since $D_2 D_2^*$ is hermitian, it follows that u_2 and u_1' must be approximately orthogonal to each other.

Table 1a Statistics of the Squared Magnitude of the Two-Dimensional Fourier Transform of the Data Matrix D_3 over 50 Trials at Limit of Weak Signal Detectability

f_1 frequency	H0		H1		weak signal SNR
	m	σ	m	σ	
.01	1.693	1.5	11.209	6.709	−13 dB
.02	1.614	1.396	9.076	5.546	−14 dB
.03	1.554	1.367	8.937	5.049	−14 dB
.04	1.525	1.381	7.281	4.314	−15 dB
.05	1.536	1.406	7.316	4.371	−15 dB
.06	1.59	1.437	9.113	5.06	−14 dB
.07	1.683	1.483	9.547	5.323	−14 dB
.08	1.805	1.548	10.149	5.677	−14 dB
.09	1.94	1.625	9.08	5.396	−15 dB
.10	2.072	1.706	9.625	5.502	−15 dB

$INR = 20$ dB $\qquad m -$ mean

$f_2 = .06 \qquad \sigma -$ standard deviation

Table 1b Statistics of the Squared Magnitude of the Two-Dimensional Fourier Transform of the Data Matrix D_3 over 50 Trials at Weak Signal SNR that is 1 dB Below the Limit of Weak Signal Detectability

f_1 frequency	H0		H1		weak signal SNR
	m	σ	m	σ	
.01	1.693	1.5	9.203	6.045	−14 dB
.02	1.614	1.396	7.493	4.991	−15 dB
.03	1.554	1.367	7.363	4.519	−15 dB
.04	1.525	1.381	6.042	3.853	−16 dB
.05	1.536	1.406	6.074	3.914	−16 dB
.06	1.59	1.437	7.523	4.541	−15 dB
.07	1.683	1.483	7.92	4.798	−15 dB
.08	1.805	1.548	8.467	5.118	−15 dB
.09	1.94	1.625	7.69	4.87	−16 dB
.10	2.072	1.706	8.19	4.97	−16 dB

$INR = 20$ dB $\qquad m -$ mean

$f_2 = 0.06 \qquad \sigma -$ standard deviation

Table 1c Statistics of the Squared Magnitude of the Two-Dimensional Fourier Transform of the Residual Matrix X at the Peak Frequency over 50 Trials at Limit of Weak Signal Detectability

f_1 frequency	H0		H1		weak signal SNR	peak frequency
	m	σ	m	σ		
.01	1.403	1.27	8.741	5.264	4	.07
.02	1.403	1.27	7.424	4.544	−3	.07
.03	1.403	1.27	6.758	4.093	−7	.07
.04	1.586	1.435	7.477	4.255	−9	.08
.05	1.586	1.435	7.35	4.221	−11	.08
.06	1.586	1.435	8.027	4.62	−12	.08
.07	1.721	1.535	8.297	5.068	−13	.09
.08	1.721	1.535	9.724	5.746	−13	.09
.09	1.813	1.588	9.031	5.491	−14	.10
.10	1.813	1.588	9.567	5.527	−14	.10

$INR = 20\ dB$ m − mean
$f_2 = 0.06$ σ − standard deviation

peak frequency− f_1 frequency at which the maximum
squared magnitude of the two-dimensional
Fourier transform occurred.

Table 1d Statistics of the Squared Magnitude of the Two-Dimensional Fourier Transform of the Residual Matrix X at the Peak Frequency over 50 Trials at Weak Signal SNR That is 1 dB Below the Limit of Weak Signal Detectability

f_1 frequency	H0		H1		weak signal SNR	peak frequency
	m	σ	m	σ		
.01	1.403	1.27	7.186	4.701	3 dB	.07
.02	1.403	1.27	6.09	4.065	−4 dB	.07
.03	1.403	1.27	5.557	3.624	−8 dB	.07
.04	1.586	1.435	6.159	3.808	−10 dB	.08
.05	1.586	1.435	6.084	3.778	−12 dB	.08
.06	1.586	1.434	6.653	4.129	−13 dB	.08
.07	1.721	1.535	6.939	4.533	−14 dB	.09
.08	1.721	1.535	8.119	5.101	−14 dB	.09
.09	1.813	1.588	7.631	4.928	−15 dB	.10
.10	1.813	1.588	8.088	4.966	−15 dB	.10

$INR = 20\ dB$ m − mean
$f_2 = .06$ σ − standard deviation

peak frequency− f_1 frequency at which the maximum

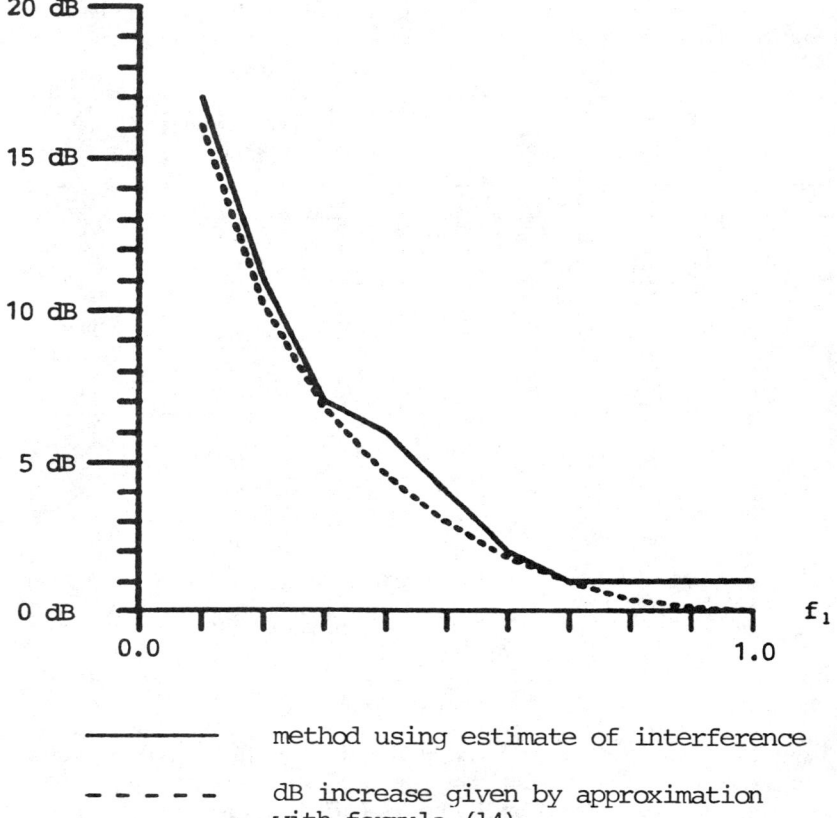

Figure 1a Decibel increase in signal power which is required to achieve signal-detectability performance of the interference-free comparison case

Since $u_1 = u_1'$, the sinusoidal vectors that will be orthogonal to u_1' will be of the form

$$e_k = \left[1, e^{j2\pi\left(\frac{k}{10}\right)}, e^{j2\pi q\left(\frac{k}{10}\right)^2}, \ldots e^{j2\pi\left(\frac{k}{10}\right)^{10}}\right]^T, \quad (12)$$

for $k = 1, 2, \ldots 9$.

Therefore, the approximate signal power in the residual matrix X can be

F_1	ΔdB_1	ΔdB_2
0.01	17	16.1
0.02	11	10.2
0.03	7	6.8
0.04	6	4.6
0.05	4	3.0
0.06	2	1.8
0.07	1	1.0
0.08	1	0.4
0.09	1	0.11
0.10	1	0

ΔdB_1 - method using estimate of interference

ΔdB_2 - dB increase given by approximation with Formula (14)

Figure 1b Decibel increase in signal power which is required to achieve signal-detectability performance of the interference-free case

obtained by projecting the spatial signal vector onto the vectors e_k as follows

$$\sum_{k=1}^{9} |S_1^T e_k|^2 \qquad (13)$$

where S_1 is the spatial signal vector or the approximate decibel increase in signal power can be obtained by the formula

$$\Delta dB_{increase} = 10 \log_{10} (\sin^2[102\pi F_1]). \qquad (14)$$

This approximation is also plotted in Figure 2. The results indicate that this is a reasonable approximation.

At frequencies close to the second bin of the DFT of the spatial signal vector, the method presented performs nearly as well as two-dimensional Fourier transforming the data matrix directly for the case when there is no interference present (see Table 1 and Figure 2a-b).

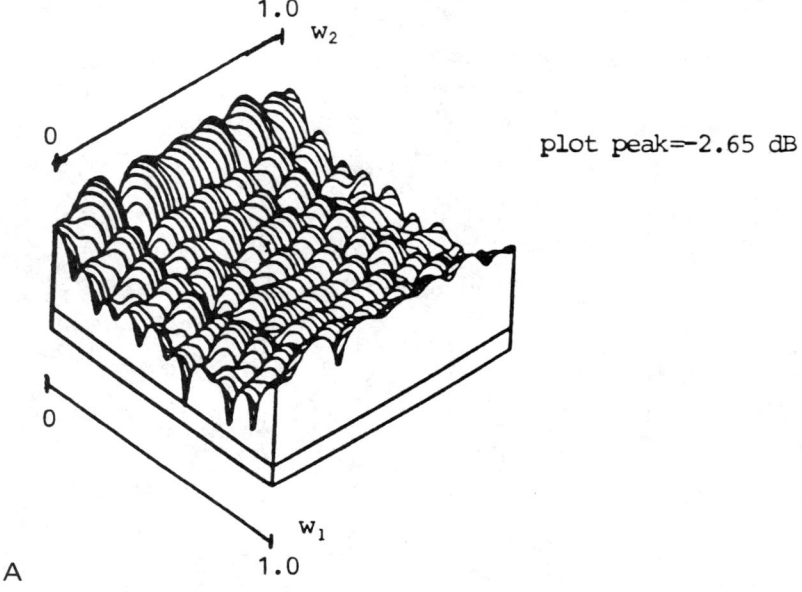

Figure 2 Log plots of the squared magnitude of the two-dimensional Fourier transform of the data matrices and the residual matrix from one simulation trial with $\begin{matrix} f_1 = .03 \\ f_2 = .06 \end{matrix}$ $INR = 20$ dB

Note: All plots have range of 50 dB.

plot peak $= -2.65$ dB

(a) Data matrix $D_3 = D_1 + S + W$ with $SNR = -4$ dB.

plot peak $= -26.3$ dB

(b) Residual matrix X with $SNR = -4$ dB.

Note: Using the same data matrix D_3 as before.

plot peak $= -26.2$ dB

(c) Data matrix $D_3 = S + W$ with $SNR = -10$ dB.

Data-Adaptive Detection

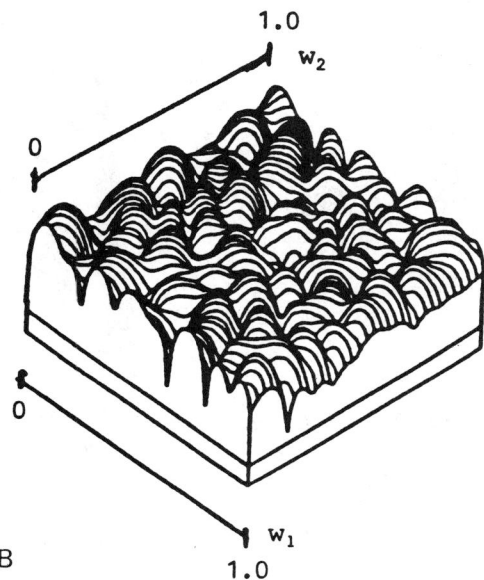

plot peak=−26.3 dB

B

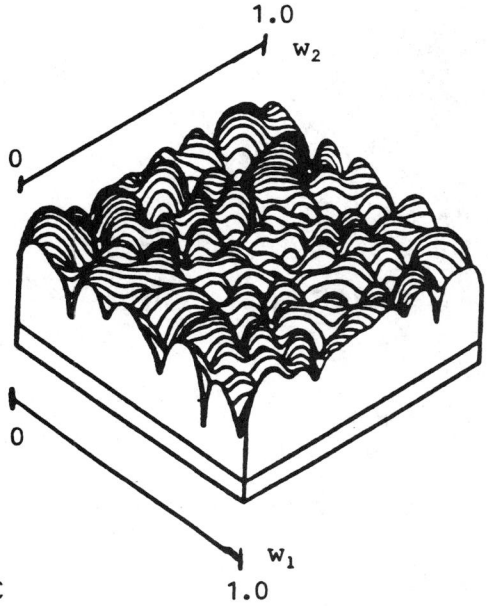

plot peak=−26.2 dB

C

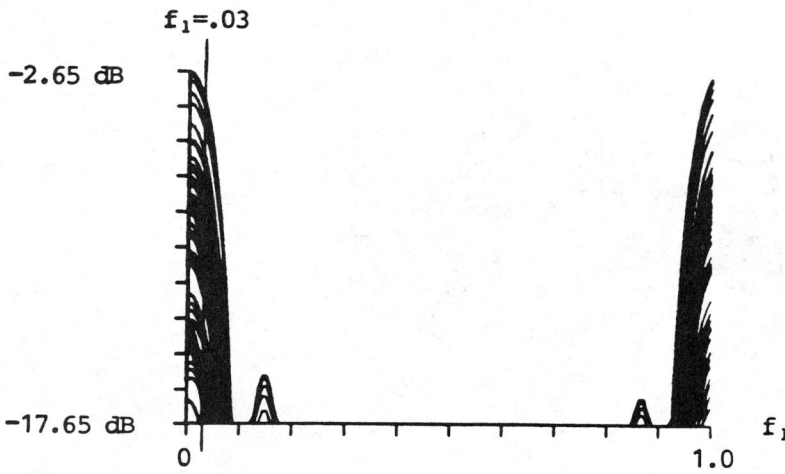

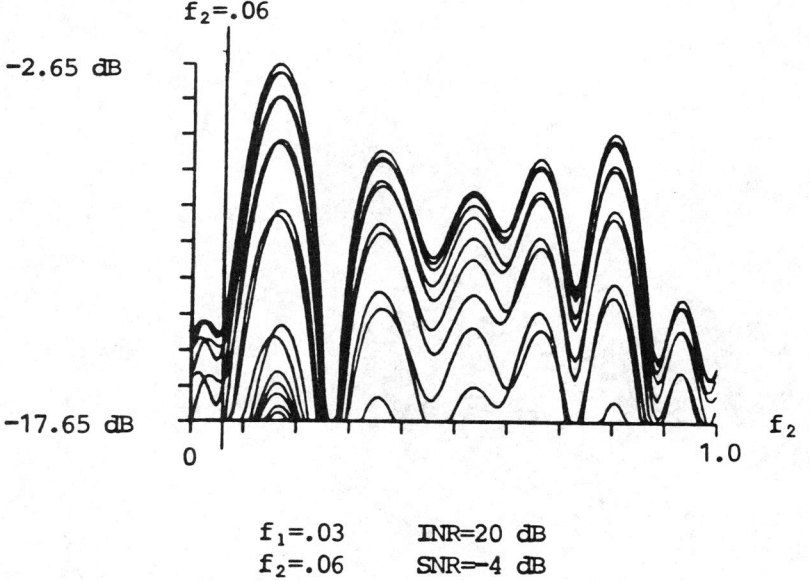

$f_1 = .03$ INR = 20 dB
$f_2 = .06$ SNR = −4 dB

Figure 3a Log plot of cuts of the squared magnitude of the two-dimensional Fourier transform of the data matrix $D_3 = D_1 + S + W$ from one trial.

Note: Same data as in Figure 2a.

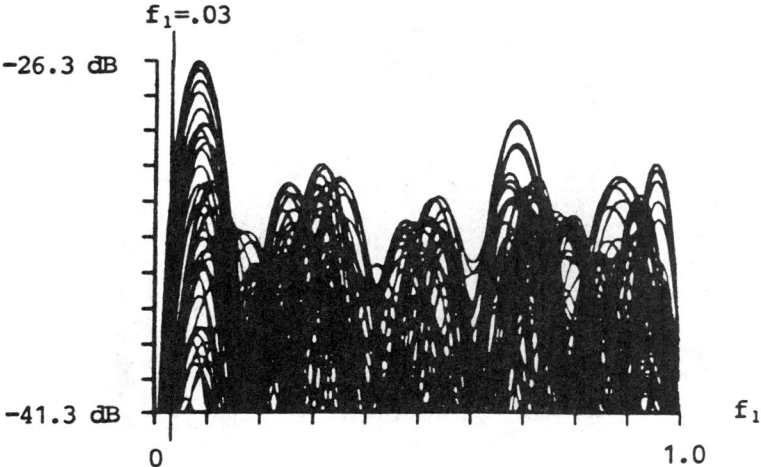

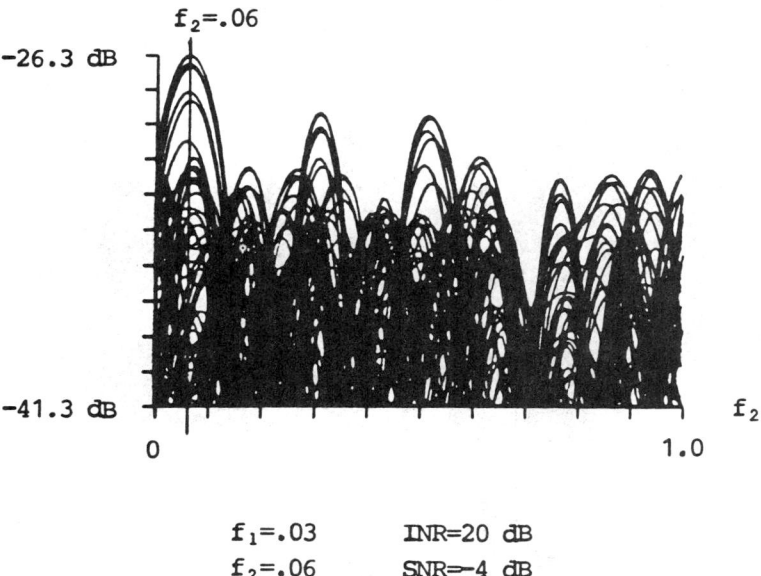

$f_1=.03$ INR=20 dB
$f_2=.06$ SNR=-4 dB

Figure 3b Log plot of cuts of the squared magnitude of the two-dimensional Fourier transform of the residual matrix X from one trial.

Note: Same data as in Figure 2b.

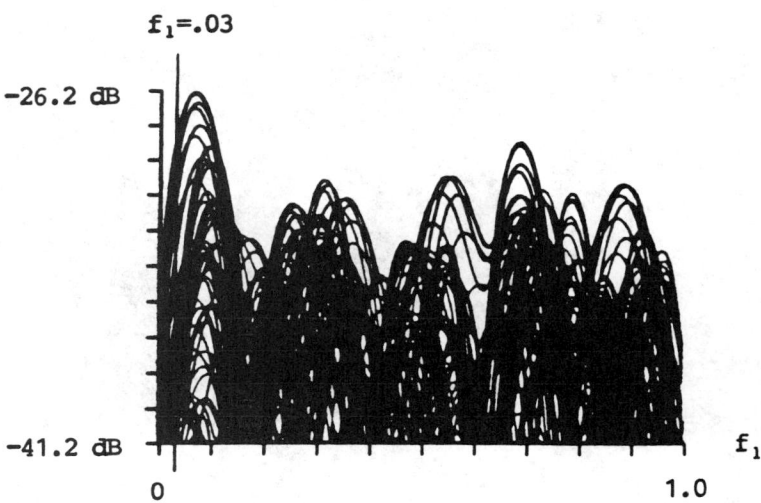

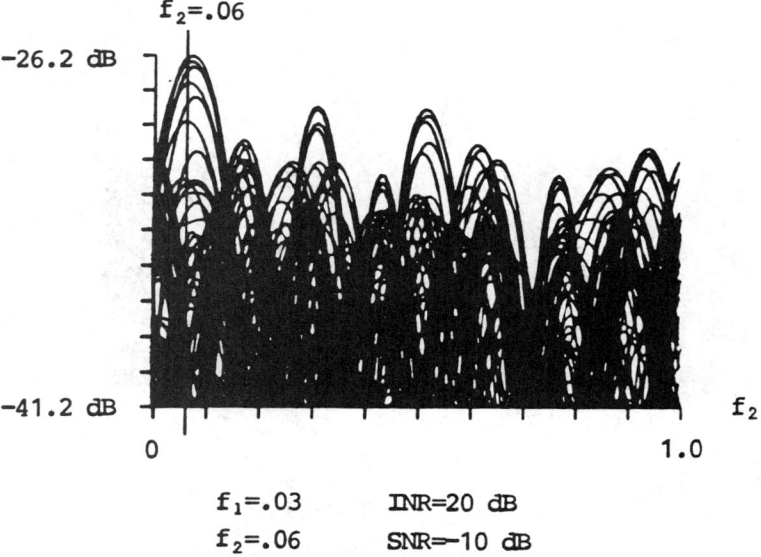

$f_1 = .03$ INR = 20 dB
$f_2 = .06$ SNR = −10 dB

Figure 3c Log plot of cuts of the squared magnitude of the two-dimensional Fourier transform of the data matrix $D_3 = S + W$ from one trial.

Note: Same data as in Figure 2c.

REFERENCES

1. D.W. Tufts, Ramdas Kumaresan, and Ivars Kirsteins, "Data Adaptive Signal Estimation by Singular Value Decomposition of a Data Matrix," Proceedings Letters, Proc. IEEE, Vol. 70, pp. 685-884, June 1982.

2. D.W. Tufts, Ivars Kirsteins, and Ramdas Kumaresan, "Adaptive Detection of a Weak Signal in Noise and Strong Interference," to appear, IEEE Transactions on Aerospace and Electronics Systems.

3. A.J. Claus, T.T. Kadota, and D.M. Romain, "Efficient Approximation for a Family of Noises for Application in Adaptive Spatial Processing for Signal Detection," IEEE Transactions on Information Theory, Vol. IT-26, pp. 588-595, Sept. 1980.

Optimal Detection in Linear Reverberation Noise*

Patrick L. Brockett

Department of Finance and Applied Research Laboratories
University of Texas
Austin, TX

1 INTRODUCTION

The problem of optimal detection of signals in stochastic noise has been solved in only a relatively few cases. Most investigators assume the signal, or the noise, or both are Gaussian. However, in many very important practical situations, e.g., sonar, radar, or satellite transmission, this assumption may not hold (cf. Girodan and Haber (1972), Kennedy (1969), Trunk and George (1970), and VanTrees (1971)). A particular case in point is Middleton's model for reverberation in randomly rough surfaces (1967a, 1967b, 1972a, 1972b). This has been called the most complete model for reverberation in the review paper of Fortuin (1967).

A brief summary of the Middleton model and its assumptions are given in Wilson (1981). In particular the following assumptions are made.

1. The total scattering process is a linear superposition of point scatterers (inhomogenities such as the case of bubbles in an otherwise homogeneous medium). These point scatterers are assumed to be independently distributed in space and time. Moreover, only primary scattering is considered.

2. Sound interacts with the scatterers via an impulse response function.

3. The number of scatterers in a given region follows a Poisson point

---
*Research supported under Contract N00014-81-K-0145, Task SRO 106 with the Office of Naval Research.

process. (This is consistent with assumption 1.)

4. The elementary scattered waveform is the same for each scatterer, and any random parameters associated with an individual scatterer have identical distributions.

5. The scattering zone is in the farfield of the projector and receivers.

Using these assumptions, it is shown in Middleton (1967a) that the reverberation process $Y(t)$ has a joint characteristic function of the form

$$\phi_{Y(t_1), \ldots, Y(t_n)}(\xi_1, \xi_2, \ldots, \xi_n)$$

$$= \exp \int_\Lambda \int_\Omega \left[\exp \left\{ i \sum_{j=1}^n \xi_j U(t_j; \lambda; \theta) \right\} - 1 \right] q(\lambda; t') F(d\theta) d\lambda \quad (1)$$

where t' is the temporal dependence of the scattering, Λ is the scattering region and, for $\lambda \in \Lambda$, $U(t_j, \lambda; \theta)$ is the elementary scattered waveform received at time t_j from point λ at time t', and $q(\lambda; t')$ is the density of scatterers at λ and time t'. Here, θ is a random parameter associated with the scattering, which has probability density $F(\theta)$. We shall assume θ is a scale factor and $U(t, \lambda, \theta) = \theta U(t, \lambda)$. In particular, θ could be constant.

The estimation of parameters in the Middleton model is considered in Wilson (1975), and in Brockett and Wilson (1982). In this paper we shall assume the parameters of the model are known and proceed to develop an optimal detector, i.e., a likelihood ratio detector for testing the hypothesis of signal plus noise versus pure noise for a class of stochastic processes which includes the Middleton model. We show that the Middleton noise process is a special type of linear process defined in Section 2, and is infinitely divisible. For two linear processes we explicitly calculate the likelihood ratio and its distribution.

2 LINEAR INFINITELY DIVISIBLE PROCESSES, AND MIDDLETON'S MODEL

In this section we shall present a brief description of the linear process. A more extensive discussion and references are given in Lugannani and Thomas (1967).

The linear process $Y(t)$ is defined by the stochastic integral

$$Y(t) = \int_B f(t, s) \, dX(s) \quad (2)$$

where $X(s)$ is a zero mean, second order, stochastically continuous process with independent increments and $B \subseteq R^k$. Additionally, $f(t, s)$ is real valued and square integrable with respect to $dV(s) = E|dX(s)|^2$. In brief, Y is an L_2 filtering of an independent increment process. We shall additionally assume f is L_2 continuous (so that $Y(t)$ is stochastically continuous). As particular cases we have the Gaussian process and the shot noise process and, as we shall see, the Middleton model for reverberation noise.

Following the method used in Papoulis (1965) for shot noise processes, one may determine the finite dimensional characteristic functions of the linear process defined in Equation (2) (cf. Lugannani and Thomas (1967) or Eastwood and Lugannani (1977)). In the case without Gaussian component, they are given by

$$\phi_{\mathbf{t}}(\mathbf{u}) = \exp\left\{\int_B \int_0^\infty \{\exp(izw) - 1 - izw\} M(ds, dz)\right\} \quad (3)$$

where $\mathbf{t} = (t_1, \ldots, t_n)$, $\mathbf{u} = (u_1, \ldots, u_n)$, and $w = u_1 f(t_1, s) + u_2 f(t_2, s) + \ldots + u_n f(t_n, s)$. The measure M is the place-jump measure of the additive process X, i.e., $M(C \times A)$ is the expected number of jumps (pulses) of the process X occurring in the set C with the magnitude (amplitude) in the Borel set A. See Gikhman and Skorokhod (1969) for a more detailed explanation of the Levy measure M and its properties. In Brockett and Tucker (1977) where B is one-dimensional and associated with the time of the pulse of X, the measure M is called the time-jump measure. We have deleted the Gaussian component of $Y(t)$ from Equation (3) because it is independent of the non-Gaussian part, it has already been treated extensively in the literature, and it does not appear explicitly in Middleton's model.

To demonstrate that Middleton's model is actually a linear process, we observe that if $X(s)$ is a Poisson point process over Λ with density $q(\lambda)$, and if the size of the pulse (jump) at λ is θ with probability distribution $F(\theta)$, and if the function f is actually the elementary waveform $U(t, \lambda)$, then the linear process given in Equation (2) has the joint characteristic function shown in Equation (3) with $M(d\lambda, dz) = q(\lambda) F(d\theta) d\lambda$. This is of the form defined in Equation (1) that is translated by a constant. Thus, the Middleton process is a particular type of linear process. Several other noise models can also be shown to be linear processes as well (cf. Kennedy (1969)).

The method we shall use to derive likelihood ratios is to embed the linear processes in the class of infinitely divisible processes, and then utilize known representation theorems for the larger class to obtain a manageable form for Equation (2).

The class of infinitely divisible stochastic processes was evidently first studied by Lee (1967), and subsequently studied by Maruyama (1970), Wright (1975), Briggs (1975), and Veeh (1981). A stochastic process is considered infinitely divisible if all of its n-dimensional marginal distributions are infinitely divisible random vectors. A second order infinitely divisible process without a Gaussian component has the following representation. By definition, for every finite subcollection $\lambda = \{t_1, t_2, \ldots, t_n\} \subseteq [a, b]$, there exists a random vector $\mathbf{c}_\lambda$, and an n-dimensional Levy measure $\overline{M}_\lambda$ such that the characteristic function of $(Y(t_1), \ldots, Y(t_n))$ is

$$\ln \phi_\lambda(\mathbf{u}) = i\mathbf{u}'\mathbf{c}_\lambda + \int \{\exp(i\mathbf{u}'\mathbf{x}) - 1 - i\mathbf{u}'\mathbf{x}\} dM_\lambda(\mathbf{x}). \quad (4)$$

Here we have used a variant of the Kolmogorov representation valid for second order, infinitely divisible vectors (cf. Lukacs (1970), p. 119).

Let $\Lambda = \{\lambda = (t_1, \ldots, t_n)\}$ denote the set of all finite subsets of $[a, b]$. The collection $\{(\mathbf{c}_\lambda, M_\lambda), \lambda \in \Lambda\}$ uniquely determines the distribution of an infinitely divisible process Y, and vice versa (Maruyama (1970), Theorems 1 and 3). Using the partial ordering on Λ by inclusion, we obtained a system of projections $\{P_\lambda, \lambda \in \Lambda\}$ from function space $R^{[a,b]}$ onto the coordinate space R^λ. The system of Levy measures $\{M_\lambda, \lambda \in \Lambda\}$ is consistent, and Maruyama shows that a measure Q may be defined on $R^{[a,b]}$ as the projective limit of the collection $\{M_\lambda, \lambda \in \Lambda\}$. The σ-algebra, and the construction of Q is similar to the usual construction in Kolmogorov's existence theorem for processes with given marginals. As we shall see, for linear processes Q is actually explicitly obtainable.

To demonstrate that a linear process is an infinitely divisible process, we manipulate the characteristic function given by Equation (3) into the form of that of Equation (4). Towards this end we first note that if $\lambda = (t_1, \ldots, t_n)$ is given, and M_λ is defined on R^λ via $M_\lambda(A) = M[\{(s, z): [zf(t_1, s), \ldots, zf(t_n, s)] \in A\}]$, then M_λ is a Levy measure on R^λ concentrated on the surface $[zf(t_1, s), \ldots, zf(t_n, s)]$, $s \in B$, $z \in R$. Moreover, the integral relationship

$$\int_B h(\mathbf{x}) dM_\lambda(\mathbf{x}) = \int_B \int_{-\infty}^{\infty} h[zf(t_1, s), \ldots, zf(t_n, s)] M(dz, ds)$$

holds for measurable h. The fact that M_λ is indeed a Levy measure on R^λ follows (after some calculations) from the square integrability of f with respect to V and from the formula $\int h(s) dV(s) = \int \int z^2 h(s) M(ds, dz)$ which relates the variance measure to the place-jump measure M.

Now, let us write $\mathbf{f}_\lambda(s) = P_\lambda f(\cdot, s) = (f(t_1, s), \ldots, f(t_n, s))$ and $\mathbf{c}_\lambda = 0$. We observe that

$$i\mathbf{u}'\mathbf{c}_\lambda + \int_{-\infty}^{\infty} \{\exp(i\mathbf{u}'\mathbf{x}) - 1 - i\mathbf{u}'\mathbf{x}\} M_\lambda(dx)$$
$$= \int \int \{\exp(iz\mathbf{u}'\mathbf{f}_\lambda(s)) - 1 - iz\mathbf{u}'\mathbf{f}_\lambda(s)\} M(ds, dz)$$
$$= \int \int \{\exp(izw) - 1 - izw\} M(ds, dz),$$

where $w = u_1 f(t_1, s) + u_2 f(t_2, s) + \ldots + u_n f(t_n, s)$ as before. Thus Equation (3) is now in the form of Equation (4). It follows that linear processes are in fact infinitely divisible processes. Moreover, we can determine the projective limit Q of the system of Levy measures $\{M_\lambda, \lambda \in \Lambda\}$. Namely, if $A \subseteq R^{[a,b]}$ is a collection of functions, then $Q(A) = M[\{(s, z): zf(\cdot, s) \in A\}]$. This follows since if $B \subseteq R^\lambda$, then

$$M_\lambda(B) = M[\{(s, z): [zf(t_1, s), \ldots, zf(t_n, s)] \in B\}]$$
$$= M[\{(s, z): zP_\lambda f(\cdot, s) \in B\}]$$
$$= Q[P_\lambda^{-1}(B)],$$

so that the λth coordinate projection of Q is M_λ.

Given the projective limiting measure Q on $R^{[a,b]}$, we obtain a pathwise representation of the linear (or, more generally, infinitely divisible) process Y as follows: let π be a Poisson random measure on $R^{[a,b]}$ with intensity measure Q, i.e., for disjoint $A_1, \ldots, A_n$, $\pi(A_1), \ldots, \pi(A_n)$ are independent Poisson random variables with means $Q(A_1), \ldots, Q(A_n)$ respectively. Then,

$$Y(t) = \underset{\epsilon \to 0}{\text{limit in Prob}} \int_{A_\epsilon} x(t)(\pi - Q)(dx) \qquad (5)$$

where $A_\epsilon = \{x: |x(t)| \geq \epsilon\}$ has joint characteristic functions shown in Equation (3). Indeed,

$$\log E\left\{\exp\left[i \sum_{j=1}^n u_j Y(t_j)\right]\right\} = \int \left\{\exp\left[i \sum_{j=1}^n u_j x(t_j)\right] - 1 - i \sum_{j=1}^n u_j x(t_j)\right\} Q(dx)$$
$$= \int \{\exp(i\mathbf{u}'P_\lambda x) - 1 - i\mathbf{u}'P_\lambda x\} Q(dx)$$

$$= \int \{\exp(i\mathbf{u}'y) - 1 - i\mathbf{u}'y\} M_\lambda(dy).$$

In the next section we use the representation given in Equation (5) to calculate the likelihood ratio for linear processes.

3 LIKELIHOOD RATIO DETECTION

Let Y_1 and Y_2 denote two linear processes which might be noise and signal-plus-noise in long range active sonar, for example. Their multivariate Levy measures M_1 and M_2 induce (via projective limits) the measures Q_1 and Q_2 on function space. Moreover, the processes Y_1, Y_2 also determine measures on function space via $\mu_i(A) = P\{Y_i(\cdot) \in A\}$ $i = 1, 2$.

According to the Neyman-Pearson lemma, the most powerful test, i.e., an optimum detector, for H_0: Y_1 vs H_1: Y_2 is a likelihood ratio test, i.e., reject H_0 if $d\mu_1/d\mu_2(X) > C_\alpha$, where C_α is the αth quantile value of the μ_1 distribution of $d\mu_1/d\mu_2$. In this section we relate the measures μ_i and Q_i $i = 1, 2$ and obtain the desired likelihood ratio and its distribution.

The main result of this section is summarized on Theorem 1. It concerns only the case where Q_1 and Q_2 are finite measures. The general result is stated and proved in Brockett (1982). The finite case is the pertinent case for using Middleton's model.

Theorem 1. (Nonstationary Compound Poisson Case)

Suppose $Y_1(t)$ and $Y_2(t)$ are two stochastically continuous infinitely divisible processes with corresponding projective limit measures Q_1 and Q_2 finite.

(a) If $Q_1 \ll Q_2$ and $\int x(Q_1 - Q_2)(dx) = 0$, then $\mu_1 = PY_1^{-1} \ll \mu_2 = PY_2^{-1}$. Moreover, using the representation stated in Equation (5), we have

$$\ln \frac{d\mu_1}{d\mu_2}[Y_1(\cdot)] = \int \ln \rho(x) \pi_1(dx) + Q_2(R^{[a,b]}) - Q_1(R^{[a,b]}), \quad (7)$$

where $\rho(x) = dQ_1/dQ_2(x)$.

(b) The μ_1 distribution of the log likelihood ratio in (a) is given via the characteristic function whose logarithm is

$$\ln \phi(u) = iu[(Q_2 - Q_1)(R^{[a,b]})] + \int [\exp\{iu \ln \rho(x)\} - 1]Q_1(dx).$$

Thus, $\ln d\mu_1/d\mu_2(Y_1(\cdot, \omega))$ is a translated compound Poisson random variable on R with intensity measure $\nu(A) = Q_1(\{x: \ln \rho(x) \in A\})$.

Proof. The proof is given in Brockett (1982). Roughly, it uses the representation stated in Equation (4) to relate Y to π, and then uses Brown's results (1971) on general Poisson point processes to obtain conditions of absolute continuity. The likelihood ratio formulae can also be derived from Brown's formulae by an appropriate correspondence between μ_i and π_i. Details are given in Brockett (1982) where it is also shown how to obtain equivalent results when a Gaussian component is also present by using the methods of Skorokhod (1964).

To apply the above results in the case of reverberation noise with Middleton's model, we assume the scatter density $q(\lambda)$ over Λ is estimated, and that there is an impulse response, i.e., θ is a constant of one, for example. The presence of a fixed object superimposes an additional density $h(\lambda)$ of scatterers in the position of the object. For example, certain highlights on an object with known geometrical configurations might constitute relative maximums of the object scatter density h. Of course h is supported only on the region containing the object. Thus the detection of the object constitutes a test of H_0: Y_1 vs H_1: Y_2 where Y_1 has Levy measure $q(\lambda)d\lambda$, and the signal-plus-noise Levy measure is $\{q(\lambda) + h(\lambda)\}d\lambda$. The elementary waveform $U(\lambda, t)$ is known (in the active sonar case) or estimatable (in the passive sonar case). Thus the projective measures Q_1 and Q_2 on function space may be determined. Namely, $Q_1(A) = M_1\{\lambda: U(\cdot, \lambda) \in A\}$. For convenience, consider $U: \Lambda \to R^{[a,b]}$ as a function space-valued mapping. We have

$$Q_1(A) = \int_{U^{-1}(A)} q(\lambda) d\lambda,$$

and similarly

$$Q_2(A) = \int_{U^{-1}(A)} [q(\lambda) + h(\lambda)] d\lambda.$$

Now let

$$B_\lambda = \{s: U(\cdot, s) = U(\cdot, \lambda)\} = U^{-1}(U(\cdot, \lambda)).$$

Then the quantity

$$\rho[U(\cdot, \lambda)] = \frac{dQ_1}{dQ_2}[U(\cdot, \lambda)] = \frac{\int_{B_\lambda} q(s) ds}{\int_{B_\lambda} [q(s) + h(s)] ds}$$

so that the intensity measure for the compound Poisson measure ν on R^1 in Theorem 1(b) is given by

$$\nu([c, d]) = Q_1(\{f: \ln \rho(f) \in [c, d]\}) = \int_D q(\lambda) d\lambda$$

where

$$D = \left\{\lambda: c + \ln \int_{B_\lambda} q(\lambda) + h(\lambda) d\lambda \leq \ln \int_{B_\lambda} q(\lambda) dx \leq d + \ln \int_{B_\lambda} q(\lambda) + h(\lambda) d\lambda\right\}.$$

Since the likelihood ratio is a translate of a compound Poisson, we know its probability measure is given by translating the measure

$$n(B) = e^{-1} \sum_{k \geq 0} \frac{\nu^{*k}(B)}{k!},$$

where $*$ represents convolution. By truncating this series after a few terms, we may approximate the probability distribution for the likelihood ratio, and hence obtain the critical value C_α needed for the likelihood ratio test.

REFERENCES

Briggs, D.V. (1975). "Densities for Infinitely Divisible Random Processes," J. Mult. Anal., 5, 278-205.

Brockett, P.L. (1982). "The Likelihood Ratio Detector for Non-Gaussian Infinitely Divisible and Linear Stochastic Processes," The University of Texas at Austin, Department of Finanace Working Paper, 81/82-2-54. To appear Annals of Statistics.

Brockett, P.L., and H.G. Tucker (1977). "A Conditional Dichotomy Theorem for Stochastic Processes with Independent Increments," J. Mult. Anal., 7, 13-27.

Brockett, P.L., and G. Wilson (1982). "Likelihood Detection Using Middleton's Model of Surface Reverberation," in preparation.

Brown, M. (1971). "Discrimination of Poisson Processes," Ann. Math. Statist., 42, 771-775.

Eastwood, L.F., Jr., and R. Lugannani (1977). "Approximate Likelihood Ratio Detectors for Linear Processes," IEEE Trans. Inf. Th., IT-23, 482-489.

Fortuin, L. (1979). "Survey of Literature on Reflection and Scattering of Sound Waves at the Sea Surface," J. Acoust. Soc. Am., 47, 1209-1228.

Gikhman, I.I., and A.V. Skorokhod (1969). *Introduction to the Theory of Random Processes* (Saunders, Philadelphia).

Girodan, A.A., and F. Haber (1972). "Modeling of Atmospheric Noise," Radio Science, 7, 1011-1023.

Kennedy, R.S. (1969). *Fading Dispersive Communication Channels* (Wiley, New York).

Lukacs, E. (1970). *Characteristic Functions* (Hafner Publisher, New York).

Lee, P.M. (1967). "Infinitely Divisible Stochastic Processes," Z. Wahrscheinlichkertstheorie Verw. Geb., 7, 147-160.

Lugannani, R., and J.B. Thomas (1967). "On a Class of Stochastic Processes which are Closed Under Linear Transformations," Information and Control, 10, 1-21.

Maruyama, G. (1970). "Infinitely Divisible Process," Th. Prob. Appl., XV, No. 1, 1-22.

Middleton, D. (1967a). "A Statistical Theory of Reverberation and Similar First Order Scattered Fields, Part I: Waveform and General Processes," IEEE Trans. Inf. Theory, 13, 372-392.

Middleton, D. (1967b). "A Statistical Theory of Reverberation and Similar First Order Scattered Fields, Part II: Moments, Spectra, and Spatial Distribution," IEEE Trans. Inf. Theory, 13, 393-414.

Middleton, D. (1972b). "A Statistical Theory of Reverberation and Similar First Order Scattered Fields, Part IV: Statistical Models," IEEE Trans. Inf. Theory, 18, 35-67.

Middleton, D. (1972a). "A Statistical Theory of Reverberation and Similar First Order Scattered Fields, Part IV: Statistical Models," IEEE Trans. Inf. Theory, 18, 68-90.

Newman, C.M. (1973). "On the Orthogonality of Independent Increment Processes," *Topics in Probability Theory* (Courant Institute of Mathematical Science).

Papoulis, A. (1965). *Probability, Random Variables, and Stochastic Processes* (McGraw-Hill, New York).

Skorokhod, A.V. (1964). *Random Processes with Independent Increments* (Izdatel'stvo "Nauka," Moscow (in Russian)).

Trunk, G.V., and S.F. George (1970). "Detection of Targets in Non-Gaussian Sea Clutter," IEEE Trans. Aerosp. Electron. Syst., AES-6, 620-628.

VanTrees, H.L. (1971). *Detection, Estimation, and Modulation Theory, Part III: Radar-Sonar Signal Processing and Gaussian Signals in Noise* (Wiley, New York).

Veeh, J.A. (1981). "Multidimensional Infinitely Divisible Measures," Ph.D. Dissertation, University of California, Irvine, Dept. of Mathematics.

Wilson, G.R. (198). "Covariance Functions and Related Statistical Properties of Acoustic Backscattering from a Randomly Rough Air-Water Interface," Applied Research Laboratories Technical Report No. 81-23 (ARL-TR-81-23), Applied Research Laboratories, the University of Texas at Austin.

Wright, A.L. (1975). "Statistical Inference for Stationary Gaussian processes," Ph.D. Dissertation, University of California, Irvine, Dept. of Mathematics.

On Some Estimation/Detection Problems from Sampled Data*

Elias Masry

*Electrical Engineering and
Computer Sciences Department
University of California, San Diego
La Jolla, CA*

1 INTRODUCTION

The nature of many signals in such diverse fields as acoustics, sonar, seismology, speech communication, and many others is intrinsically analog and continuous in time yet the processing of such signals is invariably digital. This is largely due to the availability of fast digital computers and the development of efficient signal processing algorithms. The process of analog-to-digital conversion consists of sampling and quantization and we confine ourselves here to the sampling aspects of the problem.

It is a common practice to employ periodic sampling whereby the sampling instants $\{t_k\}$ are equally-spaced. However, in many problems of signal estimation and detection, periodic sampling is not necessarily the best choice in the sense that the use of appropriate nonperiodic sampling schemes can provide significantly better performance for the discrete-time estimators and detectors. Perhaps the most illuminating example is in the area of covariance and spectral estimation of continuous-time processes from sampled data: Let $\{X(t), -\infty < t < \infty\}$ be a stationary stochastic process with covariance function $R(t)$ and spectral density function $\phi(\lambda)$. Suppose equally-spaced data $\{X_k = X(k/\rho)\}_{k=1}^{n}$ is taken, with sampling rate $\rho > 0$, and is used to form estimates $\hat{R}_n(t)$ and $\hat{\phi}_n(\lambda)$ of $R(t)$ and $\phi(\lambda)$ respectively. Due to the aliasing of the spectrum, introduced by periodic sampling, these estimates are *not* consistent, as the sample size $n \to \infty$, unless the process X is bandlimited, i.e.,

---

*Work Supported under Contract N00014-75-C-0652, Task NR 042-289 with the Office of Naval Research.

$\phi(\lambda) = 0$ for $|\lambda| > W$, $W > 0$, and the sampling rate ρ satisfies $\rho \geq \pi/W$ [5]. This requires, however, prior knowledge of the spectrum and, more significantly, high rates of sampling for broadband signals. The above situation changes dramatically if *non*equally-spaced data $\{X(t_k)\}_{k=1}^n$ are used to form the covariance and spectral estimates $\hat{R}_n(t)$ and $\hat{\phi}_n(\lambda)$; here $\{t_k\}$ is an appropriate stationary point process on the real line with mean intensity ρ. The characterization of point processes $\{t_k\}$ which introduce no aliasing (alias—free) is given in [6]; an example is the stationary Poisson point process. The asymptotic statistics of the corresponding covariance and spectral estimates, based on the data $\{X(t_k)\}_{k=1}^n$, are derived in [7][8][9]; these estimates are quadratic-mean consistent, as $n \to \infty$, for *all* values of the average sampling rate ρ. Thus, there is no minimum sampling rate (Nyquist rate) required to ensure consistency—a fact of particular importance when the process X is broadband.

In this paper we concentrate on the optimality and performance of deterministic and random sampling schemes in the context of detecting a (known) signal in additive colored noise. Unlike the alias-free sampling schemes $\{t_k\}$ employed in [7][8][9] for covariance and spectral estimation, where $\{t_k\}$ is a point process on the real line with finite mean intensity, the sampling instants in the detection problem lie in a *fixed* finite interval and become dense as the sample size n tends to infinity. Thus, in the context of the detection problem, we are concerned with deterministic and random sampling *designs*. The detection problem with sampled data is formulated in Section II. The results are presented and discussed in Section III, where an optimal detector is used, and in Section IV where a nonoptimal detector is employed.

2 FORMULATION OF THE PROBLEM

Continuous-Time Data. Consider the problem of detecting a sure signal $s(t)$ in additive noise $N(t)$ over a finite interval I which, for simplicity, is taken to be [0,1]:

$$H_1 : X(t) = s(t) + N(t),$$
$$t \in [0,1] = I,$$
$$H_0 : X(t) = N(t),$$

where the noise $N(t)$ is stationary Gaussian with mean zero and a known strictly positive definite covariance function $R(t)$. It is assumed that the detection problem is nonsingular and in fact that the solution f of the detection integral equation

$$s(t) = \int_0^1 R(t-\tau)f(\tau)d\tau = (Rf)(t), \quad 0 \leq t \leq 1,$$

is square integrable, i.e., the detection problem is strongly regular (see Root [10], Kailath [4]). Then the test statistic T is the linear operation on the continuous time observation $\{X(t), 0 \leq t \leq 1\}$:

$$T = \int_0^1 f(t)X(t)dt, \tag{2.1}$$

where the integral is defined in the Lebesgue sense for a measurable version of X.

Sampled Data. In practice the evaluation of the test statistic T is frequently carried out in a digital fashion where the observation process X is sampled at the points
$$I_n = \{t_{n,1}, \ldots, t_{n,n}\}$$
and based on the sample $\{X(t_{n,i})\}_{i=1}^n$ of size n, an appropriately defined test statistic T_n of the form

$$T_n = \sum_{i=1}^n c_{n,i} X(t_{n,i}) \qquad (2.2)$$

is used to test H_1 vs. H_0. It is a common practice to use equally spaced sampling points, $t_{n,i} = (i-1)/(n-1)$, $i = 1, \ldots, n$, which, as will be seen in the following, may not represent the best possible choice. Using the detector T_n based on a sample of n observations, instead of the continuous time detector T, naturally results in a degradation of performance. Our objective is, for each sample size n, to choose the sampling scheme I_n and the coefficients $\{c_{n,i}\}_i$ so as to minimize the resulting degradation in the performance of the detector.

The natural choice for the coefficients $\{c_{n,i}\}_i$, for each given sampling design I_n, is that for which T_n is the (optimal) test statistic for the finite sample detection problem,

$$H_1(I_n): X_i = s(t_{n,i}) + N(t_{n,i}),$$
$$i = 1, \ldots, n.$$
$$H_0(I_n): X_i = N(t_{n,i}),$$

i.e.,
$$c_{n,i} = \sum_{k=1}^n s(t_{n,k}) R_{I_n}^{-1}(k,i), \qquad (2.3)$$

where $[R_{I_n}^{-1}(i,j)]_{i,j=1}^n$ is the inverse matrix of $[R(t_{n,i} - t_{n,j})]_{i,j=1}^n$. This optimal choice of coefficients requires inverting an $n \times n$ matrix, which may become ill conditioned for large sample sizes (which may be required for good performance). Thus, while the choice of the optimal detector corresponding to each sampling design is the natural one, we also consider simpler nonoptimal detectors whose performance is comparable to that of the optimal one, but whose form (coefficients) is easier to compute. Such nonoptimal detectors have coefficients of the form

$$c_{n,i} = \frac{1}{n} c_n(t_{n,i}) \text{ or } c_{n,i} = \frac{1}{n} c(t_{n,i}) \qquad (2.4)$$

for some function $c_n(t)$ or $c(t)$, so that the ith coefficient depends only on the ith sampling point, unlike the ith coefficient (2.3) of the optimal detector which depends on *all* sampling points.

We consider two kinds of sampling schemes, deterministic and random. In deterministic sampling, the sampling points are determined according to a deterministic rule, e.g., periodic sampling. In random sampling, the sampling points are chosen according to a randomized rule: $(t_{n,1}, \ldots, t_{n,n})$ is a random vector, independent of the noise N, whose joint distribution determines the scheme; random sampling schemes can be easier to analyze and they provide a simple way of choosing (asymptotically appropriate or sometimes optimal) coefficients of the simple form (2.4).

Performance Measures. As a measure of performance of a test it is natural to use the probability of detection at a fixed false alarm rate α. If $P_d(T_n)$ and $P_d(T)$ are the probabilities of detection of the test based on T_n and T, then our goal is to choose the sampling design I_n and the coefficients $\{c_{n,i}\}_i$ so that $P_d(T_n)$ will approximate $P_d(T)$ as closely as possible. Moreover, the sequence of designs and coefficients should be chosen so that $P_d(T_n) \to P_d(T)$, as the sample size n tends to infinity. A related quantity of interest for such designs is the rate at which $P_d(T) - P_d(T_n) \to 0$, and this rate can be used to determine the required number of samples n for an acceptable probability of detection, as well as to compare different sampling designs. Within a specified class D of sampling designs, we are interested in finding the best design of size n, in the sense of maximizing $P_d(T_n)$ among all sampling designs in the class D, a goal rarely attained for deterministic designs. Alternatively, we are interested in finding an asymptotically optimal sequence of sampling designs, i.e., a sequence of designs for which the corresponding detectors T_n^* are such that

$$\frac{P_d(T) - P_d(T_n^*)}{P_d(T) - \sup P_d(T_n)} \to 1 \qquad (2.5)$$

where the supremum is taken over all sampling designs of size n in D. This notion of asymptotically optimal sequences of designs was introduced by Sacks and Ylvisaker [11]-[13] in connection with a regression problem closely related to the problem at hand.

When a random sampling scheme is used, the test statistic T_n is no longer normally distributed and it is generally difficult to find the distribution of T_n under H_0 and H_1, and thus evaluate the probability of detection. For this reason, as an alternative measure of performance of the test based on T_n we use the generalized signal-to-noise ratio (see Helstrom [3])

$$S^2(T_n) = \frac{\{E(T_n|H_1) - E(T_n|H_0)\}^2}{\text{var }(T_n|H_0)}, \qquad (2.6)$$

and attempt to find optimal designs of fixed sample size n, which would maximize $S^2(T_n)$, or asymptotically optimal designs with corresponding detectors T_n^* such that

$$\frac{S^2(T) - S^2(T_n^*)}{S^2(T) - \sup S^2(T_n)} \to 1. \qquad (2.7)$$

The generalized signal-to-noise ratio $S^2(T)$ of the continuous time optimal detector T is equal to

$$S^2(T) = \iint_0^1 R(t-\tau)f(t)f(\tau)\,dt d\tau = \int_0^1 s(t)f(t)\,dt \qquad (2.8)$$

and, as T is Gaussian, the probability of detection $P_d(T)$ at false alarm rate α is

$$P_d(T) = \Phi[S(T) - \Phi^{-1}(1-\alpha)], \qquad (2.9)$$

where Φ is the standard normal distribution function, i.e., $P_d(T)$ is a strictly increasing function of $S(T)$.

In the next section we consider sampling designs when an optimal detector, with coefficients (2.3), is used.

3 SAMPLING DESIGNS—OPTIMAL DETECTORS

Here T_n is the optimal detector, with coefficients (2.3), and the discussion is concerned exclusively with the sampling design I_n. For fairly general deterministic and random sampling designs we have $S^2(T_n) \to S^2(T)$ as $n \to \infty$ [1, Propositions 1 and 2]. For deterministic designs more precise results are available under certain smoothness conditions on the noise covariance R and we confine ourselves to this case in this section.

We first note that for deterministic designs, both criteria of performance defined in terms of probability of detection or generalized signal-to-noise ratio are *equivalent* for a fixed sample size as well as asymptotically; this is seen from the fact that the relationship between $P_d(T_n)$ and $S(T_n)$ is the same as between $P_d(T)$ and $S(T)$, i.e., $P_d(T_n) = \Phi[S(T_n) - \Phi^{-1}(1-\alpha)]$. It suffices therefore to concentrate on generalized signal-to-noise ratios.

The question of finding the best design of size n, or an asymptotically optimal sequence of designs, when an optimal detector is employed, turns out [1] to coincide with that considered in a series of papers by Sacks and Ylvisaker [11]-[13] in the context of a regression or an integral approximation problem. We summarize here the results relevant to us cast in the context of the detection problem (see [1] for details).

Optimal designs for a fixed sample size do not generally exist when the noise N is quadratic-mean differentiable and, when they do, they are usually difficult to obtain analytically. However, asymptotically optimal designs and precise rates of convergence are known at the present time for regular sequences of designs. This important class of designs was introduced by Sacks and Ylvisaker in [11]-[13] and is defined as follows: For each sample size n, the sampling points $\{t_{n,1}, \ldots, t_{n,n}\}$ are the quantiles of a probability density function h on [0,1] which is strictly positive (almost everywhere); namely, $t_{n,i}$ is the $(i-1)/(n-1)$ quantile of h,

$$\int_0^{t_{n,i}} h(t)\,dt = \frac{i-1}{n-1}, \quad i = 1, \ldots, n.$$

The resulting sequence is called a regular sequence of designs and is denoted by $RD(h)$. The corresponding partition of [0,1] into n intervals: $0 = t_{n+1,1} < \ldots < t_{n+1,n+1} = 1$ generates a regular sequence of partitions denoted by $RP(h)$. When h is the uniform density on [0,1] the design points are equally spaced, i.e., we have periodic sampling, and the partition is an interval equipartition.

The full technical conditions needed for the validity of the following results on asymptotically optimal designs can be found in [1].

(i) If the noise N has no quadratic-mean derivatives, f is continuous, and a regular sequence of designs $RD(h)$ is used with h continuous on $(0,1)$, then as $n \to \infty$,

$$n^2[S^2(T) - S^2(T_n)] \to \frac{\beta}{12} \int_0^1 \frac{f^2(t)}{h^2(t)}\,dt, \qquad (3.1)$$

where $\beta = R'(0-) - R'(0+)$. The sequence of sampling designs $\{I_n^*\}$

corresponding to the regular sequence $RD(h^*)$ with $h^*(t) = |f(t)|^{2/3}/\int_0^1 |f(\tau)|^{2/3}d\tau$ is asymptotically optimal, and the corresponding detector T_n^* satisfies

$$n^2[S^2(T) - S^2(T_n^*)] \to \frac{\beta}{12}\left\{\int_0^1 |f(t)|^{2/3}dt\right\}^3. \quad (3.2)$$

Examples of noise processes for which the above result holds include the Gauss-Markov process, the stationary process with triangular covariance $R(t) = (1-|t|)$ on $[-1,1]$ and $= 0$ elsewhere, and convex combination thereof such as $R(t) = \sum_{i=1}^{m} a_i \exp(-\mu_i|t|)$ with μ_i, $a_i > 0$, $\sum_{i=1}^{m} a_i = 1$.

(ii) If the noise N has precisely one quadratic-mean derivative, f is continuous and has at most finitely many zeros, and a regular sequence $RD(h)$ is used with h continuous on $(0,1)$, then as $n \to \infty$

$$n^4[S^2(T) - S^2(T_n)] \to \frac{\gamma}{720} \int_0^1 [f(t)/h^2(t)]^2 dt, \quad (3.3)$$

where $\gamma = r'(0-) - r'(0+)$, $r(t) = -R''(t)$. The sequence of sampling designs $\{I_n^*\}$ corresponding to a regular sequence $RD(h^*)$ with $h^*(t)$ proportional to $|f(t)|^{2/5}$ is asymptotically optimal, and the corresponding detector T_n^* satisfies

$$n^4[S^2(T) - S^2(T_n^*)] \to \frac{\gamma}{720} \{\int_0^1 |f(t)|^{2/5}dt\}^5. \quad (3.4)$$

(iii) If the noise N has exactly $k \geq 2$ quadratic-mean derivatives, then recent results of Eubank et al. [2] for very special noise processes suggest a rate of convergence of n^{-2k-2} for $S^2(T) - S^2(T_n)$ but a complete derivation of this is not available yet for general noise processes.

The results in (i) and (ii) can also be expressed in terms of probability of detection, at false alarm rate α, using the relationship

$$\frac{P_d(T) - P_d(T_n)}{S(T) - S(T_n)} \to \phi[S(T) - \Phi^{-1}(1-\alpha)]$$

where $\phi(u)$ is the standard normal density. For example, under (i) we have for a regular sequence $RD(h)$ of designs

$$n^2[P_d(T) - P_d(T_n)] \to \frac{\beta}{24S(T)}\phi[S(T) - \Phi^{-1}(1-\alpha)] \int_0^1 \frac{f^2(t)}{h^2(t)} dt \quad (3.1\text{'})$$

and the sequence of sampling designs $\{I_n^*\}$, corresponding to $RD(h^*)$ with $h^*(t)$ proportional to $[f^2(t)]^{1/3}$, is asymptotically optimal, in the sense of the probability of detection criterion (2.5), and the detector T_n^* satisfies

$$n^2[P_d(T) - P_d(T_n^*)] \to \frac{\beta}{24S(T)} \phi[S(T) - \Phi^{-1}(1-\alpha)]\left\{\int_0^1 [f^2(t)]^{1/3}dt\right\}^3. \quad (3.2\text{'})$$

Similarly under (ii).

Periodic sampling, $t_{n,i} = (i-1)/(n-1)$, $i = 1, \ldots, n$, is covered under (i) and (ii) by taking $h(t)$ to be the uniform density on $[0,1]$. It is clear then that periodic sampling is not asymptotically optimal in general (unless f^2 is constant in case (i), or $|f|$ is constant in case (ii)). The improvement in performance of the asymptotically optimal design $\{I_n^*\}$ over periodic sampling is determined, for large sample size n, by the ratio of the integrals $\left\{\int (f^2)^{1/3}\right\}^3 / \int (f^2)$ in case (i) and $\left\{\int |f|^{2/5}\right\}^5 / \int f^2$ in case (ii). By Holder's inequality, these ratios take values in $(0,1]$ and, depending on f, can in fact take any value in $(0,1]$ and thus they can be quite small, as seen in case (i) from the example $f^2(t) = (\nu + 1) t^\nu$, $t \in [0,1]$ for large ν. This improvement in performance was also demonstrated in [1], via examples, for small sample size. Unlike periodic sampling, the asymptotically optimal sequence of sampling designs depends on f and, through the integral equation of detection, is therefore dependent on both the signal s and the noise covariance R.

4 SAMPLING DESIGNS—NONOPTIMAL DETECTORS

Here the test statistic T_n based on the observation $\{X(t_{n,i})\}_{i=1}^n$ of sample size n, has the simple form

$$T_n = \frac{1}{n} \sum_{i=1}^n c_n(t_{n,i}) X(t_{n,i}) \qquad (4.1)$$

for some appropriately chosen functions $c_n(t)$. The form of these functions c_n is determined by the requirement that as the sample size n tends to infinity the performance of the detector based on T_n should converge to the performance of the continuous time optimal detector based on T.

We consider both deterministic (Subsection B) and random designs (Subsection A). When random sampling designs are used, the statistic T_n is no longer normally distributed and its distributions under H_0 and H_1 are in general difficult to evaluate explicitly. However, it is comforting to note that, for the appropriately chosen functions c_n, the statistic T_n is asymptotically normal under H_0 and H_1 and the probability of detection $P_d(T_n)$ and generalized signal-to-noise ratio $S^2(T_n)$ converge to $P_d(T)$ and $S^2(T)$, respectively, as $n \to \infty$.

A. *Random Design.* We consider two specific random designs: simple random sampling, presented in detail, and stratified sampling for which the results are discussed briefly. We adopt the generalized signal-to-noise ratio $S^2(T_n)$ as the measure of performance of the detector based on T_n.

In simple random sampling, for each sample size n, the sampling points $t_{n,1}, \ldots, t_{n,n}$ are independent and identically distributed with common probability density function $g(t)$ on $[0,1]$. The requirement that $S^2(T_n) \to S^2(T)$ as $n \to \infty$ leads to a natural choice for $c_n(t)$ [1, Proposition 3] of the form $c_n(t) = c(t)$ for all n where $c(t)$ satisfies $c(t)g(t) = f(t)$. Then the test statistic T_n becomes

$$T_n = \frac{1}{n} \sum_{i=1}^n \frac{f(t_{n,i})}{g(t_{n,i})} X(t_{n,i}), \qquad (4.2)$$

and its generalized signal-to-noise ratio can be shown to be given by

$$S^2(T_n) = \frac{S^4(T)}{S^2(T) + \frac{1}{n}[B^2 - S^2(T)]},$$

where

$$B^2 = \sigma^2 \int_0^1 c(t) f(t) \, dt, \quad \sigma^2 = R(0). \tag{4.3}$$

Thus the exact rate of convergence of $S^2(T) - S^2(T_n)$ to zero is $1/n$ and this result is obtained with no assumptions whatsoever on the noise covariance. No analogous result exists for deterministic designs. Also this result is free of dimensionality and is thus true for multiparameter detection problems as well. When the function f/g is bounded on $[0,1]$, the distributions of T_n under H_0 and H_1 are asymptotically normal and the probability of detection $P_d(T_n)$ converges to $P_d(T)$ as $n \to \infty$ [1, Proposition 3].

Furthermore, the optimal design (density g) of fixed sample size is easily obtained for *all* noise processes—in sharp contrast to deterministic designs: A variational argument shows that the optimal sampling density $\bar{g}(t) = |f(t)|/\int_0^1 |f(\tau)| d\tau$ on $[0,1]$ and the corresponding detector $\bar{T}_n$ and $S^2(\bar{T}_n)$ are given by

$$\bar{T}_n = \frac{D/\sigma}{n} \sum_{i=1}^n \text{sgn}[f(t_{n,i})] X(t_{n,i}), \quad S^2(\bar{T}_n) = \frac{S^4(T)}{S^2(T) + \frac{1}{n}[D^2 - S^2(T)]}, \tag{4.4}$$

where $D = \sigma \int_0^1 |f(t)| dt$. Finally, $\bar{T}_n$ is asymptotically normal under H_0 and H_1 and $P_d(\bar{T}_n) \to P_d(T)$ as $n \to \infty$.

Next we briefly discuss the results for stratified sampling. In stratified sampling the interval $[0,1]$ is partitioned into strata $\{A_{n,i}\}_{i=1}^n$ and the sampling points are chosen independently one from each stratum, $t_{n,i} \in A_{n,i}$. If $g_{n,i}$ is the sampling probability density within stratum $A_{n,i}$, i.e., the density of $t_{n,i}$, the natural choice (see [1]) for $c_n(t)$ satisfies $c_n(t) g_{n,i}(t) = nf(t)$ on each $A_{n,i}$. Then

$$T_n = \sum_{i=1}^n \frac{f(t_{n,i})}{g_{n,i}(t_{n,i})} X(t_{n,i}) \tag{4.5}$$

and

$$S^2(T_n) = \frac{S^4(T)}{S^2(T) + \sum_{i=1}^n \left\{ \sigma^2 \int_{A_{n,i}} f^2/g_{n,i} - V_{n,i}^2 \right\}} \tag{4.6}$$

where $\sigma^2 = R(0)$, and

$$V_{n,i}^2 = \int_{A_{n,i}} \int_{A_{n,i}} R(t - \tau) f(t) f(\tau) \, dt \, d\tau, \tag{4.7}$$

and of course we always have $S^2(T_n) \leq S^2(T)$. As for simple random sampling, the sampling densities $\bar{g}_{n,i}(t), i = 1, \ldots, n$, which maximize $S^2(T_n)$, for each fixed sample size n, are proportional to $|f(t)|$ on each $A_{n,i}$, and the resulting (optimal) detector $\bar{T}_n$ and $S^2(\bar{T}_n)$ are as follows:

Estimation/Detection Problems

$$\bar{T}_n = \frac{1}{\sigma} \sum_{i=1}^{n} U_{n,i} \, \text{sgn} \, [f(t_{n,i})] X(t_{n,i}), \quad S^2(\bar{T}_n) = \frac{S^4(T)}{S^2(T) + \sum_{i=1}^{n} (U_{n,i}^2 - V_{n,i}^2)} \quad (4.8)$$

where $U_{n,i} = \sigma \int_{A_{n,i}} |f(t)| \, dt$, and $V_{n,i}^2$ is given in (4.7). If a regular sequence of partitions $RP(h)$ is used with f/h bounded, the statistic $\bar{T}_n$ is asymptotically normal under H_0 and H_1 and $P_d(\bar{T}_n) \to P_d(T)$ and $S^2(\bar{T}_n) \to S^2(T)$ as $n \to \infty$ [1, Proposition 4]. In particular, for equispaced interval partitions we always have

$$S^2(T) - S^2(\bar{T}_n) = O\left(\frac{1}{n}\right),$$

and when f has a constant sign and the noise N is quadratic-mean differentiable we have

$$S^2(T) - S^2(\bar{T}_n) = O\left(\frac{1}{n^3}\right).$$

It turns out [14] that the rate $1/n^3$ is the best possible for stratified sampling, and asymptotically optimal sequences of stratified designs can be obtained under considerably weaker conditions on continuous f than constant sign. (It is assumed that the statistic $\bar{T}_n$ is used and one is concerned with the *choice* of the regular partition $RP(h)$ which leads to asymptotically optimal stratified design. As in the case of optimal detectors with deterministic designs, it turns out that the optimal density $h(t)$ is proportional to $|f(t)|^\nu$ where ν depends on the number of quadratic-mean derivatives of the noise N. See [1] and [14] for details. Thus, unless $|f(t)|$ is constant on $[0,1]$, equispaced interval partition does not provide an asymptotically optimal stratified design.)

B. *Deterministic Design-Midpoint Sampling.* Mid-point sampling is a deterministic sampling scheme introduced, in a somewhat more general form, by Schoenfelder [14]. The sampling points $\{t_{n,1}, \ldots, t_{n,n}\}_n$ are chosen as the mid-points of a regular sequence of partitions $RP(h)$ of $[0,1]$,

$$\{0 = x_{n+1,1} < x_{n+1,2} < \ldots < x_{n+1,n} < x_{n+1,n+1} = 1\}_n,$$

i.e., $t_{n,i} = (x_{n+1,\,i+1} + x_{n+1,i})/2$.

When h is the uniform density, mid-point sampling becomes periodic sampling with $t_{n,i} = (2i-1)/(2n)$ (the endpoints are not included).

In mid-point sampling the test statistic $T_n = (1/n) \sum_{i=1}^{n} c_n(t_{n,i}) X(t_{n,i})$ is normally distributed under H_0 and H_1 and, the two criteria of performance defined in terms of probability of detection and of generalized signal-to-noise ratio are equivalent for finite sample size as well as asymptotically and we will therefore focus on the signal-to-noise ratio $S^2(T_n)$.

The requirement that $S^2(T_n) \to S^2(T)$ as $n \to \infty$ leads [1, Proposition 5] to the choice for $c_n(t)$ of the form $c_n(t) = c(t)$ for all n where $c(t)$ satisfies $c(t)h(t) = f(t)$. The test statistic T_n and $S^2(T_n)$ are then given by

$$T_n = \frac{1}{n} \sum_{i=1}^{n} \frac{f(t_{n,i})}{h(t_{n,i})} X(t_{n,i}), \tag{4.9}$$

$$S^2(T_n) = \frac{\{\sum_{i=1}^{n} \frac{f(t_{n,i})}{h(t_{n,i})} s(t_{n,i})\}^2}{\sum_{i,j=1}^{n} \frac{f(t_{n,i})f(t_{n,j})}{h(t_{n,i})h(t_{n,j})} R(t_{n,i} - t_{n,j})}. \tag{4.10}$$

The optimal design (i.e., the optimal density h generating the regular partition $RP(h)$) which maximizes $S^2(T_n)$ is not known at the present. However, asymptotically optimal designs and the precise rate of convergence of $S^2(T_n)$ to $S^2(T)$ and $P_d(T_n)$ to $P_d(T)$ can be derived under conditions analogous to those in Section III. Using the results of Schoenfelder [14], who studied mid-point sampling in the context of estimating integrals of stochastic processes, it can be shown that the following result holds under certain technical conditions stated fully in [1].

(i) If the noise N has no quadratic-mean derivatives, f and h are continuous on $(0,1)$, then, as $n \to \infty$,

$$n^2[S^2(T) - S^2(T_n)] \to \frac{\beta}{12} \int_0^1 \frac{f^2(t)}{h^2(t)} dt, \tag{4.11}$$

$\beta = R'(0-) - R'(0+)$. The mid-point sampling design with $h^*(t)$ proportional to $|f(t)|^{2/3}$ is asymptotically optimal and the corresponding detector T_n^* satisfies

$$n^2[S^2(T) - S^2(T_n^*)] \to \frac{\beta}{12} \{\int_0^1 |f(t)|^{2/3} dt\}^3. \tag{4.12}$$

(ii) If the noise N has k quadratic-mean derivatives it is expected that the analog of the results (ii) - (iii) of Section III should hold with identical rates of convergence; the derivations are not available yet. The constant and rate of convergence in (4.11), (4.12) are identical to the corresponding quantities in (3.1) and (3.2) where an *optimal* detector is used. Thus under the present conditions, the mid-point sampling design is both the simplest (in view of the form of the test statistic T_n of (4.9) for each sample size) and asymptotically best. The computational results obtained in [1] show that, even for small sample size, the performance of the nonoptimal detector (4.9) compares favorably with that of the optimal detector with coefficients (2.3). We finally remark that mid-point design with equispaced partition is included in (4.11), with $h(t) = 1$ on $(0,1)$, and is not asymptotically optimal in general. Mid-point design using the partition $RP(h^*)$ can perform, for large n, appreciably better than that using equispaced partition.

5 COMMENTS

It is clear from the above results that equally-spaced designs and partitions are not asymptotically optimal in general regardless of whether an optimal or nonoptimal detectors are used. Moreover, certain nonoptimal detectors (such as those associated with mid-point design) can perform as well as the optimal detectors, at least for large sample size n. Finally, the merit of the simple random sampling, which has the slowest rate (n^{-1}) of convergence, is the simplicity of the corresponding detector and the fact that this rate holds with no restrictive assumptions on the covariance R of the noise.

REFERENCES

[1] S. Cambanis and E. Masry, "Sampling designs for the detection of signals in noise," *IEEE Trans. Information Theory*, Vol. IT-29, pp. 83-104, 1983.

[2] R.L. Eubank, P.L. Smith and P.W. Smith, "Uniqueness and eventual uniqueness of optimal designs in some time series models," *Ann. Statist.*, Vol. 9, pp. 486-493, 1981.

[3] C.W. Helstrom, "Detection theory and quantum mechanics," *Information and Control*, Vol. 10, pp. 254-291, 1967.

[4] T. Kailath, "RKHS approach to detection and estimation problems—Part I: Deterministic signals in Gaussian noise," *IEEE Trans. Information Theory*, Vol. IT-17, pp. 530-549, 1971.

[5] L.H. Koopmans, *The Spectral Analysis of Time Series*. New York: Academic Press, 1974.

[6] E. Masry, "Alias-free sampling: An alternative conceptualization and its applications," *IEEE Trans. Information Theory*, Vol. IT-24, pp. 317-324, 1978.

[7] E. Masry, "Poisson sampling and spectral estimation of continuous-time processes," *IEEE Trans. Information Theory*, Vol. IT-24, pp. 173-183, 1978.

[8] E. Masry, "Discrete-time spectral estimation of continuous-time processes—The orthogonal series method," *Ann. Statist.*, Vol. 8, pp. 1100-1109, 1980.

[9] E. Masry, "Nonparametric covariance estimation from irregularly-spaced data," *Advances in Appl. Prob.*, Vol. 15, pp. 113-132, 1983.

[10] W.L. Root, "The detection of signals in Gaussian noise," in A.V. Balakrishnan, Ed., *Communication Theory*, pp. 160-191, New York: McGraw-Hill, 1968.

[11] J. Sacks and D. Ylvisaker, "Designs for regression problems with correlated errors," *Ann. Math Statist.*, Vol. 37, pp. 66-89, 1966.

[12] J. Sacks and D. Ylvisaker, "Designs for regression problems with correlated errors III," *Ann. Math. Statist.*, Vol. 41, pp. 2057-2074, 1970.

[13] J. Sacks and D. Ylvisaker, "Statistical designs and integral approximation," *Proc. Twelfth Biennial Seminar of the Canadian Math. Congress*, pp. 115-136, Canadian Math. Congress, Montreal, 1970.

[14] C. Schoenfelder, "Random designs for estimating integrals of stochastic processes," *Institute of Statistics Mimeo Series* No. 1201, University of North Carolina at Chapel Hill, November 1978.

[15] C. Schoenfelder and S. Cambanis, "Random designs for estimating integrals of stochastic processes," *Ann. Statist.*, Vol. 10, pp. 526-538, 1982.

A Technique For Improving Detection and Estimation of Signals Contaminated by Under Ice Noise*

Roger F. Dwyer

Surface Ship Sonar Department
Naval Underwater Systems Center
New London, CT

1 INTRODUCTION

The recent analyses of FRAM II data have shown that strong non-Gaussian noise exists at times in the frequency domain. This paper presents the results of processing FRAM II data in the frequency domain using data partitioning. We also present the frequency domain envelope distribution at three frequencies. These results suggest that data modeling should also be considered as a signal processing technique. However, here we shall concentrate on the data partitioning results and leave to a later time the discussion of the results of data modeling.

2 MATHEMATICAL PRELIMINARIES

Let the received data $x(i, q) = x[(i + (q - 1)M)h]$, $i = 0, 1, 2, \ldots, M - 1$, $q = 1, 2, \ldots, n$ where h is the interval between successive samples, be composed of an additive mixture of signal, interference, and Gaussian noise of the form

$$x(i, q) = m(i, q) s(i, q, a_1) + c(i, q) I(i, q, a_2) + N(i, q), \quad (1)$$

where

$s(i, q, a_1)$ is the transmitted or radiated signal,
$I(i, q, a_2)$ is a narrowband interference (ice noise),

---
*Work supported under Work Request N0001483WR30054, Task NR 042-499 with the Office of Naval Research.

$N(i, q)$ is independent Gaussian noise,
$m(i, q)$ represents fading or multipath effects of the signal,
$c(i, q)$ represents fading or multipath effects of the interference, and
a_t is a random variable taking on two values, $a_t = 0$, $a_t = 1$ with probabilities $P(a_t = 0) = 1 - L_t$, $P(a_t = 1) = L_t$ respectively, for $t = 1, 2$.

The random variables a_1, a_2 model transient or frequency modulation components of the signal and interference, respectively, in the sense that L_1 and L_2 represent the probability of the signal or interference being in a particular frequency location measured over the observation interval. The parameters a_1, a_2, m, and c render the data non-Gaussian in the frequency domain. In general these parameters are also functions of frequency; however, for the sake of notational simplicity, we shall not explicitly show this dependence. The medium can also cause the received signal to be frequency modulated; we will assume that the frequency modulation caused by the medium is incorporated in the parameters a_1 and a_2.

The discrete fourier transform (DFT) is defined as

$$X(q, F_p) = \sqrt{h/M} \sum_{i=0}^{M-1} x(i, q) \exp(-jF_p i)$$
$$= X^r(q, F_p) - j X^I(q, F_p), \qquad (2)$$

where $j = \sqrt{-1}$, $F_p = 2\pi f_p h$, and $f_p = p/Mh$ Hz. The components $X^r(q, F_p)$ and $X^I(q, F_p)$ signify real and imaginary parts, respectively. Temporal weighting may also be included but is not treated here.

If we assume that the components of Equation (2) are mutually independent, the corresponding spectrum estimate is

$$P(F_p) = L_1 E(m^2(q)) S_s(F_p) + L_2 E(c^2(q)) S_I(F_p) + R_N, \qquad (3)$$

where we have assumed slow fading with respect to the length of the DFT so that $m(i, q) = m(q)$ and $c(i, q) = c(q)$, and $S_s(F_p)$ is the spectrum estimate at the pth frequency of the signal, $S_I(F_p)$ corresponds to the pth frequency spectrum estimate of the interference, and R_N represents the independent Gaussian noise spectrum estimate. The components L_1 and L_2 were assumed to be multiplicative in the frequency domain.

The problem to be addressed here concerns processing the data in the frequency domain (as given by equation (2)) in the presence of strong non-Gaussian interference. This problem arose as a consequence of the FRAM II data analysis study, although it was addressed on a theoretical level in References 1 and 2.

In this paper we will present under ice ambient noise data which are dominated by non-Gaussian interference in both the time and frequency domains. Then we will show that by processing the data using partitioning techniques, significant performance improvements in terms of signal-to-noise ratio (SNR) are possible in the under ice environment, when non-Gaussian noise occurs assuming that the signal level is much smaller than the interference. This assumption is not always met in practice. Reference 3 discusses a frequency domain processing method that treats the interference-free case.

The theory of data partitioning methods was introduced by Ching and Kurz. Here we will present only the FRAM II data partitioning results. Anyone interested in the theoretical aspects of partitioning should consult References 4 and 5.

Another approach to optimum processing is data modeling. The frequency domain envelope distribution of FRAM II data is also presented for three frequencies at three different resolutions.

3 FRAM II DATA RESULTS

Since the FRAM II data results of under ice ambient noise have been reported in detail in Reference 6, we will only summarize the results needed in this paper.

The data analyses are composed of time and frequency domain statistical measurements. The time domain data were filtered, sampled, and grouped into records of 1024 samples each. The mean, variance, skew, and kurtosis were then estimated for each record. Over time intervals consisting of hundreds of records, the cumulative distribution function (CDF) of the energy (square of the data samples) was estimated and was shown, for the most part, to be non-Gaussian but with nonstationary behavior over successive intervals. The time domain data were then transformed into the frequency domain via a 1024 point fast Fourier transform (FFT). Frequency domain statistics were then compiled for each frequency cell for both the real and imaginary parts.

Figure 1 shows the statistical moments for the time domain data, which were filtered through a 2500 Hz lowpass filter and then sampled at a 10,000 Hz rate. Therefore each record represents a time interval of about 0.1 seconds, giving an overall data length of 10 minutes. Some important observations about these data are the variability of the variance over time and the significant deviation from the Gaussian assumption based on the skew and kurtosis estimates. We found, by processing the data in bands, that the variability in the variance was due to higher frequency (greater than 750 Hz) components. The kurtosis is especially important because it indicates deviations from the Gaussian distribution by values greater or less than 3. The values greater than 3 pertain (in many cases) to distributions that are more peaked than the Gaussian distribution whereas values less than 3 correspond (in many cases) to distributions that are less peaked.[†] For example, a purely sinusoidal signal with uniformly distributed phase has a kurtosis of 1.5. None of the records in Figure 1 has a kurtosis value of 1.5 although some are near 2. The additive Gaussian noise is most likely contributing to this result since it is known that kurtosis is SNR dependent (Reference 3).

Figure 2 compares the power spectral density (PSD) (top curve), real skew (middle curve), and the real kurtosis for the under ice data which have been processed with a 10 Hz resolution and averaged over 1000 consecutive

---

[†]There may be pathological distributions where this interpretation is not justified (12). However, deviations from 3 pertain to non-Gaussian distributions.

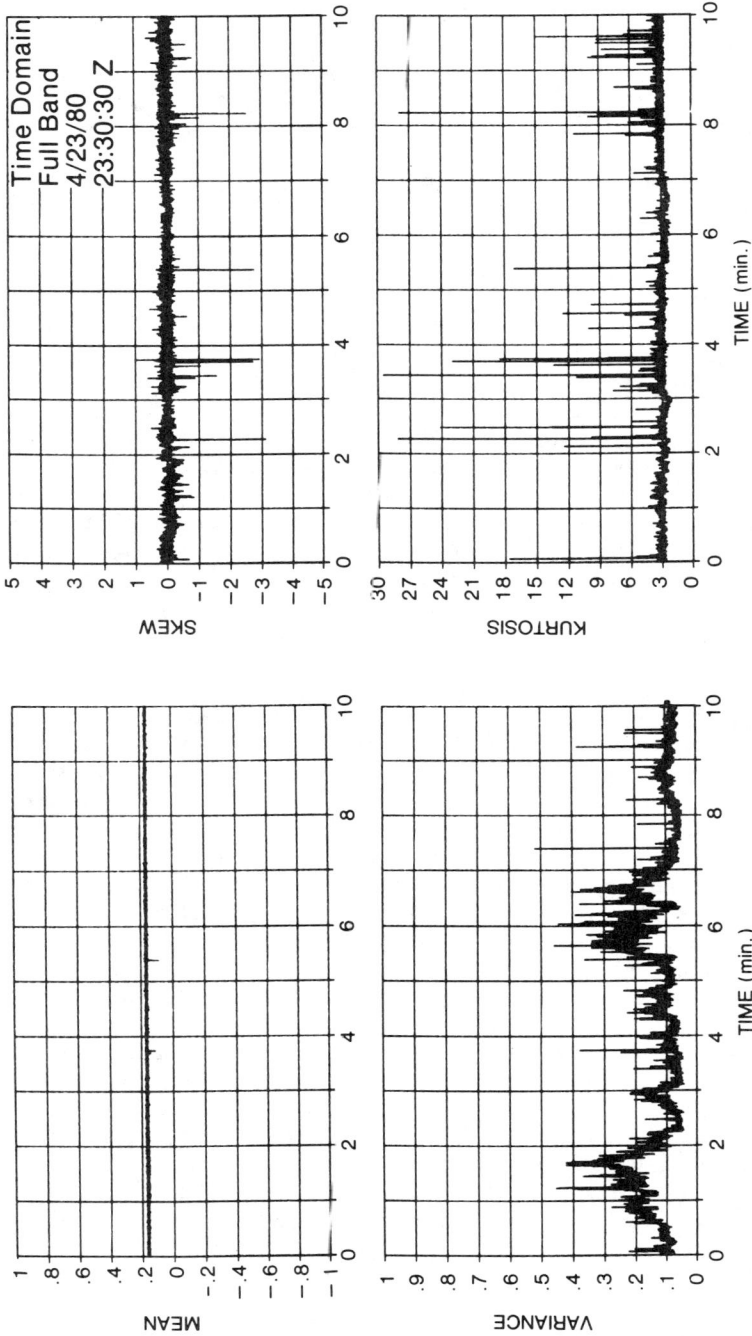

Figure 1 Time domain statistical moments for 2500 Hz band

Improving Detection

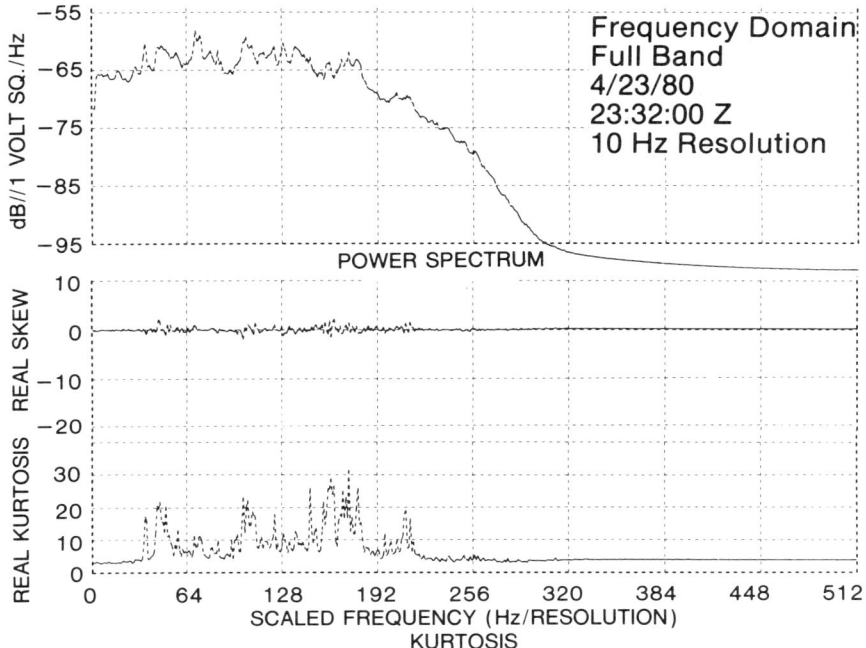

Figure 2 Frequency domain statistical moments

FFT's. This gives a total time interval of 1.7 minutes. The data clearly indicate non-Gaussian noise based on the frequency domain skew and kurtosis. Over a relatively flat portion of the band as seen in the PSD we estimated the amplitude CDF using the 1000 consecutive FFT's for both the real and imaginary parts. The results (Figure 3) show significant deviation from a Gaussian distribution (dashed curve) for both the real and imaginary parts. Later in our discussion we will compare these results with the original data after it has been processed by partitioning.

Another data set of under ice noise is given in Figures 4 and 5. The data were first filtered by a 100 Hz bandpass filter centered at 350 Hz, were sampled at 2000 Hz, and were processed in records of 1024 samples, each giving an interval of approximately 0.5 second. The important observations in Figure 4 are that the variability in the variance is greatly reduced in this band and that many records deviate from a Gaussian distribution based on the kurtosis estimate. The frequency domain results (Figure 5) show that many frequency locations also significantly deviate from a Gaussian distribution based on the frequency domain kurtosis estimate. In addition, a 60 Hz tonal and some of its harmonics are present in the PSD estimate in Figure 5. The corresponding kurtosis estimate shows values significantly less than 3. A theoretical discussion explaining the significance of these results for a signal propagating in a medium with fading or multipath effects is given in Reference 3.

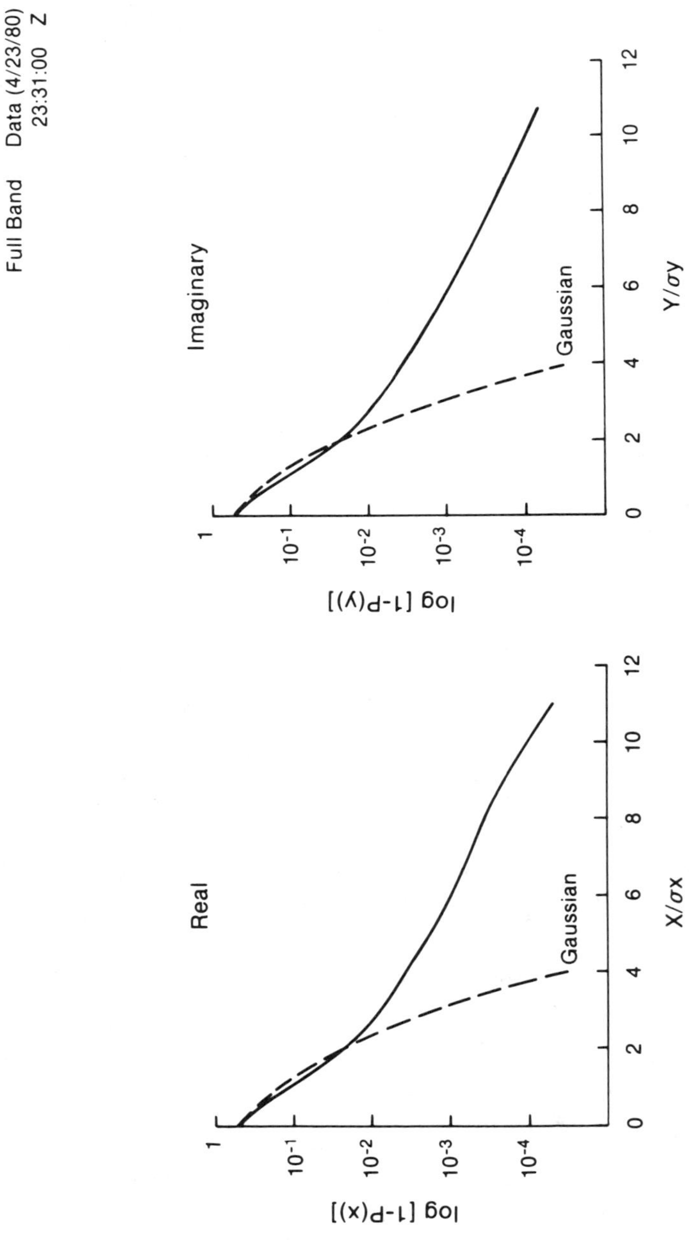

Figure 3 Frequency domain amplitude distribution

Improving Detection

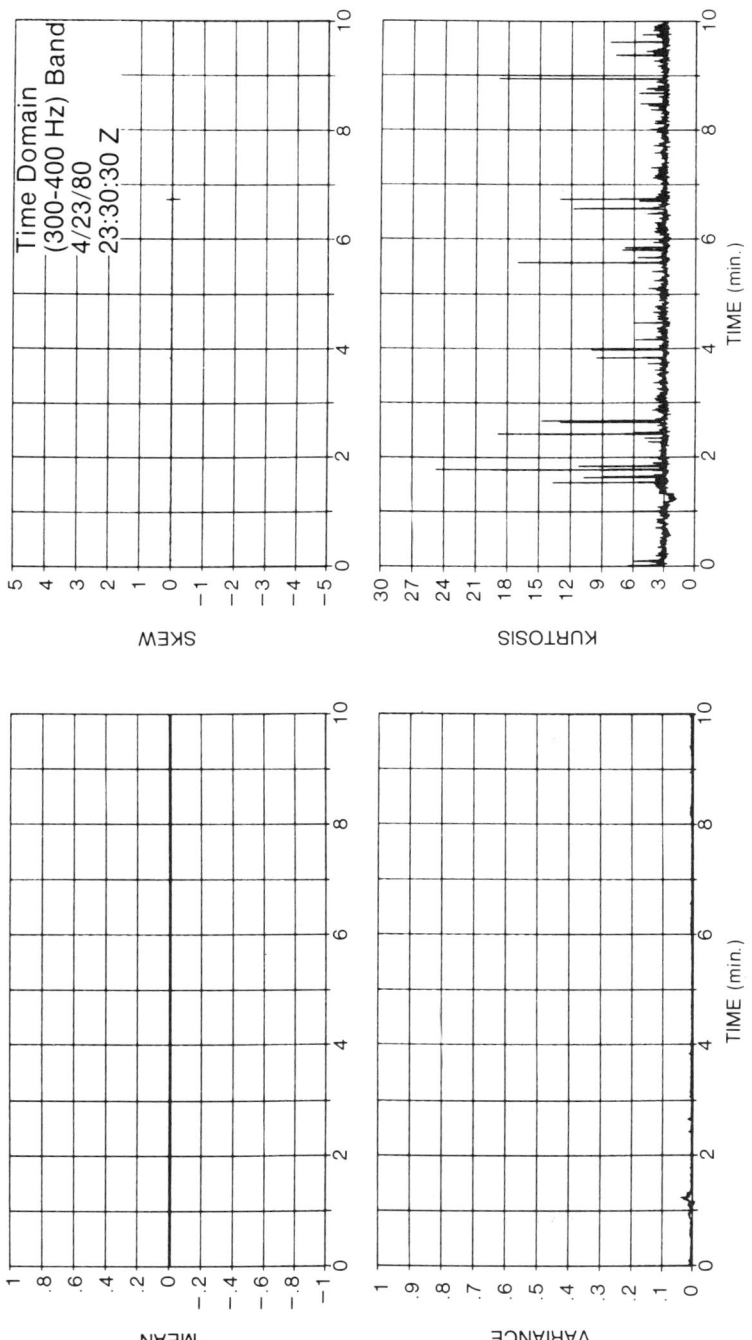

Figure 4 Time domain statistical moments for 300-400 Hz band

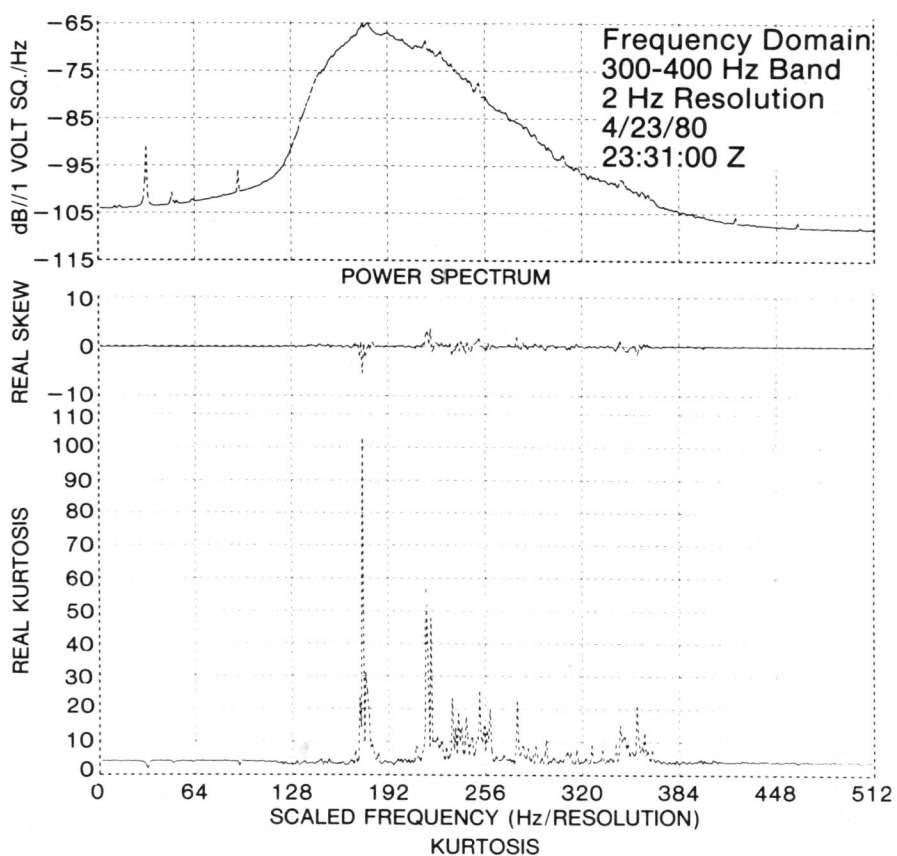

Figure 5 Frequency domain statistical moments for 2 Hz resolution

4 FRAM II PROCESSING RESULTS

We shall discuss the results of processing signals in the frequency domain here. Optimum techniques for processing time domain signals in non-Gaussin noise can be found in References 7 through 10 and 13 and 14. These frequency domain processing techniques, which we shall discuss, represent a new methodology for extracting signals embedded in narrowband non-Gaussian noise. References 1 and 2 develop, on a theoretical level, optimum partitioning techniques in sufficient detail that we can concentrate here on the results of processing FRAM II data. First, the results of processing the 10 Hz resolution frequency domain data by partitioning the real and imaginary parts separately will be given. Then we shall present the frequency domain envelope distribution results in 2, 6, and 10 Hz resolution cells. These results suggest that noise

FREQUENCY DOMAIN: AMPLITUDE DISTRIBUTION

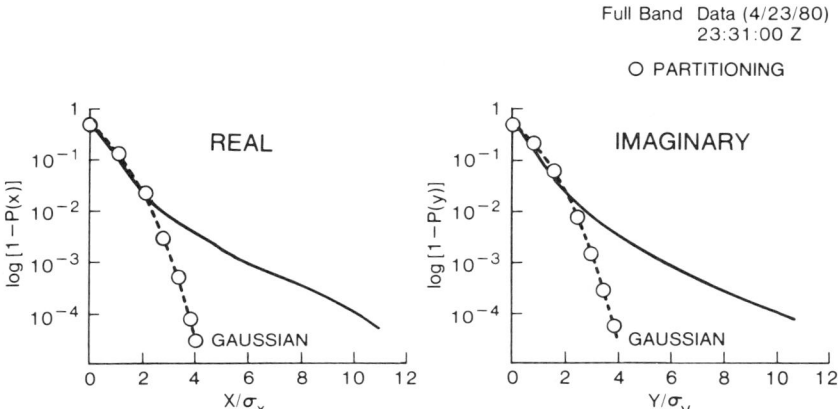

Figure 6 Frequency domain data and partitioned data

modeling should also be considered for optimum signal processing in the frequency domain.

Figure 6 compares the amplitude CDF of the partitioned data with that of the original data for both the real and imaginary parts. The circles correspond to the partitioned results. Here the data is partitioned by recursively estimating two quantities which partition the data into three intervals. Of the three intervals the outer two are assigned zero weight. A linear region is maintained in the center interval. The interesting feature is that partitioning for this data set produces nearly a Gaussian distribution. This result can be explained by noting that partitioning is equivalent to a nonlinear transformation that significantly reduces the high amplitude excursions and leaves the smaller amplitudes unchanged. It should also be pointed out that this result of a Gaussian output can be predicted from Reference 11. The performance improvement can be significant, as shown by the following simple example. Suppose it is desired to set the threshold so that the false alarm rate per FFT is .0001. Then, assuming a small SNR, the partitioned detector would have approximately a 12 dB processing advantage* over a conventional (linear) detector, as shown in Figure 6.

Another way of comparing performance is based on the asymptotic relative efficiency, which is essentially a ratio of the output SNR of two detectors as the signal approaches zero and the integration time approaches infinity. As shown in Reference 9, this turns out to be a ratio of output variances for the two detectors under noise-only conditions. Figure 7 compares, for the 10 Hz resolution case, the CDF of the partitioned and original data for the real part only. The circles represent the partitioned data. Again note the close approximation to a Gaussian distribution after partitioning. We calculated the output variance for both data sets and found that partitioning improves performance in

---

*Includes gain due to reduction of variance.

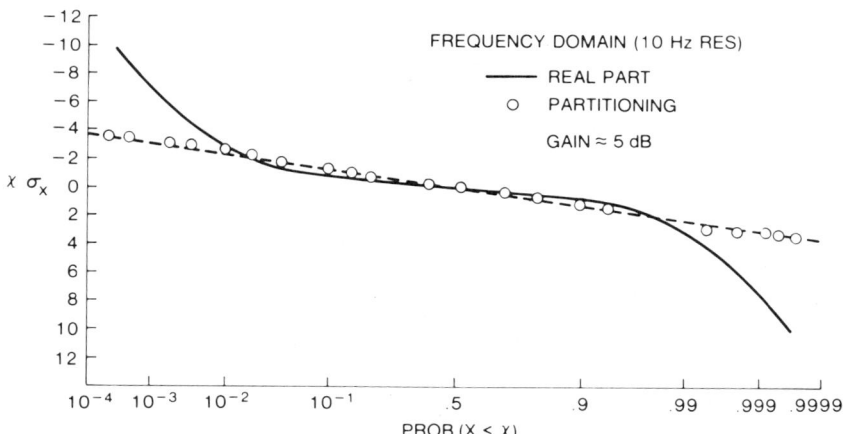

Figure 7 **Frequency domain data and partitioned data with gain**

this case by about 5 dB in the sense that its variance is decreased by 5 dB over the variance of the original data.

Another approach to optimum processing of the frequency domain FRAM II data is to model the data. Figure 8 shows a plot of the envelope distribution at the output of an FFT at three frequencies and for three different resolutions. We considered FFT's with 10, 6, and 2 Hz resolutions with time-resolution products (TRP) of 1000, 1000, and 750, respectively. The vertical scale represents the envelope (normalized by its standard deviation) in dB, and the horizontal axis is the exceedance probability. In order to better visualize the tail behavior of the envelope distribution, we included a small number of adjacent bins in the estimate. These estimates followed approximately the curves shown in Figure 8. In this way we were able to extend the tail region and observe its trend. The solid line in the figure represents a Rayleigh distribution. As can be seen, the data deviate from the Rayleigh distribution for all three cases considered. This suggests the possibility of modeling the envelope distribution of FRAM II data in the frequency domain.

5 SUMMARY

We have presented the FRAM II data analysis results and have shown that the ambient noise was at times highly non-Gaussian in the time and frequency domains. The frequency domain results are new, but we have confirmed them by a theoretical analysis and by comparison with other data. For the small signal case, two techniques were considered for improving detection and estimation of signals contaminated with under ice noise.

In the first technique, we employed an adaptive partitioning method which approximates the optimum nonlinearity, as derived by an expansion of the

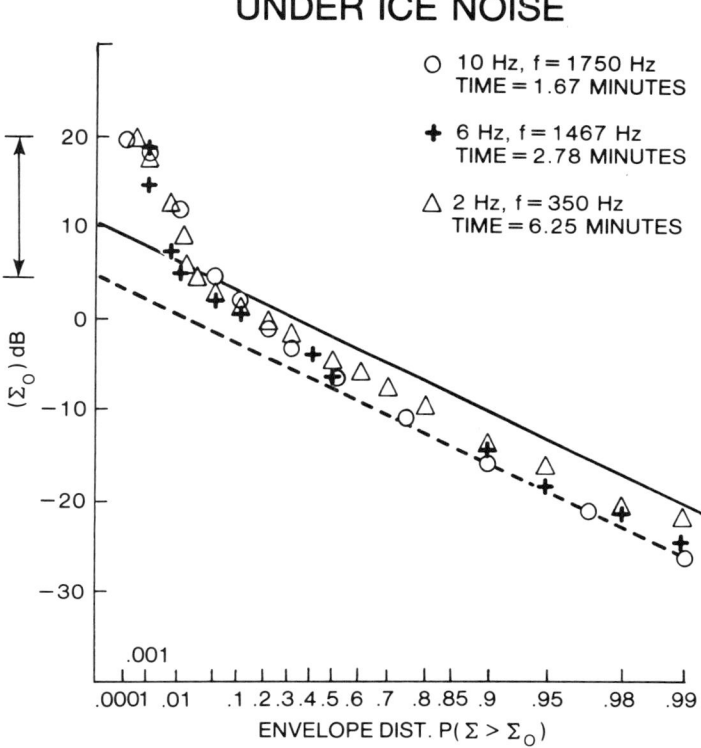

Figure 8 Frequency domain envelope distribution

loglikelihood ratio, for the particular noise distribution prevailing at the time a decision is to be made. Partitioning transformed the CDF of the FRAM II data into a Gaussian distribution and thereby decreased the threshold needed to maintain a prescribed false alarm rate. A particular example showed that, for a false alarm rate of .0001, a 12 dB improvement in performance could be achieved. We also calculated the variance of the original and partitioned data and found that for this example a gain of 5 dB in performance could be achieved based solely on reduction of the variance. This improvement applies to both detection and estimation of signals in the under ice noise environment.

Another approach to optimum processing was considered. We plotted the envelope distribution, at the output of the FFT for three frequencies for three different resolutions. All three cases deviated from the Rayleigh distribution which suggests that data modeling should be considered as a signal processing approach in the frequency domain.

Future studies will concentrate on determining the rate of occurrence of arctic non-Gaussian noise so that the overall significance of these results can be addressed.

ACKNOWLEDGMENT

The author wishes to thank I. Dyer and A. Baggeroer of MIT and F. Dinapoli and E. Hug of NUSC for making the FRAM II data available for analyses. The support and encouragement of D. Tufts of the University of Rhode Island and S. Schwartz of Princeton University during this research are appreciated.

REFERENCES

1. R. Dwyer, "Robust Sequential Detection of Narrowband Acoustic Signals in Noise," *Proceedings of IEEE International Conference on Acoustics, Speech and Signal Processing,* Washington, D.C., 2-4 April 1979.

2. R. Dwyer, "A Robust Post-Processing Technique for Detection of Narrowband Signals," NUSC Technical Memorandum 791090, to be published.

3. R. Dwyer, "A Statistical Frequency Domain Signal Processing Method," *Proceedings of the 16th Annual Conference on Information Sciences and Systems, held at Princeton University,* 17-19 March 1982.

4. Y. Ching and L. Kurz, "Nonparametric Detectors Based On M-Interval Partitioning," *IEEE Transitions on Information Theory* IT-18, pp. 241-250, 1972.

5. L. Kurz, "Nonparametric Detectors Based on Partition Tests," in *Nonparametric Methods in Communications: Selected Topics,* edited by Papantoni-Kazakos and D. Kazakos, Marcel Dekker, New York, 1977.

6. R. Dwyer, *FRAM II Single Channel Ambient Noise Statistics,* NUSC Technical Document 6583, 25 November 1981.

7. R. Dwyer and L. Kurz, "Sequential Partition Detectors," *Journal of Cybernetics,* Vol. 8, pp. 133-157, 1978.

8. R. Dwyer and L. Kurz, "Sequential Partition Detectors with Dependent Sampling," *Journal of Cybernetics,* Vol. 10, pp. 211-232, 1980.

9. R. Dwyer, "Robust Sequential Detection of Weak Signals in Undefined Noise Using Acoustical Arrays," Journal of Acoustic Society America, Vol. 67, March 1980.

10. R. Dwyer, "Detection of Partitioned Signals by Discrete Cross-Spectrum Analysis," *IEEE Conference Proceedings of ICASSP,* 1980.

11. L. Kurz, "A Method of Digital Signalling in the Presence of Additive Gaussian and Impulsive Noise," *IRE International Convention Record,* Part 4, pp. 161-173, 1962.

12. A. Nuttall, personal communication.

13. J. Miller and J. Thomas, "Detectors for Discrete-Time Signals in Non-Gaussian Noise," IEEE Trans. Inf. Theory **IT-18**, 241-250 (1972).

14. R. Martin and S. Schwartz, "Robust Detection of a Known Signal in Nearly Gaussian Noise," IEEE Trans. Inf. Theory **IT-17** (1), 50-56 (1971).

Estimation in the Presence of Noise of a Signal Which is Flat Except for Jumps*

Yi-Ching Yao**

Statistics Center
Massachusetts Institute of Technology
Cambridge, MA

1 INTRODUCTION

We consider the problem of estimating a signal which is a step function when one observes the signal plus Gaussian noise. Optimal linear and nonlinear estimates are derived and compared.

This problem is a simplified version of a more general one, applications of which appear in many fields such as seismology, tomography, image processing, econometric modelling, regression analysis and tracking problems. In these problems, the unknown underlying structure is a function, of one or more variables, which is discontinuous or has discontinuous derivatives. It is desired to estimate these nonsmooth functions (signal processes). They can be measured either directly with measurement error or indirectly through various transformations. There are two important and relevant problems:

(1) Can one estimate such signals efficiently?

(2) Can one detect whether or where a process changes its character?

We shall restrict ourselves to the simple case where the signal processes are flat except for jumps and can be measured directly. In other words, in discrete time denote the signal process by $\mu_1, \mu_2, \ldots, \mu_T$ and let $\mu_{n+1} = \mu_n$ except for occasional changes. Let the observations $X_n = \mu_n + \epsilon_n$, $n = 1, 2, \ldots, T$ where the ϵ_n are measurement noise. We shall concentrate on estimating the signal process (i.e., the first problem) and pay little attention to detecting change points.

---

*Work supported by ONR Contract N00014-75-C-0555, task NR-609-001.
**Current Affiliation: Department of Statistics, Colorado State University, Fort Collins, CO

If the change points were known, we could estimate μ_n by the average of the data points between the two surrounding change points. If jumps are not large, it is hard to tell when jumps take place and to take appropriate action. Moreover, if measurement noise has a heavy-tailed distribution, outliers may be disguised as jumps.

In order to develop insight for estimating the signal from the observations, we take a Bayesian point of view and consider a simple model. To be specific, we will characterize the underlying problem through the following special assumptions, which form the discrete time version of a model of Duncan [1, p. 255].

(1) The sequence of the change points forms a discrete renewal process with identically geometrically distributed interarrival times.

(2) The distinct heights of the signal are mutually independent from a common Gaussian distribution.

(3) The measurement noise is Gaussian white noise.

With respect to these assumptions, some basic questions arise:

(Qa) How well can we do if we know the change points?

(Qb) How well can we do if we use the best linear estimate?

(Qc) How well can we do with the best nonlinear estimate?

(Qd) If the parameters are not known, can we estimate them from the empirical data?

(Qe) What if the model is not satisfied?

(Qf) What about the analogous continuous time problem?

The first three of these questions are studied in great detail here. Results on the others will be presented in a forthcoming report. In Section 3, we summarize our results and describe some open questions. Proofs of the details in Section 2 will also appear in a forthcoming report (see [7,8]).

2 A SPECIAL BAYESIAN MODEL

In this section the following Bayesian model is studied in great detail:

(1) Let $\mathbf{J} = (J_1, J_2, \ldots, J_{T-1})$ be a Bernoulli sequence indicating when *changes* take place. i.e.

$$J_n = 1 \text{ if there is a change between } n \text{ and } n+1, \quad (2.1)$$
$$= 0 \text{ otherwise.}$$

where $Pr(J_n = 1) = p$, for $1 \leq n \leq T-1$. For convenience, define $J_0 \equiv J_T \equiv 1$.

(2) Let $Y_1, Y_2, \ldots, Y_T$ be i.i.d. $N(\theta, \sigma^2)$. Define the *signal process* $\{\mu_n\}$ recursively as follows

$$\mu_1 = Y_1$$
$$\mu_{n+1} = (1 - J_n)\mu_n + J_n Y_{n+1}, \quad n = 1, 2, \ldots, T-1 \quad (2.2)$$

(3) Let the *observation process* $\{X_n\}$ be given by

$$X_n = \mu_n + \epsilon_n, \quad n = 1, 2, \ldots, T \tag{2.3}$$

where the *noise* $\{\epsilon_n\}$ is i.i.d. $N(0, \sigma_\epsilon^2)$. The processes $\{J_n\}, \{Y_n\}$ and $\{\epsilon_n\}$ are mutually independent.

Throughout this paper the parameters p, θ, σ^2, and σ_ϵ^2 are assumed known and without loss of generality, θ and σ_ϵ^2 are set equal to 0 and 1, respectively.

In Section 2.1 the minimum variance linear estimates of μ_n are presented and their average mean squared error is expressed in a closed form. In Section 2.2 a characterization of the Bayes estimates of μ_n is presented which has a reasonable interpretation. In Section 2.3 the estimates $E(\mu_n | \mathbf{X}, \mathbf{J})$ where $\mathbf{X} \equiv (X_1, X_2, \ldots, X_T)'$ are derived and their asymptotic average mean squared error as $T \to \infty$ is found. Finally in Section 2.4 the three types of estimates are compared.

2.1 The Minimum Variance Linear Estimates (MVLE)

The minimum variance linear estimates depend only on the first and second moments of the signal and observation processes. The process $\{\mu_n\}$ has the same covariance structure as an AR(1) with parameter $\rho \equiv 1 - p$. In other words, we may regard the linear estimation problem as the estimation of an AR(1) in the presence of white noise.

It is easy to derive that $\tilde{\mu}_n$, the MVLE of μ_n, satisfies

$$\tilde{\mu}_n = \mathbf{e}_n'(\mathbf{I} - \mathbf{M}^{-1})\mathbf{X} \tag{2.4}$$

where $\mathbf{e}_n = (0, 0, \ldots, 1, 0, \ldots, 0)'$ is the nth natural coordinate vector,

$\mathbf{I} =$ the $T \times T$ identity matrix, and

$\mathbf{M} = (M_{ij})_{T \times T}, M_{ij} = \delta_{ij} + \sigma^2 \rho^{|i-j|}$.

Several explicit expressions have been derived for the asymptotic behavior of the minimum mean squared errors as $T \to \infty$ in linear filtering, prediction and interpolation of weakly stationary discrete time processes corrupted by additive noise under very general conditions [5]. In contrast, for finite T, explicit expressions have seldom been found. The following proposition presents a closed-form representation for AMSE $(\tilde{\mu}_n)$, *the average of the mean squared errors of $\tilde{\mu}_n$*. The proof can be found in [6].

Proposition 2.1

$$AMSE(\tilde{\mu}_n) \equiv T^{-1} \sum_{n=1}^{T} E(\tilde{\mu}_n - \mu_n)^2$$

$$= 1 + T^{-1}\sigma^{-2} f_T'(-\sigma^{-2}) / f_T(-\sigma^{-2})$$

$$= 1 + \sigma^{-2} u_+'(-\sigma^{-2}) / u_+(-\sigma^{-2}) + o(1), (T \to \infty).$$

where

$$f_T(\lambda) = a(\lambda)(u_+(\lambda))^T + b(\lambda)(u_-(\lambda))^T,$$
$$u_\pm(\lambda) = [1-\rho^2-\lambda(1+\rho^2) \pm \sqrt{(1-\rho^2-\lambda(1+\rho^2))^2 - 4\rho^2\lambda^2}]/2,$$
$$a(\lambda) = [(1-\lambda)^2 - \rho^2 - (1-\lambda)u_-]/(u_+^2 - u_+u_-),$$
$$b(\lambda) = [(1-\lambda)u_+ - (1-\lambda)^2 + \rho^2]/(u_+u_- - u_-^2).$$

2.2 The Bayes Solution—the Minimum Variance Nonlinear Estimates

The Bayes solution can be computed by brute force with a number of operations of the order of 2^T. In this section, we present a characterization which has a reasonable interpretation and also provides a way to compute the solution with $0(T^3)$ operations.

In the following, we consider the Bayes estimates of μ_n based on (1) the past and present data, $E(\mu_n | X_1, \ldots, X_n)$, (2) the future data, $E(\mu_n | X_{n+1}, \ldots, X_T)$, and (3) all of the data, $E(\mu_n | X_1, X_2, \ldots, X_T)$.

Here are convenient notations:

(1) $X_i^j \equiv (X_i, X_{i+1}, \ldots, X_j)$ $(i \leq j)$
(2) $S_0 \equiv 0$, $S_n \equiv \sum_{k=1}^{n} X_k$ (cumulative sums)
(3) $L(Y) \equiv$ the distribution of random variable Y
(4) $f_{\mu_n}(z|X_i^j) \equiv$ the conditional probability density of μ_n at z given X_i^j
(5) "$f(x,z) \propto g(x,z)$ in z" means that there exists $c(x)$ such that $f(x,z) = c(x)g(x,z)$ for all x,z.

An Expression for $L(\mu_n | X_1^n)$

Proposition 2.2

$$f_{\mu_{n+1}}(z|X_1^{n+1}) \propto \phi(X_{n+1}-z)[(1-p)f_{\mu_n}(z|X_1^n) + pf_\mu(z)] \text{ in } z, 1 \leq n \leq T-1$$

where ϕ is the standard normal density and

$$f_\mu(x) = (2\pi\sigma^2)^{-1/2}\exp(-x^2/2\sigma^2) \tag{2.8}$$

is the density of the prior signal distribution.

Proposition 2.2 is an updating formula for computing $L(\mu_n | X_1^n)$, $n = 1, 2, \ldots, T$. Since $L(\mu_1|X_1) = N(S_1(1+\sigma^{-2})^{-1}, (1+\sigma^{-2})^{-1})$, we can demonstrate by use of induction and Proposition 2.2

Proposition 2.3

$$L(\mu_n|X_1^n) = \sum_{k=1}^{n} A_k^{(n)} \cdot N\left[\frac{S_n - S_{n-k}}{k+\sigma^{-2}}, \frac{1}{k+\sigma^{-2}}\right]$$

where

$$A_k^{(n)} = \frac{p(1-p)^{k-1}\alpha_{n-k+1}}{\sqrt{1+k\sigma^2}\alpha_{n+1}}$$

Estimation in Noise

$$\exp\left[\frac{(S_n - S_{n-k})^2}{2(k + \sigma^{-2})}\right] \quad (n = 1, 2, \ldots, T; k = 1, \ldots, n)$$

and α_n ($n = 1, 2, \ldots, T+1$) are defined recursively by $\alpha_1 = 1$, and

$$\alpha_{n+1} = \sum_{k=1}^{n} \alpha_{n-k+1} \frac{p(1-p)^{k-1}}{\sqrt{1+k\sigma^2}} \exp\left[\frac{(S_n - S_{n-k})^2}{2(k + \sigma^{-2})}\right].$$

Therefore, we have

Proposition 2.4

$$E(\mu_n | X_1^n) = \sum_{k=1}^{n} A_k^{(n)} \cdot \frac{S_n - S_{n-k}}{k + \sigma^{-2}}$$

An Expression for $L(\mu_n | X_{n+1}^T)$

As for Proposition 2.2, we can derive

Proposition 2.5

$$f_{\mu_{n-1}}(z|X_n^T) \propto (1-p)\,\phi\,(X_n - z)f_{\mu_n}(z|X_{n+1}^T)$$

$$+ pf_\mu(z)\int_{-\infty}^{\infty} \phi\,(X_n - z')f_{\mu_n}(z'|X_{n+1}^T)\,dz' \quad \text{in } z, 2 \leq n \leq T.$$

Since $L(\mu_T) = N(0, \sigma^2)$, we can derive by use of induction and Proposition 2.5

Proposition 2.6

$$L(\mu_n | X_{n+1}^T) = \sum_{k=0}^{T-n} B_k^{(T-n)} \cdot N\left[\frac{S_{n+k} - S_n}{k + \sigma^{-2}}, \frac{1}{k + \sigma^{-2}}\right]$$

where

$$B_k^{(T-n)} = \frac{p(1-p)^k \beta_{T-n-k}}{\sqrt{1+k\sigma^2}\,\beta_{T-n}} \exp\left[\frac{(S_{n+k} - S_n)^2}{2(k + \sigma^{-2})}\right] + (1-p)\delta_{nT}$$

$(T - n = 0, 1, \ldots, T-1; k = 0, \ldots, T-n)$

and β_{T-n} ($n = T, T-1, \ldots, 1$) are defined recursively by $\beta_0 = 1$ and

$$\beta_{T-n} = \sum_{k=0}^{T-n-1} \beta_{T-n-1-k} \frac{p(1-p)^k}{\sqrt{1+(k+1)\sigma^2}} \exp\left[\frac{(S_{n+k+1} - S_n)^2}{2(k+1+\sigma^{-2})}\right].$$

Expressions for $L(X_n | X_1^T)$ and $E(\mu_n | X_1^T)$

Proposition 2.7

$$f_{\mu_n}(z|X_1^T) \propto f_{\mu_n}(z|X_1^n)f_{\mu_n}(z|X_{n+1}^T)/f_\mu(z) \quad \text{in } z.$$

This states that the "two-sided" conditional density of the signal is proportional to the product of the two "one-sided" conditional densities divided by its prior density. The idea of using forward and backward recursions has been introduced in the engineering literature, e.g., [2, Appendix] [3] [4]. The proof of this proposition is based on the Markov property of $\{\mu_n\}$.

From Propositions 2.3, 2.6 and 2.7, we can derive

Proposition 2.8

$$L(\mu_n|X_1^T) = \sum_{1 \leq i \leq n \leq j \leq T} C_{ij} \cdot N\left[\frac{S_j - S_{i-1}}{j-i+1+\sigma^{-2}}, \frac{1}{j-i+1+\sigma^{-2}}\right]$$

where

$$C_{ij} = C'_{ij}/D, \quad D = \sum_{1 \leq i \leq n \leq j \leq T} C'_{ij}$$

and

$$C'_{ij} = \alpha_i \beta_{T-j} \frac{(1-p)^{j-i}}{\sqrt{1+(j-i+1)\sigma^2}} \exp\left[\frac{(S_j - S_{i-1})^2}{2(j-i+1+\sigma^{-2})}\right].$$

Note: D is independent of n and therefore equal to α_{T+1}/p.

Therefore, we have

Proposition 2.9

$$E(\mu_n|X_1^T) = \sum_{1 \leq i \leq n \leq j \leq T} C_{ij}(S_j - S_{i-1})/(j-i+1+\sigma^{-2})$$

Remarks:

(1) Since, for $i \leq n \leq j$,

$$L(\mu_n|X_1^T, J_{i-1}=1, J_i = J_{i+1} = \ldots = J_{j-1} = 0, J_j = 1)$$

$$= N\left[\frac{S_j - S_{i-1}}{j-i+1+\sigma^{-2}}, \frac{1}{j-i+1+\sigma^{-2}}\right],$$

one can see from Proposition 2.8 that

$$C_{ij} = Pr(J_{i-1}=1, J_i = \ldots = J_{j-1}=0, J_j=1|X_1^T).$$

Thus $\{C_{(i+1)j}: 0 \leq i < n \leq j \leq T\}$ represents the conditional distribution of the two change points surrounding time n. Hence,

$$Pr(J_n = 1|X_1^T) = \sum_{k=0}^{n-1} Pr(J_k = 1, J_{k+1} = \ldots = J_{n-1}=0, J_n=1|X_1^T) = \sum_{k=0}^{n-1} C_{(k+1)n}$$

In particular,

$$Pr(\text{No change in } [1,T]|X_1^T) = C_{1T}$$

can be used to test whether changes have ever happened.

(2) $$E(\mu_n|X_1^T) = \sum_{1 \leq i \leq n \leq j \leq T} C_{ij}(S_j - S_{i-1})/(j-i+1+\sigma^{-2}) = \sum_{k=1}^{T} d_k^{(n)} X_k$$

Estimation in Noise 173

where $d_k^{(n)} = \sum_{\substack{1 \le i \le \min(n,k) \\ \max(n,k) \le j \le T}} C_{ij} / (j - i + 1 + \sigma^{-2})$, $1 \le k \le T$.

So, $0 < d_1^{(n)} < d_2^{(n)} < \ldots < d_{n-1}^{(n)} < d_n^{(n)} > d_{n+1}^{(n)} > \ldots > d_T^{(n)} > 0$

and

$$\sum_{k=1}^T d_k^{(n)} = \sum_{1 \le i \le n \le j \le T} \frac{j-i+1}{j-i+1+\sigma^{-2}} C_{ij} < 1.$$

Thus, the Bayes estimate $E(\mu_n | X_1^T)$ is a sample dependent weighted average of the observations X_k and the prior mean 0, and the weights $d_k^{(n)}$ attain their maximum at $k = n$ and decrease strictly as k moves away from n on either side.

(3) The number of operations required to compute α_n, β_{T-n}, C_{ij} ($1 \le n \le T$, $1 \le i \le j \le T$) is $0(T^2)$. The number of operations to compute $E(\mu_n | X_1^T)$ is $0\,(n(T-n))$. So the total number of operations to compute $E(\mu_n | X_1^T)$ for all n is $0\,(T^3)$.

2.3 $E(\mu_n | \mathbf{X}, \mathbf{J})$: The Estimates of μ_n Given the Change Points

In this section, we consider $E(\mu_n | \mathbf{X}, \mathbf{J})$ which can be used to see how much additional information for estimating μ_n is obtained from the knowledge of the change points.

Define $[r_n(\mathbf{J}), s_n(\mathbf{J})]$ to be the largest integral interval containing n which contains no change ($1 \le r_n(\mathbf{J}) \le n \le s_n(\mathbf{J}) \le T$). Since $\mathbf{X}$ and $\boldsymbol{\mu}$ are Gaussian conditional on $\mathbf{J}$, the minimum variance estimate of μ_n given $\mathbf{X}$ and $\mathbf{J}$ is the linear estimate

$$E(\mu_n | \mathbf{X}, \mathbf{J}) = \sum_{k=r_n(\mathbf{J})}^{s_n(\mathbf{J})} X_k / (s_n(\mathbf{J}) - r_n(\mathbf{J}) + 1 + \sigma^{-2}).$$

The following proposition gives an explicit expression for the asymptotic behavior of AMSE $(E(\mu_n | \mathbf{X}, \mathbf{J}))$ as $T \to \infty$.

Proposition 2.10

$$\text{AMSE}\,(E(\mu_n|\mathbf{X},\mathbf{J})) \equiv T^{-1} \sum_{n=1}^T E[E(\mu_n|\mathbf{X},\mathbf{J}) - \mu_n]^2$$

$$= p - p^2 \sigma^{-2} (1-p)^{-\sigma^{-2}-1} \int_0^{1-p} \frac{x^{\sigma^{-2}}}{1-x} dx + o(1), \quad (T \to \infty).$$

2.4 Comparison Among the Three Types of Estimates, $\tilde{\mu}_n$, $E(\mu_n|\mathbf{X})$, and $E(\mu_n|\mathbf{X}, \mathbf{J})$.

In this section, the performance of $\tilde{\mu}_n$, $E(\mu_n|\mathbf{X})$, and $E(\mu_n|\mathbf{X},\mathbf{J})$ is compared in terms of their average mean squared errors for $T = 20$.

The AMSE of $\tilde{\mu}_n$ is calculated from Proposition 2.1 while that of $E(\mu_n|\mathbf{X})$ (or $E(\mu_n|\mathbf{X},\mathbf{J})$) is estimated by simulation with 200 (or 2000, respectively) replications for each one of 48 cases where $p \in \{0.05, 0.1, 0.2, 0.4, 0.6,$

(a) $\sigma = 4$

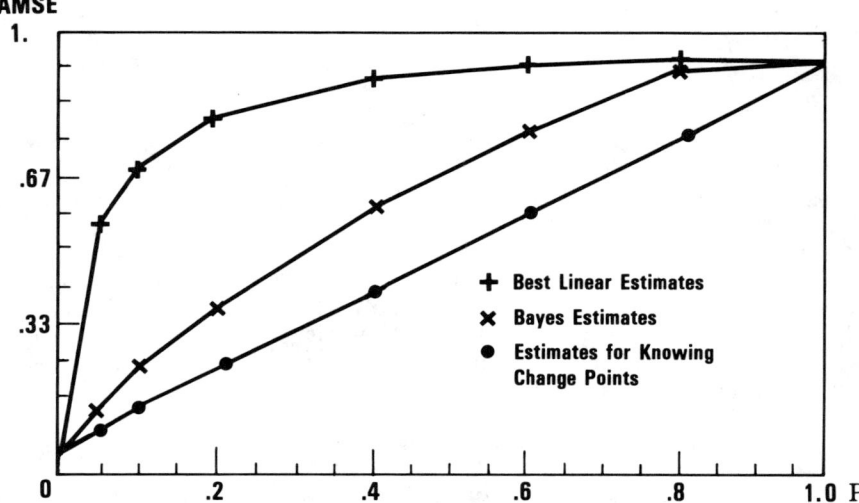

(b) $p = 0.1$

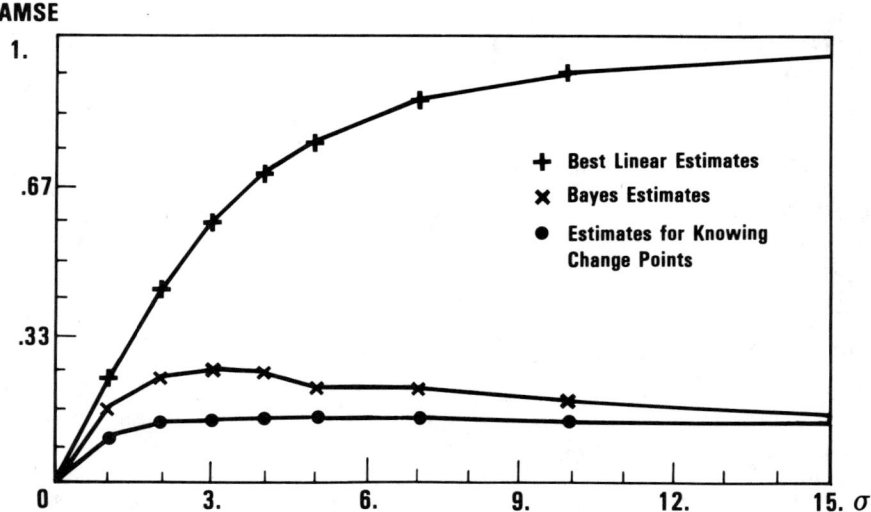

Figure 1 AMSE as a function of p and σ

0.8} and $\sigma \epsilon$ {1, 2, 3, 4, 5, 7, 10, 15}. The estimated numbers for AMSE ($E(\mu_n|\mathbf{X})$) (or AMSE ($E(\mu_n|\mathbf{X},\mathbf{J})$)) have precision (standard deviation) about 0.02 (or 0.002, respectively) or less. Figure 1 presents partial simulation results.

Estimation in Noise

Remarks

(1) It can be shown that the AMSE of $\tilde{\mu}_n$ and $E(\mu_n|\mathbf{X},\mathbf{J})$ are increasing as p or σ^2 increases. So is the AMSE of $E(\mu_n|\mathbf{X})$ as p increases. However, from the simulation results, it appears that as σ^2 increases, AMSE $(E(\mu_n|\mathbf{X}))$ first increases and then decreases and eventually approaches AMSE $(E(\mu_n|\mathbf{X},\mathbf{J}))$. One explanation is that when σ^2 is large enough $\mathbf{J}$ can be well estimated from $\mathbf{X}$, and this information can offset the loss of the relatively small prior information about μ_n.

(2) From the simulation results, it appears that $E(\mu_n|\mathbf{X})$ is only slightly worse than $E(\mu_n|\mathbf{X},\mathbf{J})$ in every case. However, $\tilde{\mu}_n$ is very poor when σ^2 is moderately large and p is small. Actually, if we allow $T \to \infty$ and fix σ^2 it is not difficult to show that

$$\text{AMSE}(\tilde{\mu}_n) = \sqrt{\frac{\sigma^2}{2}}\, p^{1/2} + o(p^{1/2}) \qquad (p \to 0^+)$$

$$\text{AMSE}(E(\mu_n|\mathbf{X},\mathbf{J})) = p + o(p) \qquad (p \to 0^+).$$

In other words, $\tilde{\mu}_n$ is very inefficient compared with $E(\mu_n|\mathbf{X},\mathbf{J})$ when p is small. The asymptotic behavior of AMSE $(E(\mu_n|\mathbf{X}))$ as $p \to 0^+$ is not known.

(3) Since σ^2 can be regarded as the signal to noise ratio, it is interesting to consider relative mean squared error, i.e., mean squared error divided by the energy of the signal. In other words, σ^{-2} AMSE is used to replace AMSE. As a matter of fact, the σ^{-2} AMSE for the case that $L(\mu_n) = N(0,\sigma^2)$ and $L(\epsilon_n) = N(0,1)$ is the same as the AMSE for the case that $L(\mu_n) = N(0,1)$ and $L(\epsilon_n) = N(0,\sigma^{-2})$. Therefore, it is not hard to show that the σ^{-2} AMSE of $\tilde{\mu}_n$ and $E(\mu_n|\mathbf{X})$ and $E(\mu_n|\mathbf{X},\mathbf{J})$ are decreasing as the signal to noise ratio σ^2 increases.

3 SUMMARY AND OPEN QUESTIONS

In this paper, we propose a special discrete time Bayesian model to describe step function signals which are corrupted by white noise. The best linear estimates, the Bayes estimates and the estimates with known change points are derived, evaluated and compared analytically and numerically. It is found that the best linear estimates are insensitive to sudden changes in signals and therefore cannot perform efficiently. On the other hand, the Bayes estimates are strongly related to the estimation of the change points. The total number of operations to compute the Bayes estimates is $0(T^3)$ where T is the fixed time span.

It is worthwhile to note that the results in Section 2.2 can be easily extended to the multivariate case. Also, the criterion of mean squared error can be relaxed since the posterior distributions of the signal states can be computed.

However, the estimation problem becomes much more difficult when the noise is colored and/or the domain of the signal is of higher dimension. We list three open questions as the end of this paper.

(1) How should one compute efficiently the Bayes solution of a modified version of the model where the additive noise is colored (serially dependent)?

(2) What is a good formulation and an efficient solution for a two-dimensional model with reasonable properties to deal with the problem of image processing? Here "reasonable property" is not well-defined but properties such as rotational invariance seem desired.

(3) How does one deal with problems where signals cannot be measured directly and are observed through transformations, as is the case in CAT scanners?

REFERENCES

[1] Barnard, G.A. (1959). Control charts and stochastic processes. *J. Roy. Stat. Soc.* B 21, 239-271 (with discussion).

[2] Forney, G.D. (1973). The Viterbi algorithm. *IEEE Proc.*, 61, 268-278.

[3] Fraser, D.C. (1967). *A New Technique for the Optimal Smoothing of Data.* Sc. D. Dissertation, M.I.T., Cambridge, MA.

[4] Mayne, D.Q. (1966). A solution of the smoothing problem for linear dynamic systems. *Automatica*, 4, 73-92.

[5] Snyders, J. (1972). Error formulae for optimal linear filtering prediction and interpolation of stationary time series. *Ann. Math. Statist.*, 43, 1935-1943.

[6] Yao, Y.C. (1981). *The Linear Theory of Estimating Means in Time Series Subjected to Changes in Geometrically Distributed Time Intervals.* Tech. Rept. ONR21, Statist. Center, M.I.T., Cambridge, MA.

[7] Yao, Y.C. (1982), *Estimation in the Presence of Noise of a Signal which is Flat except for Jumps—Part I, A Bayesian Study.* Tech. Rept. ONR25, Statist. Center, M.I.T., Cambridge, MA.

[8] Yao, Y.C. (1983). *Estimation in the Presence of Noise of a Signal which is Flat except for Jumps—Part II, The Empirical Bayes Approach.* Tech. Rept. ONR27, Statist. Center, M.I.T., Cambridge, MA.

Part III: Data Analysis and Modeling

Cross Validated Spline Methods for Direct and Indirect Sensing Experiments*

Grace Wahba

*Department of Statistics
University of Wisconsin
Madison, WI*

1 OVERVIEW OF CROSS VALIDATED SPLINE METHODS FOR DIRECT AND INDIRECT SENSING EXPERIMENTS

For several years, we have been developing data analysis techniques for a fairly general class of (direct and indirect) sensing experiments. The model may be described as follows: f is some unknown function of one or more variables, which is assumed to be in a (Hilbert) space H of "smooth" functions, and one observes $\{y_i\}$

$$y_i = L_i f + \epsilon_i \quad i = 1, 2, \ldots, n, \tag{1.1}$$

where the data functionals $L_1, \ldots, L_n$ are n bounded linear functionals on H and the ϵ_i are independent, zero mean measurement errors, with variances $w_i \sigma^2$, $i = 1, 2, \ldots, n$. The parameter σ^2 may be unknown. Examples of L_i are

(1) $L_i f = f(P_i)$
(2) $L_i f = \int_\Omega K(P_i, P) f(P) dP$
(3) $L_i f = \Sigma\, a_\alpha(P_i)\, \dfrac{\partial^{\alpha_1 + \ldots \alpha_d}}{\partial x_1^{\alpha_1} \ldots \partial x_d^{\alpha_d}}\, f(x_1, \ldots, x_d)$

---

*Work supported under Contract N00014-77-C-0675, Task NR 609-003 with the Office of Naval Research.

Example (1) corresponds to the "direct sensing" problem, example (2) to a Fredholm integral equation of the first kind given discrete, noisy data, and example (3) corresponds to a partial differential equation. Example (2) is pervasive in many problems in geophysics, meteorology and medicine. Usually one desires to estimate $f(P)$, for all P in some set S. Sometimes only moments $\int w_i(P) f(P) dP$ are desired, in other applications derivatives of f may be desired. Side information may frequently be available, such as

$$\begin{array}{ll} 0 \leq f(P) & \text{positivity} \\ a(P) \leq f(P) \leq b(P) & \text{boundedness} \\ 0 \leq f'(P) & \text{Monotonicity} \\ 0 \leq f''(P) & \text{convexity} \end{array} \quad (1.2)$$

etc. More generally we have considered the case where $f \in C \subset H$, where C is any closed convex set in H which is the intersection of a "smoothly varying" family of hyperplanes See [2]. H can be chosen so that there exists a C with the requisite properties for each of the examples in Eq. (1.2).

The general approach to the estimation of f we have developed and tested goes as follows: the estimate, call it f_λ, is the solution to the minimization problem: find $f \in H$ to minimize

$$\frac{1}{n} \sum_{i=1}^{n} (y_i - L_i f)^2 / w_i + \lambda J_{m,\delta}(f), \quad (1.3)$$

subject to $f \in C$. Here $J_{m,\delta}$ is a seminorm on H, or "penalty function" indexed by the parameters m and δ (to be described in a moment), and λ is the bandwidth or smoothing parameter. Some examples of J are

(1) $J_m(f) = \int_0^1 (f^{(m)}(P))^2 dP$

(2) $J_{2,\delta}(f) = \int_{-\infty}^{\infty} \int_{-\infty}^{\infty} (f_{xx}^2 + 2\delta f_{xy}^2 + \delta^2 f_{yy}^2) dx dy$ or, in d dimensions

$$J_m(f) = \sum_{\alpha_1 + \alpha_2 + \ldots \alpha_d = m} \frac{m!}{\alpha_1! \alpha_2! \ldots \alpha_d!} \int \cdots \int \left| \frac{\partial^m f}{\partial x_1^{\alpha_1} \ldots \partial x_d^{\alpha_d}} \right|^2 dx_1 \ldots dx_d.$$

(3) $J_m(f) = \int_{\text{sphere}} (\Delta^m f)^2 dP$

where Δ is the Laplace-Beltrami operator on the sphere

$$\Delta f = \frac{1}{\cos^2 \phi} f_{\lambda\lambda} + \frac{1}{\cos \phi} (\cos \phi f_\phi)_\phi, \quad \lambda = \text{longitude}, \phi = \text{latitude}.$$

Another example appears in Section 3 below, where the parameter δ plays a different role. The important bandwidth parameter λ, and sometimes m and δ, can be estimated by the method of generalized cross validation (GCV). The GCV method has recently been extended to cover problems where the constraints $f \in C$ are imposed [21,24]. In example (1), if the $L_i f$ are point evaluation functionals, that is, $L_i f = f(P_i)$, then f_λ is a polynomial smoothing spline, in example (2) f_λ will be a thin plate spline [11,13] and in example (3) f_λ will be a spline on the sphere [20].

f_λ can be shown to be a Bayes estimate, with a certain prior determined by H and J [1,6]. On the other hand, it can be shown that f_λ is a natural generalization of the output of a low pass filter, see [7]. If the $L_i f$'s are derivatives, and the limit is taken as $\lambda \to 0$, then spline collocation methods for the solution to differential equations results [10].

Table 1 gives a list of some of the problems involved in estimating f by cross validated spline methods, and relevant references of the author, her collaborators, and students. In Table 1, $\hat{\lambda}$ is the GCV estimate of λ.

Table 1   Problems in the Estimation of f, with References

(1) Explicit representation of the minimizer of (1.3) in various contexts [1,4,5,11,13,18,20,22].

(2) Generalized cross validation methods for choosing λ, m, and δ [4,7,8,21,24].

(3) Use of prior information to choose H and J [22,23].

(4) Smoothing of multidimensional scattered data [11,13].

(5) Smoothing splines on the sphere [20,22].

(6) Numerical methods for computing $f_{\hat{\lambda}}$ with large data sets [14,21,25].

(7) Quadrature methods tailored to specific integral equation problems [19,28].

(8) Optimal design (choice of $t_1, \ldots, t_n$) [10,17].

(9) Convergence properties of $f_{\hat{\lambda}}$ [3,4,7,12].

(10) Convergence properties when the constraints are discretized [2].

(11) Confidence intervals for $f_{\hat{\lambda}}$ [26].

(12) Relation of $f_{\hat{\lambda}}$ to Bayes estimates and Weiner filtering [1,6].

(13) Nonlinear data functionals [29].

Four especially interesting areas of application are

(1) Smoothing of scattered, noisy data in two and higher dimensions with thin plate splines [11,13]. Transportable code is available from the Madison, WI Academic Computing Center.

(2) Approximate solution of Abel's Integral Equations, as they occur in stereology and computerized tomography [16,28]. These equations have a singular kernel.

(3) Equations of radiative transfer as they occur, for example in the estimation of vertical temperature profiles from satellite-observed radiances [29]. These equations are mildly nonlinear.

(4) As an alternative to logistic regression [27].

The integral equation methods have found use in a number of engineering applications, for example in the determination of adsorption energy distributions from an adsorption isotherm experiment [30] and in the recovery of aerosol size distributions from Marple impactor data [31].

For lack of space we have omitted explicit mention of recent related work by others, in particular by Cox; Chow, Geman and Wu; Ragozin; Rosenblatt; Lii; Silverman; Speckman; and Utreras.

In the remainder of this paper we briefly review recent numerical results in two interesting areas: Section 2: deconvolution with positivity constraints and Section 3: estimation of wind horizontal divergence and vorticity from scattered noisy wind vector measurements.

2 DECONVOLUTION WITH NONNEGATIVITY CONSTRAINS

We numerically studied the convolution equation

$$y_i = \int_0^1 k\left(\frac{i}{n} - s\right) f(s) ds + \epsilon_i, \quad i = 1, 2, \ldots, n \tag{2.1}$$

$$f(s) \geq 0, \quad 0 \leq s \leq 1$$

with $J(f) = \int_0^1 (f''(s))^2 ds$. The results appear in [24]. The cross validated spline method prescribes the estimate of the solution, call it f_λ, say, as the minimizer of

$$\frac{1}{n} \sum_{i=1}^n \int_0^1 k\left(\frac{i}{n} - s\right) f(s) ds + \lambda \int_0^1 (f''(s))^2 ds \tag{2.2}$$

in the nonnegative quadrant of a certain Sobolev space W_2^2 (which we will not discuss further).

For n of moderate size (say 32-256) a good, readily computable approximation to the minimizer f_λ of (2.2) may be found by approximating f_λ by a trigonometric polynomial of the form

$$f_\lambda(x) = \alpha_0 + \sum_{\nu=1}^{n/2} \alpha_\nu \cos 2\pi\nu x + \sum_{\nu=1}^{n/2-1} \beta_\nu \sin 2\pi\nu x$$

and finding the α's and β's to minimize Eq. (2.2) subject to the discretized positivity constraints

$$f\left(\frac{i}{n}\right) \geq 0, \quad i = 1, 2, \ldots, n.$$

For fixed λ, a quadratic programming problem with n unknowns and n linear inequality constraints results. GCV for constrained problems may be used to choose λ; for each trial value of λ a q.p. must be solved, a good starting guess can be obtained by applying GCV to the unconstrained problem. (Thus, the method is not "cheap," however it can be very cheap compared to the cost of some experiments.)

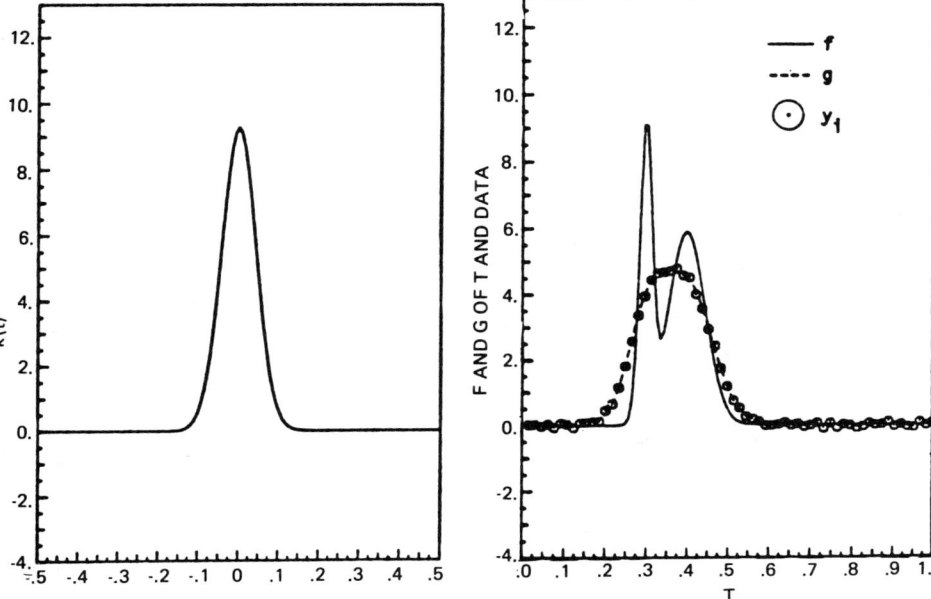

Figure 1.1 The convolution kernel $k(t)$

Figure 1.2 f, g, data, f_λ and $f_{\lambda_c}^c$

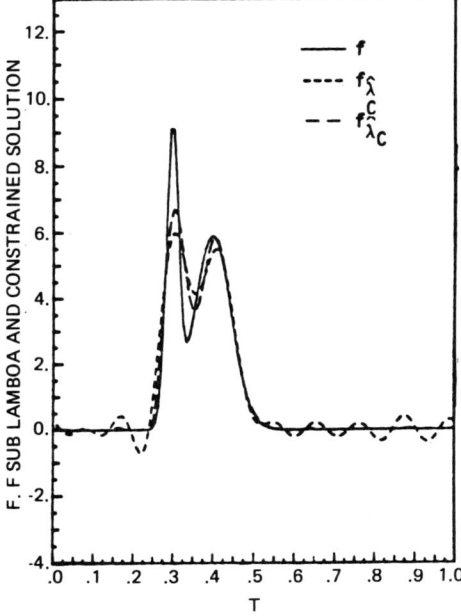

Figure 1.3 f, g, data, f_λ and $f_{\lambda_c}^c$

We give here a single numerical example, from [24]. Figure 1.1 gives a plot of the convolution kernel $k(t)$ (assumed periodic). Figure 1.2 gives a plot of the (true) test solution $f(x)$, $0 \leq x \leq 1$, "exact" data $g(x)$,

$$g(x) = \int_0^1 k(x-s)f(s)ds \quad 0 \leq x \leq 1$$

and simulated measured data

$$y_i = g\left(\frac{i}{n}\right) + \epsilon_i, \quad i = 1, 2, \ldots, n$$

for $n = 64$. The ϵ_i were independent normally distributed pseudorandom variables with common standard deviation $\sigma = 0.05$. The fact that there are two distinct peaks in the true f is not at all obvious from the measured data. Figure 1.3 gives the true f (again) and it gives $f_{\hat{\lambda}}$, which is the cross validated spline estimated of f obtained *without* imposing positivity constraints. Figure 1.3 also gives $f_{\hat{\lambda}_C}^C$, which is the estimate for f with the nonnegativity constraints imposed. $\hat{\lambda}_C$ is the GCV estimate of λ for constrained problems.

The unconstrained cross-validated spline estimate is not bad, however, spurious oscillations are clearly visible. The constrained solution not only eliminates the spurious oscillations, it enhances the resolution of the two peaks. This innocuous looking problem is illposed to a high degree. If there were no measurement error, there would be only about 20 linearly independent pieces of information in the 64 dimensional data vector recorded to 8 significant figures. The nonnegativity constraints are adding important information. For further information concerning the relative degree of illposedness, see [21]. Unsophisticated estimation methods will give garbage.

3 VECTOR FIELDS ON THE SPHERE

In meteorology and geophysics measurement of vector fields are made at discrete locations around the world and it is desired to estimate the global vector field. In meteorology, for example, the wind field at the 500 millibar height (height at which the pressure is 500 millibars) and other heights, is measured at the worldwide radiosonde network. It is desired to estimate the vorticity (curl) ζ and the divergence D of this wind field on a regular grid, to be used as initial conditions for the differential equations of numerical weather forecasting. The following is abstracted from [22].

Letting $U(P)$ and $V(P)$ be the east and north wind respectively at $P = (\lambda, \phi)$, the data are

$$U_i = U(P_i) + \epsilon_i^U, \quad V_i = V(P_i) + \epsilon_i^V, \quad (3.1)$$

where ϵ_i^U and ϵ_i^V are measurement errors, and it is desired to estimate ζ and D on a regular grid. ζ and D are related to U and V by

$$\zeta = \frac{1}{a \cos \phi} \left[-\frac{\partial}{\partial \phi}(U \cos \phi) + \frac{\partial V}{\partial \lambda} \right] \quad D = \frac{1}{a \cos \phi} \left[\frac{\partial U}{\partial \lambda} + \frac{\partial}{\partial \phi}(V \cos \phi) \right], \quad (3.2)$$

where a is the radius of the earth. There exists (by Helmoltz Theorem) two functions $\Psi(P)$ and $\Phi(P)$, called the stream function and the velocity potential respectively, with the following properties:

$$U = \frac{1}{a}\left\{-\frac{\partial \Psi}{\partial \phi} + \frac{1}{\cos \phi}\frac{\partial \Phi}{\partial \lambda}\right\}, \quad V = \frac{1}{a}\left\{\frac{1}{\cos \phi}\frac{\partial \Psi}{\partial \lambda} + \frac{\partial \Phi}{\partial \phi}\right\}; \quad (3.3)$$

$$\zeta = \Delta \Psi, \quad D = \Delta \Phi. \quad (3.4)$$

We estimate ζ and D from U_i and V_i by solving a minimization problem of the form: find Ψ and Φ in an appropriate function space to minimize

$$\frac{1}{n}\sum_{i=1}^{n}\left\{-\frac{1}{a}\frac{\partial \Psi}{\partial \phi}(P_i) + \frac{1}{a\cos \phi_i}\frac{\partial \Phi}{\partial \lambda}(P_i) - U_i\right\}^2$$

$$+ \frac{1}{n}\sum_{i=1}^{n}\left\{\frac{1}{a\cos \phi_i}\frac{\partial \Psi}{\partial \lambda}(P_i) + \frac{1}{a}\frac{\partial \Phi}{\partial \phi}(P_i) - V_i\right\} \quad (3.5)$$

$$+ \lambda\left[J^{(1)}(\Psi) + \frac{1}{\delta}J^{(2)}(\Phi)\right]$$

and then obtaining ζ and D analytically as $\Delta \Psi$ and $\Delta \Phi$. The parameter λ is the usual bandwidth parameter and the parameter δ controls the relative amount of energy in the divergent and nondivergent part of the wind. An approximation to the minimizer of Eq. (3.5) may be obtained by first approximating Ψ and Φ as a finite number of spherical harmonics

$$\Psi = \sum_{l=1}^{N}\sum_{s=-l}^{l}\alpha_{ls}Y_l^s, \quad \Phi = \sum_{l=1}^{N}\sum_{s=-l}^{l}\beta_{ls}Y_l^s. \quad (3.6)$$

The spherical harmonics $\{Y_l^s\}$ are the eigenfunctions of the Laplace-Beltrami operator Δ,

$$\Delta Y_l^s = -l(l+1)Y_l^s,$$

and approximating Ψ and Φ this way is analogous to the approximation of f in Section 2 by a finite number of sines and cosines. It can be shown that if $J^{(1)}$ and $J^{(2)}$ are isotropic seminorms (isotropic = unchanged under arbitrary rotations of the coordinate system) then for Ψ and Φ of (3.6),

$$J^{(1)}(\Psi) = \sum_{l=1}^{N}\sum_{s=-l}^{l}\frac{\alpha_{ls}^2}{\lambda_{ls}^{(1)}}, \quad J^{(2)}(\Phi) = \sum_{l=1}^{N}\sum_{s=-l}^{l}\frac{\beta_{ls}^2}{\lambda_{ls}^{(2)}} \quad (3.7)$$

for some $\{\lambda_{ls}^{(1)}, \lambda_{ls}^{(2)}\}$. For more details, and in particular for a discussion concerning the choice of the $\{\lambda_{ls}\}$ from historical data, see [22].

By substituting Eq. (3.7) into Eq. (3.5), for fixed λ, δ, the minimization problem reduces to finding the $\{\alpha_{ls}, \beta_{ls}\}$ which minimize a quadratic form. The parameters λ and δ are estimated by GCV.

To see how well this method may be expected to do on real meteorological data, realistic "true" 500 millibar stream function-velocity potential pairs (Ψ, Φ) were generated and the "true" wind fields at 114 North American weather stations determined using Eq. (3.3). Realistic measurement errors (s.d. in each component of 2.5 m/s) were added. Figure 2.1 gives the simulated wind data, and Fig. 2.2 gives the estimated wind field, which is obtained by using Eq. (3.3) in conjunction with the estimated (Ψ, Φ). Figures 2.3 and 2.4 give the "true" and estimated vorticity and divergence respectively. The

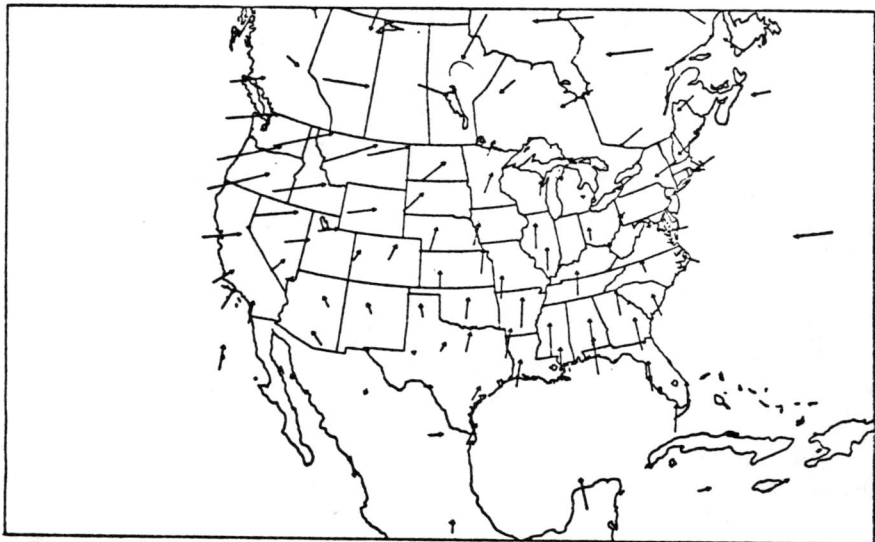

Figure 2.1 Simulated wind data

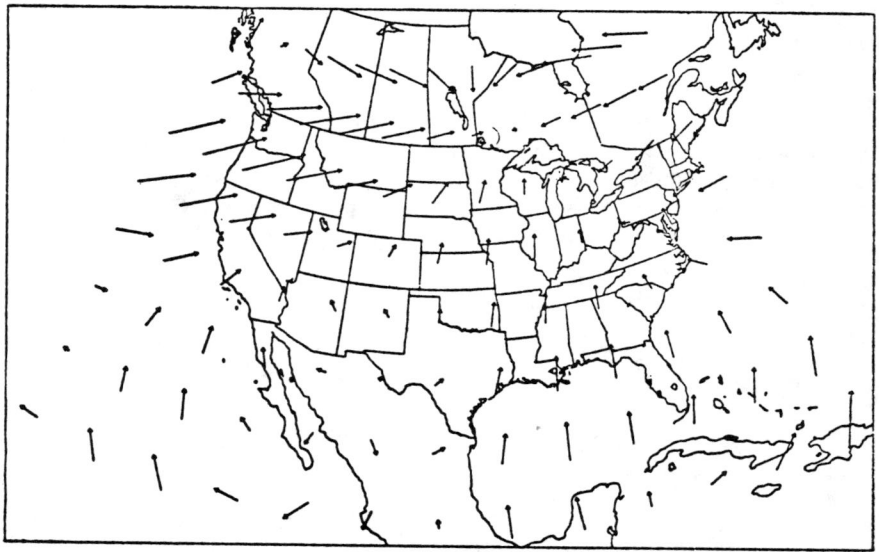

Figure 2.2 Estimated wind field

Cross Validated Methods

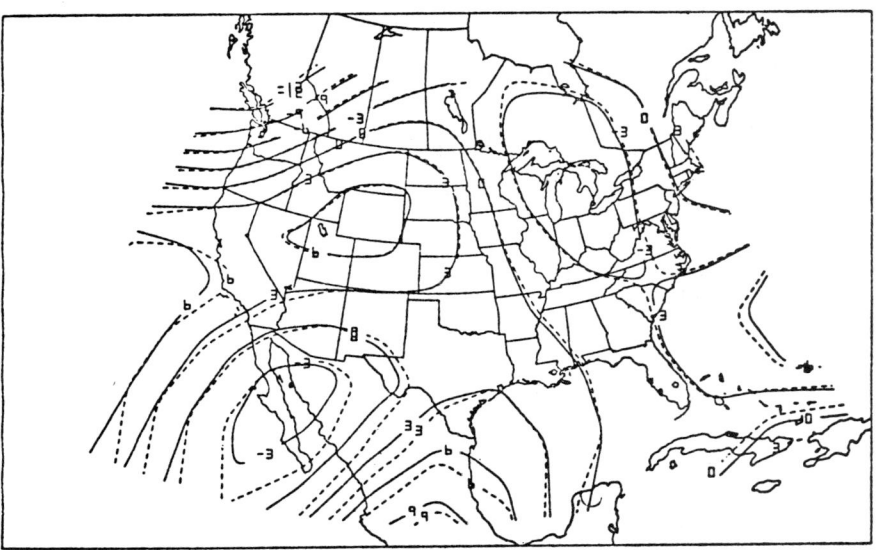

Figure 2.3 Model and estimated vorticity, $\times 10^{-9}$/sec

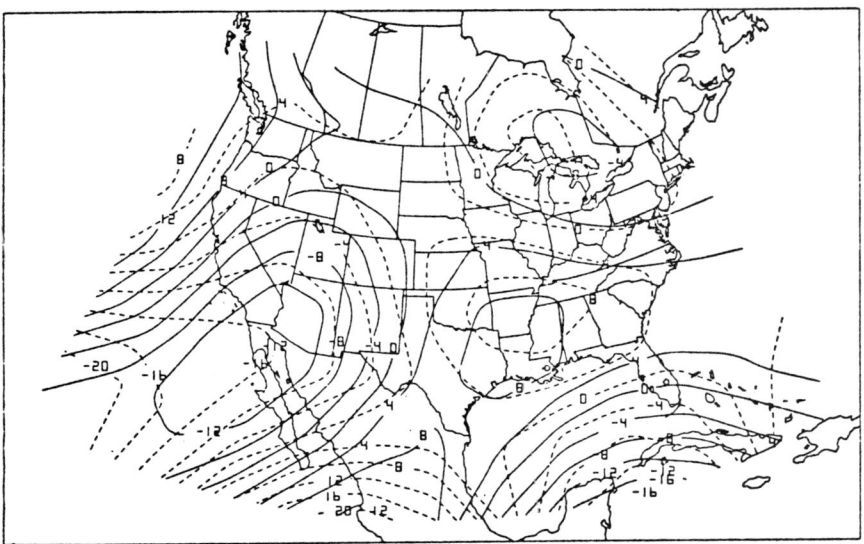

Figure 2.4 Model and estimated divergence, $\times 10^{-6}$/sec

results appear to be excellent when compared to previous estimation methods. One of the results of this experiment was that the estimates were sensitive to changes in δ as well as λ and, the GCV method gives a good estimate of an optimal δ as well as λ. In a problem like this, it can be very important to parameterize J in a physically meaningful way, as well as in a mathematically well posed way. Methods for doing this are discussed in [22], see also [15].

REFERENCES

[1] Some results on Tchebycheffian spline functions (with George S. Kimeldorf). *J. Math. Anal. and Applic.*, *33*, 1, 82-95 (1971).

[2] On the minimization of a quadratic functional subject to a continuous family of linear inequality constraints. *SIAM J. Control*, *11*, 1, 64-79 (1973).

[3] Smoothing noisy data by spline functions. *Numer. Math.*, *24*, 383-393 (1975).

[4] Practical approximate solutions to linear operator equations when the data are noisy. *SIAM J. Numerical Analysis*, *14*, 4, 651-667 (1977).

[5] A survey of some smoothing problems and the method of generalized cross validation for solving them. Applications of Statistics, P.R. Krishnaiah, ed., 507-523, North Holland (1977).

[6] Improper priors, spline smoothing and the problem of guarding against model errors in regression. *J. Roy. Stat. Soc. Ser. B.* *40*, 3, 364-372 (1978).

[7] Smoothing noisy data with spline functions: estimating the correct degree of smoothing by the method of generalized cross-validation (with Peter Craven). *Numer. Math.*, *31*, 377-403 (1979), see also IMSL subroutine ICSSCV.

[8] Generalized cross-validation as a method for choosing a good ridge parameter (with G. Golub and M. Heath). *Technometrics*, *21*, 215-223 (1979).

[9] Smoothing and ill posed problems. in "Solution Methods for Integral Equations with Applications," M. Goldberg, ed., 183-194, Plenum Press (1979).

[10] Determination of an optimal mesh for a collocation-projection method for solving two-point boundary value problems (with Manohar Athavale). *J. Approximation Theory*, *28*, 1, 38-48 (1979).

[11] How to smooth curves and surfaces with splines and cross-validation, in "Proceedings of the 24th Conference on the Design of Experiments." U.S. Army Research Office, Report 79-2, 167-192 (1979).

[12] Convergence rates of "Thin Plate" smoothing splines when the data are noisy, in "Smoothing Techniques for Curve Estimation." T. Gasser and M. Rosenblatt, eds., Lecture Notes in Mathematics No. 757, 232-246, Springer-Verlag (1979).

[13] Some new mathematical methods for variational objective analysis using splines and cross-validation (with J. Wendelberger), *Monthly Weather Review 108*, 36-27 (1980).

[14] Spline bases, regularization, and generalized cross-validation for solving appoximation problems with large quantities of noisy data, in "Approximation Theory III." W. Cheney, ed., 905-912, Academic Press (1980).

[15] Data-based optimal smoothing of orthogonal series density estimates. *Ann. Statist.*, 9, 1, 146-156 (1981).

[16] A new approach to the numerical inversion of the Radon transform with discrete, noisy data, in *Mathematical Aspects of Computerized Tomography.* G.T. Herman and F. Natterer, eds., 198-203, Lecture notes in Medical Informatics, Springer-Verlag, 1981.

[17] Design problems for optimal surface interpolation (with C. Micchelli), in *Approximation Theory and Applications.* Z. Ziegler, ed., 329-348, Academic Press.

[18] On the estimation of functions of several variables from aggregated data (with N. Dyn). *SIAM J. Math. Anal.*, 13, 1, 134-152 (1982).

[19] Numerical experiments with the thin plate histospline. *Commun. Statist. Theor. Meth.* A10(24), 2475-2514 (1981).

[20] Spline interpolation and smoothing on the sphere. *SIAM J. Scientific and Statistical Computing,* 2, 1, 5-16 (1981).

[21] Ill posed problems: numerical and statistical methods for mildly, moderately, and severely ill posed problems with noisy data, University of Wisconsin-Madison Department of Statistics, Technical Report No. 595, February 1980, to appear in "Proceedings of the International Conference on Ill Posed Problems." M.Z. Nashed, ed.

[22] Vector splines on the sphere, with application to the estimation of vorticity and divergence from discrete, noisy data. In "Multivariate Approximation Theory." Vol. 2, W. Schempp and K. Zeller, eds., 407-429, Birkhauser Verlag (1982).

[23] Variational methods in simultaneous optimum interpolation and initialization. In "The Interaction Between Objective Analysis and Initialization," D. Williamson, ed., Pub. No. 127, Atmospheric Analysis and Prediction Division, National Center for Atmospheric Research, Boulder, CO, 178-185 (1982).

[24] Constrained regularization for ill posed linear operator equations, with applications in meteorology and medicine, in "Statistical Decision Theory and Related Topics III." Vol. 2, S.S. Gupta and J.O. Berger, eds., 383-418, Academic Press (1982).

[25] Computational methods for generalized cross-validation with large data sets (with D.M. Bates). In "Treatment of Integral Equations by Numerical Methods." C.T.H. Baker and G.F. Miller, eds., 283-296, Academic Press (1982).

[26] Bayesian confidence intervals for the cross validated smoothing spline. University of Wisconsin-Madison Statistics Department, Technical Report No. 645, July 1981, to appear in *J. Roy. Stat. Soc. B*.

[27] Multivariate thin plate spline estimates for the posterior probabilities in the classification problem (with M. Villalobos). University of Wisconsin-Madison Statistics Department. Technical Report No. 686, July 1982, to appear, *Commun. Statist. Theor. Meth.*, A11, 13 (1984).

[28] Nychka, D. (1982), Solving integral equations with noisy data: an application of smoothing splines in pathology. In "Proceedings of Interface XIV," Karl Heiner, ed., Springer-Verlag.

[29] O'Sullivan, F., Remote sensing of temperature profiles in the atmosphere, to appear in "Proceedings of a Workshop on Density Estimation and Function Smoothing." Held at Texas A&M, March 11-13, 1982, E. Parzen, ed.

[30] Crump, J.G., and Seinfeld, J.H. (1982), A new algorithm for inversion of aerosol size distribution data. *Aerosol Science and Technology, 1:* 15-34.

[31] Merz, P. (1980), Determination of adsorption energy distribution by regularization and a characterization of certain adsorption isotherms. *J. Comput. Phys., 38*, 64-85.

Tools for Large Data Set Analysis*

Leo Breiman

Department of Statistics
University of California
Berkeley, CA

Jerome Friedman

Stanford Linear Accelerator Center
Stanford University
Stanford, CA

1 INTRODUCTION

In an ONR sponsored research project we are simultaneously developing philosophy and software tools for large data set analysis. The philosophical issues concern such questions as: What is a large data set? What are the goals in large data set analysis? Given these goals what is an appropriate set of tools? Since the answers to these questions have determined the course of our software development, this paper will first discuss these questions and then describe some of the specific tools under development.

The class of large data sets that we are interested in dealing with are partially defined by these properties.

1. Many variables.
2. Mixed variable types, i.e., ordered, categorical, periodic, etc.
3. Inhomogeneous.
4. Nonstandard.

Property 3 can be interpreted as meaning that the relationships between variables changes over different parts of the data base. A consequence is that the idea of constructing a single parametric model for the data set is not useful. The data is not multivariate normal. It cannot be transformed to multivariate normality. Because of the larger sample size, the lack of homogeneity becomes discernable and dominant.

---

*Supported by ONR Contract No. N00014-82-K-0054

Property 4 uses the word nonstandard for data bases that do not have a cases by variable matrix form. For instance, the files for each case may be of variable dimensionality, the data may have a hierarchical structure, etc.

Large sample size has the interesting implication that classical statistical methods are of dubious value. These were developed to squeeze the maximum possible information out of small data set. But in a large data set, for instance, there is an endless stream of hypotheses involving very minor differences that are "statistically significant at the 5% level." In fact, from the point of view that considers the essence of statistics to consist of fitting models and then drawing conclusions from the models, large data set analysis is not statistics. From a more general point of view that statistics consists in the art and science of gaining information and insights about the state of nature through the gathering and analysis of quantitative data, it is an interesting and important problem.

The goal, in our formulation, is: to explore and quantify the relationship between variables. Given the nature of the data sets of interest to us, some specifications for appropriate tools are:

(A) Uses a minimum of distributional assumptions.
(B) Can cope with nonhomogeneous data sets.
(C) Computationally efficient for data bases of the order of 100 megabytes.

We have been working in three areas:

I. *Preliminary Screening;* cleaning the data base, flagging outliers and anomalies, filling in missing data.
II. *Microanalysis;* looking at pieces of the data base, and contrasting different pieces.
III. *Macroanalysis;* trying to assess overall relationships, i.e., putting the pieces together.

Large data base are usually marked by numerous errors, missing values, and outliers. Often, adequate data quality auditing procedures have not been imposed. The result is that a major effort is necessary to get the data base into shape for any analyses.

We have coded a procedure called DATA-GIN which automatically flags outliers and imputes missing values in a standard structure data base. This program has been interfaced into the "S" statistical package developed at Bell Telephone Labs. The other software we are developing will also be interfaced into "S". The programs will also be made available in the form of stand-alone FORTRAN subroutines.

2 DESCRIPTION OF DATA-GIN

DATA-GIN and many of the other programs we are developing make repeated calls on CART. CART is a tree-structured classification and regression method that makes a minimum of distribution assumptions. It works through a recursive binary splitting of the data, at each node, finding that split among many candidate splits that maximizes a goodness-of-split criteria. A brief description of CART can be found in Breiman [1]. A monograph on CART by Breiman, Friedman, Olshen and Stone is in its 3rd (and final) draft.

Since CART functions through successive splits of the data base, it is well suited to dealing with nonhomogeneous pieces. Another CART characteristic is that by successive use of subsamples, the tree can be built utilizing a successively larger proportion of the data with only a few passes through the data base. It works equally well on ordered, categorical and mixed data types.

If any variable, say x_j, is designated as the response variable, CART builds a tree for predicting x_j. If x_j is ordered, the tree gives a predicted value

$$\hat{x}_f = f_j(x_1, \ldots, x_{j-1}, x_{j+1}, \ldots x_M).$$

If some of the variables, $x_1, \ldots, x_{j-1}, x_{j+1}, x_N$ are missing in a particular case, CART gives a predicted value of x_j based only on the nonmissing values. If x_j is categorical, say, $x_j \in \{1, \ldots, I\}$, then CART gives predicted values $\hat{p}_i$ for the probability that x_j falls into the ith category.

DATA-GIN builds a prediction tree $\hat{x}_j = f_j(x_1, \ldots, x_{j-1}, x_{j+1}, \ldots, x_M)$ for each variable x_j, $j = 1, \ldots, N$. An ordered variable x_j in a case is flagged if $|x_j - \hat{x}_j|$ is too large. For categorical variables x_j is flagged if $x_j = i$ and $\hat{p}_i$ is too small.

If x_j is missing, it is filled in by $\hat{x}_j$, if ordered. If categorical, it is filled in by that category i' such that $\hat{p}_{i'} = \max_i \hat{p}_i$. The output flags all outliers, missing values, and gives overall summaries. The interactive version allows inspection of individual cases and comparison to predicted values.

Once DATA-GIN is built for a particular data base, then all future entries into the data base can be run through and checked. Because of the binary tree structure, the storage requirements for the complete DATA-GIN are modest, and new cases can be run through quickly. DATA-GIN has a characteristic that is typical of the tools we are interested in developing. It lacks the elegance and precision of maximum likelihood imputation of missing data assuming some model—say, multivariate normal. But it is more generally applicable, both in terms of data distribution and variable types. It will give sensible results for many large data bases having standard structure.

3 MICROANALYSIS

In a large nonhomogeneous data base, part of the analytic process is looking at interesting pieces of the data base, contrasting selected pieces between themselves and perhaps with the overall characteristics of the data base. To do this, an elementary data base management system is necessary. James Reeds is building ours and (a poor pun) is calling it HORSE.

HORSE pulls out the subset of the data base corresponding to any Boolean combination of keys at a prescribed sampling rate. For instance, the command

$$5 \leq x_1 < 8 \text{ and } X_3 = \text{red or } X_7 > 10, \ 1/10$$

creates a file consisting of every tenth case from the subset of the data base whose variables satisfy the stated Boolean conditions. Three different sets of conditions will create three different files. The command ALL 1/1000 creates a file of every 1000th case from the entire data base. Once a file has been created then all the standard statistical descriptor programs can be run on it. Since HORSE is interfaced to "S," in particular, any of the "S" programs can be run on the file. However, HORSE was originally conceived as the front end of a program called DESTREE.

DESTREE is one of the important tools in analyzing the relationship between variables is the capability of contrasting pieces of the data set with one or more selected variables restricted to a given range. For instance, typical questions might be:

> In what characteristics do white males over 21 in Mobile, Alabama differ most markedly from white males over 21 in New York City and if we add the group of white males over 21 in Alameda County, California, what characteristics most separate the three groups? If we add a subsample of the entire white male over 21 population, what characteristics of the 4 groups differ most?

The standard method of answering such questions has been to make numbers of tables, i.e. 2×2, $2 \times n$, etc. and inspect differences of means. For high dimensional data, this is an ineffective and tedious procedure.

The approach used in DESTREE is to contrast J pieces of the data base by considering the problem as a J class classification problem and invoking CART. CART goes to work and finds the best split on the variable that best discriminates between the groups. For instance, in the above question, the best split among the two original groups of white males over 21 may be on income, say at the point $12,000/year. CART performs this split on both groups and then proceeds to see which split best discriminates between the white male over 21 less than $12,000/year income subsets in the two groups and similarly for the over $12,000/year income subsets. The analyst then has available those variables and split points that most discriminate the groups, together with a measure of how well the groups can be discriminated between. In the 2nd phase of developing DESTREE, an interactive graphical version is planned that will permit operator intervention in the selection of split points and variables being split on, and display simultaneous histograms of a given variable in the different classes. We consider DESTREE as a general purpose descriptive tool in the microanalysis stage. It offers flexible descriptive capabilities considerably more powerful than currently available.

4 MACROANALYSIS

In this phase of the analysis the goal is to put it all together—that is, to get an overall view of how the variables in the data relate to each other.

The obvious problem is that with, say, a 100 megabyte data base, extensive exploration and analysis is difficult and expensive. The preponderance of the data is highly redundant. The difficult problem is how to remove redundancy and work with a much smaller data set that carries most of the information present in the original data set.

Some suggestions that are current are:

(1) **Subsample:** that is, take every 100th or 1000th case, or whatever small fraction is needed to reduce sample size to manageable proportions. The main difficulty we see in this procedure is that it may reduce information without reducing redundancy.

For instance, a large group of cases in the data base may be similar to each other, and constitute a high density region. Subsampling will produce a large number of these similar cases. A few cases that are unique in some ways have a smaller change of being included.

(2) **Cluster:** A "flow-through" clustering algorithm is used to create clusters. Then each cluster is represented by its center, appropriately weighted.

There are some notable difficulties. First, in most data bases of our knowledge the cases are not clustered. They are spread out like a cloud, thicker in some places and thinner in others. In such situations, clustering is an artificial device whose performance is unknown. Second, clustering assumes a distance between points. The definition of this distance is almost always ad hoc, and different definitions can lead to considerably different clusters. Furthermore, the center of the cluster is representative of the cluster only in terms of the given distance.

(3) **Key cut and describe:** The idea is to cut the data base into disjoint pieces by conditions on certain preselected variables, i.e., the pieces could be $a_1 \leq X_1 < b_1$, $a_3 \leq X_3 \leq b_3$, $a_{10} \leq X_{10} \leq b_{10}$ etc. With each piece store summary statistics such as size, means, covariance matrix. Then use only these summary statistics in the macroanalysis.

There are obvious drawbacks. Since there is no guarantee that the pieces are homogeneous, then there is no implied guarantee that the summary statistics give an adequate description of it. Further, there is no guarantee that the ensemble of summary statistics can give a good description of the overall data base structure. Still, this approach does directly tackle the problem of redundancy, and could be a start in a promising direction.

Our preliminary thinking on this problem has led to the design of a program called **DELEGATE SAMPLING**. The question it addresses is how to draw a subsample from the data base that reduces redundancy and preserves information. The idea is that such a subsample, appropriately weighted, would be used in the macroanalysis.

If one knows the density $f(\underline{x})$ of the data distribution, then one method of reducing redundancy is *inverse density sampling*. Given a data point $\underline{x}_k$, include it in the subsample with probability $c/f(\underline{x}_k)$, and if it is included, weight it by $f(\underline{x}_k)/c$. Inverse density sampling takes relatively fewer points from high density areas, and samples more completely from low density areas. A simple computation gives the result that an inverse density subsample is spread approximately uniformly over the data base. Since the density is unknown, an inverse density sampling scheme necessarily involves density estimation. However, the usual density estimates, i.e., kernel nearest neighbor, etc. are computationally unfeasible for a large data set.

DELEGATE SAMPLING uses an adaptive tree-structured histogram type of density estimate. Suppose all data in a node N of the tree is confined to the ranges $a_j \leq x_j \leq b_j$, $j = 1, \ldots, M$. For the jth variable, compute the empirical distribution function $F_j(x)$ using only the data in the node. Let $U_j(x) = (x - a_j)/b_j - a_j$ be the uniform distribution over the range. Define

$$\Delta_j = \sup_{a_j \leq x \leq b_j} |F_j(x) - U_j(x)|.$$

Take the best splitting variable to be the ith if $\Delta_i = \max_j \Delta_j$ and suppose that

$$\Delta_i = F_j(x_i^*) - U_j(x_i^*)|.$$

Then split the node into the two pieces N_L, N_R

$$N_L = \{\underline{x} \in N; x_i \leq x_i^*\}, N_R = \{\underline{x} \in N; x_i > x_i^*\}.$$

This process is then repeated on the descendant nodes. It is initiated on a $1/n$ subsample $\{\underline{x}'\}$ of the data with a_j the minimum (maximum) value of the jth variable in the full data set. At a given point in the tree construction, when the sample size/node has dropped below a given threshold, $1/n'$ subsample is selected, $n' << n$, and dropped through the partially constructed tree. Then the node splitting is continued using the augmented sample in each current node. The process is continued until $\max_j \Delta_j$ falls below a threshold value. Then the tree is trimmed upward to reduce ineffective splits.

The density in each terminal node is estimated as the ratio of the proportion of the total sample used in construction that is in the node to the original volume occupied by the sample. This procedure produces rectangular regions in data space such that all the univariate marginals are nearly uniform over the region. The assumption is that in consequence, the multivariate distribution over the rectangle will be nearly uniform. After the tree density estimator is constructed, the entire data set is run through it, and delegates randomly selected from each node, the number being inversely proportional to the node density and each one weighted directly proportional to the node density. A minor modification of the program allows categorical variables to be treated along with the ordered variables.

Clearly, the above method will not produce refined density estimates. It is not meant to. The purpose is to produce delegate points, appropriately weighted, that can be used in macroanalysis of the data base. In this context, the difference between a refined density estimation method and the method outlined above are probably second order.

5 FOR FURTHER CONSIDERATION

'An important problem, currently lacking for an effective soluItion, is how to cut the data base into "homogeneous" pieces. A good solution to this problem could significantly assist in the analysis as follows:

(i) Since each piece is "homogeneous" the analysis of each such piece is relatively straightforward. The difficulties of changes in the relationships between variables does not have to be dealt with.

(ii) Summarizing the statistical properties of each homogeneous piece would largely reduce redundancy, while preserving the relevant statistical information.

REFERENCE

[1] Breiman, L. "Growing trees to analyze high dimensional data" Proceedings, 3rd Annual Department of Energy Statistical Symposium, Albuquerque, New Mexico, November 1978.

Data Analysis and Modeling of Arctic Sea Ice Subsurface Roughness: A Summary

Donald P. Gaver and Patricia Jacobs

Department of Operations Research
Naval Postgraduate School
Monterey, CA

1 INTRODUCTION

The spatial pattern of the sea ice cover in the Arctic has been of considerable scientific interest to geophysicists and oceanographers for some time. Its presence importantly affects the environment for naval and other military operations, and for oil and mineral exploration. In particular, naval submarine operations are influenced by the existence of deep downward projections ("ice keels") from the surface canopy, by acoustic wave reflections from the underside of that canopy, and by the apparently random incidence of essentially open regions in the ice pack ("leads" or "polynyas") that permit access to the surface from below.

In this paper statistical methods are used to characterize and summarize features of the Arctic ice pack related to those mentioned above. The analysis is based on a particular set of data furnished by Dr. Peter Wadhams of Scott Polar Research Institute, Cambridge, England, to whom we are grateful. A previous analysis of these data has been reported by Wadhams and Horne [1980], hereafter abbreviated *WH*. While the approaches of earlier investigators have lead to simple one-parameter exponential distributions as summaries of data describing (a) spatial intervals between keel occurrences, and also (b) keel depths, our definitions and data analysis suggest that both keel spacings and keel depths are longer-tailed than the exponential. We further suggest simple parametric forms to summarize the observed statistical behavior.

2 THE DATA

The data we analyze were obtained by upward-looking sonar aboard the submarine U.S.S. GURNARD during the period April 7-10, 1976, from beneath the Beaufort Sea ice canopy. The route followed by the GURNARD was from a point north of Barter Island (just over 70° N) to 75-76° N, thence southeasterly to a point 72-73° N, and finally westerly to a point northeast of Pt. Barrow. For a detailed map see *WH*. The data—ice drafts, measured from below the ice to the surface—were taken over a 1400 km transect length. Data tapes were initially cleaned and processed at the Arctic Submarine Laboratory, Naval Undersea Center, San Diego; they were later further processed at Scott Polar Research Institute, and observations which were taken at intervals of 1.3-1.5 m were referred by interpolation to a nominal 1.0 m spacing. Furthermore, the data file was split into sections, each of which make up about 50 km of data. There were 27 such sections, with a gap appearing between two of them. More detail is available in *WH*.

Certainly the data set referred to can provide considerable information concerning the underside of Arctic ice. However, there are limitations in the inferences that even a sophisticated analysis of these particular measurements can produce. For instance, the data were obtained during a relatively short period of time in one year, so there is no opportunity to assess month-to-month or season-to-season variability. In order to obtain more information, more data must be subjected to analysis. Our purpose here is to suggest methods of analysis that may be useful when such data become available.

3 DATA ON KEEL SPACINGS AND KEEL DEPTHS: DEFINITIONS

The raw data on ice drafts were transformed into data on keel spacings and magnitudes by the simple expedient of constructing an imaginary line, L, at a constant depth d(feet) below sea surface, and then measuring distances (spacings) between successive up- and down-crossings of L, denoted generically by x, and the maximum depth (keel depth, relative to d) achieved between a down-crossing and the first subsequent up-crossings, denoted by y. Figure 1 should clarify this definition, which differs somewhat from that of *WH*: it permits the occurrence of more small spacings than does theirs.

Data on spacings and keels and were initially obtained for three levels: d = 30 (feet), 40, 50. These depths are apparently of interest from a submarine operational view point, but are too deep to be of great interest to acousticians. Further analysis of crossings at smaller depths is in progress.

4 EXPLORATORY ANALYSIS OF SPACINGS

A data segment running from (lat. 71.140, long. 144.225) to (lat. 74.328, long. 144.378) was selected, and graphical displays of spacings at d = 30 ft. were constructed. There were

    (a) *Serial plot:* x_i (spacing i, in order) vs i. This suggested fewer *long* spacings for i = 1, 2, ... 200, i.e., some nonstationary behaviour.
    (b) *Serial histograms:* Successive groups of spacings (n = 73, in order)

ARCTIC SEA-ICE DATA APPEARANCE

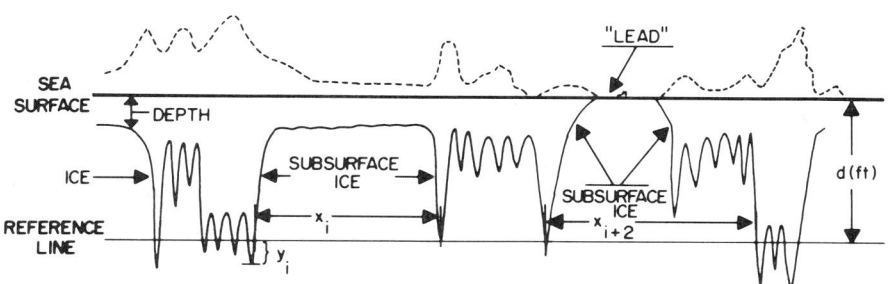

x_i = DISTANCE BETWEEN UP-CROSSING AND SUBSEQUENT DOWN CROSSING OF REFERENCE LEVEL (PARALLEL TO SURFACE) AT DEPTH d (e.g. 30', 40', 50').

y_i = KEEL DEPTH, WITH RESPECT TO REFERENCE LEVEL, DEPTH d.

Figure 1 Arctic sea-ice date appearance

were histogrammed; exponential-*like* positive skewness revealed itself.

(c) *Serial box plots:* supplementary to (b); long spacings occurred late along the track.

(d) *Histograms of log (spacings):* A symmetrizing transformation, see Tukey [1977], McNeil [1977] reveals *two* roughly symmetrical "bumps" or regions of data concentration; (see Fig. 2). The lower bump is associated with crossings occurring across the bottom of wide keels; the upper bump represents distinct, rather large keel spacings of operational significance (e.g., submarine operations).

The data associated with the upper bump was selected for more detailed statistical analysis. Specifically, the first 200 observations of the series were set aside. The median of the remaining data was calculated, and those data points less than the median were also set aside. The median was subtracted from the remaining spacings, and the remainders are then analyzed in what follows; these remainders are henceforth referred to as the upper half of the data. They are the meaningfully large spacings between subsequent keels.

5 TOWARDS A PARAMETRIC DESCRIPTION OF SPACING

An exponential distribution summarization of spacings data has been suggested by Hibler [1972], and adopted by *WH* also for purposes of discussion.

Model 1: $X \sim$ Exponential. Let X be a random variable representing a typical spacing. Then if x is any positive number, the simple exponential model is that ($P\{\ \}$ denotes probability of the enclosed event)

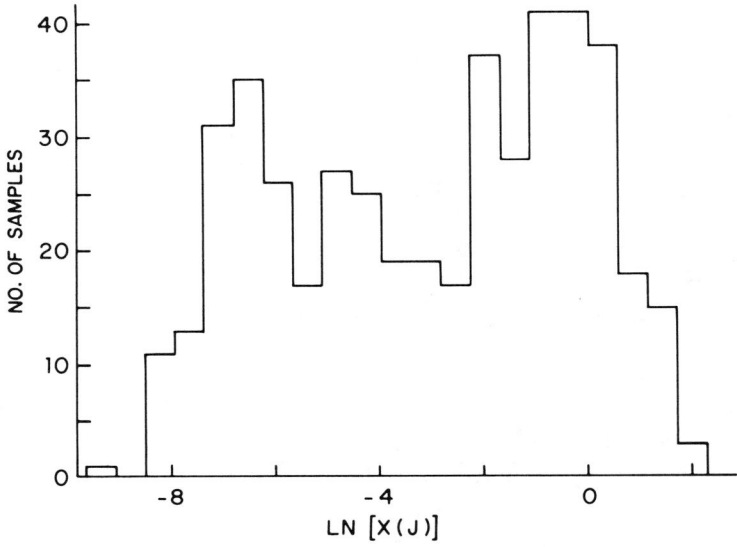

Figure 2 Histogram of log (spacings)

$$P\{X > x\} = e^{-\mu x}, \tag{5.1}$$

so the probability density of spacings is

$$f_x(x; \mu) = e^{-\mu x}\mu \quad x \geq 0 \tag{5.2}$$

and μ is the rate of occurrence of spacings per unit distance; equivalently,

$$E[X] = \frac{1}{\mu}. \tag{5.3}$$

This says that the population average is $1/\mu$; the maximum likelihood estimate of μ in model (5.1) is simply $\hat{\mu} = (\bar{x})^{-1}$. A time series (lagged correlation and spectrum) analysis of successive spacings at $d = 30$ ft shows only very weak dependence between spacings; such dependence will be ignored.

An informal but informative check for the suitability of the exponential model is to examine a plot of the order statistics $x_{(j)}$ of the (upper half of the) data to the corresponding expected exponential order statistics: a straight line relationship signifies that the simple exponential fits well. Figures 3 and 4 suggest a systematic upward bow in the data, indicating a systematically longer-than-exponential right tail. Such an effect is also present for $d = 40$ ft, and $d = 50$ ft, but the curvature becomes progressively less noticeable as the depth, d, increases. Numerical summaries (moments) also suggest systematic longer-than-exponential tails (See Table 1.)

The occurrence of a near-exponential distribution of spacings between deep keels is perhaps not surprising, in that ice structure formation appears to have a random nature without much long-term order. It is the near-independence of ice structure sizes in neighboring parts of the pack that is probably responsible for the near-exponentiality of the spacing observed. See

Table 1 Moment and quantile summaries of spacings in excess of median (70 meters) at depth $d = 30$ ft

Mean: $\bar{x} = 0.929$ (km)

$s^2 \equiv \hat{v} \text{ ar } [X] = 1.485$, $s = 1.218$

Coeff. of variation $= \dfrac{s}{\bar{x}} = 1.311$ (1.0)*

Skewness $= \hat{\gamma}_1 = 2.678$ (2)*

Kurtosis $= \hat{\gamma}_2 = 9.853$ (6)*

Lower Quartile $\equiv \underline{Q} = 0.155$

Median $= 0.487$

Upper Quartile $\equiv \overline{Q} = 1.279$

*Numbers in parentheses are characteristic of an Exponential model.

Gaver and Jacobs (1981) for situations in which the exponential arises as a limiting distribution.

6 MODIFICATION OF THE EXPONENTIAL MODEL: THE SCULPTURED EXPONENTIAL

As an alternative to the exponential fit (the upper half of) the $d = 30$ ft spacings data were next represented by a modified or *sculptured* exponential.

Model 2: Linearly Sculptured Exponential. Let

$$X = AZ(1 + CZ), \qquad (6.1)$$

where X represents a spacing, Z is a unit exponential *basic* r.v. and A and C are constants, A being a scale and C reflecting departure from exponentiality. The term $(1 + CZ)$ "sculptures" Z by leaving small values of Z virtually unchanged $(1 + CZ \approx 1$ for Z small), but expanding large values $(1 + CZ \approx CZ$ for Z large). We can represent the order statistics, $X_{(j)}$, as follows:

$$X_{(j)} = AZ_{(j)}(1 + CZ_{(j)}), \qquad (6.2)$$

that is, the size-ordered X-values, $X_{(1)} < X_{(2)} < \ldots < X_{(n)}$ are easily represented in terms of those for Z, $Z_{(j)}$. Note that quantiles or percent points, $x(\alpha)$, of X can be written in terms of those of Z, $z(\alpha)$:

$$x(\alpha) = Az(\alpha)(1 + Cz(\alpha)), \quad (0 \leqslant \alpha \leqslant 1). \qquad (6.3)$$

Thus sculpturing gives a simple representation of the inverse distribution of X in terms of that of Z. Furthermore, explicit formulas for the distribution and density functions of Model 2, (6.1), can be derived.

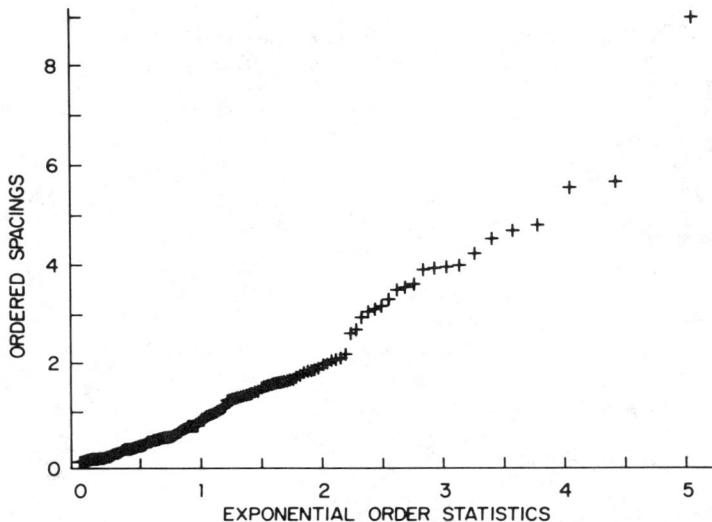

Figure 3 Exponential model diagnostic plot

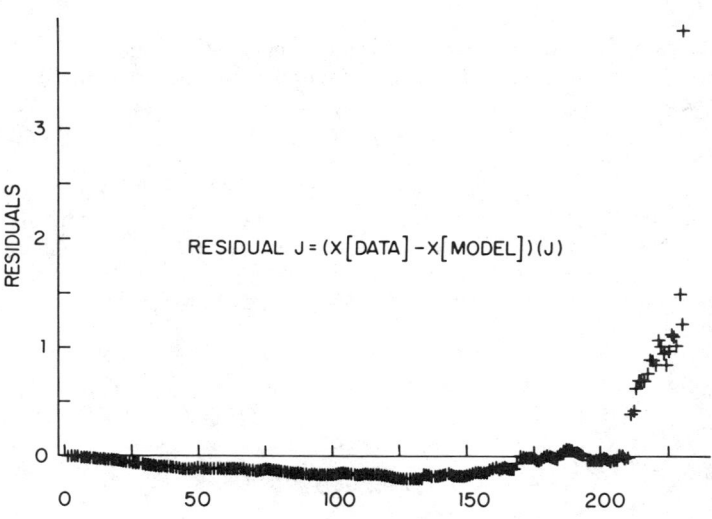

Figure 4 Exponential model diagnostic plot (spacing)

Arctic Sea Ice

$$F_X(x; A, C) = P\{X \leq x\} = 1 - \exp\left[-\frac{2x}{A + \sqrt{A^2 + 4ACx}}\right] \quad (6.4)$$

is the distribution function, and

$$f_X(x; A, C) = \exp\left[-\frac{2x}{A + \sqrt{A^2 + 4ACx}}\right] \cdot \frac{1}{\sqrt{A^2 + 4ACx}} \quad (6.5)$$

is the density.

Model 3: Exponentially Sculptured Exponential. A further alternative to the simple exponential (5.1) is of the form

$$X = AZe^{CZ}, A, C \geq 0. \quad (6.6)$$

The sculpturing term e^{CZ} again leaves small values of Z (exponential) nearly unchanged, but considerably extends large Z values. Once again by monotonicity we have

$$X_{(j)} = AZ_{(j)}e^{CZ_{(j)}} \quad (j = 1, 2, \ldots, n) \quad (6.7)$$

and

$$x(\alpha) = Az(\alpha)e^{Cz(\alpha)}, \quad (0 \leq \alpha \leq 1) \quad (6.8)$$

for order statistics and quantiles of X from (6.6).

Results are reported in the following tables and figures. For more detail see Gaver and Jacobs [1982].

Expressions for the moments of this model come by differentiating the moment-generating function of Z. Details are omitted.

7 FITTING THE SCULPTURED EXPONENTIAL MODEL FOR SPACING

A number of ways of fitting models or representations such as (6.1) and (6.6) suggest themselves; among these are the following.

(A) Maximum likelihood: possible for (6.1), using (6.5).
(B) Moment-matching: possible for (6.1), and also (6.6).
(C) Quartile matching: feasible for any sculptured representation.
(D) Hybrid methods: e.g., constrained likelihood fit by requiring $E[X] = \bar{x}$ and allowing C in (6.1) to be determined by maximum likelihood.
(E) Generalized nonlinear least-squares, robustified if necessary: it is proposed to regress $x_{(j)}$ on $Az(j)[1 + Cz(j)]$ where e.g., $z(j) = \ln(1 - j/(n+1))$, the approximate expected value of the basic r.v. Z, and an appropriate covariance matrix is utilized.

We report the results of applying several of these methods to fit Models 2 and 3 to spacings (at $d = 30$ ft) data; the results are then diagnostically examined.

Maximum Likelihood Fitting: Linear Sculpturing

The results of a maximum likelihood fit of the spacings data by model (6.1) appear in Table 2. Although the agreement of the lower moments is satisfactory for the maximum likelihood fit of (6.1), that of the higher moments (skewness and kurtosis) is less so; this points to the apparent sensitivity of the m.l.e. to extreme values, assigning an unreasonably high C ("correction") value. Nevertheless, a Kolmogorov-Smirnov test of goodness of fit yields a value of 0.65 for the fitted sculptured model, while a Kolmogorov-Smirnov value of 1.73 is found for a simple exponential fit; the sculptured model is seen by this test as providing a substantially improved fit. Furthermore, a diagnostic plot of observed (x_j) vs predicted $(\hat{x}_j(\hat{A}, \hat{C}))$ order statistics provides a more satisfactory straight-line fit than does an exponential model. See Figs. 5 and 6. The sculptured model fitted by maximum likelihood to the particular set of data under discussion appears to predict a somewhat longer far right tail than is consistent with visual examination of the data.

Moment Matching: Linear Sculpturing

An alternative to the maximum likelihood method is that of matching moments. It is customary to equate the two lowest moments, e.g., sample and model mean and variance, when fitting two-parameter models. There is theoretical justification in the present case for matching sample and model

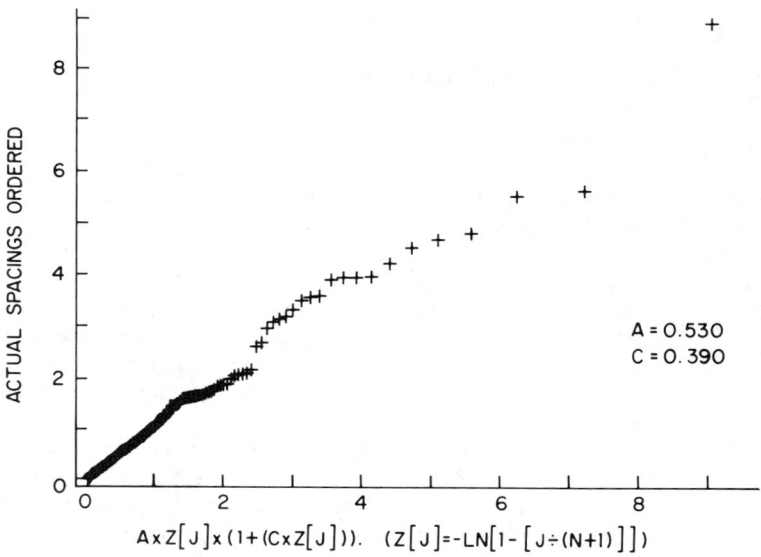

Figure 5 Sculptured model diagnostic plot (maximum likelihood fit)

Arctic Sea Ice

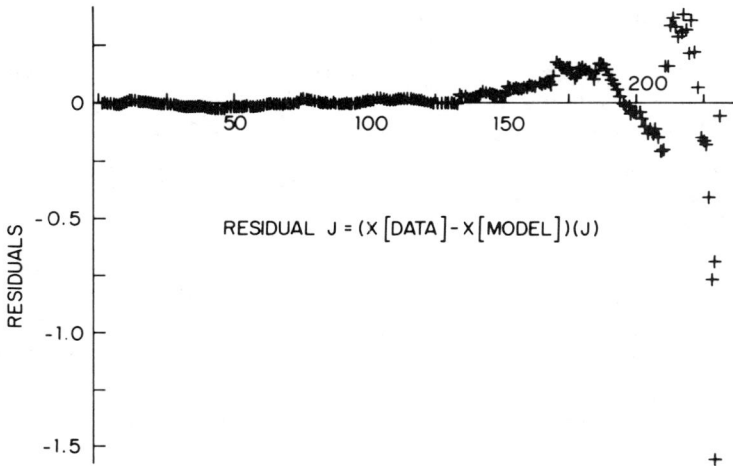

Figure 6 Sculptured model diagnostic plot (MLE for spacings)

means and coefficients of variation (C.V.); see Stephens [1978] and Shapiro and Wilk [1972]. However, mean and skewness matching has also been used.

When applied to spacings at $d = 30$ ft the skewness-matching method produces estimates that differ noticeably from those given by maximum likelihood as seen in Table 2: $\tilde{A}$ is larger, and $\tilde{C}$ is smaller than the corresponding maximum likelihood estimates. In several respects the skewness-match is to be preferred: it agrees best with the data evidence in the far tail, i.e., at the upper quartile and with respect to the kurtosis measures of model and data.

Table 2 Fits of Models to Spacing Data ($n = 231$)

Estimates	Raw Data	Sculptured Exponential (6.1)			Gamma	
		M.L.E.	C.V.	Skew	M.L.E.	Moment
		$\hat{A} = 0.53$	$\tilde{A} = 0.67$	$\tilde{A} = 0.827$	$\hat{A} = 0.70$	$\tilde{A} = 0.582$
		$\hat{C} = 0.39$	$\tilde{C} = 0.193$	$\tilde{C} = 0.062$	$\hat{B} = 1.33$	$\tilde{B} = 1.60$
		S.E. (.09)* (.16)*			S.E. (.15)* (.60)*	
		Corr $(\hat{A}, \hat{C}) = -.84*$			Corr $(\hat{A}, \hat{B}) = -.71*$	
$\hat{E}[X]$	0.929	0.95	0.929	0.929	0.929	0.929
$\hat{C}.V.[X]$	1.31	1.50	1.31	1.12	1.311	1.196
Skew [X]	2.68	4.57	3.71	2.68	2.622	2.390
Kurt [X]	9.85	40.32	26.01	12.54	10.32	8.58
Q	0.155	0.170	0.203	0.242	0.127	0.172
Med.	0.487	0.467	0.53	0.598	0.478	0.541
$\bar{Q}$	1.29	1.13	1.18	1.24	1.256	1.276

*( ) represent large-sample standard errors calculated from the likelihood (Fisher information)

Moment Matching: Exponential Sculpturing

The model (6.6) was also fit to the data by choosing C to match the coefficient of variation and choosing A to match the mean of the data. The values of the estimated parameters and predicted moments are as follows.

Table 3   Coefficient of Variation - Match fit of Spacings by the Sculptured Model (6.6)

$$\tilde{A} = 0.721 \qquad \tilde{C} = 0.119$$

Estimates	Raw Data	Model
$\hat{E}[X]$	0.929	0.929
$\hat{CV}.[X]$	1.31	1.31
$\widehat{\text{Skew}}[X]$	2.68	4.53
$\widehat{\text{Kurt}}[X]$	9.84	53.64
Lower Quartile, $\underline{Q}$	0.155	0.215
Median	0.487	0.543
Upper Quartile, $\overline{Q}$	1.29	1.18

The large model kurtosis is a surprise, but otherwise the fit is good.

8 THE GAMMA MODEL FOR SPACINGS

The gamma distribution is an alternative, and classical, model for spacings that includes the exponential as a special case.

The gamma has been fitted to the 30 ft spacings data by maximum likelihood and also by matching the first two moments. The results are summarized in Table 2.

It turns out that the m.l.e.-fitted gamma and the skewness-fitted sculptured exponential both tend to predict *smaller* extreme right tails than indicated by the raw data, in contrast to the m.l.e.-fitted sculptured exponential.

9 STATISTICAL PROPERTIES OF THE ICE KEELS

Next consider briefly the distributional properties of keels; see Figure 1. Again the data give evidence of a systematically longer-than-exponential right tail.

Two theoretical models were fitted to the raw data: the sculptured exponential, (6.1), and the gamma. Several methods of fitting were employed: lower moment-matching and maximum likelihood. The adequacy of the fits was assessed by numerical and graphical methods; Table 4 summarizes the results.

10 CONCLUSIONS

The larger spacings between ice keels along the Gurnard track, and the keel depths below 30 ft reference line, both exhibit systematic longer-than-

Table 4   Fits of Models to Ice Keel Depths Beyond 30 ft
($n = 365$)

Estimates	Raw Data	Sculptured Exponential			Gamma	
		M.L.E.	C.V.	Skew	M.L.E.	Moment
		$\hat{A} = 4.04$ $\hat{C} = 0.37$ S.E. (.52)* (.12)* (Corr $(\hat{A}, \hat{C})$ = -.83	$\tilde{A} = 5.27$ $\tilde{C} = .155$	$\tilde{A} = 6.65$ $\tilde{C} = .019$	$\hat{A} = .73$ $\hat{B} = 9.43$ S.E. (.83)* (.04)* Corr $(\hat{A}, \hat{B})$ = -.72	$\tilde{A} = .63$ $\tilde{B} = 11$
$\hat{E}[X]$	6.90	7.03	6.90	6.90	6.90	6.90
$\hat{C}.V.[X]$	1.26	1.49	1.26	1.04	1.17	1.26
Skew $[X]$	2.22	4.51	3.47	2.22	2.34	2.53
Kurt $[X]$	5.55	39.32	22.4	7.90	8.20	9.57
$\underline{Q}$	1.20	1.29	1.58	1.92	1.36	1.08
Med.	3.60	3.52	4.04	4.67	4.11	3.77
$\overline{Q}$	8.60	8.47	8.88	9.46	9.48	9.44

K.S. for raw data to be exponential = 2.3.

exponential tails. We introduce a new class of distributions ("sculptured exponentials") for representing such measurements, and compare the result to those obtained using a gamma. The quality of the data summary depends both upon the analytical representation used and the fitting technique employed. Graphical assessments of fits are given.

In other work we utilize our representations to study the magnitudes of extreme (maximal) ice keels.

ACKNOWLEDGMENT

The authors wish to thank L. Uribe for his programming assistance. Figures 2-6 are copies of graphs produced by an experimental APL package GRAFST2 which the Naval Postgraduate School is using under a test agreement with IBM Watson Research Center, Yorktown Heights, NY. We are grateful to Professor P. A. W. Lewis of the Naval Postgraduate School, and to Dr. P. D. Welch and Dr. P. Heidelberger of IBM for making this excellent data-analytic implement available to us. The research of the authors was partially supported by the Office of Naval Research and the Naval Postgraduate School Foundation.

REFERENCES

[1] Cramér, H. and Leadbetter, M.R. (1967). *Stationary and Related Stochastic Processes.* John Wiley and Sons, Inc. New York, NY.

[2] Gaver, D.P. and Acar, M. (1979). Analytical hazard representations for use in reliability, mortality, and simulation studies. *Commun. Statist.-Simula. Computat.*, B 8(2), pp. 91-111.

[3] Gaver, D.P. and Jacobs, P.A. (1981). On combinations of random loads. *S.I.A.M. J. Applied Math.*, Vol. 40, pp. 454-466.

[4] Gaver, D.P. (1982). Sculptured distributions and some applications. Report in preparation.

[5] Gaver, D.P. and Jacobs, P.A. (1982). Data Analysis and Modeling of Arctic Sea Ice Subsurface Roughness. Naval Postgraduate School Tech. Report.

[6] Hibler, W.D. III, Weeks, W.F. and Moch, S.J. (1972). Statistical aspects of sea-ice ridge distributions. *J. of Geophysical Research*, Vol. 77, pp. 5954-5970.

[7] McNeil, D.R. (1977). *Interactive Data Analysis.* Wiley-Interscience, New York, NY.

[8] Shapiro, S.S. and Wilk, M.B. (1972). Analysis of variance test for the exponential distribution (complete samples). *Technometrics*, Vol. 14, pp. 355-370.

[9] Stephens, M.A. (1978). On the W test for exponentiality with origin known. *Technometrics*, Vol. 20, pp. 33-36.

[10] Tukey, J.W. (1977). *Exploratory Data Analysis.* Addison-Wesley Pub. Co., Reading, MA.

[11] Wadhams, P. and Horne, R.J. (1980). An analysis of ice profiles obtained by submarine sonar in the Beaufort Sea. *J. of Glaciology*, Vol. 25, pp. 401-424.

[12] Wilk, M.B. and Gnanadesikan, R. (1968). Probability plotting methods for the analysis of data. *Biometrika*, Vol. 55, pp. 1-17.

Probability Density Functions of Ocean Acoustic Noise Processes*

Frederick W. Machell and Clark S. Penrod

Applied Research Laboratories
University of Texas
Austin, TX

1 INTRODUCTION

A common assumption in the processing of ocean acoustic data is that the first order probability density function governing the underlying noise process is the Gaussian density. One reason for this assumption is that more often than not it leads to analytically tractable results. Frequently, signal processors side-step the need for considering non-Gaussian processes by appealing to the central limit theorem. In this paper, examples of non-Gaussian ocean acoustic processes are presented in addition to some of the effects that filtering has on the densities of these processes. As will be seen, appealing to the central limit theorem can lead to the wrong conclusion in some situations.

The primary tool used for characterization of the noise processes is density estimation. In previous work, the finite sample performance of fixed and variable kernel density estimation techniques was examined in considerable detail.[1] In that study, artificially generated data with Gaussian and Laplacian densities were used to compare the finite sample performance of fixed and variable kernel density estimation techniques. The main conclusion from that work was that the standard fixed kernel method performed best in terms of accuracy per computational unit. All of the density estimates presented here were generated with the fixed kernel algorithm.

---

*Work supported under Contract N00014-81-K-0145, Task SRO 106 with the Office of Naval Research.

The paper opens with two examples of broadband non-Gaussian acoustic processes; density estimates for these processes are displayed in this section. In Section III, the effects of filtering these processes are discussed and the corresponding density estimates are presented.

2 DENSITIES OF BROADBAND ACOUSTIC PROCESSES

Two examples of broadband acoustic processes are presented in this section. The densities in these examples deviate from the Gaussian density in different ways. In the first example, the resulting density is more peaked than Gaussian (leptokurtic), while the density in the second example is flatter than Gaussian (platykurtic). In both examples, small values of skewness (less than 10^{-1} in magnitude) were measured over the observation interval.

Seismic Data

The first data set analyzed was collected in the Gulf of Mexico during a time period when seismic exploration for oil was underway. A representative segment of the time series of the instantaneous amplitude recorded during this time period is displayed in Figure 1(a). A series of strong impulses can be observed in the time series corresponding to the arrival of energy from the explosive source.

This data set fails statistical tests for normality due to large values of kurtosis. A typical value of kurtosis for these data is 5.5 as opposed to the Gaussian value of 3. The test results are confirmed by the density estimate which results from the fixed kernel algorithm. This estimate is displayed in Figure 1(b) with the Gaussian density having the same mean and variance as the sample mean and sample variance of the data. A similar plot is given in Figure 1(c) with the Laplacian density that best fits the data set. The estimate is clearly more peaked than the Gaussian fit, while it agrees with the Laplacian fit quite closely. This result should come as no surprise, as the Laplacian density has a kurtosis value of 6. Moreover, the Laplacian density is commonly used to model impulsive noise[2] similar to this data. As a final note, the Laplacian density belongs to the generalized Gaussian family of densities with an exponential decay rate of 1 (see Ref. 3 for further discussion of the generalized Gaussian density).

Merchant Data

The second example is a data set recorded in an ocean area of high merchant shipping density. A time series plot of pressure spectrum levels in tenth octave bands centered about various frequencies is displayed in Figure 2(a). The time period selected for analysis (highlighted in the figure) was one in which a merchant vessel passed over the recording system.

The data from this time period failed statistical tests for normality due to small values of kurtosis. A typical value of kurtosis for this data is 2.3 indicating a density that is flatter than the Gaussian density. The statistical tests again are corroborated by the density estimates which result from the fixed kernel algorithm. The estimate is seen to be significantly flatter than the Gaussian

Probability Density Functions 213

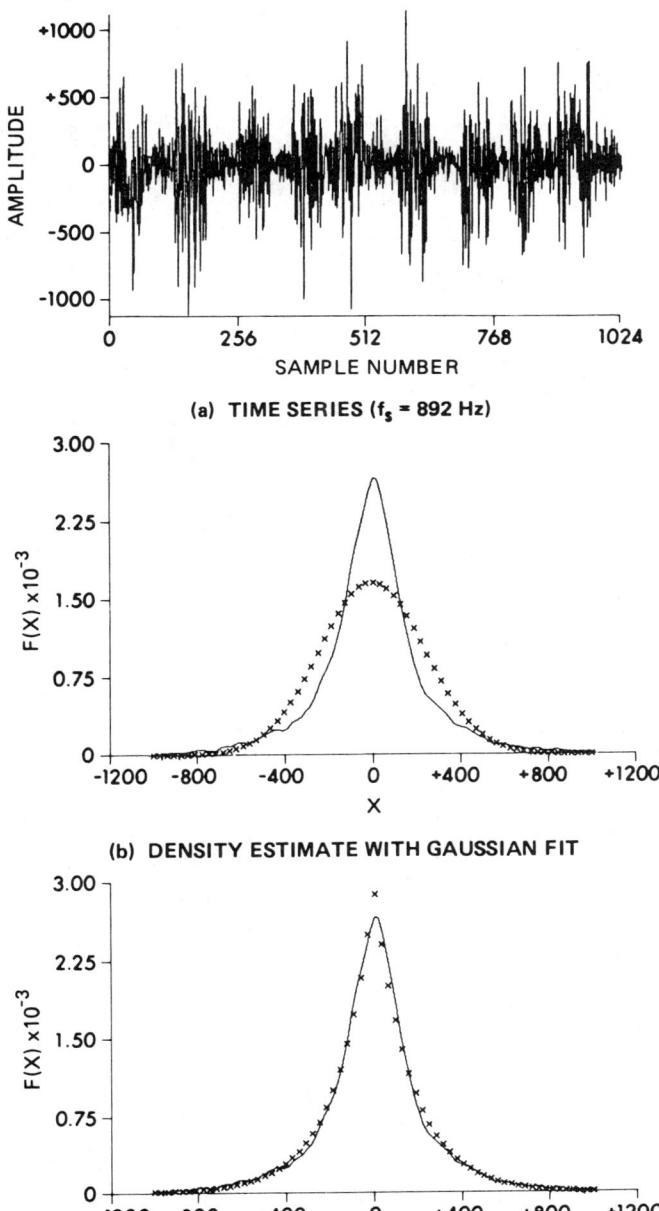

(a) TIME SERIES (f_s = 892 Hz)

(b) DENSITY ESTIMATE WITH GAUSSIAN FIT

(c) DENSITY ESTIMATE WITH LAPLACIAN FIT

Figure 1 Seismic data

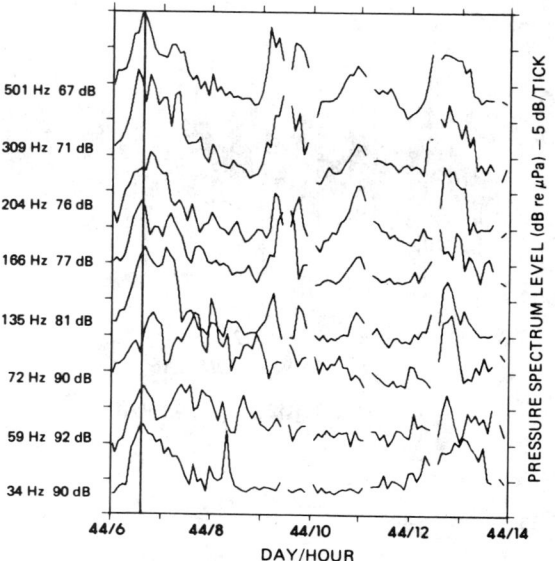

(a) TIME SERIES OF SPECTRA (f_s = 1250 Hz)

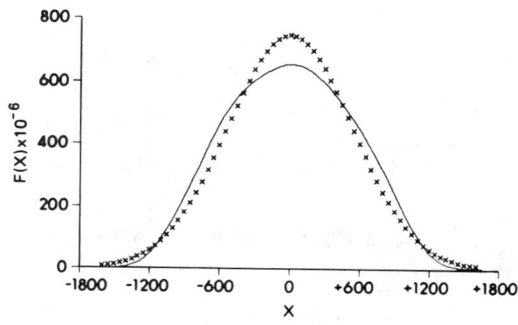

(b) DENSITY ESTIMATE WITH GAUSSIAN FIT

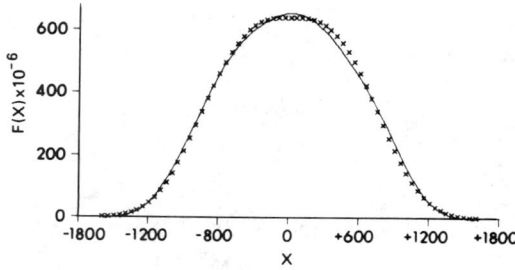

(c) DENSITY ESTIMATE WITH GENERALIZED GAUSSIAN FIT (c = 3)

Figure 2 Merchant data

Probability Density Functions 215

density having the same mean and variance as the data (Figure 2(b)). In fact, the estimate agrees closely with the generalized Gaussian density with an exponential decay rate of 3 (Figure 2(c)). This density has a kurtosis of about 2.4.

3 DENSITIES OF FILTERED DATA

In this section we examine the effects of filtering on the densities of the data described above. The fast Fourier transform (FFT) is used to produce narrowband time series by selecting a specified bin from successive transforms. With an FFT of length 32, 16 complex time series are generated with each time series containing energy concentrated in a small band of frequencies. From these time series, the frequency bands containing the strongest non-Gaussian components can be determined.

Note that the real and imaginary parts of each FFT bin are linear combinations of the data. One might expect that if the transform is long enough and the data are independently generated, a central limit effect should occur; i.e., the filtered output is more nearly Gaussian than the input. A similar phenomenon might be expected if another type of linear filter is used, such as a linear phase, finite impulse response filter. The following example shows that if the independence assumption is violated the opposite of a central limit effect can occur; i.e., the filtered output is actually less Gaussian than the input.

Seismic Data

Consider what happens when the raw seismic data (sampled at the Nyquist rate) is filtered with the FFT. In this case, the filtered time series has larger values of kurtosis than the input time series. A plot of kurtosis versus frequency for the real and imaginary FFT outputs appears in Figure 3(a). The sample autocorrelation function of the data (Figure 3(b)) provides a possible explanation for this phenomenon. The large degree of correlation displayed indicates sample-to-sample dependence in the data, and hence the central limit theorem does not apply.

The kurtosis spectrum in Figure 3(a) provides some additional information about the data. Extreme values of kurtosis are exhibited at the high frequencies with moderately high levels of kurtosis occurring at other frequencies. The reason for this frequency dependence is easily explained. The explosive source produces strong non-Gaussian contributions across a wide frequency range. In contrast, the Gaussian component due to the background ambient noise possesses spectral content which is relatively large at frequencies below 100 Hz and decreases rapidly at higher frequencies, resulting in increased domination by the non-Gaussian component. Another interesting property of the kurtosis spectra is the large difference in kurtosis between the real and imaginary parts of the FFT at the high frequencies. The reason for such a large difference remains unknown.

The corresponding density estimates show the dependence on frequency of the statistics of the process. Figures 3(c)-(f) display density estimates for FFT bins corresponding to frequencies of 56, 139, 251, and 390 Hz. For comparison, the Laplacian density which best fits the filtered data is also shown.

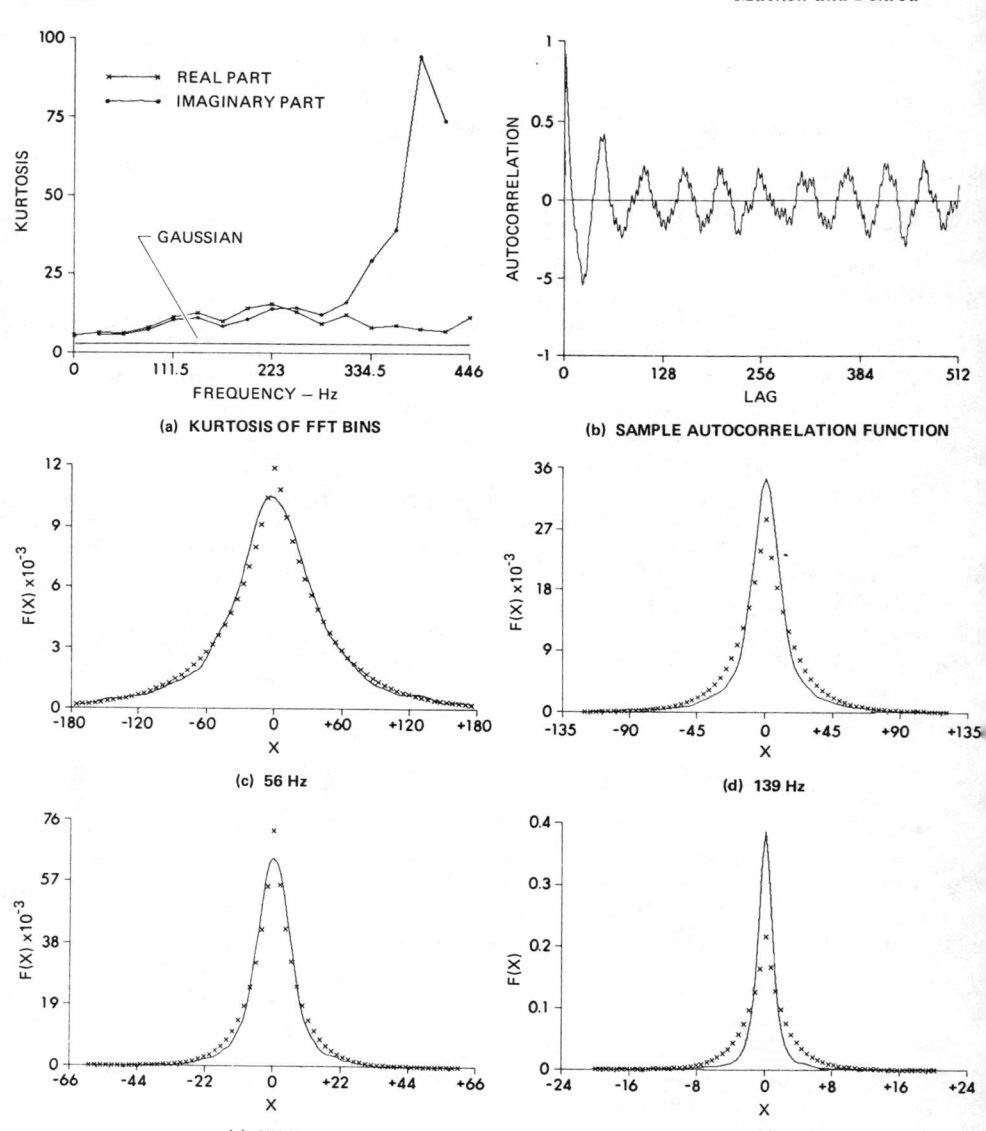

Figure 3 Filtered seismic data sampled at Nyquist rate

From this analysis, one can see that the performance of a system in this kind of noise will largely depend on the frequency band selected for processing the data.

Now consider what happens when the filtering procedure is repeated on the seismic data after being resampled to obtain independence. For convenience, every 128th sample is selected. In this case, the filtered data is closer to

Probability Density Functions

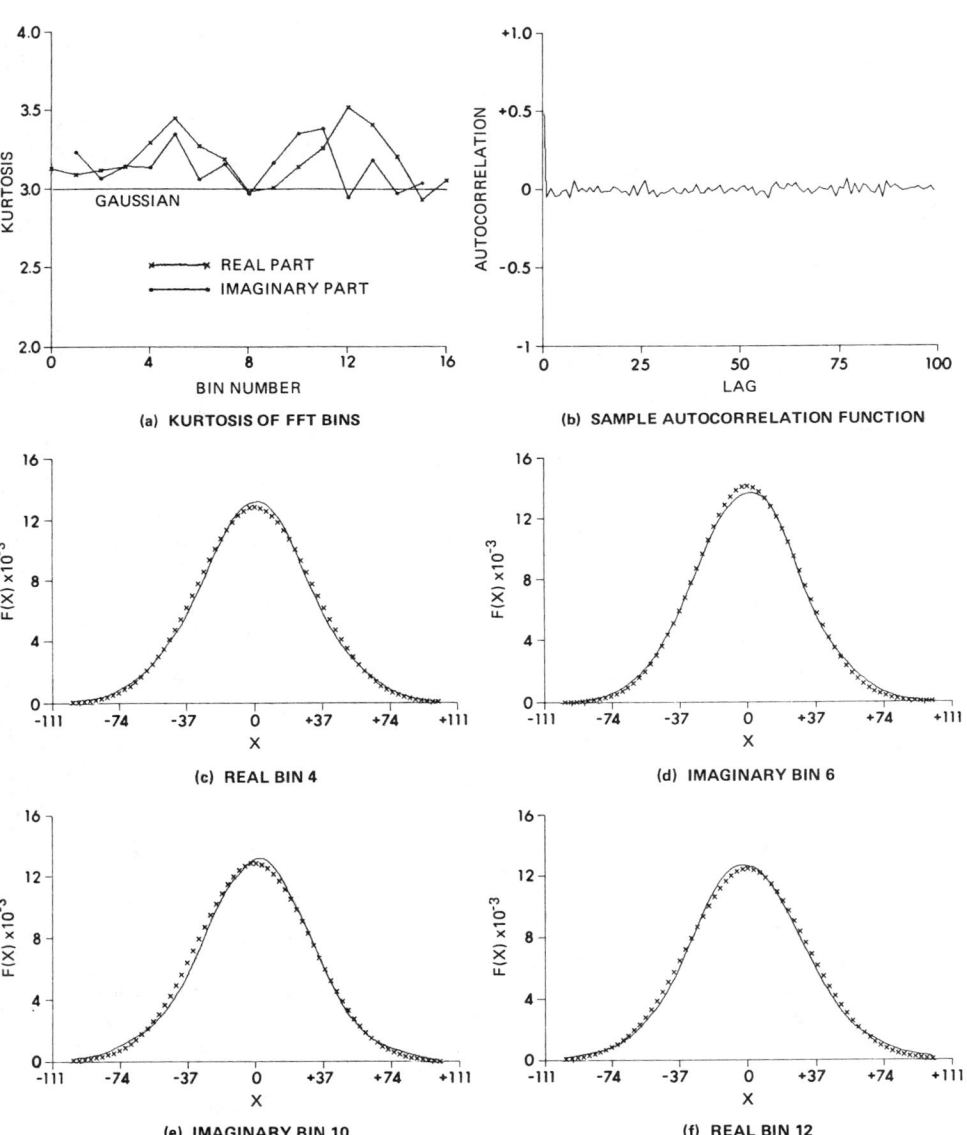

Figure 4 Filtered seismic data Ith by 128

Gaussian than the input as might be expected in consideration of the central limit theorem. The kurtosis values for the real and imaginary FFT bins fall close to the Gaussian value of 3 as depicted in Figure 4(a). Note that the sample autocorrelation function (Figure 4(b)) very nearly resembles a delta function, a necessary condition for independence. Filtering the resampled data produces the anticipated central limit effect.

The density estimates confirm the notion that a central limit effect has occurred. Density estimates corresponding to four of the FFT bins are displayed in Figure 4(c)-(f). The Gaussian density having the same mean and variance as the data is also shown. As can be seen, the estimates show only minor deviations from the Gaussian fit. Furthermore, the statistics of the data now exhibit a much smaller frequency dependence as a result of undersampling the process.

Merchant Data

In this section the effects of filtering on the density of the merchant data are examined. Figure 5(a) displays the values of kurtosis versus frequency for the real and imaginary FFT bins when the merchant data are sampled at the Nyquist rate (1250 Hz). The low frequencies tend to have smaller kurtosis than Gaussian, while the high frequencies tend to have kurtosis values slightly above Gaussian. The low kurtosis values are due to the low frequency tones associated with the merchant vessel nearby mixing with the Gaussian background noise (a sinusoidal distribution has a kurtosis of 1.5). At the high frequencies, the merchant contributes mostly broadband energy and the kurtosis spectra show no significant features at these frequencies.

The sample autocorrelation function in Figure 5(b) reflects the dominance of the low frequency spectrum by the merchant. The observed oscillations correspond to a frequency of about 88 Hz which falls within the bandwidth of bin 2, the bin showing the lowest value of kurtosis. Density estimates for bins centered at frequencies of 78, 234, 352, and 469 Hz are displayed in Figure 5(c)-(f) along with the Gaussian density that best fits the data. The estimate for 78 Hz reflects the low value of kurtosis associated with the 88 Hz tone. Note the decrease in variance with increasing frequency characteristic of ambient noise data.

Consider what happens when the filtering procedure is applied to the merchant data after being resampled to obtain independence. Every 128th sample is selected so that the resulting sampling frequency is about 10 Hz. The plot of kurtosis for the real and imaginary FFT bins in Figure 6(a) shows values that are closely distributed about the Gaussian value of 3. The sample autocorrelation function in Figure 6(b) reveals virtually no correlation in the data, and as with the undersampled seismic data, a central limit effect has occurred.

Density estimates for four of the FFT bins are displayed in Figure 6(c)-(f) along with the Gaussian density with the same mean and variance as the data. The estimates agree very closely with the Gaussian fit reinforcing the idea that the central limit theorem has entered the picture. Another thing to notice about the density estimates is that the variance of the FFT bins does not change much from one bin to the next, a consequence of the whitening effect of undersampling the process.

4 CONCLUSION

In summary, commonplace examples of non-Gaussian processes in the ocean acoustic environment have been exhibited. Density estimates for these processes have been displayed which agree closely with the generalized Gaus-

Probability Density Functions

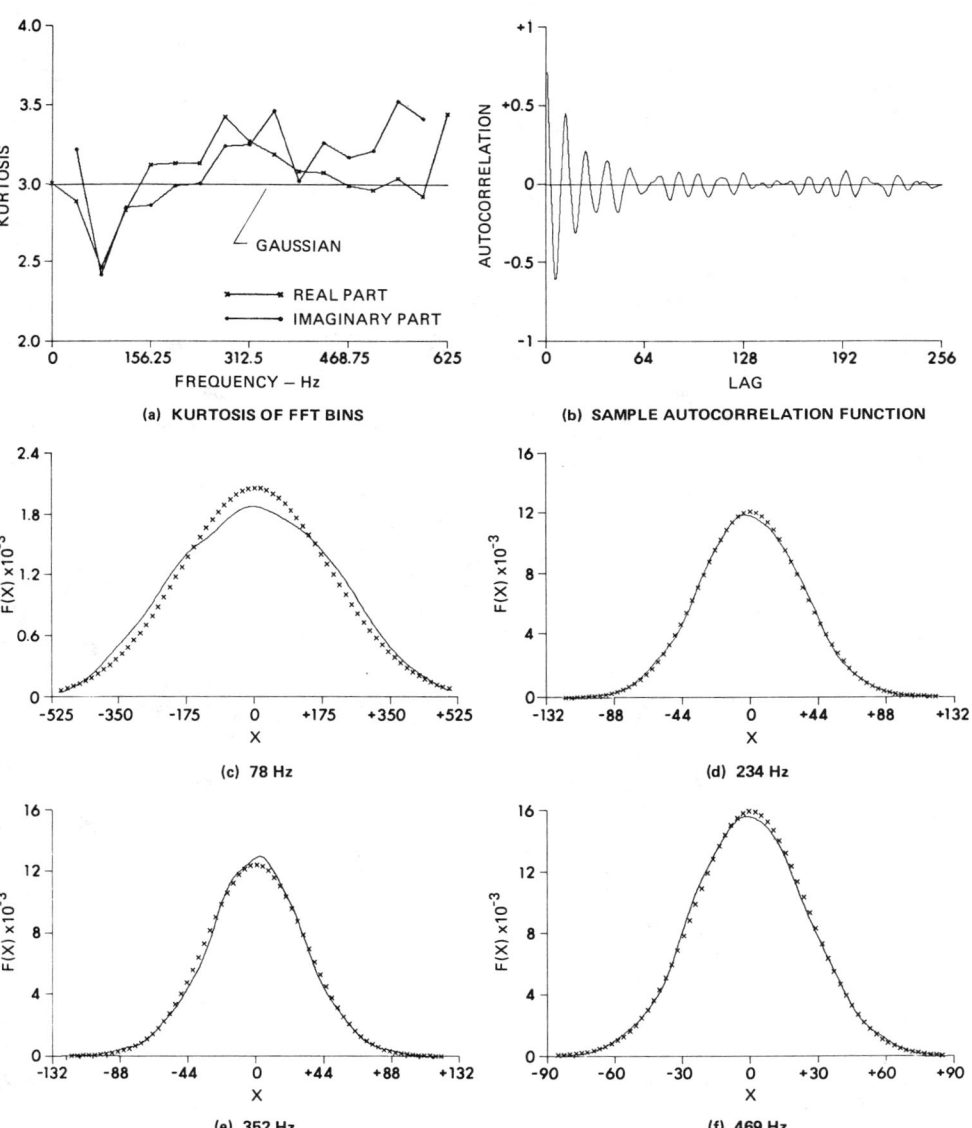

Figure 5 Filtered merchant data sampled at Nyquist rate

sian density for different exponential decay rates. The effects of filtering on the densities of these processes have been examined, showing the dependence on frequency of the statistical characterization of the processes. Filtering has been shown to produce a central limit effect when the processes are sampled to obtain independence; however, with dependent samples the opposite effect may result from filtering, as with the seismic data.

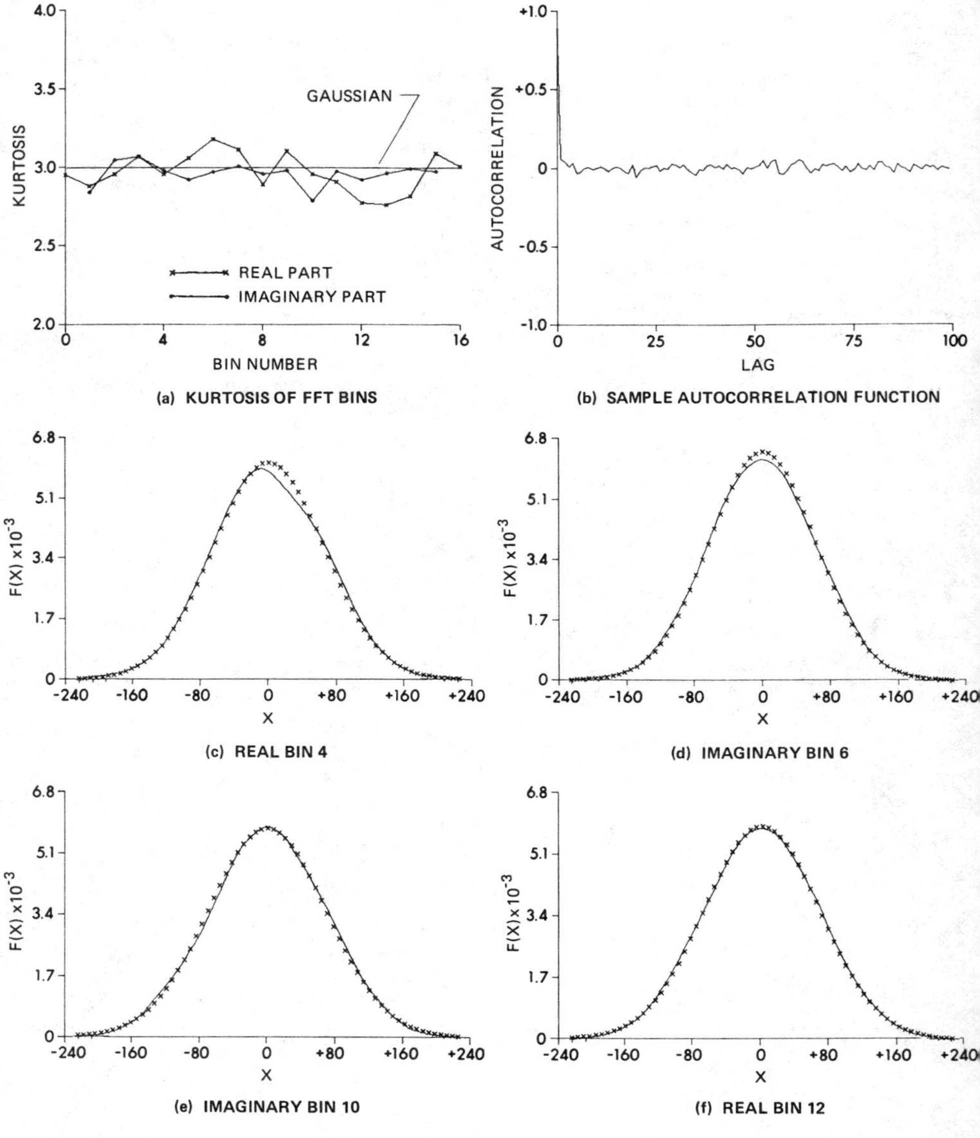

Figure 6 Filtered merchant data Ith by 128

ACKNOWLEDGMENTS

The authors are grateful for the support of the Statistics and Probability Program of the Office of Naval Research, under Contract N00014-81-K-0145, and for numerous helpful discussions with Professors Terry Wagner and Luc Devroye.

REFERENCES

1. C.S. Penrod, F.W. Machell, and T.J. Wagner, "Empirical Finite Sample Performance of Fixed and Variable Kernel Density Estimates," submitted to The Journal of the American Statistical Association.

2. R.J. Marks II, G.L. Wise, and D.G. Haldeman, "Further Results on Detection in LaPlace Noise," in Proceedings of the 20th Midwest Symposium on Circuits and Systems, 1977.

3. J.H. Miller and J.B. Thomas, "Detectors for Discrete-Time Signals in Non-Gaussian Noise," IEEE Transactions on Information Theory *IT-18*, No. 2, 241-250 (1972).

Experimental and Modeled Density Estimates of Underwater Acoustic Returns*

Gary R. Wilson and Dennis R. Powell

*Applied Research Laboratories,
University of Texas
Austin, TX*

1 INTRODUCTION

In underwater acoustic applications it is often assumed that the noise field is a Gaussian random process. In many instances this assumption is a reasonable one and usually leads to tractable analytical solutions of signal processing problems. However, it is suspected that significantly non-Gaussian noise fields can be encountered in underwater acoustics, and it is known that non-Gaussian fields can result in significant performance degradations for processors optimized for the Gaussian noise field.[1] To overcome these performance degradations, it is usually not enough to know that the noise field is non-Gaussian; certain statistical measures of the noise field may also be required. Typically one of the more useful statistical measures is the probability density function. For example, detection algorithms based on a likelihood ratio usually require a knowledge of the signal and noise densities. Previous analyses of the statistical properties of underwater acoustic fields have generally been limited to determining if the data are non-Gaussian and have stopped short of characterizing the observed probability densities. The purpose of this paper is to provide univariate density estimates of experimental non-Gaussian underwater acoustic data. Two estimation techniques are compared, a nonparametric technique using kernel functions, and a parametric technique based on Middleton's class A noise model.

---

*Work supported under Contract N00014-81-K-0145, Task SRO 106 with the Office of Naval Research.

2 DATA VALIDATION

The surface reverberation data consisted of 13 ensembles of 500 samples each. Each ensemble represented a random sampling of the reverberation. Each ensemble characterized the reverberation at a different time relative to the beginning of the pulsed transmission. The 500 samples were generated by repetitively projecting a pulsed signal at a wind roughened surface and recording the backscattered returns. The bottom reverberation data consisted of 54 ensembles of 57 samples each. Random samples were generated by insonifying independent parts of the bottom on each transmission from a moving platform. Both the surface and bottom reverberation were generated using reasonably high resolution sonar systems; the pulse length was 100 μsec in both cases and the horizontal transmit beamwidths were 3° for the bottom reverberation and 15° for the surface reverberation.

The ambient noise was recorded in a shallow water coastal region and is dominated by acoustic energy from snapping shrimp. The ambient noise was sampled to produce independent samples, and for the convenience of later processing the samples were grouped into 48 ensembles of 625 samples each.

Techniques to estimate the probability density function of experimental data require random samples of the data. Random samples are those that are independent and identically distributed (homogeneous). Thus each of the ensembles was tested for randomness, independence, and homogeneity prior to density estimation. Since the underlying probability distributions of the data are unknown, several of the more powerful nonparametric tests were used. The runs up and down test and the Kendall rank correlation test were used to test for randomness and independence. The Kolmogorov-Smirnov two-sample test and the Wilcoxon rank sum test were used to test for homogeneity. In addition, the ensembles were tested for normality to determine that they were non-Gaussian. The tests for normality included Pearson's test of skew and D'Agostino's test, among others. Descriptions of these tests and their applicability to sonar data can be found elsewhere.[2-6]

Two-sided tests were applied to all the data. The test statistic outputs for each ensemble tested have been converted to *p*-values. A *p*-value is the probability of exceeding the test statistic value under the null hypothesis (homogeneity, for example). A small *p*-value indicates that the null hypothesis of the test can be rejected with some confidence. The results of these tests for the 54 ensembles of bottom reverberation are shown in Figure 1. The *p*-value is itself a random variable that is uniformly distributed in the interval [0,1] under the null hypothesis. Independent *p*-values can thus be combined to give an overall probability that the particular hypothesis being tested is valid for the entire set of ensembles.[7] This combination probability is designated in Figure 1 as P_c. Edgington's normal curve method was used to combine the probabilities.[8] If P_c is not low, say not less than 0.05, then the null hypothesis can be accepted for the entire set of ensembles even though some ensembles may have small *p*-values. For some tests, a single *p*-value cannot be calculated, but only an upper and lower bound on the *p*-value. For these tests only upper and lower bounds on P_c are given. As can be seen from Figure 1, the bottom reverberation ensembles could be accepted as random, independent, and homogeneous. However, the Gaussian assumption could not be accepted. Similar results were obtained from analyses of surface reverberation and ambient noise data.

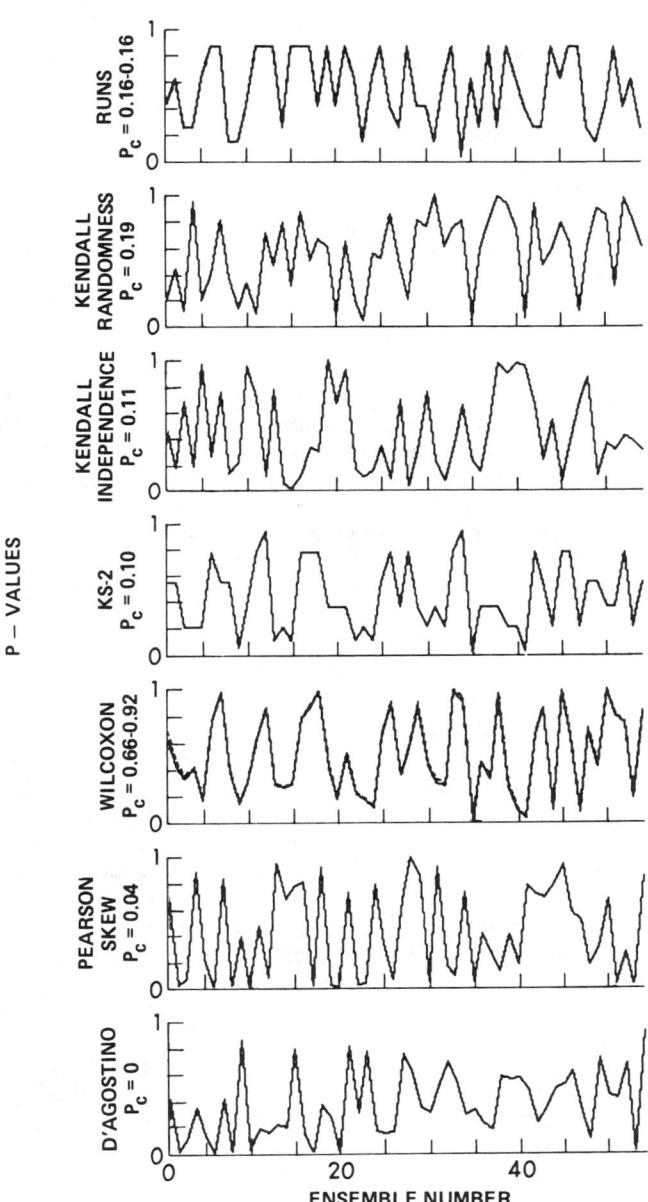

Figure 1 p-values for several hypothesis tests applied to bottom reverberation ensembles

Although the ensembles as a whole tested to be non-Gaussian, it was not clear that all the ensembles of each type of data were produced from the same underlying non-Gaussian distribution. For this reason, a k-sample homogeneity test[9] was applied to the ensembles of each type of data. The results of this test indicated that the ensembles of each data set could be considered as representing the same underlying distribution.

3 DENSITY ESTIMATES

Nonparametric density estimates using kernel functions were applied to the three sets of data. The kernel estimate is given by

$$f_n(x) = \frac{1}{n} \sum_{i=1}^{n} \frac{1}{h(n)} K\left(\frac{x - X_i}{h(n)}\right),$$

where $X_1, X_2, \ldots, X_n$ are i.i.d. random variables, $h(n)$ is a sequence of kernel widths, and $K(x)$ is a kernel function satisfying certain regularity conditions. The kernel function used in this study is a quartic kernel.[10] Both fixed kernel widths and variable kernel widths were used.[11] The density estimates were applied to each ensemble and the resulting estimates from each data set were averaged together to produce an average density for each data set.

Figure 2 displays the average density estimate for the surface reverberation data using fixed width kernels. Both linear and log plots of the density are shown. The kernel estimate is compared to a Gaussian density of the same mean and variance. As can be seen, the surface reverberation density is nearly Gaussian, but with slightly heavier tails. The heavier tails indicate that there is a slightly greater probability of observing a large amplitude reverberation return than would be expected if the reverberation were truly Gaussian. For a detector designed for a Gaussian noise field, this would result in an actual false alarm rate greater than the design false alarm rate. However, for these data the difference in false alarm rates would not be significant in most cases. In Figure 3, the density estimate of the bottom reverberation is compared to a Gaussian density. One limitation of the fixed width kernels is their erratic behavior in the extreme tails when only a few or perhaps no samples are within the kernel width. This behavior is obvious in the log plot of Figure 3. The bottom reverberation density is also nearly Gaussian but with apparently heavier tails. In contrast to the reverberation densities, the density of the ambient noise is significantly non-Gaussian. In Figure 4 the ambient noise is compared to a Laplace density.

A comparison of the sample moments and the moments computed from the density estimates provide one indication of how well the density estimates represent the true (but unknown) underlying density based on a finite number of samples. Table I compares the sample and kernel estimates of the mean, variance, skew, and kurtosis for the three data sets. The 95% confidence intervals for the mean and variance are also provided. The moments are in generally good agreement. However, the kernel estimate of the variance of the bottom reverberation differs significantly from the sample estimate.

To determine the effect of kernel width on the moments derived from fixed width kernel density estimates, Gaussian samples of a known mean and

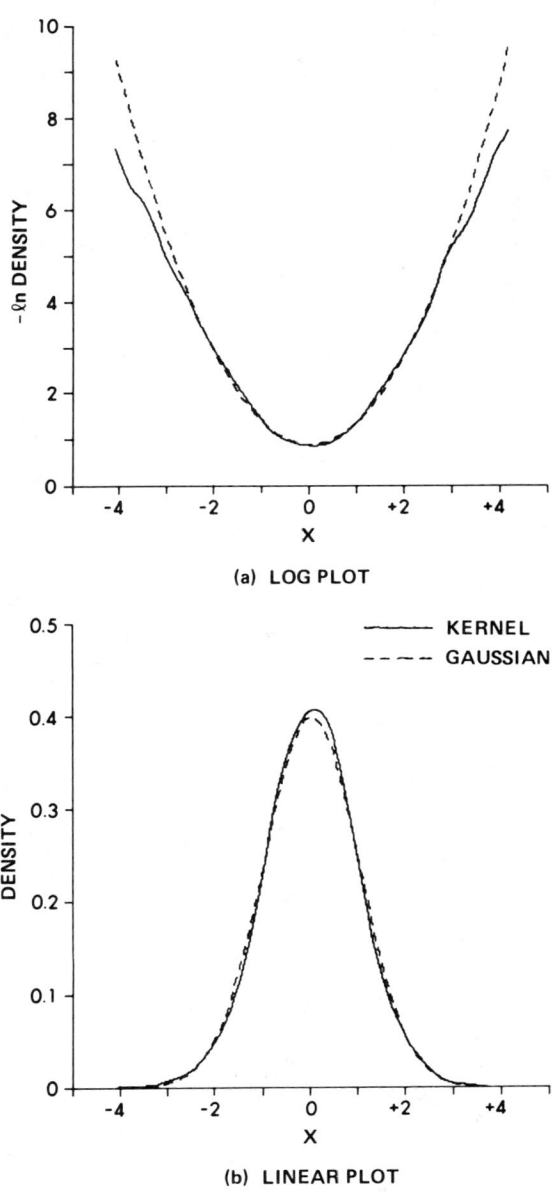

Figure 2 Density estimate of surface reverberation compared to a Gaussian density

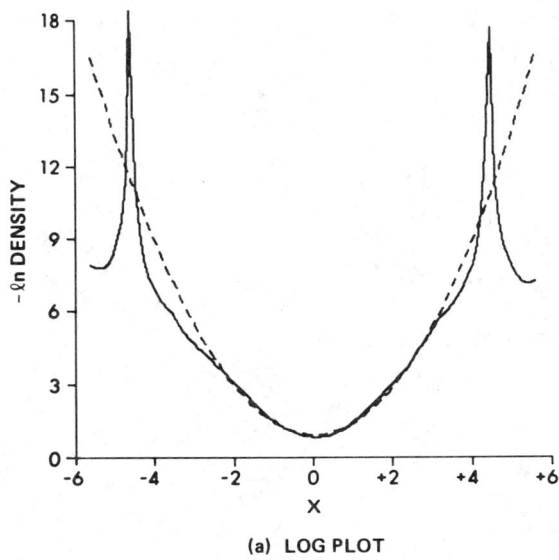

(a) LOG PLOT

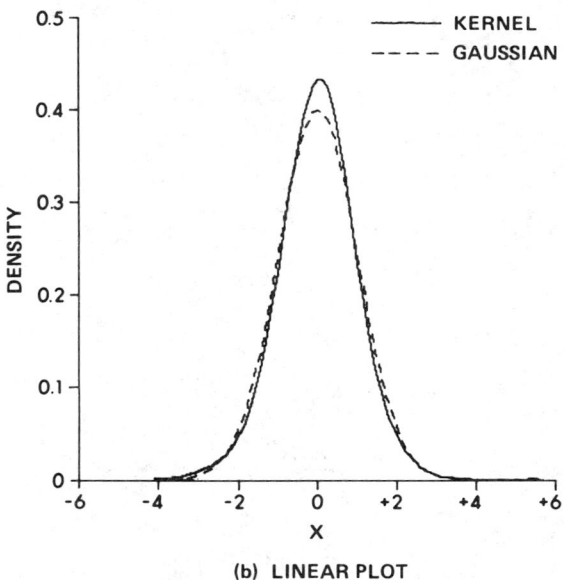

(b) LINEAR PLOT

Figure 3 Density estimate of bottom reverberation compared to a Gaussian density

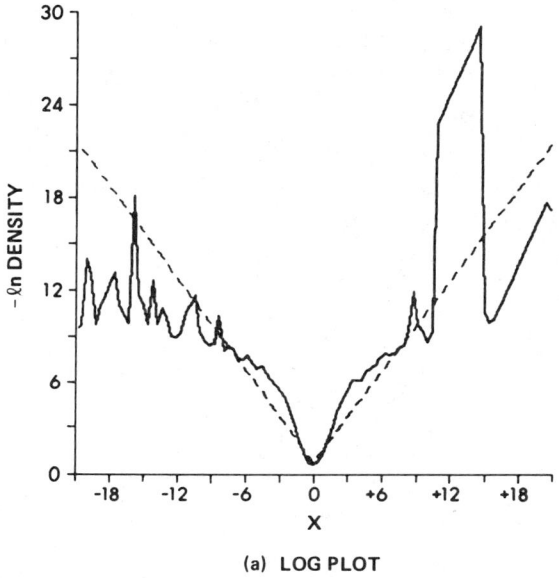

(a) LOG PLOT

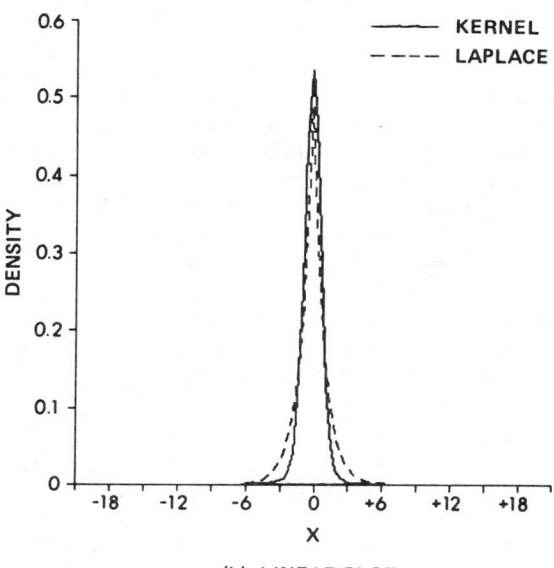

(b) LINEAR PLOT

Figure 4 Density estimate of biological noise compared to a Laplace density

Table I Comparison of Sample and Kernel Estimates of Moments

Statistic	Sample Estimate	Kernel Estimate	95% Confidence Interval
Surface Reverberation			
Mean	0	0.0	(−0.02, 0.02)
Variance	1	1.05	(0.96, 1.04)
Skew	−0.03	−0.03	—
Kurtosis	3.4	3.3	—
Bottom Reverberation			
Mean	0	0.01	(−0.04, 0.04)
Variance	1	1.18	(0.93, 1.07)
Skew	−0.07	0.03	—
Kurtosis	4.4	4.2	—
Biological Noise			
Mean	0	0.0	(−0.01, 0.01)
Variance	1	1.06	(0.92, 1.08)
Skew	−1.54	−1.40	—
Kurtosis	44.4	43.4	—

variance were generated and the density of the Gaussian samples was estimated using various kernel widths. The mean, variance, skew, and kurtosis were computed from these density estimates and plotted as a function of kernel width. Figure 5 contains the plots of these moments compared to their 98% confidence intervals. Except for small kernel widths, the fixed width kernel density estimate overestimates the variance even though the other moments are in reasonable agreement with their expected values. These small kernel widths did not provide acceptably smooth density estimates.

A variable width kernel estimate in which the kernel width is determined for each data point by the k nearest neighbors to that data point[11] was also applied to the Gaussian data. The variable width kernel estimates tend to provide smoother estimates in the tails than the fixed width kernel estimates. However, the tails are apparently overestimated. Subsequently, the variance is overly large, which in turn affects the kurtosis. These results are shown in Figure 6 as a function of the smoothing parameter k/n. Comparison of the finite sample performance of fixed and variable width kernel estimates has confirmed these observations by demonstrating that the variable width data centered kernel is more biased but has less variance on the tails than the fixed width kernel.[11]

To summarize, based on the criteria of moments, the fixed width kernel provided a better estimate of the density of the experimental data than the variable width kernel. The variable width kernel estimates tended to place too much probability in the tails. A detection threshold based on this estimate

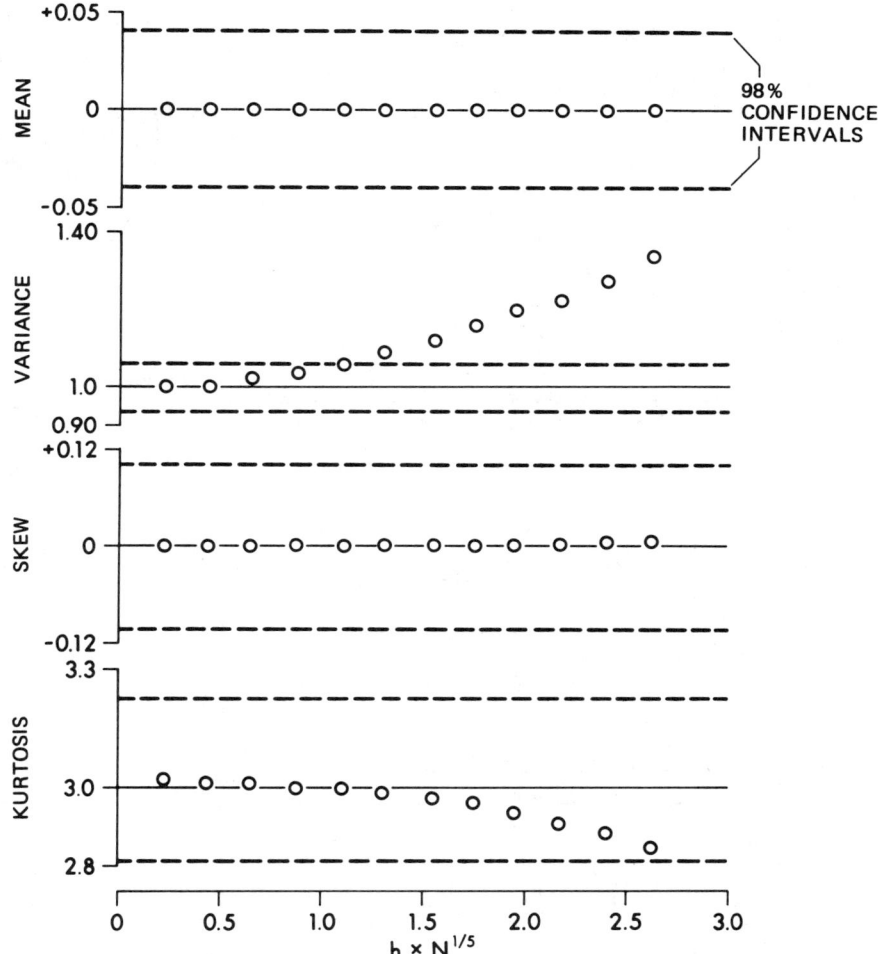

Figure 5 Fixed width kernel moment estimates of Gaussian data as a function of kernel width

would result in an operating point significantly different than expected. However, the fixed width kernel estimates were not entirely satisfactory either. It was not possible to choose a kernel width that simultaneously provided smooth density estimates, especially in the tail, and acceptable estimates of the moments in every case. Erratic estimates of the tails can result in unacceptable likelihood ratios based on these density estimates. Thus the proper estimation technique to use in practice will depend on the application for which the estimates will be used. In some cases, neither the fixed nor the variable width kernel density estimates may be appropriate.

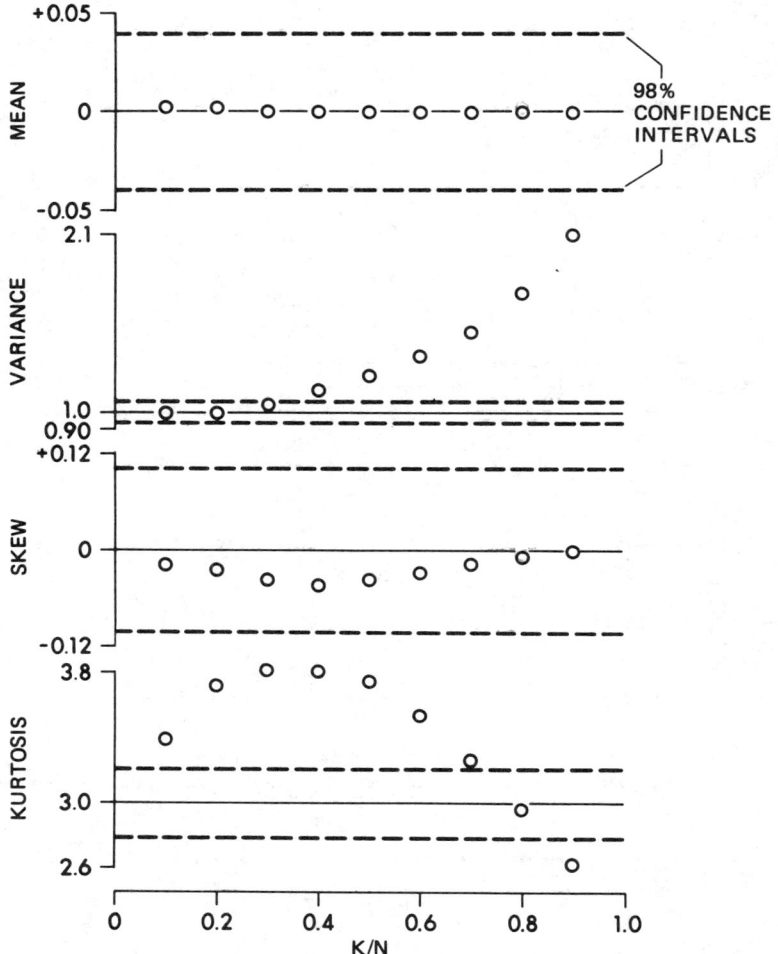

Figure 6 Variable width kernel moment estimates of Gaussian data as a function of smoothing parameter

4 NOISE MODEL

An impulsive interference model developed by D. Middleton[12] was implemented and applied to the three data sets. Only the class A noise model was implemented. Class A noise is defined by Middleton as noise that does not produce significant ringing in the receiver due to transient effects of the impulsive interference. Class A noise generally has a bandwidth less than the receiver bandwidth. Class B noise on the other hand results in significant build-up and decay of a transient, and usually occurs when the noise has a bandwidth greater than the receiver bandwidth. The different responses of the receiver to the noise result in statistically different receiver outputs, requiring different noise models. The reverberation data had bandwidths commensurate with the receiver bandwidths, while the ambient noise bandwidth was larger than the receiver bandwidth. Thus these data do not completely fit the assumptions of the class A model; nevertheless the class A model was easier to implement and provided a reasonable starting point. Surprisingly, the class A model provided reasonable density estimates of these data. Apparently the spectral distribution of the noise energy was such that significant ringing of the receiver did not occur.

The density of the instantaneous amplitude of class A noise is given by

$$f_Z(z) = e^{-A} \sum_{m=0}^{\infty} \frac{A^m}{m!} \phi(z, 0, \sigma_m^2).$$

$\phi(z, 0, \sigma_m^2)$ is the normal density function with mean 0 and variance σ_m^2. σ_m^2 is given by the expression

$$\sigma_m^2 = \frac{\frac{m}{A} + \Gamma'}{1 + \Gamma'},$$

where A is the impulsive index and Γ' is the ratio of the intensities of the Gaussian and non-Gaussian components. The parameters A and Γ' are estimated from the data by graphical means and by a method of moments.[13]

The model density is compared to the fixed width kernel density estimate of the surface reverberation in Figure 7. The two estimates agree well. The Γ' parameter value of 7.0 indicates that the impulsive component is very small, as expected. The model density is compared to the bottom reverberation density estimate in Figure 8. Again there is good agreement, with the model predicting the heavier than normal tails reasonably well.

The reverberation was nominally class A noise in both cases, but the assumption of class A noise for the ambient noise is more difficult to justify. The bandwidth of the receiver was nominally 25 kHz, but the spectrum of the acoustic energy of snapping shrimp is known to extend to at least 150 kHz.[14] Nevertheless, the class A model provided a surprisingly close estimate of the density of the ambient noise, as can be seen in Figure 9. Part of the explanation may be that even though the snapping shrimp spectrum extended to 150 kHz, the majority of the energy was below 25 kHz,[14] reducing the amount of ringing in the receiver and allowing the snapping shrimp noise to be represented by the class A noise model. The ambient noise was more impulsive than the reverberation data as evidenced by the lower value of the impulsive index A (Figure 9). The impulsive index was 0.005 for the ambient noise and 0.1 and 0.086 for the surface and bottom reverberation, respectively. The relative intensity of the non-Gaussian component of the ambient noise was also larger than the non-Gaussian component of the reverberation since the parameter Γ' was smaller for the ambient noise (2.84 for the ambient noise compared to 7.0 and 3.97 for the surface and bottom reverberation, respectively).

Table II is an extension of Table I that includes the estimates of the mean, variance, skew, and kurtosis of the three data sets based on the model densities. The model estimates of the even moments are in very good agreement with the sample estimates. Since the model density is a symmetric den-

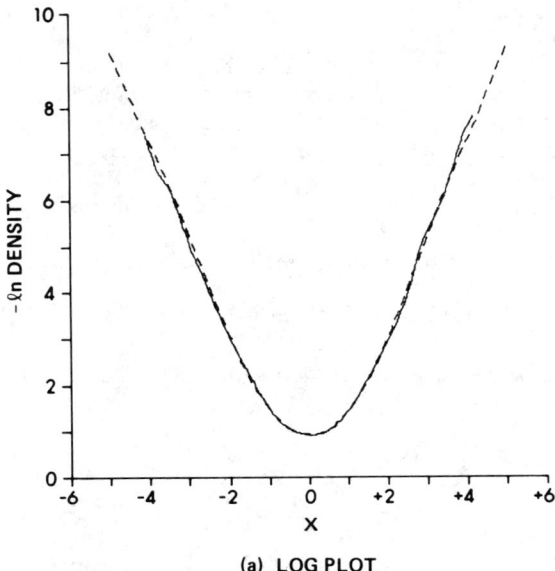

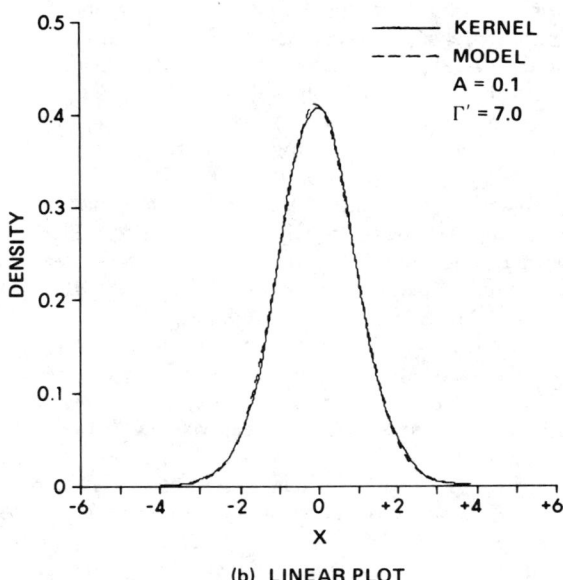

Figure 7 Comparison of kernel density estimate and model density of surface reverberation

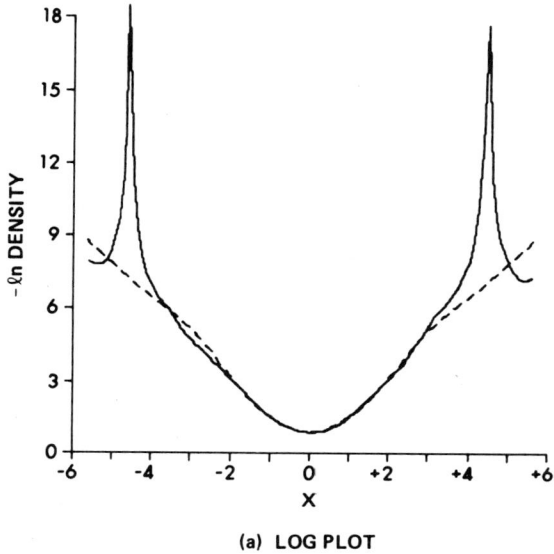

(a) LOG PLOT

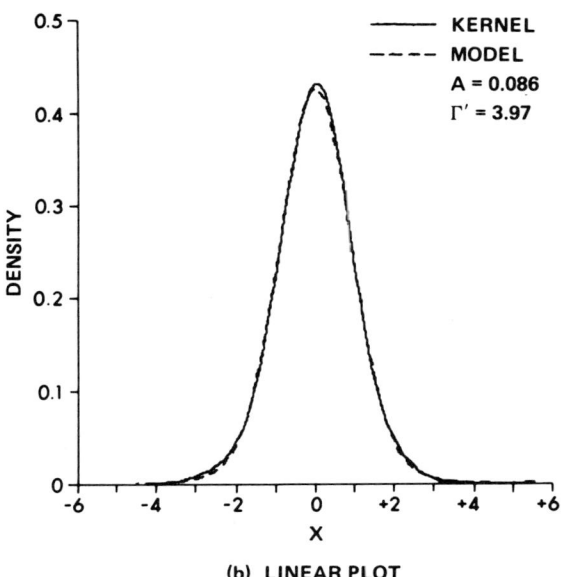

(b) LINEAR PLOT

Figure 8 Comparison of kernel density estimate and model density of bottom reverberation

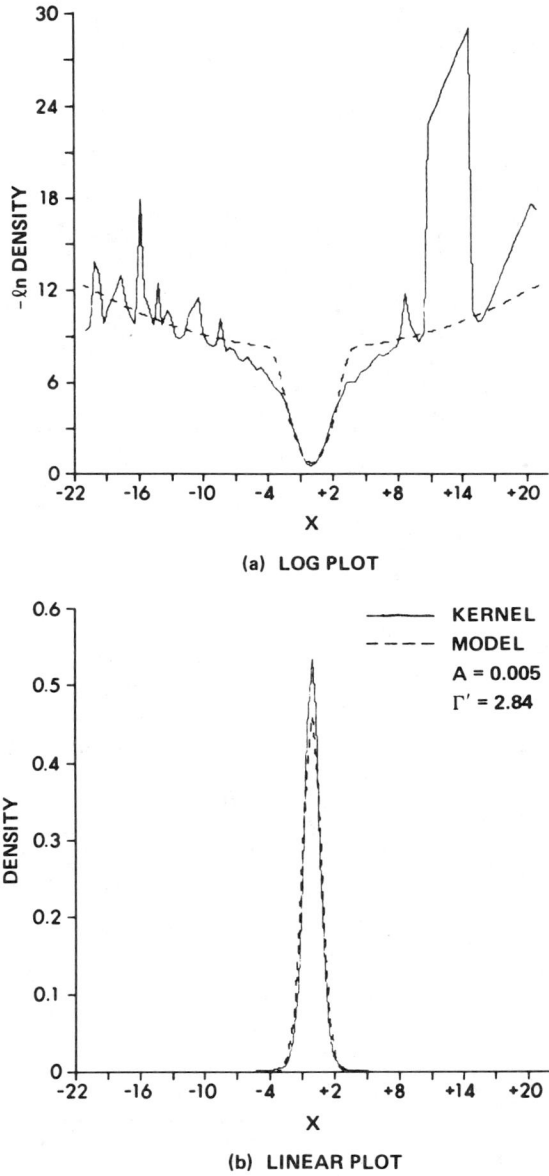

Figure 9 Comparison of kernel density estimate and model density of biological noise

Table II Comparison of Sample, Kernel and Model Estimates of Moments

Statistic	Sample Estimate	Kernel Estimate	Model Estimate	95% Confidence Interval
Surface Reverberation				
Mean	0	0.0	0	(−0.02, 0.02)
Variance	1	1.05	1	(0.96, 1.04)
Skew	−0.03	−0.03	0	—
Kurtosis	3.4	3.3	3.4	—
Bottom Reverberation				
Mean	0	0.01	0	(−0.04, 0.04)
Variance	1	1.18	1	(0.93, 1.07)
Skew	−0.07	0.03	0	—
Kurtosis	4.4	4.2	4.4	—
Biological Noise				
Mean	0	0.0	0	(−0.01, 0.01)
Variance	1	1.06	1	(0.92, 1.08)
Skew	−1.54	−1.40	0	—
Kurtosis	44.4	43.4	43.7	—

sity, the odd moments are all zero. Thus the model estimates of the skew do not agree with the sample estimates. However, the sample skew was usually rather small (except for the biological noise). The model estimates of the moments provided better agreement with the sample estimates than did the kernel estimates with the exception of the skew.

5 SUMMARY

Density estimates of statistically validated ensembles of experimental surface reverberation, bottom reverberation, and ambient biological noise were made. The ensembles were tested for normality and found to be non-Gaussian. Two density estimation methods were used—nonparametric kernel methods and a parametric method based on Middleton's class A noise model. The reverberation densities were only slightly different from a Gaussian density, with somewhat heavier tails. The biological noise density was significantly non-Gaussian. The kernel and class A density estimates were in reasonably good agreement. However, the class A estimates were smoother in the tails and provided somewhat better agreement with the sample moments. These results demonstrate that reasonable density estimates of experimental underwater acoustic noise processes can be made. Furthermore, although these results do not represent an exhaustive description of underwater acoustic noise processes, they nevertheless provide a first indication of the density function of some commonly encountered processes. As such, these results can be used by sonar system designers to evaluate the effects that these types of non-Gaussian noise fields have on detectors that are based on the assumption of a Gaussian noise field, such as increased false alarm rate. These experimental density estimates can also be used to guide the design of detectors with improved performance in these types of realistic noise environments.

REFERENCES

1. Arthur D. Spaulding and David Middleton, "Optimum Reception in an Impulsive Interference Environment—Part I: Coherent Detection," IEEE Trans. Commun., IT-COM 25, 910-923 (1977).

2. D. Middleton, "Acoustic Modeling, Simulation, and Analysis of Complex Underwater Targets, II. Statistical Evaluation of Experimental Data," Applied Research Laboratories Technical Report No. 69-22 (ARL-TR-69-22), Applied Research Laboratories, The University of Texas at Austin (1969).

3. Claude Horton, Sr., "Modern Methods in Data Collection and Analysis of Surface Reverberation and of the Associated Water Waves," paper presented at The 86th meeting of the Acoustical Society of America, Los Angeles, California, 30 October-2 November 1973.

4. Charles R. Baker, "Some Statistical Tests for the Analysis of Sonar Data," Report No. B-74-3, Department of Statistics, University of North Carolina, Chapel Hill, North Carolina (1974).

5. Marshall E. Frazer, "Some Statistical Properties of Lake Surface Reverberation," J. Acoust. Soc. Am., 64, 858-868 (1978).

6. Gary R. Wilson, "A Statistical Analysis of Surface Reverberation," accepted for publication in The Journal of the Acoustical Society of America (1983).

7. Robert Rosenthal, "Combining Results of Independent Studies," Psychol. Bull., 85, 185-193 (1978).

8. Eugene S. Edgington, "A Normal Curve Method for Combining Probability Values from Independent Experiments," J. Psychol., 82, 85-89 (1972).

9. J. Kiefer, "K-Sample Analogues of the Kolmogorov-Smirnov and Cramer-V Mises Tests," Ann. Math. Stat., 30, 420-447 (1959).

10. B. Silverman, "Choosing the Window Width When Estimating a Density," Biometrika, 65, 1-11 (1978).

11. C.S. Penrod, F.W. Machell, and T.J. Wagner, "Empirical Finite Sample Performance of Fixed and Variable Kernel Density Estimates," submitted for publication in The Journal of the American Statistical Association (1982).

12. David Middleton, "Statistical-Physical Models of Electromagnetic Interference," IEEE Trans. Elec. Comp., EC-19, 106-127 (1977).

13. David Middleton, "Procedures for Determining the Parameters of the First-Order Canonical Models of Class A and Class B Electromagnetic Interference," IEEE Trans. Elec. Comp., EC-21, 190-208 (1979).

14. M. Ward Widener, "Ambient-Noise Levels in Selected Shallow Water off Miami, Florida," J. Acoust. Soc. Am., 42, 904-905 (1967).

Part IV: Array Processing and Target Tracking

Coherent Array Processing*

Melvin J. Hinich

Departments of Government and Economics
University of Texas
Austin, TX

1 INTRODUCTION

Suppose that an array of sensors is *receiving* coherent plane wave signals from a distant source. Delay-and-sum beamforming is the usual approach to coherent processing of the received signals. The output of the beamformer is then analyzed to obtain an estimate of the source's bearing.[1] For sonar arrays, accurate bearing estimation of single or multiple targets is the major goal. It is necessary to have a good array design and efficient signal processing to achieve this goal.

There is a close relationship between beamforming and frequency-wavenumber Fourier analysis, which has computational advantages over beamforming when the noise is spatially coherent. This relationship, described in the first part of this paper, is used to calculate the array response for any array geometry, and to mitigate the jamming effects of a coherent interfering signal. The final section deals with bearing estimation of a broadband wave when the noise is spatially correlated.

2 SPATIAL FOURIER TRANSFORM APPROACH TO BEAMFORMING

To begin the exposition, consider a simple one dimensional problem: a linear array of M sensors and a single frequency plane wave signal in complex variable form. Let θ_0 denote the wave's direction of arrival *with respect to the array axis*, let c denote the wave's velocity, and $A = |A| \exp(i\phi)$ its complex amplitude.

---

*Work supported under Contract N00014-82-K-0281, Task NR 042-315 with the Office of Naval Research.

A plane wave signal at the k th sensor is

$$s(t,x_k) = A \exp\left[i\omega_0\left(t - \frac{x_k \cos \theta_0}{c}\right)\right], \quad (1.1)$$

where x_k is the location of the k th sensor ($x_1 < x_2 < \ldots < x_M$). The signal in a beam pointed at angle θ (and $-\theta$) is

$$B(t,\theta) = \sum_{k=1}^{M} s(t + \tau_k, x_k), \quad (1.2)$$

where the k th delay is $\tau_k = \dfrac{x_k \cos \theta}{c}$. A linear array cannot identify between θ_0 and $-\theta_0$, so arbitrarily assume that $\theta_0 > 0$.[2]

Since the wavenumber component on the array axis is $\kappa_0 = (\omega_0/c) \cos \theta_0$, it follows from (1.1) and (1.2) that for $\kappa = (\omega_0/c) \cos \theta$

$$B(t,\theta) = A \exp(i\omega_0 t) \sum_{k=1}^{M} \exp[i(\kappa - \kappa_0)x_k]$$

$$= \sum_{k=1}^{M} s(t,x_k) \exp(i\kappa x_k). \quad (1.3)$$

In other words, beamforming a linear array is the same as computing a one dimensional spatial Fourier transform of the M array channels. The spatial frequency $\kappa = (\omega_0/c) \cos \theta$ corresponds to the look angle θ.

In actual practice, a beam is computed from a finite record of the M channels, the beam output is filtered in a narrow band about ω_0, and the filtered signal is squared and summed to give the average energy in the beam for the data set. In frequency-wavenumber analysis, each channel is filtered, and then the spatial Fourier transform is computed. If the received signal is a wave, the magnitude of this transform has a peak of height $M|A|$ when $\kappa = \kappa_0 = (\omega_0/c) \cos \theta_0$.

A three dimensional plane wave signal, defined with respect to a given coordinate system, is

$s(t,x,y,z) =$

$A \exp\{i\omega_0[t - (x \cos \theta_0 \sin \gamma_0 + y \sin \theta_0 \sin \gamma_0 + z \cos \gamma_0)/c]\}$ (1.4)

where θ_0 is the propagation angle with respect to the x-axis in the (x,y) plane, and γ_0 is the angle with respect to the z-axis. If the medium is homogeneous between the source and receiver, there is no refraction of propagation, and thus θ_0 and γ_0 are the *bearing* and *elevation* of the source, respectively. Let (x_k, y_k, z_k) denote the coordinates of the k th sensor of an array in the space. The signal in a beam pointed at bearing θ and elevation γ is

$$B(t,\theta,\gamma) = \sum_{k=1}^{M} s(t + \tau_k, x_k, y_k, z_k) \quad (1.5)$$

where the k th delay is $\tau_k = (x_k \cos \theta \sin \gamma + y_k \sin \theta \sin \gamma + z_k \cos \gamma)/c$. Defining $\kappa_x = (\omega_0/c) \cos \theta \sin \gamma$, $\kappa_y = (\omega_0/c) \sin \theta \sin \gamma$, and $\kappa_z = (\omega_0/c) \cos \gamma$, it follows from (1.4) and (.15) that

Coherent Array Processing

$$B(t,\theta,\gamma) = \sum_{k=1}^{M} s(t,x_k,y_k,z_k) \exp[i(\kappa_x x_k + \kappa_y y_k + \kappa_z z_k)], \quad (1.6)$$

and $|B(t,\theta_0,\gamma_0)| = M|A|$. Thus $B(t,\theta,\gamma)$ is the three dimensional spatial Fourier transform of the data.

3 ARRAY RESPONSE FUNCTIONS

The response of an array can easily be expressed as a function of a wavenumber. Let us start with a linear geometry. Suppose all array channels are filtered in a narrow band about ω_0. The dependence on time can be dropped from all expressions involving the signal. For example, the plane wave is $s(x) = A \exp[-i(\omega_0/c)x \cos \theta_0]$. Then the filtered beam at angle θ is

$$B(\theta) = \sum_{k=1}^{M} s(x_k) \exp(i\kappa x_k) = A \sum_{k=1}^{M} \exp[i(\kappa - \kappa_0)x_k] \quad (2.1)$$

where $\theta = \cos^{-1}[(c/\omega_0)\kappa]$. The function $R(\kappa) = \sum_{k=1}^{M} \exp(i\kappa x_k)$ is called the array's *complex response*. Its amplitude is $|R(\kappa)|$, and the beam power pattern for a plane wave input is $|R(\kappa - \kappa_0)|^2$ as a function of κ or θ.

For example, suppose the array has equally spaced sensors. Normalize the distance unit so that $d = 1$ is the distance between adjacent response. For an array of length N composed of two clusters of $M/2$ equally spaced sensors with sensor spacing $d = 1$,[3]

$$|R(\kappa)| = 2|\cos[(N - M/2)\kappa/2]\sin(M\kappa/4)/\sin(\kappa/2)|. \quad (2.2)$$

Unless the x_k are all irrational numbers, which is impossible to achieve, then $R(\kappa)$ is a periodic function of κ. The replicas of $R(\kappa)$ for κ outside its principle domain are called *grating lobes* in antenna theory, and can lead to spatial aliasing.[4] Suppose that $x_k = n_k$ for integers n_k and $d = 1$. Set $n_1 = 0$, and let D be the greatest common divisor of $\{n_2, \ldots, n_M\}$. From Hinich and Weber,[5] $R(\kappa)$ is periodic with period $2\pi/D$. Its principle domain is either $-\pi/D < \kappa < \pi/D$ or $0 \leq \kappa < 2\pi/D$, depending on the convention one is using. This result implies that a real wave signal has no aliases in the array's principle domain if its wavelength $\lambda_0 = 2\pi c/\omega_0$ is greater than $2D$. For example, if $x_2 = 11$, $x_3 = 17$, and $x_4 = 30$, there is no aliasing if $\lambda_0 > 2$.

The response function for a three dimensional array is

$$R(\underline{\kappa}) = \sum_{k=1}^{M} \exp(i\underline{\kappa}^T \underline{v}_k), \quad (2.3)$$

where $\underline{\kappa} = (\kappa_x, \kappa_y, \kappa_z)^T$ and $\underline{v}_k = (x_k, y_k, z_k)^T$. The filtered beam for bearing θ and elevation γ is simply $B(\theta,\gamma) = R(\underline{\kappa} - \underline{\kappa}_0)$ when the input signal is a wave with $\theta = \theta_0$ and $\gamma = \gamma_0$. In terms of the wavenumber components,

$$\theta = \tan^{-1}(\kappa_y/\kappa_x)$$

and $\quad (2.4)$

$$\gamma = \tan^{-1}(\kappa_y/\kappa_z \sin \theta).$$

Thus the phase velocity does not need to be known to obtain (θ,γ) from $\underline{\kappa}$ if the array is truly three dimensional.

4 SIGNAL PLUS NOISE

The major advantage of the frequency-wavenumber approach over beamforming is the ease with which spatially correlated (coherent) noise is prewhitened. Spatially correlated noise makes the signal-to-noise ratio spatially dependent. Simple beamforming is then statistically inefficient. Large scale filtering operations are needed to prewhiten the received signals by time delay methods to obtain efficient bearing estimates.[6] Calculations of optimal bearings are relatively easy in the wavenumber domain if the nois's wavenumber spectrum has been estimated.

Once again it is easier to explain the method for a linear array and a single frequency wave. Suppose that the signal at x_k is a plane wave plus stationary, zero-mean Gaussian noise denoted $\epsilon(t,x_k)$. The filtered signal is

$$s(x_k) = A \exp[-i(2\pi/\lambda_0)x_k \cos\theta_0] + \epsilon(x_k), \qquad (3.1)$$
$$= A \exp(-i\kappa_0 x_k) + \epsilon(x_k)$$

where $\epsilon(x_k)$ denotes the filtered noise and $\lambda_0 = 2\pi c/\omega_0$ is the wavelength. If the noise is spatially correlated, then the *wavenumber spectrum* $S_\epsilon(\kappa)$ of the noise is not flat. For large M, the expected value of $M^{-1}|\sum_{k=1}^{M} \epsilon(x_k) \exp(i\kappa x_k)|^2$ is $S_\epsilon(\kappa)$.[7] Assume that the noise field has been sampled sufficiently long so that $S_\epsilon(\kappa)$ can be considered as *known* for κ in its principle domain.

For large M, the expected value of $|B(\theta)|^2/M$ is

$$E[M^{-1}|B(\theta)|^2] = |A|^2 M^{-1}|R(\kappa - \kappa_0)|^2 + S_\epsilon(\kappa) \qquad (3.2)$$

from (2.1). Since $E[M^{-1}|B(\theta_0)|^2 = M|A|^2 + S_\epsilon(\kappa_0)$, the normalized beam pattern has a peak of order $M|A|^2$ against a background of order $S_\epsilon(\kappa)$ for the look angle θ_0, provided that $M|A|^2 \gg S_\epsilon(\kappa)$ for κ in a band about κ_0.

From now on in the paper, the beam output (the spatial Fourier transform) will be expressed as a function of κ rather than θ, i.e. as $B(\kappa)$. In practice $B(\kappa)$ is computed for a finite number of κ's. Suppose that for an array geometry $\{x_k = n_k d\}$, $B(\kappa)$ is computed for the grid values $\{\kappa_l = 2\pi l/Nd: l = 0, 1, \ldots, N-1\}$ where $N > M$. This is easily done using the fast Fourier transform FFT to compute $\sum_{k=0}^{N-1} \delta(k) s(x_k) \exp(i2\pi kl/N)$, where $\delta(k) = 1$ if there is a sensor at the point kd, and $\delta(k) = 0$ otherwise.

It is often useful to have a test statistic to determine the statistical signficance of the maximum energy peak in the beam output. Let the null hypothesis be $A = 0$, i.e. the peak is due to noise alone. Consider the test statistic

$$X = \max_{l=0,\ldots,N-1} \frac{2|B(\kappa_l)|^2}{M\, S_\epsilon(\kappa_l)}, \qquad (3.3)$$

which is analogous to the statistic of the Fisher test for the presence of a sinusoid in additive noise.[8] The distribution of X is needed to compute the threshold for an α-level test of the null hypothesis. This distribution is hard to obtain since the beam outputs $B(\kappa_l)$ are correlated. But if the sidelobes rapidly diminish as M increases, as is the case for the optimal array or an equally spaced array, the distribution of X is approximately the same as the distribution of the maximum of M uncorrelated χ_2^2 variates. The cumulative distribution function (cdf) of this maximum is $[F(x)]^M$, where $F(x)$ is the cdf of a χ_2^2 variate, i.e. $F(x) = Pr(\chi_2^2 < x)$. The null hypothesis of noise alone is then rejected at the α-level if $X > x_0$, where x_0 satisfies

$$\alpha = 1 - [F(x_0)]^M. \tag{3.4}$$

Note that in (3.3), $|B(\kappa_l)|^2$ is divided by the noise wavenumber spectrum. This division is the spatial analogy of prewhitening. If the beam outputs are to be visually inspected for the presence of a wave, then the *equalized power pattern* $P(\kappa_l) = |B(\kappa_l)|^2/S_\epsilon(\kappa_l)$ should be plotted for $l = 0, \ldots, N - 1$.

Suppose that $P(\kappa_l)$ has a peak at $\kappa_{\hat{l}}$ that is statistically significant. Then $\hat{\theta}_0 = \cos(\lambda_0 \hat{l}/Nd)$ is the natural estimate of θ_0. If $(2\pi/\lambda_0)\cos\theta_0$ falls between two grid points, there is a *quantization error* in $\hat{\theta}_0$ of order $1/Nd$.

When M or the signal-to-noise ratio (SNR) is large, the root mean-square error of $\hat{\theta}_0$ due to noise is approximated by

$$\text{rmse } \hat{\theta}_0 \cong \frac{\lambda_0}{2\pi(2\rho M)^{1/2}\sigma \sin\theta_0} \tag{3.5}$$

where $\sigma = [M^{-1}\sum_{k=1}^{M}(x_k - \bar{x})^2]^{1/2}$ is a measure of the array's aperture, ρ is the power SNR in a narrow band about ω_0, and $\bar{x} = M^{-1}\sum_{k=1}^{M} x_k$.[9] Since $\lim_{M \to \infty} M^{1/2}$ rmse $\hat{\theta}_0$ is equal to the Cramer-Rao bound for the asymptotic variance of a consistent estimator of θ_0 for Gaussian noise, $\hat{\theta}_0$ is approximately maximum-likelihood for large M. For a large aperture array with many sensors, this bearing error component is often less than the quantization error due to the grid width of the discrete wavenumbers (or beam angles).

Now consider the problem of estimating the complex amplitude A.[10] If the noise is *spatially uncorrelated*, the maximum likelihood estimator of A is

$$\hat{A} = \frac{1}{M} B(\hat{\kappa}_0). \tag{3.6}$$

Since $E[B(\hat{\theta}_0)] = AR(0) = AM$, it follows that $\hat{A}$ is unbiased, i.e. $\hat{E}[A] = A$. The complex variance of $\hat{A}$ is $E|\hat{A} - A|^2 = \sigma_\epsilon^2/M$, where $\sigma_\epsilon^2 = E[\epsilon^2(x_k)]$ is the variance of the noise.

If $S_\epsilon(\kappa)$ is not flat, $\hat{A}$ is still unbiased.[11] Its complex variance is approximately $E|\hat{A} - A|^2 = M^{-1}S_\epsilon(\kappa_0)$. Thus $\hat{A}$ is a precise estimator of A when M is large, and is useful in removing the effect of a coherent jamming signal.

Planar array processing is a simple straightforward extension of the linear case, using two indices and two sums. Let $(\hat{\kappa}_{x0}, \hat{\kappa}_{y0})$ denote the wavenumbers on a two dimensional grid $\{\kappa_{xl} = 2\pi l/N, \kappa_{yk} = 2\pi k/N\}$ that jointly maximize the equalized power pattern $|B(\kappa_{xl}, \kappa_{yk})|^2/S_\epsilon(\kappa_{xl}, \kappa_{yk})$. The estimator of A using a planar array is $\hat{A} = \frac{1}{M} B(\hat{\kappa}_{x0}, \hat{\kappa}_{y0})$. The variance of A is approximately $M^{-1}S_\epsilon(\kappa_{xl_0}, \kappa_{yk_0})$.

The estimator of θ_0 is $\hat{\theta}_0 = \pm\tan^{-1}(\hat{\kappa}_{y0}, \hat{\kappa}_{x0})$ where the sign is determined by the signs of $\hat{\kappa}_{x0}$ and $\hat{\kappa}_{y0}$. When M is large, $\hat{\theta}_0$ is approximately maximum-likelihood[12] and

$$\text{rmse } \hat{\theta}_0 \cong \frac{\lambda_0}{2\pi(2\rho M)^{1/2}\sigma} \tag{3.7}$$

if $\sigma_y^2 = M^{-1} \sum_{k=1}^{M} (y_k - \bar{y})^2 = \sigma^2$, i.e. the y-aperture is equal to the x-aperture.

5 BLOCKING A JAMMING SIGNAL

Suppose the filtered signal for the k th sensor is

$$s(x_k) = A \exp[-i(2\pi/\lambda_0)x_k \cos \theta_0]$$
$$+ A_J \exp[-i(2\pi/\lambda_J)x_k \cos \theta_J] + \epsilon(x_k), \tag{4.1}$$

where A_J, λ_J, and θ_J are the amplitude, wavelength, and direction, respectively, of a wave that is interfering with the wave of interest. If $\lambda_J \neq \lambda_0$, then all or most of the jamming energy can be filtered out. Thus let $\lambda_J = \lambda_0$, and assume that A_J and θ_J are unknown.

If $|A_J| \gg |A|$, then the beam pattern will be dominated by the jammer. If this is the case, then θ_J is estimated by the angle $\hat{\theta}_J = \cos^{-1}[\lambda_J l_J/N]$ such that $|B(\kappa_{l_J})|^2/S_\epsilon(\kappa_{l_J})$ is a maximum for $l = 0, \ldots, N - 1$. The amplitude A_J is estimated using (3.6) with l_J in place of l_0.

The response of the jammer can be then subtracted from $B(\kappa_l)$, and the adjusted beam pattern defined by

$$|\tilde{B}(\kappa_l)|^2 = |B(\kappa_l) - \hat{A}_J R(\kappa_l - l_J)|^2 \tag{4.2}$$

can be used to estimate the bearing θ_0.[13]

If $|A_J|$ is of the order of $|A|$, we need to know θ_J (if $\theta_J \neq \theta_0$) to block the jammer. If so, compute (4.2) with l_J as the integer that makes θ_{l_J} closest to θ_J.

6 BROADBAND SIGNAL PROCESSING

Until now the signal has been assumed to be a single frequency plane wave. To exposite the processing of a broadband wave, let the array be linear and let

$$s(t,x_k) = s(t - c^{-1}x_k \cos \theta_0) + \epsilon(t,x_k) \tag{5.1}$$

where $s(t)$ is a bandlimited signal whose upper frequency is ω_u. Once again the noise is assumed to be Gaussian and to be stationary in time and space.

All signals are eventually transient. Select the time origin so that $s(t) = 0$ for $t < 0$ and $t > T$, where T is the signal duration. Let H be the largest integer less than or equal to $T\omega_u/2\pi$. The signal has the simple Fourier representation

$$s(t) = \sum_{j=-H}^{H} A(\omega j) \exp i\omega jt \tag{5.2}$$

where $\omega j = 2\pi j/T$ and

$$A(\omega j) = \frac{1}{T} \int_0^T s(t) \exp(-i\omega jt) dt. \tag{5.3}$$

Assume that the signal has zero mean, i.e. $A(0) = 0$.

Suppose that $s(t, x_k)$ is sampled at times $t_n = n\Delta (n = 0, \ldots, N_T - 1)$ where $\Delta = \pi/\omega_u$ and $N_T = [T/\Delta]$.[14] If $\cos \theta_0 > 0$, part of the leading edge of the signal is lost for $k > 1$. The trailing edge is lost when $\cos \theta_0 < 0$. These end-effects are negligible if $x_M/c \ll T$. Then from (5.2),

$$s_j(x_k) = N_T^{-1} \sum_{n=0}^{N_T-1} s(n\Delta, x_k) \exp(-i\omega_j n\Delta)$$

$$= A(\omega_j) \exp(-i\kappa_j x_k) + \epsilon_j(x_k) \quad (5.4)$$

where $\kappa j = (\omega_j/c) \cos \theta_0$ and $\epsilon_j(x_k)$ is a zero mean complex Gaussian variate. Its variance for large N_T is $E|\epsilon_j(x_k)|^2 \simeq N_T^{-1} S_\epsilon(\omega_j)$, where $S_\epsilon(\omega)$, the power spectrum of the noise, is independent of x_k. Thus the discrete Fourier transform of the received signals yields H single frequency waves plus filtered noise. For each $j = 1, \ldots, H$, compute

$$B(\kappa_l, \omega_j) = \sum_{k=1}^{M} s_j(x_k) \exp i\kappa_l x_k \quad (5.5)$$

for the κ_l grid discussed in Section 3. Concentrating on bearing estimation let $\hat{\theta}_j$ denote the look angle associated with the maximum $|B(\kappa_l, \omega_j)|^2/S_\epsilon(\kappa_l)$. The maximum-likelihood estimator of θ_0, denoted $\hat{\theta}_0$, is approximated by

$$\sum_{j=1}^{H} \sigma_j^{-2} \hat{\theta}_j / \sum_{j=1}^{H} \sigma_j^{-2}$$

where σ_j^2 is the large sample variance of $\hat{\theta}_j$.[15] Using the large sample approximation (3.5) with ρ replaced by $\rho(\omega_j) = N_T|A_j|^2/S_\epsilon(\omega_j)$ and λ_0 by $\lambda_j = 2\pi c/\omega_j$,

$$\hat{\theta}_0 \simeq \sum_{j=1}^{H} \omega_j^2 \rho(\omega_j) \hat{\theta}_j, \quad (5.6)$$

and its root-mean-square error is

$$\text{rms } \hat{\theta}_0 \simeq \frac{c \left(\sum_{j=1}^{H} \omega_j^2 \rho(\omega j) \right)^{-1/2}}{\sqrt{M} \, \sigma \sin \theta_0}. \quad (5.7)$$

The SNR $\rho(\omega_j)$ is estimable since $A(\omega_j)$ is precisely estimated by $M^{-1} B(\hat{\kappa}_0, \omega_j)$ when M is large.

Since a maximum-likelihood estimator has minimum mean-square error when the array size is large, $\hat{\theta}_0$ is *optimal* in a mean-square sense for nonsparse arrays. There is no need for ad-hoc bearing estimators for such arrays.

This paper has shown the connection between beamforming and frequency-wavenumber spectral analysis using discrete time and space measurements. A designer of a robust and effective array processing system must have a broad understanding of the relationships between physical models of propagating waves, the capabilities and limitations of existing hardware, and the statistical properties of estimators of the key parameters in the signal models. It is hoped that this paper will help promote such understanding.

REFERENCES

1. B.D. Steinberg, *Principles of Aperture and Array System Design,* Section 5.4 (Wiley, New York, 1976). His use of symbols differs from mine. For example, he uses θ_0 to denote the direction of arrival from the array normal. See also C.S. Clay and H. Medwin, *Acoustical Oceanography,* Section 5.3.2 (Wiley, New York, 1977).

2. Since there is a going versus coming ambiguity for real waves, it is impossible to distinguish between a θ in $\pi/2 < \theta < \pi$ and in $0 < \theta < \pi/2$. If the direction of energy propagation is measured, this ambiguity vanishes.

3. Carter shows that this design minimizes the asymptotic bearing variance for a linear array with a given minimum sensor spacing. See G.C. Carter, "Variance bounds for passively locating an acoustic source with a symmetric line array," *J. Acoust. Soc. Am. 62,* 922-926 (1977). This design is also optimal for range estimation. M.J. Hinich, "Passive Range Estimation Using Subarray Parallax," *J. Acoust. Soc. Am* 65 (5), 1229-1239 (1979).

4. Aliasing is defined in M.J. Hinich, "Processing spatially aliased arrays," *J. Acoust. Soc. Am. 64* (3), 792-794 (1978). Also see Steinberg, Section 5.2, op. cit.

5. M.J. Hinich and W.E. Weber, "Determination of the Nyquist frequency for unequally spaced data," ONR Tech Report 17 (revised). Virginia Tech (1980).

6. G.C. Carter and C.H. Knapp, "Time delay estimation," in *Proceedings of the 1976 IEEE Conference on Acoustics, Speech and Signal Processing* (IEEE, New York, 1976), pp. 357-360. W.J. Bangs and P.M. Schultheiss, "Space-time processing for optimal parameter estimation," in *Signal Processing,* ed. J.W.R. Griffiths, P.L. Stocklin, and C. van Schooneveld (Academic, New York, 1973), pp. 577-590.

7. D. Brillinger, *Time Series, Data Analysis and Theory,* Section 4.4 (Holt, Rinehart, and Winston, New York, 1975).

8. P. Bloomfield, *Fourier Analysis of Time Series: An Introduction,* Chapter 5 (Wiley, New York, 1976).

9. Levin heuristically derives the maximum-likelihood bearing estimator and its properties for a general three-dimensional array, assuming Gaussian noise. M.J. Levin, "Least-qsquares array processing for signals of unknown form," *Radio Electron. Engr. 29,* 213-222 (1965). These results are rigorously derived for a uniformly spaced array by Hinich and Shaman, M.J. Hinich and P. Shaman, "Parameter estimation for an r-dimensional plane wave observed with additive independent Gaussian errors," *Ann. Math. Statist. 43,* 153-169 (1972). The Cramer-Rao bound for the asymp-

totic bearing variance for a linear array is given by V.H. McDonald and P.J. Schultheiss, "Optimum passive bearing estimation in a spatially incoherent noise environment," *J. Acoust. Soc. Am.* **46**, 37-43 (1969). These rigorous derivations match Levin's results.

10. This simple result is derived in M.J. Hinich, "Frequency-wavenumber array processing," *J. Acoust. Soc. Am.* **69**, 732-737 (1981).

11. The maximum-likelihood estimator of A for correlated noise is given by J. Capon, R.J. Greenfield, and R.J. Kolker, "Multidimensional maximum-likelihood processing of a large aperture seismic array," *Proc. IEEE* **55**, 192-211 (1967).

12. C.S. Clay, M.J. Hinich, and P. Shaman, "Error analysis of velocity and direction measurements of plane waves using thick large aperture arrays," *J. Acoust. Soc. Am.* **53**, 1161-1166 (1973), and M.J. Hinich, "Estimating signal and noise using a random array," *J. Acoust. Soc. Am.* **71**, 97-99 (1982).

13. This is equivalent to steering a null at θ_j, and then estimating θ_0. See V.C. Anderson and P. Rudnick, "Rejection of a coherent arrival of an array," *J. Acoust. Soc. Am.* **45**, 406-410 (1969).

14. $[x]$ denotes the integer closest to x.

15. If $\hat{\theta}_1$ and $\hat{\theta}_2$ are maximum-likelihood estimators of θ_0 computed from two independent samples, the $(\sigma_1^{-2} + \sigma_2^{-2})^{-1}(\sigma_1^{-2}\hat{\theta}_1 + \sigma_2^{-2}\hat{\theta}_2)$ is maximum-likelihhod for the combined sample. Its variance is $(\sigma_1^{-2} + \sigma_2^{-2})^{-1}$. The large sample variance of a consistent estimator is equal to its mean square error when the sample size is large. Thus σ_j^2 is the right hand side of (3.5) with ρ replaced by $\rho(\omega_j)$ and λ_0 by $\lambda_j = 2\pi c/\omega_j$.

On Nonlinear Filtering and Tracking*

R. R. Mohler, W. J. Kolodziej, and R. S. Engelbrecht

Department of Electrical and Computer Engineering

H. D. Brunk

Departments of Statistics and Mathematics Oregon State University Corvallis, OR

1 INTRODUCTION

It is the *long-range objective* of the research, which is summarized here, to develop a working theory and an efficient methodology for nonlinear signal processing as appropriate for sonar tracking in a noisy underwater environment. The intent is to use all possible information which may be available about the target dynamics and its statistical characteristics about the ocean medium and about ambient noise sources. In general, this may mean working with "raw" data (as opposed to previously filtered data such as through a correlation detector). Ideally, any prefiltering would evolve from a generalized estimation process.

The basis for this research evolved from the principal investigator's NSF-sponsored research on bilinear systems, BLS (e.g., see Mohler, 1972; Mohler and Ruberti, 1973; Mohler and Kolodziej, 1980). Foundations of the nonlinear filtering approach used here are presented by Mohler and Kolodziej (1980, 1981) and by Lipster and Shiryayev (1978).

In some cases the system equations originally may be linear in the unobservable states, nonlinear only in the observable states, and take (or may be

---

*Work supported under Contract N00014-81-K-0814, Task NR 608-003 with the Office of Naval Research.

approximated by) the general stochastic form of

$$dx_1 = A(y,t)x_t dt + B(y,t)dt + G(y,t)dw^1$$

$$dy_t = C(y,t)x_t dt + D(y,t)dt + R(y,t)dw^2 \qquad (1)$$

where $w^1(\cdot)$, $w^2(\cdot)$ are independent vector Wiener processes; $A(\cdot)$, $B(\cdot)$, $C(\cdot)$, $D(\cdot)$, $G(\cdot)$, $R(\cdot)$ are appropriate matrix functionals in general. If it is assumed that the initial unobservable state, x_0, is conditional Gaussian with respect to the initial observable state, y_0, then under certain broad assumptions the optimal m.s.e. filter takes the form of

$$dm_t = (Am_t + B)dt + \Lambda dv_t,$$
$$d\Gamma_t = (A\Gamma_t + \Gamma_t A + GG^* - \Lambda\Lambda^*)dt,$$
$$\Lambda = (GR^* + \Gamma_t C^*)R_1^{-1}, \qquad (2)$$
$$dv_t = R_1^{-1}[dy_t - (Cm_t + D)dt],$$
$$R_1^2 = R(y,t)R^*(y,t),$$

where

$$m_t = E[x_t/\gamma_t], \quad \Gamma_t = \text{cov}(x_t/\gamma_t),$$

E denotes statistical expectation operator, Γ_t conditional covariance, * transposition, and γ_t is the σ-algebra generated by the observations on $[0,t]$.

If the system (e.g., target and observations) does not take this form directly, it is reasonable to approximate the system with nonlinear terms based on or adapted to the observation, y_t.

The methods which are summarized here are of a more general nature than the linear-model approach (or extended linear filtering) with Gaussian noise. They are developed to handle certain "bilinear" forms, for which the rigid-body motion equations are a good example or may be so approximated. Equations (1) in their general form may include modeling of a broad class of non-Gaussian noise sources, since certain non-Gaussian (and possibly non-stationary) processes may be modeled by Wiener noise passed through an appropriate nonlinear filter.

A very formal representation of the filter is given in Figure 1 to realize a functional understanding of its design.

In a crude fashion, the relative roles of the "target" model and "observation" model (as well as transmitted signals in active cases) are given in terms of $A(\cdot)$, $B(\cdot)$ and $D(\cdot)$, $C(\cdot)$, respectively, with $\Lambda(\cdot)$ affected by both models. Consequently, it is not too difficult to realize the roles: of "optimal" signal transmission for active sonar, of "optimal" observer maneuvers relative to target motion, of target parameters and of the medium's statistical description. These aspects are presently under investigation.

2 TRACKING-FILTER PROCESS

Two similar, but new, approaches to nonlinear filtering are presented here and tracking simulations have been made on three different computing facilities: CDC CYBER 70, HP 1000 and EAI Hybrid (100-680-690 combination) for

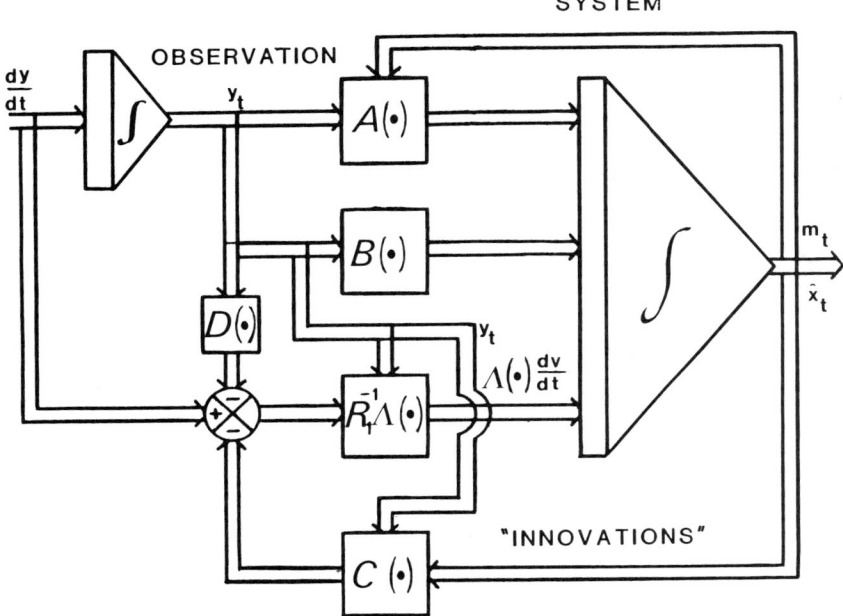

Figure 1 Nonlinear filter

comparison. Both methods depend on the method reviewed in Section 1, and are being compared along with a comparison of computers.

2.1 Nonlinear Filtering—Examples

The first approach approximates the general nonlinear model by a control model of a "bilinear" form. The "best" model approximation of this form is then computed by the appropriate optimal stochastic control policy (see Mohler and Kolodziej, 1981). The final step, with the feedback control a function of the estimated states $\hat{x}_t$, requires computation of the optimal (m.s.e.) state estimator by the method of Section 1.

Example 1. Consider a process with an absolute-value detector so that

$$dx_t = f|x_t|dt + \sigma_1 dw_t^1,$$

$$dy_t = hx_t dt + \sigma_2 dw_t^2, \tag{3}$$

where w_t^1, w_t^2 are independent Wiener processes, f, h, σ_1, σ_2 are appropriate constants and $t \in [0,T]$.

The approximation yields

$$dx_t \simeq \hat{u}_t x_t dt + \sigma_1 dw_t^1, \tag{4}$$

$$dy_t \simeq x_t dt + \sigma_2 dw_t^2,$$

where $\hat{u}$ is selected to minimize

$$E[\int_0^T (f|x_t| - \hat{u}_t x_t)^2 dt]. \tag{5}$$

The optimal (m.s.e.) filter for (4), with x_0 conditionally Gaussian w.r.t., y_0 has the form of

$$d\hat{x}_t = \hat{u}_t \hat{x}_t dt + \Gamma_t h (\sigma_2)^{-1} dv_t, \tag{6}$$

$$d\Gamma_t = [2\hat{u}_t \Gamma_t + \sigma_1^2 - h^2 \Gamma_t^2 (\sigma_2)^{-2}] dt,$$

where

$$dv_t = (\sigma_2)^{-1}(dy_t - h\hat{x}_t) dt$$

is the innovation process. From stochastic dynamic programming, it is found that an approximating optimal control is given by

$$\hat{u} \approx \tilde{u} = f \left[\frac{2\hat{x}\Gamma \exp(-\hat{x}^2/2\Gamma)}{\sqrt{2\pi\Gamma}\ (\hat{x}^2 + \Gamma)} + 2 \operatorname{erf}(-\hat{x}/\sqrt{2\Gamma}) \right]. \tag{7}$$

The EKF is obtained by $\tilde{\tilde{u}} = f \operatorname{sgn} \hat{x}$, with $\tilde{u} \to \tilde{\tilde{u}}$ as $\Gamma \to 0$, which holds for the general form of nonlinear system.

A comparison of the EKF and the two-step nonlinear filter, TNF, is given by Table I. Here, Q_1 and Q_2 are the mean-square "distances" of the TNF and the EKF, respectively.

Table 1 TNF-EKF Filter Comparison

Parameter Values				
	f	3	3	7
	h	1	1	1
	σ_1	5	8	8
	σ_2	1	2	2
	Q_1	2.21	7.47	36.7
	Q_2	2.45	8.98	39.2
% Improvement		11	21	7

While the TNF obviously is more accurate than the EKF, further consideration should be made to computational complexity for any given application.

Example 2. A simple second-order, linear target track is considered here with a nonlinear observation such that

$$dx_1 = x_2 dt,$$

$$dx_2 = -\alpha x_2 dt + \sigma_1 dw_t^1, \tag{8}$$

Nonlinear Filtering/Tracking

$$dy_1 = \left(x_1 + \frac{x_1 x_2}{c}\right) dt + \sigma_2 dw_t^2,$$

$$dy_2 = x_2 dt + \sigma_3 dw_t^3. \tag{9}$$

w_t^1, w_t^2, w_t^3 are independent Wiener processes. Here, α, a drag coefficient, c, acoustic velocity and standard deviations, σ_1, σ_2, σ_3, are assumed constant. The product term in the first observation equation arises due to target motion during transmission.

To fit the above control-approximation approach, the best bilinear observation approximation is obtained such that u_t minimizes

$$J(u) = E \int_0^T (x_1 x_2 - u_t x_1)^2 dt. \tag{10}$$

The filtering problem may be solved in two stages again. First, however, the KBF, $\hat{x}_1 = E[x_1/y_2]$, is obtained from the linear, y_2, observation. Then the nonlinear conditional-Gaussian filter, 2) NCF, improves the estimate by using y_1 observation.

In this manner, the KBF yields

$$d\hat{x}_1 = \hat{x}_2 dt + \frac{P_3}{\sigma_3} dv,$$

$$d\hat{x}_2 = -\alpha \hat{x}_2 dt + \frac{P_2}{\sigma_3} dv, \tag{11}$$

$$dv = \frac{1}{\sigma_3} (dy_2 - \hat{x}_2 dt)$$

where P_2, P_3 are the standard time-variant solutions to components of the Riccati equation. The NCF yields

$$d\overset{*}{x}_1 = \overset{*}{x}_2 dt + \frac{\Gamma_1}{\sigma_2}\left[1 + \frac{\hat{u}}{c}\right] dv,$$

$$d\overset{*}{x}_2 = -\alpha \overset{*}{x}_2 dt + \frac{\Gamma_3}{\sigma_2}\left[1 + \frac{\hat{u}}{c}\right] dv,$$

$$dv = \frac{1}{\sigma_2}\left\{dy_1 - \overset{*}{x}_1\left[1 + \frac{\hat{u}}{c}\right] dt\right\},$$

$$d\Gamma_1 = \left\{2\Gamma_3 - \frac{\left[1 + \frac{\hat{u}}{c}\right]^2}{\sigma_2^2} \Gamma_1^2\right\} dt \tag{12}$$

$$d\Gamma_2 = \left\{\Gamma_2 - \alpha\Gamma_3 - \frac{\left[1 + \frac{\hat{u}}{c}\right]^2}{\sigma_2^2} \Gamma_1 \Gamma_3\right\} dt,$$

$$d\Gamma_3 = \left\{ -2\alpha\Gamma_2 + \sigma_1^2 - \frac{\left(1 + \frac{\hat{u}}{c}\right)^2}{\sigma_2^2} \Gamma_3^2 \right\} dt.$$

The optimal control, computed after filter stage 1, in order to obtain (12), is

$$\hat{u} = \hat{x}_2 + \frac{2\hat{x}_1 P_3}{\hat{x}_1^2 + P_1}, \quad (13)$$

by solution of the Bellman-Hamilton-Jacobi stochastic equation.

A comparison of this TNF and KBF is given in Figure 2, and indicates the improved accuracy of nonlinear tracking. Obviously, more meaningful (higher-order, evasively controlled and nonlinear) target dynamics and observations need to be analyzed in the future.

2.2 Active Sonar Tracking

While a simplified tracking problem, which may represent either the passive or the active case, is presented in 2.1, certain front-end processing is assumed in the observation equation. However, it is the eventual intent of this project to analyze the raw-data formulation to whatever extent is possible. Ideally, optimal detection may evolve as part of the filter signal-processing.

It is the purpose of the observation model to describe effectively the target state or at least position and possibly velocity. And, in the active-sonar case this is accomplished by measuring the time-delay, T_d, and time-compression, σ (or Doppler frequency shift) of echoes which are distortions of a transmitted signal due to random scatterings, reverberations, inhomogeneities target-radiated noise, ambient noise, etc.—though some, such as ambient noise may not be significant in the active frequency range of interest. Obviously, approximations are required to make the problem amenable to solution by even nonlinear processing.

While it is known that multiplicative noise distortion is present, it might first be modeled with the additive noise, so that a somewhat simplified model of the received signal takes the form of

$$R_t = a(t)s[(t - T_d)(1 - \delta)] + \sigma n_t, \quad (14)$$

where $s(t)$ is the transmitted signal, $T_d \approx 2r_t/c_t$, $\delta \approx 2\dot{r}_t/c_t$, r_t is range of target from receiver-transmitter, c_t is acoustic velocity, and n_t is additive noise with perhaps constant σ. Subscript t refers to random functions of time. Multiple echoes would result in a summation of terms similar to (14).

As a simple example assume the range is defined by

$$\ddot{r} = -\alpha\dot{r}^2 + u, \quad (15)$$

where α is a drag-force coefficient and u is a thrust-control term which may or may not be known or statistically known to the "observer." For convenience, $s(t)$ is assumed to be a simple square pulse, width T_0 and unity amplitude, which is repeated every T seconds, so that

$$y_t = s(t - r/c) + \sigma n_t, \quad (16)$$

Nonlinear Filtering/Tracking

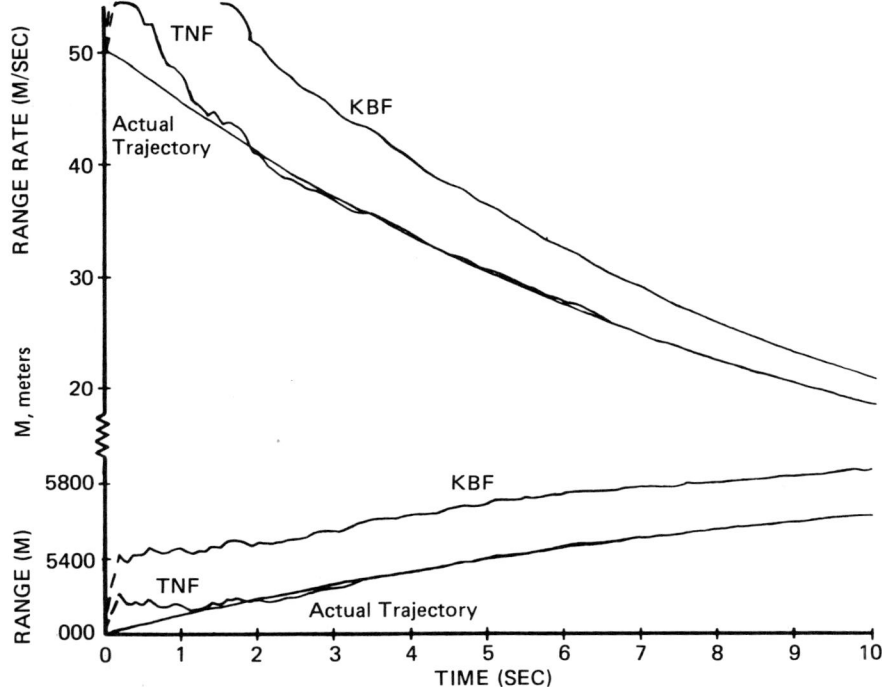

Figure 2 Rigid body tracking

where σ is assumed zero, c is constant, and n_t is a Gaussian random process of mean zero and unity variance.

The target model is described by state variables $x_1 = t - r/c$, $x_2 = \dot{r}$ for convenience. Then approximation of the nonlinear drag term, by again a bilinear stochastic feedback of the observation, results in the following recursive filter:

$$\frac{d\hat{x}_1}{dt} = -\frac{\hat{x}_2}{c} + 1 + P_1 \beta v,$$

$$\frac{d\hat{x}_2}{dt} = -\alpha(P_2 + \hat{x}_2^2) + u + P_3 \beta v,$$

$$\frac{dP_1}{dt} = -\frac{2}{c} P_3 - \beta^2 P_1^2, \qquad (17)$$

$$\frac{dP_2}{dt} = -4\alpha \hat{x}_2 P_2 - \beta^2 P_3^2,$$

$$\frac{dP_3}{dt} = -\frac{P_2}{c} - 2\alpha \hat{x}_2 P_3 - \beta^2 P_1 P_3,$$

where the innovations are

$$v = \frac{1}{\sigma}[y - \text{erf}(\hat{\alpha}) + \text{erf}(\hat{\alpha} - T_0/\sqrt{P_1})],$$

$$\beta = \frac{\exp(-0.5\hat{\alpha}^2) - \exp[-0.5(\hat{\alpha} - T_0/\sqrt{P_1})^2]}{\sqrt{2\pi P_1}\,\sigma},$$

$$|\hat{\alpha}| = \min_{n \text{ (positive integer)}} \left|\frac{\hat{x}_1 - nT}{\sqrt{P_1}}\right|,$$

and

$$\text{erf}(\alpha) = \frac{1}{\sqrt{2\pi}} \int_0^\alpha \exp(-0.5\zeta^2)\,d\zeta.$$

A hybrid simulation showed this filter to track the simple nonlinear target model very accurately. But again, more meaningful models of the target and its distorted observation need to be studied. Comparisons will be made with the more ad-hoc adaptive techniques of Pryor (1977), Chan, et al. (1980), Moose (1982) and Singer and Behnke (1971).

3 NOISE AND DISTORTION CHARACTERISTICS

Traditionally, it is assumed that the sea provides a linear medium, that transmitted sonar signals are of narrow frequency band, that ocean turbulence and seismic activity are significant from 1 to 10 hertz, that shipping noise is significant in the 10-300 hertz range, that above 300 hertz wind and sea-surface activity are significant, that noise is Gaussian, etc. Consequently, ambient noise is very directionally dependent (i.e., low frequencies, shipping, important in horizontal direction, and high frequencies, seastate, in the vertical). But these assumptions, particularly linearity, Gaussian and narrow-band aspects are being investigated by this project.

3.1 Ambient Noise

The periodogram estimate of noise power spectral density (psd) historically has been the basis of experimental ocean-related research. For example, researchers at Woods Hole several decades ago found jagged estimates as a consequence of too-narrow aperture filters in their measurements. It was found that "window" shaping or ad-hoc filtering techniques improved the estimates. Next, digital techniques led to improved estimates via correlation and Fourier transforms. Then, Fast Fourier Transform algorithms, along with ad-hoc filters, led to direct psd estimates which are computed efficiently and accurately in many cases.

While the periodogram estimate has traditionally been used by engineers to define psd, it is known to be a bad estimate in most cases. Indeed, it has been shown that in the limits, used for its definition, it has a variance larger than that of the original process itself, even in the Gaussian case. However, most of the ad-hoc filtering decreases this variance, but is usually designed for Gaussian noise of particular frequency ranges. In fact, it has been established that non-Gauss continuous processes as well as point processes do exist in the environment of interest here.

Wahba (1980) proposed a novel approach to estimating the spectral density of a zero mean, stationary, Gaussian time series. This method exploits the asymptotic independence of the values of the log periodogram (cepstrum) to transform the problem into an ordinary regression problem. Suppose the spectral density, f, is to be estimated over the interval $(0, 1/2)$, and that the time series has been observed at $2N$ equally spaced time points. Let I_j denote the periodogram evaluated at frequency $j/2N$. The random variables $(\log I_j)$ are (asymptotically, as $N \to \infty$) independent, with known variances. Their means after appropriate translation are the values $g(j/2N)$ of $g = \log f$; thus g is the regression function to be estimated.

Wahba obtains her estimates by starting with the sample Fourier coefficients of the log periodogram, and damping the coefficient at a given frequency by a "filter function" of a certain class that depends on two parameters. These parameters are to be determined from the data to minimize (approximately) expected integrated squared error.

Brunk (1980) developed the application of Bayes least squares linear methods to the estimation of regression functions represented by orthogonal expansions. In the present setting, it leads to estimates of a general class that includes Wahba's. A significant difference is that the Bayes approach provides an interpretation of the filter function in terms of prior variances of the coefficients in the expansion of the regression function. From this point of view, one should expect the investigator to assign "prior" variances in such a way as to reflect his opinion concerning the regression function to be estimated. If, for example, on the basis of his knowledge and experience, he expects a very smooth spectral density, this opinion can be reflected in his assignment of variances to the coefficients: coefficients of rapidly varying functions in the orthogonal series are expected to be small. In particular, these assignments, and hence the filter functions, should be independent of sample size; indeed, they should be completely independent of the data.

Current investigations suggest that this is a promising approach. It is to be applied in the near future to records of ambient noise from a NUWES Nanoose tape. In this connection it should be pointed out that the asymptotic theory on which the approach is based does not at all require that the time series be Gaussian; the (I_j) have the same asymptotic joint distribution whenever the stationary time series is purely nondeterministic, a moving average of uncorrelated innovations with zero means and the same variance.

This is a logical first step in a study of distributions of ambient noise. A further step will be to estimate the common distribution of the random variables of the innovations process, assumed independent and identically distributed. In principle, having estimated the spectral density, one has the coefficients of the moving average representation of the (purely nondeterministic) time series as Fourier coefficients of the square root of the spectral density. Then Box-Jenkins methods yield the individual random variables of the innovations process. These constitute a simple random sample from their common distribution, which can then be estimated in any of a number of ways. A Bayes least squares linear method is proposed which is applied under the assumption that the common density of the variables in the innovations process has itself an expansion in an appropriate orthonormal series; as, for example, a series of Walsh functions.

One tape of ambient noise data taken at Nanoose (NUWES) is already being analyzed with the proposed methodology and fitted with rational power spectra for tracking analysis.

A preliminary analysis of this data suggests with only psd information available, surface craft noise (from this particular data) could be modeled approximately by $\frac{ds}{dt} = \frac{s}{T} + \sigma n$. For this data (reference number 463-1301, Nanoose 7-28-8) $T = 0.384 \, (10)^{-3}$ sec, $20 \log \sigma = 90.88$ dB. n is normalized white noise. An added harmonic appears at about 650 hz. Propeller noise spectrum is a bit more complex (with the same harmonic appearing in the data) but approximately modeled by white noise passed through a linear filter of transfer function form:

$$k \left(\frac{a_1 s^2 + 2\zeta_1 a_1 s + 1}{s^2(a_2 s^2 + 2\zeta_2 a_2 s + 1)} \right).$$

Here, $a_1 = 0.89 \, (10)^{-3}$, $a_2 = 0.48 \, (10)^{-4}$, $\zeta_1 \approx \zeta_2 \approx 0.7$, $20 \log k \approx 79$ dB.

3.2 Acoustic Channel

The sonar system senses the acoustic pressure field which is a function of time and space coordinates. The medium traditionally is described by a linear system with impulse response dependent likewise on two time variables and three space variables which may include multiplicative noise, time smear, frequency smear and frequency wave number or angular-sector smear.

At short ranges a spherical spreading model (r^{-2} signal-energy dependence in received signal) and at long ranges a cylindrical spreading model (r^{-1} dependence) is assumed in deriving the received signal and observation model (with $r^{-3/2}$ sometimes used for interaction with the seabed such as for shallow-water ranges). However, absorption losses, with exponential range dependence, usually dominate at long ranges.

The sound velocity c in ocean water is a function of several physical parameters, but of these only temperature exhibits enough variability to produce a sizeable random effect ($dc/dT \doteq 4.2$ meters/sec-degree C). If we ignore the more-or-less permanent temperature gradient with water depth which produces such important effects like sound refraction and deep-water focussing, the main variation of temperature is due to internal waves. These result in disc-shaped regions of water with different temperature relatively to their surrounding (typically 0.1-1.0 degrees Celsius, several kilometers horizontal and several hundred meters vertical in extent). These temperature "blobs" move about slowly (1 to 10 cm/sec) and thus generate corresponding slow random variations in local sound velocity. The overall effect of these sound-speed inhomogeneities on the amplitude and phase of an acoustic signal received at a distance $|\vec{r}_s - \vec{r}|$ from a point source, and consisting of a single frequency ω, is concisely given in terms of the medium's so called "strength parameter" Φ and its "diffraction parameter" Λ, as used in the analysis of wave propagation through turbulent media (Tatarskie, 1964):

$$\Phi^2 = E \left[\frac{\omega}{c_0} \int_0^{|\vec{r}_s - \vec{r}|} c(x) dx \right]^2,$$

where

x = distance from point source,

c_0 = average sound velocity (1.5 km/sec),

$E(\cdot)$ = expectation operator,

$$\Lambda \cong \frac{|\vec{r}_t - \vec{r}| c_0}{6\omega l_z^2},$$

and l_z = vertical extent of temperature "blobs."

With $H_0(\omega)$ the deterministic part, the acoustic medium may be approximated by a transfer function of form:

$$\frac{H(\omega)}{H_0(\omega)} = A(t) \exp(i\phi(t)).$$

The dependence of $A(t)$ and $\phi(t)$ on Φ and Λ is complicated in general. However for the regions in the Φ-Λ plane which are far from the boundaries $\Phi = 1$, $\Lambda = 1$, and $\Phi\Lambda = 1$, the following approximate first-order distributions are obtained:

Table 2   Medium Inhomogeneity Distortion

Region	$A(t)$:	$\Phi(t)$:
$\Phi \ll 1$ $\Lambda \ll 1$	lognormal $\ln A(t) = \text{normal}\left(0, \frac{\Lambda\Phi^2}{4}\right)$	normal $(0, \Phi^2)$
$\Phi \ll 1$ $\Lambda \gg 1$	Rice-Nakagami Mean: 1 Variance: $\exp(\Phi^2)$	normal $(0, \Phi^2/2)$
$\Phi \gg 1$ $\Phi\Lambda \ll 1$	lognormal $\ln A(t) = \text{normal}\left(0, \frac{\Lambda\Phi^2}{4}\right)$	uniform $(0, 2\pi)$
$\Phi \gg 1$ $\Phi\Lambda \gg 1$	Rayleigh	uniform $(0, 2\pi)$

To a first approximation, the following relations hold between the parameters Φ and Λ, and the range $|\vec{r}_s - \vec{r}|$ (km) and frequency ω (hz):

$$\Lambda \simeq 0.1 \frac{|\vec{r}_s - \vec{r}|}{\omega}$$

$$\Phi \simeq 1.5 \times 10^{-3} \omega \sqrt{|\vec{r}_s - \vec{r}|}.$$

4 CONCLUSIONS

A new approach to nonlinear filtering is presented which seems effective for sonar target tracking. Preliminary studies and simulation lay the foundation for continued research. More meaningful target models, observation models and noise characterizations are being developed along with new signal processing methodologies. General extensions, where convenient, are being made to aerospace and immune defense systems.

ACKNOWLEDGMENT

The investigators wish to acknowledge the computational and analytical services of R. Bucolo, R. Rathja, T. Halawani, J. Platt and M. Lee. The support of ONR through Contract No. N00014-81-K-0814 is gratefully acknowledged.

REFERENCES

Brunk, H.D. (1980). Bayesian least squares estimates of univariate regression functions. *Comm. Stat. Theor. Method.*, $A9$:1/01-1136.

Chan, Y.T., Plant, J.B. and Bottomley, J.R.T. (1980). A Kalman Tracker with a Simple Input Estimator. *IEEE Trans. Aerospace and Electron. Sys.*, $AES\text{-}18$:235-240.

Lipster, R.S. and Shiryayev, A.N. (1978). *Statistics of Random Processes II — Application.* Springer-Verlag, New York.

Mohler, R.R. (1973). *Bilinear Control Processes.* Academic, New York.

Mohler, R.R. and Ruberti (eds.) (1972). *Theory and Applications of Variable Structure Systems.* Academic, New York.

Mohler, R.R. and Kolodziej (1980). An overview of bilinear stochastic systems. *IEEE Trans. Sys. Man. Cyb.*, 10:913-920.

Mohler, R.R. and Kolodziej (1981). Optimal control of a class of nonlinear stochastic systems. *IEEE Trans. Auto. Contr.*, $AC26$:1048-1053.

Moose, R.L. (1982). Adaptive target tracking of underwater maneuvering targets using passive targets. ONR Annual Report. (Also see this volume.)

Pryor, C.M. (1977). An adaptive Kalman filter tracker for multi-mode range/Doppler sonar. *Aspects of Signal Processing, Part I (G. Tacconi, ed.)*:265-278. D. Reidel, Dordrecht-Holland.

Singer, R.A. and Behnke, K.W. (1971). Real-time tracking filter evaluation and selection for tactical targets. *IEEE Trans. Aerosp. and Electron. Sys.*, $AES\text{-}7$:100-110.

Tatarskie, V.I. (1961). *Wave Propagation in a Turbulent Medium.* McGraw-Hill, New York.

Wahba, G. (1980). Automatic smoothing for the log periodogram. *J. Amer. Statist. Assn.* 72:122-132.

Generalized Search Optimization*

Lawrence D. Stone

Daniel H. Wagner Associates
Sunnyvale, CA

1 RECENT DEVELOPMENTS IN OPTIMAL SEARCH FOR MOVING TARGETS

Prior to 1977 there were few results on optimal search for moving targets. Reviewing Stone [1975] one can see that the theory for solving stationary target problems was well developed, but there were few moving target problems that had been solved. For a special class of problems involving conditionally deterministic motion with a factorable Jacobian, one could find optimal plans. This special class of moving target problems can be reduced to stationary target problems and solved by those methods. For other types of moving target problems, Markovian motion was usually assumed. Even here one sees that the typical problem for which one could obtain a solution involved a discrete-time, two-cell model which was solved using dynamic programming. For problems involving "large" numbers of cells (e.g., ten!), the curse of dimensionality prevents numerical solution by this technique. Necessary and sufficient conditions had been obtained for Markov processes and exponential detection functions (Saratsalo [1973]), but no one had been able to apply these conditions to find optimal plans.

In 1977 Brown applied the Karush-Kuhn-Tucker conditions to the problem of finding optimal plans for targets that move in discrete space and time when the detection function is exponential. By writing these conditions in a suitable form, he observed that the optimal plan for the moving target problem had the interesting property that if one selected a time t and conditioned on failure at all times other than t (both before and after t), then the optimal plan allocated the effort at time t so as to maximize the detection probability for the *stationary* target problem one obtained from the above conditioning. Since there are very efficient methods for finding optimal plans for stationary targets,

---

*This paper was written with the support of ONR contract N00014-80-C-0766.

especially when the detection function is exponential (see Example 2.2.8 of Stone [1975]), Brown took advantage of this fact to devise an iterative algorithm which maximizes the probability of detecting the target some time in the interval $[0, \overline{T}]$. This algorithm applies to target motions which are modeled by a mixture of discrete time and space Markov chains and is very efficient. See Brown [1980].

In Stone et al, [1978], algorithms were devised for arbitrary discrete time target motions and exponential detection functions. Stone [1979] generalized the necessary and sufficient conditions of Brown [1980] to target motions which are modeled by an arbitrary stochastic process with any mixture of discrete or continuous space or time. This generalization also applies to the wider class of regular detection functions (See Stone [1975]). Recently Arnold [1982] has shown that optimal search plans exist for a wide class of search problems in which the search density is bounded.

All of the above results apply to the problem of finding a plan which maximizes the probability of detecting a target by time $\overline{T}$. In Stromquist and Stone [1981], the necessary and sufficient conditions for optimal detection search were generalized to a wider class of constrained optimization problems which include problems not related to search as well as numerous search related ones. In the following sections, we state the necessary and sufficient conditions of Stromquist and Stone [1981] and outline how these conditions can be used to produce algorithms to find numerical solutions. The combination of the necessary and sufficient conditions and the algorithms derived from them constitute the Generalized Search Optimization technique.

2 NECESSARY AND SUFFICIENT CONDITIONS FOR CONSTRAINED OPTIMIZATION

In this section we describe the necessary and sufficient conditions which form the basis of the GSO. These conditions apply to a general class of constrained optimization problems.

Let

Y be a σ-finite measure space with measure ν

T be a σ-finite measure space with measure τ

$Z = Y \times T$ have the product measure $\mu = \nu \times \tau$.

In search problems the space Y is usually the search space, e.g., Euclidean two-space, and T is the time interval of the search.

Constraints. Let $c: Z \to (0, \infty)$ be μ-measureable, $M: T \to [0, \infty)$ be τ-measurable and $B \epsilon (0, \infty)$. Define

Ψ = the set of real-valued μ-measurable functions $\psi: Z \to [0,B]$ such that

$$\int_Y c(y,t)\psi(y,t)d\nu(y) \leq m(t) \text{ for a.e. } t \epsilon T \quad (1)$$

Let Ψ_0 be the subset of Ψ for which equality holds in Equation (1).

Functional. Let P be a real valued functional on Ψ.

Gateaux Differential. In order to state the necessary and sufficient conditions of the Generalized Search Optimization technique, we must introduce the notion of a Gateaux differential. Define the *Gateaux differential* of P at ψ in the direction h as

$$P'[\psi, h] = \lim_{\epsilon \to 0} (P[\psi + \epsilon h] - P[\psi])/\epsilon \tag{2}$$

if it exists. If P' exists and is given by

$$P'[\psi, h] = \int_Z d(\psi, y, t) h(y, t) d\mu(y, t), \tag{3}$$

then we say $d(\psi, \cdot, \cdot)$ is the *kernel* of the Gateaux differential at ψ.

Constrained optimality. We say that $\psi^* \epsilon \Psi$ is *optimal within* Ψ if $P[\psi^*] \geq P[\psi]$ for $\psi \epsilon \Psi$.

Theorems 1 and 2 below are taken from Stromquist and Stone [1981].

THEOREM 1. *If P has a Gateaux differential with kernel d, then a necessary condition for ψ^* to be optimal within Ψ is the existence of a measureable function $\lambda: T \to [-\infty, \infty]$ such that for a.e. $(y, t) \epsilon Z$*

$$d(\psi^*, y, t) \geq \lambda(t) c(y, t) \text{ if } \psi^*(y, t) = B$$
$$= \lambda(t) c(y, t) \text{ if } 0 < \psi^*(y, t) < B$$
$$\leq \lambda(t) c(y, t) \text{ if } \psi^*(y, t) = 0. \tag{4}$$

THEOREM 2. *If the conditions of Theorem 1 are satisfied, P is a concave functional, and $\psi \epsilon \Psi_0$, then condition (4) is sufficient for ψ^* to be optimal within Ψ.*

For concave functionals condition (4) gives a necessary and sufficient condition for a constrained optimum. In the remainder of this section we outline an algorithm which can be applied to discrete time problems to find optimal allocations ψ^*. The outline is intended to cover a wide range of problems which fit the framework given in this section. As a result the directions necessarily lack detail. For example, step (v) below requires the developer of the algorithm to find a function which satisfies condition (4) for one time t and also satisfies the constraint in Equation (5). In the typical detection search problem, step (v) amounts to solving a stationary search problem. Even though this may appear formidable to the the reader, it is often the case that when the specific functions are substituted into the algorithm, the way to proceed is clear. In any case this outline provides a method of attack which has proven successful in a number of problems some of which are discussed in Section 3. Having developed an algorithm along the lines of this outline one must still prove that the algorithm converges.

GSO Algorithm for discrete time. Suppose that time is discrete with $t = 0, 1, \ldots, \bar{T}$. The algorithm proceeds as follows:

(i) Make an initial guess $\psi_0 \epsilon \Psi$ for an allocation.
(ii) Set $n = 0$.
(iii) Set $s = n [mod(\bar{T}+1)]$.
(iv) Compute $d(\psi_n, \cdot, \cdot)$.
(v) Solve for $f^*: Y \to [0, \infty)$ to satisfy

$$\int_Y c(y, s) f^*(y) d\nu(y) = m(s) \tag{5}$$

and so that the function ψ_{n+1} defined by

$$\psi_{n+1}(\cdot,t) = \begin{cases} \psi_n(\cdot,t) & \text{for } t \neq s \\ f^* & \text{for } t = s, \end{cases}$$

satisfies condition (4) for $t = s$.

(vi) Compute the generalized Washburn upper bound (See Washburn [1981] and Stromquist and Stone [1981].). If this shows $P[\psi_{n+1}]$ to be within the desired tolerance of the optimal payoff $P[\psi^*]$, then stop. Otherwise go to step (vii).

(vii) Set $n = n+1$ and go to step (iii).

Step (v) is often accomplished by choosing a value for $\lambda(s)$ and solving for the function f for which ψ defined by

$$\psi(\cdot,t) = \begin{cases} \psi_n(\cdot,t) & \text{for } t \neq s \\ f & \text{for } t = s, \end{cases}$$

satisfies condition (4) for $t=s$. This is often easy to do but usually leads to functions f that do not satisfy Equation (5). However, since f is often a monotone function (pointwise) of the value of $\lambda(s)$, one can usually perform a binary search to determine the value of $\lambda(s)$ that yields an f which satisfies (5) to any accuracy desired.

Another method of performing step (v) is to determine an equivalent optimization problem whose solution will yield f^*. This was done by Brown [1980] in his original algorithm for maximizing the probability of detecting a target with discrete space and time Markov motion. In this case step (v) is equivalent to finding an allocation f^* to maximize the probability of detecting a stationary target.

3 APPLICATIONS OF GSO

In this section we illustrate how the Generalized Search Optimization technique can be applied to problems in search and other areas.

Detection Search. Suppose that a target is moving through the plane Y according to a discrete time stochastic process $X = \{X_t : t=0, \ldots, \overline{T}\}$. We have $m(t)$ effort available at time t and the function $\psi \in \Psi$ specifies $\psi(y,t)$ the search effort density placed at point y at time t for $y \in Y$ and $t = 0, \ldots, \overline{T}$. We take $T = \{0, \ldots, \overline{T}\}, Z = Y \times T, \nu$ to be Lesbegue measure, and τ to be counting measure. We let $c(y,t) = 1$ for $y \in Y, t \in T$.

For each sample path ω of the process X, the probability of detecting the target by time t given it follows that path is a function of the weighted total effort density

$$\zeta(\psi,\omega,t) = \sum_{s=0}^{t} W(X_s(\omega),s)\psi(X_s(\omega),s)$$

which accumulates by time t on the target over the course of the path. The weight $W(y,s)$ represents the relative detectability or sweep width against the target if it is located at point y at time s. There is a function $b:[0,\infty] \to [0,1]$ such that $b(\zeta(\psi,\omega,t))$ is the probability of detecting the target by time t given it follows the path ω and search plan ψ is executed. Letting E denote expectation over the sample paths of X, we define

$$P[\psi] = E[b(\zeta(\psi,\cdot,\bar{T}))]$$

to be the probability of detecting the target by time $\bar{T}$ with plan ψ. In the remainder of this discusion, we will suppress the variable ω.

The optimal detection problem is to find a plan $\psi^* \in \Psi$ such that $P[\psi^*] \geq P[\psi]$ for all $\psi \in \Psi$. Such a plan is called $\bar{T}$-optimal.

For this problem the Gateaux differential has a kernel provided that the function b has a bounded nonnegative derivative b'. Let E_{yt} denote expectation conditioned on $X_t = y$, and let p_t be the probability density function for X_t. The Gateaux differential of P is given by

$$P'[\psi,h] = \sum_{t=0}^{\bar{T}} \int_Y d(\psi,y,t) h(y,t) dy$$

where

$$d(\psi,y,t) = E_{yt}[b'(\zeta(\psi,\bar{T}))] W(y,t) p_t(y), \quad y \in Y, t \in T.$$

The hypotheses of Theorems 1 and 2 hold so that condition (4) is necessary and sufficient for $\psi^* \in \Psi_0$ to be a $\bar{T}$ optimal search plan.

The conditions of Brown [1980] are the special case of condition (4) that one obtains by taking $b(z) = 1 - exp(-z)$ for $z \geq 0, Y = \{1, \ldots, J\}$, and ν to be counting measure. In this case

$$d(\psi,y,t) = E_{yt}[exp(-\sum_{s \neq t} W(X_s,s)\psi(X_s,s))]exp(-W(y,t))\psi(y,t)) W(y,t) p_t(y)$$

and one can see that g_t defined by

$$g_t(y) = E_{yt}[exp(-\sum_{s \neq t} W(X_s,s)\psi(X_s,s))]p_t(y) \text{ for } y \in Y,$$

is proportional to the probability density of the target's location at time t given failure to detect the target by the search applied before and after time t but not during time t. Condition (4) becomes

$$g_t(y) \exp(-W(y,t)\psi^*(y,t)) W(y,t) = \lambda(t) \text{ if } \psi^*(y,t) > 0$$
$$< \lambda(t) \text{ if } \psi^*(y,t) = 0$$

which is equivalent to the conditions for $\psi^*(\cdot,t)$ to be an optimal allocation for the stationary search problem with target location density proportional to g_t.

Multistate Target Search. A generalization of the detection search described above is the multistate target search. In this case the target's motion and state are represented by a stochastic process $(X,S) = \{(X_t,S_t): t=0, \ldots, \bar{T}\}$ where X_t is the target's position at time T and S_t is the target's state at time t. The target may change state as well as location stochastically, and the target's state can affect the target's motion as well as its detectability. As an example consider a case where there are K states and the sweep width is a function of location, state, and time so that cumulative effort ζ becomes

$$\zeta(\psi,\overline{T}) = \sum_{t=0}^{\overline{T}} W(X_t,S_t,t)\psi(X_t,t)$$

and

$$P[\psi] = E[b(\zeta(\psi,\overline{T}))]$$

as before. Observe that effort cannot be allocated to states but only to locations. Let E_{ykt} denote expectation conditioned on $(X_t,S_t) = (y,k)$ and let $P_t(y,k) = Pr\{X_t=y, S_t=k\}$. Assuming that b has a bounded derivative b', Discenza and Stone [1981] show that P has a Gateaux differential with kernel

$$d(\psi,y,t) = \sum_{k=1}^{K} E_{ykt}[b'(\zeta(\psi,\overline{T}))] W(y,k,t) p_t(y,k) \qquad (6)$$

for $y \in Y$, $t = 0, \ldots, \overline{T}$.

Using the above definition of d, one can show that the necessary and sufficient conditions obtained by Discenza and Stone [1981] are also a special case of condition (4).

Two special cases of multistate search are survivor search and defensive search. In survivor search, the target may be a person missing at sea or lost in the wilderness. The state represents the condition of the survivor, e.g., in a boat, in a life raft, in the water, or dead. By setting the sweep width equal to its appropriate value for each state and to zero if the person is dead, the problem of finding a plan to maximize P becomes that of finding a plan that maximizes the probability of finding the target alive by time $\overline{T}$.

For defensive search, as defined by Brown, we are trying to detect an attacker before it launches a weapon. In this case the process has two states, weapon launched or not launched. Once the attacker launches a weapon the sweep width is set to zero and the target remains in the launched state for the remainder of the problem. In this case maximizing P is maximizing the probability of detecting the attacker before it launches an attack.

Surveillance. Tierney and Kadane [1982] have developed a technique for solving surveillance problems which builds on the optimal detection search results discussed above. The surveillance problem is to maximize the probability of being in contact (i.e., having a detection on the target) at time $\overline{T}$. In contrast to the detection search problem a detection before time $\overline{T}$ does not end the problem. It merely helps to obtain a detection at time $\overline{T}$. For problems where the target's motion is modeled by a discrete time and space Markov chain, Tierney and Kadane have shown that the optimal surveillance problem can be solved by solving a series of optimal detection search problems. In their method, one starts at time $\overline{T}$ and works his way backward in time in a fashion similar to dynamic programming. At each time t, one must solve what Tierney and Kadane call a general detection search problem given knowledge of the target's position at time t.

In the general detection problem, the searcher receives a payoff or return $r(j,t)$ if he detects the target in cell j at time t. The search stops the first time the target is detected, and the objective of the general detection problem is to maximize the expected payoff. Suppose that $\overline{T}$ is the time horizon, i.e., $r(j,t) = 0$ if $t > \overline{T}$. Let $\zeta(\psi,-1) = 0$. Then the expected payoff P using plan ψ is given by

$$P[\psi] = E[\sum_{t=0}^{\bar{T}} r(X_t,t)[b(\zeta(\psi,t)) - b(\zeta(\psi,t-1))]$$
$$= E[\sum_{t=0}^{\bar{T}} [r(X_t,t) - r(X_{t+1},t+1)]b(\zeta(\psi,t))] \quad (7)$$

and the Gateaux differential has the kernel $d(\psi,\cdot,\cdot)$ given by

$$d(\psi,j,t) = \sum_{s=t}^{\bar{T}} E_{jt}[\{r(X_s,s) - r(X_{s+1},s+1)\}b'(\zeta(\psi,s))]p_t(j)W(j,t) \quad (8)$$

for $j = 1, \ldots, J$ and $t = 1, \ldots, T$.

Equation (7) is a special case of Equation (16) in Stromquist and Stone [1981]. As noted there, conditions (4) are necessary but not sufficient for optimality in this problem. If one makes the additional assumption that $r(j,\cdot)$ is a decreasing function for $j = 1, \ldots, J$, then the conditions become sufficient. In the detection problems which arise in solving the surveillance problem, however, the functions $r(j,\cdot)$ are likely to be increasing rather than decreasing.

Optimal Resource Extraction. In Lipshutz and Stone [1981], the GSO technique is applied to finding the optimal extraction rate of a nonrenewable resource with uncertain reserves.

Let F be the distribution function of the total resource available, and let $r(t,v,w)$ be the instantaneous return we obtain at time t when extracting at the rate v given that w amount of resource has already been extracted. Let $Z:[0,\bar{T}] \to [0,\infty)$ denote an extraction plan, i.e., $Z(t)$ is the rate at which resources are extracted at time t. The time horizon $\bar{T}$ may be finite or infinite, and we can impose a constraint B on the rate at which resources can be extracted. Let δ be the discount rate for income received in the future. The expected discounted return, $R(Z)$, using plan Z is given by

$$R(Z) = \int_0^{\bar{T}} e^{-\delta t} r(t,Z(t),X(t))[1 - F(X(t))]dt$$

where

$$X(t) = \int_0^t Z(s)ds \text{ for } t > 0.$$

The objective is to find Z^* to maximize R.

Lipshutz and Stone use Theorem 1 to find necessary conditions for an optimal extraction plan and to obtain a version of the result of Loury [1978] which states that for an optimal extraction plan, the discounted marginal return at time t is equal to the expected value of the discounted average return at the time extraction stops. The version obtained by Lipshutz and Stone applies to more general return functions r and time horizons $\bar{T}$ than considered by Loury.

For discrete time problems, Lipshutz and Stone use the necessary conditions to develop an algorithm for computing optimal extraction plans.

Other Applications. The GSO technique has also been applied by Stone [1982] to find optimal consumption/investment plans and by Weisinger [1982] to find investment plans which maximize terminal wealth.

REFERENCES

Arnold, L.K., 1982. Existence of Search Plans when Feasible Plans are Uniformly Bounded, *Math. Opns. Res.* (To Appear).

Brown, S.S. 1980. Optimal Search for a Moving Target in Discrete Time and Space. *Opns. Res.* 28, 1275-1289.

Discenza, J.H., and L.D. Stone, 1981. Optimal Survivor Search with Multiple States. *Opns. Res.* 29, 309-323.

Lipshutz, R.J. and L.D. Stone, 1982. Optimal Resource Extraction with Uncertain Reserves. (Submitted for Publication).

Loury, G.C. 1978. The Optimum Exploitation of an Unknown Reserve. *Review of Economic Studies.* 45, 621-636.

Saratsalo, L. 1973. On the Optimal Search for a Target whose Motion is a Markov Process. *J. Appl. Prob.* 10, 847-856.

Stone, L.D. 1975. *Theory of Optimal Search.* Academic Press, New York.

Stone, L.D. 1979. Necessary and Sufficient Conditions for Optimal Search Plans for Moving Targets. *Math. Opns. Res.* 4, 431-440.

Stone, L.D. 1982. Application of the Generalized Search Optimization Technique to Optimal Consumption/Investment Plans. Daniel H. Wagner, Associates memorandum to Office of Naval Research. (670).

Stone, L.D., S.S. Brown, R.P. Buemi and C.R. Hopkins. 1978. *Numerical Optimization of Search for a Moving Target.* Daniel H. Wagner, Associates Report to Office of Naval Research.

Stromquist, W.R., and L.D. Stone. 1981. Constrained Optimization of Functionals with Search Theory Applications. *Math. Opns. Res.* 6, 518-529.

Tierney, L. and J.B. Kadane. 1982. Surveillance Search for a Moving Target *Opns. Res. (To Appear).*

Washburn, A.R. 1981. An Upper Bound Useful in Optimizing Search for a Moving Target. *Opns. Res.* 29, 1227-1230.

Weisinger, J.W. 1982. Search Theory and Maximizing the Utility of Terminal Wealth. Daniel H. Wagner, Associates Memorandum to Office of Naval Research. (670)

The Distribution of the Random Lighted Portion of a Curve in a Plane Shadowed by a Poisson Random Field of Obstacles*

Shelemyahu Zacks

Department of Mathematical Sciences
State University of New York
Binghamton, NY

Micha Yadin

Faculty of Industrial Engineering and Management
Technion, Israel Institute of Technology
Technion City, Haifa, Israel

1 INTRODUCTION

The class of problems discussed in the present paper has a wide range of applications in solving problems of target detection, disturbance in communication and other problems of incomplete visibility. In naval scenario the problem of random visibility measures, lines of sight, etc., can be described as follows. Suppose an observer (source of light) is located at a given point, 0, in the plane (origin) and a curve C in the plane specifies the path of a target (ship or submarine). If there are no obstacles between the observation point, 0, and the target-curve, C, then every point on C is visible from 0 (the whole curve is in the light). However, in reality, obstacles may appear in an obstacle-field (region) at random. The obstacle-field is between the origin, 0, and the target-curve, C, and the random obstacles cast shadows on C. The portions of C which are under shadows are invisible from 0. It is important to study the stochastic structure of the visible portions of the target-curve, C, in order to determine probabilities of certain events which are associated with these visible portions. In a series of papers by Yadin and Zacks [10, 11, 12], we have studied the problem of determining the moments of the visibility measure on C (the total visible portions), and approximating its distribution, when the field of obstacles is a Poisson random field of disks (in the plane) and spheres (in a three-

---
*Research supported by Contract N00014-81-K-0407, Task NR 042-276, with the Office of Naval Research.

dimensional space). Disks or spheres were considered as plausible models. The methodology that was developed can treat more general objects, like ellipses, triangles, etc. Every disk in the Poisson random field is characterized by three parameters ρ, θ and y. The distance from 0, ρ, and the orientation angle, θ, are the polar coordinates of the center of the disk, and y is its diameter. We consider stochastic models in which the diameters of disks are random variables having distributions $G(y|\rho,\theta)$, that may depend on the location coordinates (ρ,θ). The Poisson field may or may not be homogeneous. For the purpose of determining the moments of the visibility measure on C, we have to develop explicit formulae for the visibility probabilities of single points, and the simultaneous visibility of an ordered set of n points on C. These functions depend both on the geometry of the problem and on the intensity characteristics of the Poisson field. In cases of homogeneous Poisson fields, with distributions of disk diameters, $G(y)$, independent of locations (ρ,θ) (standard cases) the determination of the visibility probability functions can often be done by geometrical methods. In nonstandard cases we resort to analytical methods. In the present paper we present the analytical method and illustrate it in the case of circular target and annular field (region). The treatment of the case of straight line target parallel to a trapezoidal field, by geometrical method can be found in [12]. The problem can be attacked also by reducing it to a vacancy or coverage problem, as done in [10].

General target-curves and obstacle fields can be treated in a piece-wise manner, by combining elementary geometrical configurations of the type studied in the papers mentioned above.

The present paper consists of five sections. In Section 2 we discuss the stochastic model of Poisson shadowing random fields and visibility probabilities of points in the plane. In Section 3 we present the method of determining visibility probabilities of points on star-shaped target curves. Section 4 is devoted to the moments of visibility measures. Finally, in Section 5 we provide specific developments and some numerical illustration.

The literature on shadowing processes is quite limited. Chernoff and Daly [2] studied the distribution of length of shadows of disks on a line. The shadowing problem, however, is a special case of the general coverage problem on which there is extensive literature. In particular we refer to the studies of Robbins [5, 6], Ailam [1], Greenberg [4], Siegel [7, 8] and the monograph of Solomon [9] which summarizes many of the important results.

2 POISSON RANDOM FIELDS AND VISIBILITY PROBABILITIES IN THE PLANE

Consider a countable set of disks, D, scattered on the plane. Each disk is specified by a vector (ρ,θ,y), where $\rho, 0 \leqslant \rho < \infty$ and $\theta, -\pi \leqslant \theta < \pi$ are the polar coordinates of its center with respect to an origin 0 (observation point, source of light); and $y, 0 \leqslant y < y^*$, is its diameter. Let $\mathcal{S}_0$ denote the collection of Borel subsets of the space

$$S_0 = \{(\rho,\theta,y): 0 \leqslant \rho < \infty, -\pi \leqslant \theta < \pi, 0 \leqslant y < y^*\}. \quad (2.1)$$

For every $B \epsilon \mathcal{S}_0$, let $N\{B\}$ denote the number of disks such that $(\rho,\theta,y)\epsilon B$ (satisfying condition B). If $N\{B\}$ is a random variable for every $B \epsilon \mathcal{S}_0$ then the elements of D are called *random* disks. In particular, the elements of D are called *Poisson random disks* if for every $B \in \mathcal{S}_0, N\{B\}$ has a Poisson distribution with mean

$$\nu\{B\} = \mu \int \int_B \int dG(y|\rho, \theta) H(d\rho, d\theta), \qquad (2.2)$$

where $\mu, 0 < \mu < \infty$, is an intensity parameter; $H(\rho, \theta)$ and $G(y|\rho, \theta)$ are, respectively, the sigma-finite measure of the location coordinates (ρ, θ) and the conditional c.d.f. of the diameter, y, given (ρ, θ).

The special case in which $h(d\rho, d\theta) = \rho \, d\rho \, d\theta$ and $G(y|\rho, \theta) = G(y)$, is called the *standard case*. In the standard case the centers of the disks are uniformly distributed on the plane and their diameters are independent of their location. In this case

$$\nu\{B\} = \mu \int_0^{y^*} H_B(y) \, dG(y), \qquad (2.3)$$

where μ is the mean number of disks per unit area; $H_B(y)$ is the area of the region in which disks of diameter y, satisfying condition B, are centered.

A natural requirement for shadowing processes is that the source of light (the origin) in uncovered. We therefore introduce the *structural condition*

$$C_0 = \{(\rho, \theta, y); \frac{y}{2} \leq \rho < \infty, -\pi \leq \theta < \pi, 0 < y < y^*\} \qquad (2.4)$$

and assume that all random disks satisfy C_0. In special cases one considers more stringent structural conditions, specified by sets C contained in C_0. For example, in the previous paper of Yadin and Zacks [10] the structural condition is

$$C = \{(\rho, \theta, y); \frac{y}{2} \leq \rho < 1 - \frac{y}{2}; -\pi \leq \theta < \pi, 0 < y \leq 1\}.$$

Let P be a point in the plane. P is said to be *visible* (in light) if the line segment $\overline{OP}$ does not intersect any random disk. The set of all visible points in direction s, $-\pi \leq s < \pi$, starting at the origin, is called a *line of sight*, L_s. Let $P = (r,s)$ be a point in the plane, in orientation s and distance r from the origin. A random disk (ρ, θ, y) intersects the line segment $\overline{OP}$ if, and only if, (ρ, θ, y) belongs to

$$B(r,s) = \{(\rho, \theta, y); (\rho, \theta) \in B(r,s,y), 0 < y \leq y^*\}, \qquad (2.5)$$

where the set $B(r,s,y)$ is the set of all points having distances from $\overline{OP}$ smaller than $y/2$. (See Figure 1) A disk which intersects a line segment $\overline{OP}$ is said to cast *shadow* on P. Accordingly, a point $P = (r,s)$ is visible if $N\{B(r,s) \cap C\} = 0$. Hence, a line of sight L_s has magnitude

$$||L_s|| = \sup \{r; N\{B(r,s) \cap C\} = 0\}. \qquad (2.6)$$

Notice that $||L_s||$ is a random variable.

Let $P = (r,s)$ be a point in the plane. Under the Poisson randomness assumption, the probability that P is visible is

$$Q(r,s) = P[N\{B(r,s) \cap C\} = 0]$$
$$= \exp\{-\nu\{B(r,s) \cap C\}\}, \qquad (2.7)$$

where $\nu\{B(r,s) \cap C\}$ is obtained according to (2.2). From this function one can immediately obtain the distribution of $\|L_s\|$. Indeed,

$$P\{\|L_s\| > l\} = Q(l,s), 0 \leq l < \infty. \tag{2.8}$$

Notice that in the standard case

$$\nu\{B(r,s) \cap C_0\} = \mu\xi r, \tag{2.9}$$

where ξ is the expected value of a random diameter, Y, whose distribution is $G(y)$. Indeed, for $Y = y$, the area of $B_0 = B(r,s) \cap C_0$ is $H_{B_0}(y) = ry$, as can be implied from Fig. 1. Accordingly, the distribution of $\|L_s\|$ in the standard case is exponential with mean $1/\rho$, where $\rho = \mu\xi$.

Assume that the points (s,θ,y) in C have orientation coordinates in $[s',s'']$. Define, for any s in $[s',s'']$,

$$B_+(s) = \{(\rho, \theta, y): \frac{y}{2 \sin(\theta - s)} < \rho < \infty; s < \theta < s''; 0 < y < y^*\} \tag{2.10}$$

and

$$B_-(s) = \{(\rho, \theta, y): \frac{y}{2 \sin(s - \theta)} < \rho < \infty; s' < \theta < s, 0 < y < y^*\}. \tag{2.11}$$

$B_+(s)$ and $B_-(s)$ are sets of all points in C, on the right and on the left of the ray with orientation s, which do not intersect it.

Accordingly, if $s_1 < s_2$, $(B(r(s_1),s_1) \cup B(r(s_2),s_2)) \cap C = C - (((B_-(s_1) \cup B_+(s_2)) \cap C) \cup ((B_+(s_1) \cap B_-(s_2)) \cap C)$, where $A - B$ is the complement of B with respect to A. Thus, the probability that the points P_1 and P_2 are simultaneously visible is

$$P(s_1,s_2) = \exp\{-[\nu\{C\} - \nu\{(B_-(s_1) \cup B_+(s_2)) \cap C\}$$
$$- \nu\{(B_+(s_1) \cap B_-(s_2)) \cap C\}]\}. \tag{2.12}$$

Generally, if $s' \leq s_1 < \ldots < s_n \leq s''$ are the orientation coordinates of n points then the probability that they are simultaneously visible is

$$P(s_1, \ldots, s_n) = \exp\left\{-\left[\nu\{C\} - \nu\{(B_-(s_1) \cup B_+(s_n)) \cap C\}\right.\right.$$
$$\left.\left. - \sum_{i=1}^{n-1} \nu\{(B_+(s_i) \cap B_-(s_{i+1})) \cap C\}\right]\right\}. \tag{2.13}$$

3 VISIBILITY PROBABILITIES ON STAR-SHAPED CURVES

Consider a star-shaped curve in the plane, C, such that each ray originating at the origin, 0, intersects C at most once. The curve C is specified by a continuous positive function, $r(s)$, on the domain $[s',s'']$, where $-\frac{\pi}{2} \leq s' < s'' \leq \pi/2$. We furthermore require that $r(s)$ will have almost always continuous derivative.

The field structure is specified by the following assumptions on the random nature of the distributions of the disks. First, we assume that the random diameters, $Y_1, Y_2, \ldots$ of the disks are distributed over the interval $[a,b]$, where

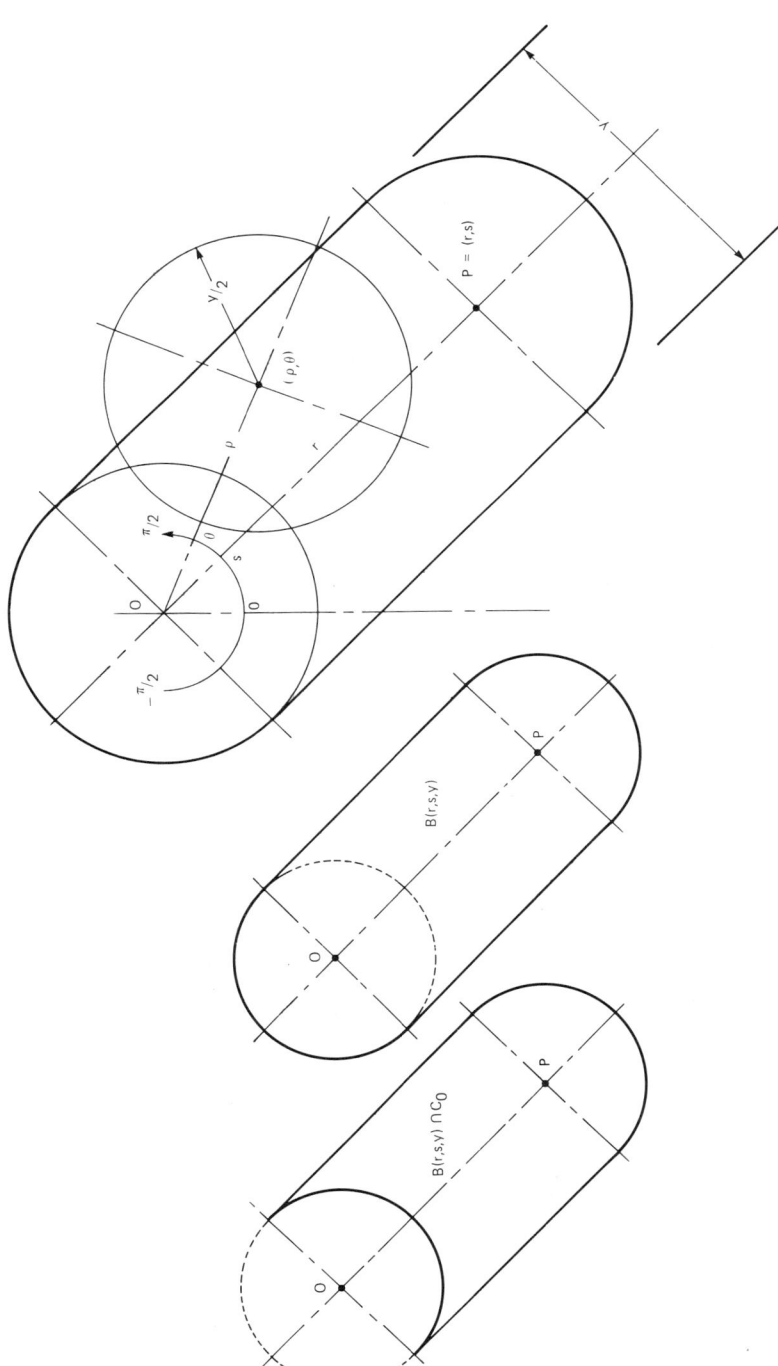

Figure 1 The geometry of disks casting shadows on a point

$0 \leqslant a < b < \infty$. Furthermore, we specify two star-shaped curves, U and W, between the origin, 0, and C, so that the centers of the disks are distributed between U and W, and no disk can either cover the origin or intersect C. More specifically, let

$$U = \{u(\theta); \theta' \leqslant \theta \leqslant ''\} \text{ and } W = \{w(\theta); \theta' \leqslant \theta \leqslant \theta''\},$$

where $[s', s''] \subset [\theta', \theta''] \subset [-\frac{\pi}{2}, \frac{\pi}{2}]$. The region between U and W is the obstacle-field, which is specified by

$$C_1 = \Big\{(\rho, \theta, y); \theta' \leqslant \theta \leqslant \theta'',$$

$$\frac{b}{2} \leqslant u(\theta) \leqslant \rho \leqslant w(\theta) \text{ and for every} \quad (3.1)$$

$$(\xi(s), s) \text{ s.t.} |(\xi(s), s)) - (w(\theta), \theta)| = b/2,$$

$$w(\theta) \cos(\theta - s) + [(b/2)^2 - w^2(\theta) \sin^2(\theta - s)]^{1/2} < r(s)\Big\},$$

as can be seen from Figure 2. The conditions in C_1 insure that no disk can either cover the origin or intersect C. Notice that the expected number of disk centers in C_1 is

$$\nu\{C_1\} = \mu \int_{\theta'}^{\theta''} \int_{w(\theta)}^{u(\theta)} H(d\rho, d\theta) \quad (3.2)$$

and

$$\nu\{C\} = \nu\{C_1\} - \nu\{B_-(s') \cap C\} - \nu\{B_+(s'') \cap C\}. \quad (3.3)$$

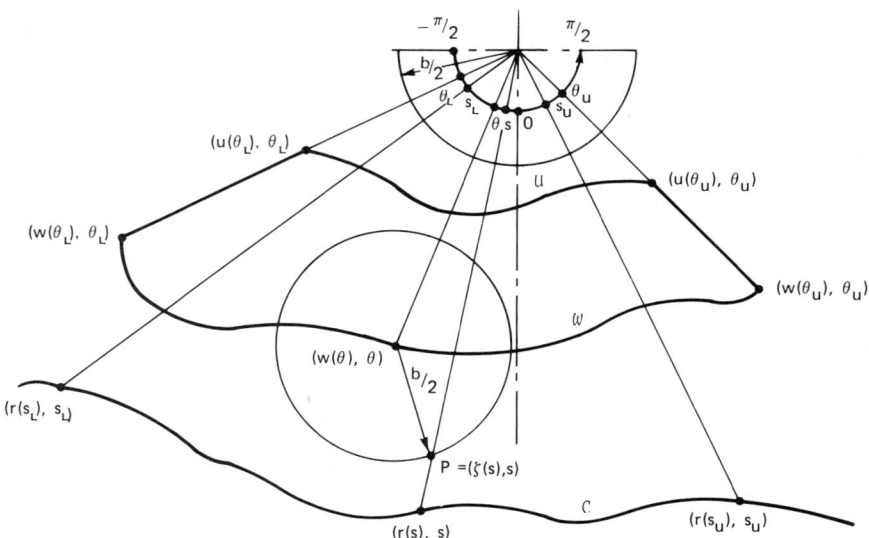

Figure 2 The geometry of disks casting shadows on a curve

Given a point $P = (r(s), s)$ on C, let $\mu K_-(s,t)$ and $\mu K_+(s,t)$ denote the expected numbers of disks $(\rho, \theta, y) \in C_1$ with $s - t \leq \theta \leq s$ and $s \leq \theta \leq s + t$, respectively, which do not intersect the line segment $\overline{OP}$. These functions are given by

$$K_-(s,t) = \int_{s-t}^{s} \int_{u(\theta)}^{w(\theta)} G(y(\rho, s - \theta) | \rho, \theta) H(d\rho, d\theta),$$

and

$$K_+(s,t) = \int_{s}^{s+t} \int_{u(\theta)}^{w(\theta)} G(y(\rho, \theta - s) | \rho, \theta) H(d\rho, d\theta), \tag{3.4}$$

where

$$y(\rho, \tau) = \begin{cases} 2\rho \sin \tau, & \text{if } \tau < \pi/2 \\ 2\rho, & \text{if } \tau \geq \pi/2 \end{cases} \tag{3.5}$$

Indeed, when $|\theta - s| < \pi/2$, a disk centered at (ρ, θ) does not intersect $\overline{OP}$ if its diameter is smaller than $2\rho \sin |\theta - s|$. If, however, $|\theta - s| \geq \pi/2$, the diameter of the disk should not exceed 2ρ.

Thus, according to (3.3) and (3.4),

$$\nu\{C\} = \nu\{C_1\} - \mu[K_-(s', s' - \theta) + K_+(s'', \theta'' - s'')]. \tag{3.6}$$

Similarly, for all $s \in [s', s'']$

$$\nu\{B_-(s) \cap C\} = \mu[K_-(s, s - \theta') - K_-(s', s' - \theta')] \tag{3.7}$$

$$\nu\{B_+(s) \cap C\} = \mu[K_+(s, \theta'' - s) - K_+(s'', \theta'' - s'')].$$

Furthermore, for $s' \leq s_1 < s_2 \leq s''$,

$$\nu\{B_+(s_1) \cap B_-(s_2) \cap C\} \tag{3.8}$$

$$= \mu\left[K_+\left(s_1, \frac{s_2 - s_1}{2}\right) + K_-\left(s_2, \frac{s_2 - s_1}{2}\right)\right].$$

The value $t = \dfrac{s_2 - s_1}{2}$ was selected in (3.8) to avoid the possibility that a disk centered at (ρ, θ) with $s < \theta < s + t$, which does not intersect $\overline{OP_1}$ will nevertheless intersect $\overline{OP_2}$, and vice versa. For the purpose of evaluating (3.4) we introduce the auxiliary functions

$$K_-(s, d\theta, \nu) = \int_0^\nu G(y(\rho, s - \theta) | \rho, \theta) H(d\rho, d\theta) \tag{3.9}$$

and

$$K_+(s, d\theta, \nu) = \int_0^\nu G(y(\rho, \theta - s) | \rho, \theta) H(d\rho, d\theta).$$

The K-functions are thus given by

$$K_-(s,t) = \int_{s-t}^{s} [K_-(s, d\theta, w(\theta)) - K_-(s, d\theta, u(\theta))] \tag{3.10}$$

and

$$K_+(s,t) = \int_{s}^{s+t} [K_+(s, d\theta, w(\theta)) - K_+(s, d\theta, u(\theta))].$$

Notice that $G(y | \rho, \theta) = 0$ for all $y < a$ and $G(y | \rho, \theta) = 1$ for all $y \geq b$. Accordingly, from (3.9)

$$K_+(s,d\theta,\nu) = \int_{A(\theta-s,\nu,a)}^{A(\theta-s,\nu,b)} G\,(y(\rho,\,\theta-s)|\rho,\,\theta)H(d\rho,\,d\theta)$$
$$+ \int_{A(\theta-s,\nu,b)}^{\nu} H(d\rho,\,d\theta), \qquad (3.11)$$

where

$$A(\tau,\,\nu,\,x) = \begin{cases} \min\left\{\nu, \dfrac{x}{2\sin\tau}\right\}, & \text{if } \tau < \pi/2 \\ \min\left\{\nu, \dfrac{x}{2}\right\}, & \text{if } \tau \geqslant \pi/2 \end{cases} \qquad (3.12)$$

The function $K_-(s,d\theta,\nu)$ can be obtained from (3.11) by replacing $\theta-s$ by $s-\theta$. In the standard case we substitute in the above formula $H(d\rho,\,d\theta) = \rho\,d\rho\,d\theta$ and $G(y|\rho,\,\theta) = G(y)$.

4 MOMENTS OF THE VISIBILITY MEASURE AND AN APPROXIMATION TO ITS DISTRIBUTION

Define the indicator function $I(s)$, $s' \leqslant s \leqslant s''$ so that $I(s) = 1$ if $(r(s),s)$ is a visible point and $I(s) = 0$ otherwise. The total measure of the visible part of C is given by

$$V\{C\} = \int_{s'}^{s''} I(s)[r^2(s) + (r'(s))^2]^{1/2} ds. \qquad (4.1)$$

$V\{C\}$ is the *visibility measure* of C. In coverage problems it is known also as the measure of vacancy (Ailam [1]).

The first moment (expected value) of $V(C)$ is

$$E\{V(C)\} = \int_{s'}^{s''} E\{I(s)\}l(s)\,ds, \qquad (4.2)$$

where $l(s) = [r^2(s) + (r'(s))^2]^{1/2}$. Moreover,

$$E\{I(s)\} = Q(r(s),s)$$
$$= \exp\{-\nu\{C_1\} + \mu[K_-(s,s-\theta') + K_+(s,\theta''-s)]\}. \qquad (4.3)$$

$\nu\{C_1\}$ is given by (3.2). In the standard case it assumes the form

$$\nu\{C_1\} = \frac{\mu}{2}\int_{\theta'}^{\theta''}[w^2(\theta) - u^2(\theta)]d\theta. \qquad (4.4)$$

Generally, for every $n \geqslant 2$, the n-th moment of $V\{C\}$ is given by

$$E\{V^n\{C\}\} = \int_{s'}^{s''}\cdots\int_{s'}^{s''} E\{\prod_{i=1}^{n} I(s_i)\}\prod_{i=1}^{n} l(s_i)\,ds_i$$

$$= n! \int\cdots\int_{s'\leqslant s_1\leqslant\cdots\leqslant s_n\leqslant s''} P(s_1,\ldots,s_n)\prod_{i=1}^{n} l(s_i)\,ds_i \qquad (4.5)$$

Indeed, $E\{\prod_{i=1}^{n} I(s_i)\}$ is the probability that all the n points are simultaneously visible. According to (2.13), (3.6), (3.7), and (3.8) we obtain, for every $s' \leqslant s_1 \leqslant \ldots \leqslant s_n \leqslant s''$,

$$P(s_1, \ldots, s_n) = \exp\{-\nu\{C_1\}\}\exp\left\{\mu K_-(s_1, s_1 - \theta')\right.$$

$$+ \mu K_+(s_n, \theta'' - s_n) + \mu \sum_{i=1}^{n-1}\left[K_+\left(s_i, \frac{s_{i+1} - s_i}{2}\right)\right.$$

$$\left.\left.+ K_-\left(s_{i+1}, \frac{s_{i+1} - s_i}{2}\right)\right]\right\} \tag{4.6}$$

The n-th moment of $V\{C\}$, for $n \geqslant 1$, can be computed according to (5.5) and (4.6) in the following recursive manner. Define first

$$\Psi_0(s) = \exp\{\mu K_-(s, s - \theta')\} \tag{4.7}$$

and for $j = 1, 2, \ldots, n - 1$ define

$$\Psi_j(s) = \int_{s'}^{s} l(y)\Psi_{j-1}(y)\exp\left\{\left[K_+\left(y, \frac{s-y}{2}\right) + K_-\left(s, \frac{s-y}{2}\right)\right]\right\}dy. \tag{4.8}$$

Then, for each $n \geqslant 1$,

$$E\{V^n\{C\}\} = n!\exp\{-\nu\{C_1\}\} \cdot \int_{s'}^{s''} l(y)\Psi_{n-1}(y)\exp\{\mu K_+(y, \theta'' - y)\}dy \tag{4.9}$$

Finally, let $L\{C\}$ denote the length of C. The distribution of $V\{C\}/L\{C\}$ is concentrated on [0, 1], with jumps at the two end points 0 and 1 and absolutely continuous elsewhere. It follows that

$$\lim_{n \to \infty} E\{V^n\{C\}\}/L^n\{C\} = P\{V\{C\} = L\{C\}\}. \tag{4.10}$$

5 SOME SPECIAL CASES

The K-Functions

In the present section we provide explicit expression for the K-functions corresponding to annular regions, under the assumptions of a standard Poisson field and uniform distribution $G(y)$ on $[a,b]$, i.e.,

$$G(y) = \begin{cases} 0 & \text{, if } y < a \\ \frac{y - a}{b - a} & \text{, if } a \leqslant y < b \\ 1 & \text{, if } b < y \end{cases} \tag{5.1}$$

Under this assumption one obtains from (3.11) and (3.12) that $K_+(s, d\theta, \nu) = \Lambda(\theta - s, \nu, a, b)d\theta$, where

$$\Lambda(\tau, \nu, a, b) =$$

$$\begin{cases} 0 & , \text{if } \tau < \sin^{-1}\left(\dfrac{a}{2\nu}\right) \\ \dfrac{1}{b-a}\left[\dfrac{2}{3}\nu^3 \sin\tau - \dfrac{a}{2}\nu^2 + \dfrac{1}{24}\dfrac{a^3}{\sin^2\tau}\right] & , \sin^{-1}\left(\dfrac{a}{2\nu}\right) \leq \tau < \sin^{-1}\left(\dfrac{b}{2\nu}\right) \\ \dfrac{1}{2}\left[\nu^2 - \dfrac{a^2+ab+b^2}{12\sin^2\tau}\right] & , \sin^{-1}\left(\dfrac{b}{2\nu}\right) \leq \tau < \pi/2 \\ \dfrac{1}{2}\left[\nu^2 - \dfrac{a^2+ab+b^2}{12}\right] & , \dfrac{\pi}{2} \leq \tau. \end{cases} \quad (5.2)$$

We define now the boundary functions $u(\theta)$ and $w(\theta)$ for annular regions and provide explicit formulae for determining $K_+(s,t)$ and $K_-(s,t)$.

The Annular Region

In the annular region the boundary functions are parallel circular arcs, i.e.,

$$u(\theta) = u, \text{ for all } \theta' \leq \theta \leq \theta'' \quad (5.3)$$

and

$$w(\theta) = w, \text{ for all } \theta' \leq \theta \leq \theta'',$$

where $\dfrac{b}{2} \leq u < w$. Moreover, according to (3.10) and (5.2), $K_+(s,t) = K_-(s,t) = K^*(t,w) - K^*(t,u)$, where

$K^*(t,\nu) = \int_0^t \Lambda(\tau,\nu,a,b) d\tau$

$$= \begin{cases} 0 & , \text{if } t < \sin^{-1}\left(\dfrac{a}{2\nu}\right) \\ K_1(t,\nu) & , \text{if } \sin^{-1}\left(\dfrac{a}{2\nu}\right) \leq t \leq \sin^{-1}\left(\dfrac{b}{2\nu}\right) \\ K_2(t,\nu) & , \text{if } \sin^{-1}\left(\dfrac{b}{2\nu}\right) \leq t \leq \pi/2 \\ K_3(t,\nu) & , \text{if } \pi/2 < t \end{cases} \quad (5.4)$$

in which

$$K_1(t,\nu) = \dfrac{1}{b-a}\left\{\dfrac{\nu^2}{3}((4\nu^2-a^2)^{1/2} - 2\nu\cos t) - \dfrac{a}{2}\nu^2\left(t - \sin^{-1}\left(\dfrac{a}{2\nu}\right)\right)\right.$$
$$\left. + \dfrac{a^2}{24}((4\nu^2-a^2)^{1/2} - a\operatorname{ctn}(t))\right\}, \quad (5.5)$$

$$K_2(t,v) = K_1\left[\sin^{-1}\left[\frac{b}{2v}\right], v\right] + \frac{v^2}{2}\left[t - \sin^{-1}\left[\frac{b}{2v}\right]\right]$$
$$- \frac{a^2 + ab + b^2}{24}\left[\frac{1}{b}(4v^2 - b^2)^{1/2} - \text{ctn}(t)\right], \quad (5.6)$$

and

$$K_3(t,v) = K_2\left[\frac{\pi}{2}, v\right] + \left[t - \frac{\pi}{2}\right]\left[\frac{v^2}{2} - \frac{a^2 + ab + b^2}{24}\right]. \quad (5.7)$$

The Moments

In the case under consideration the target curve, C, is also circular over $[s',s'']$, where $-\frac{\pi}{2} < \theta' < s' < s'' < \theta'' < \frac{\pi}{2}$. Accordingly, the function describing C is $r(s) = r$ for all $s' \geq s \leq s''$. Thus, $l(s) = r$ for all $s \in [s',s'']$ and the moments of $V\{C\}$ can be determined according to (4.7) to (4.9) in the following manner:

Let

$$\lambda = \exp\left[-\frac{\mu}{2}(w^2 - u^2)(\theta'' - \theta')\right], \quad (5.8)$$

$$\Psi_0(s) = \exp\{\mu[K^*(s - \theta', w) - K^*(s - \theta', u)]\}, \quad (5.9)$$

and

$$H(s) = \exp\{\mu[K^*(\theta'' - s, w) - K^*(\theta'' - s, u)]\} \quad (5.10)$$

for $s' \leq s \leq s''$.

Define recursively, for every $j \geq 1$,

$$\Psi_j(s) = \int_{s'}^{s} \Psi_{j-1}(y) \exp\left\{2\mu\left[K^*\left(\frac{s-y}{2}, w\right) - K^*\left(\frac{s-y}{2}, u\right)\right]\right\} dy. \quad (5.11)$$

The n-th moment of $V\{C\}$ is then

$$\mu_n = \lambda n! r^n \int_{s'}^{s''} \Psi_{n-1}(s) H(s) ds. \quad (5.12)$$

Let P_1 be the probability that $V\{C\} = L\{C\}$, which is the probability that C is completely visible. According to (3.2) and (3.6)

$$P_1 = \lambda \exp\{\mu[K^*(s' - \theta', w) - K^*(s' - \theta', u) +$$
$$K^*(\theta'' - s'', w) - K^*(\theta'' - s'', u)]\} \quad (5.13)$$

As in the previous study of Yadin and Zacks [10], the distribution of $V\{C\}/r(s'' - s')$ is approximated by a mixture of a beta-distribution with a discrete distribution concentrated on 0 and 1. The mixed distribution of the normalized visibility measure, $V^*\{C\} = V\{C\}/r(s''-s')$, has a c.d.f.

$$F^*(x) = \begin{cases} 0 & \text{, if } x < 0 \\ P_0 + (1-P_0-P_1)\dfrac{1}{B(\alpha,\beta)}\int_0^x y^{\alpha-1}(1-y)^{\beta-1}dy & \text{, if } 0 \leq x < 1. \\ 1 & \text{, if } x > 1 \end{cases} \quad (5.14)$$

Since the value of P_1 is known, we determine $F^*(x)$, in the various cases, be equating the first three moments of (5.14) to the moments of $V^*\{C\}$. The n-th moment of (5.14) is

$$\tilde{\mu}_n = P_1 + (1-P_0-P_1)\frac{\alpha(\alpha+1)\ldots(\alpha+n-1)}{(\alpha+\beta)(\alpha+\beta+1)\ldots(\alpha+\beta+n-1)}, n \geq 1 \quad (5.15)$$

Accordingly, if $c_n = \mu_n^* - P_1$, where $\mu_n^* = \mu_n/r^n(s''-s')^n$ is the n-th normalized moment of $V\{C\}$, then

$$\alpha = \frac{2c_2^2 - c_3(c_1+c_2)}{c_1 c_3 - c_2^2},$$

$$\beta = \frac{(c_1-c_2)(c_2-c_3)}{c_1 c_3 - c_2^2} \quad (5.16)$$

and

$$P_0 = 1 - P_1 - \frac{c_1(\alpha+\beta)}{\alpha}.$$

The numerical determination of the moments μ_n was performed in the following manner. The interval $[s',s'']$ was partitioned to M sub-intervals of length $\Delta = (s''-s')/M$. The grid points for the computation are then $s_j = s' + j\Delta, j = 0, 1, \ldots, M$. The functions $\Psi_0(s)$ and $G(s)$ were computed exactly for each $j = 0, \ldots, M$. The functions $\Psi_k(s), k \geq 1$, were then computed at the grid points s_j, according to the numerical integration formula

$$\Psi_k(s_j) = \begin{cases} 0 & \text{, if } j = 0 \\ \dfrac{\Delta}{2}(\Psi_{k-1}(s_0)\exp\{2\mu\tilde{K}(\dfrac{\Delta}{2})\} + \Psi_{k-1}(s_1)) & \text{, if } j = 1 \\ \dfrac{\Delta}{2}(\Psi_{k-1}(s_0)\exp\{2\mu\tilde{K}(j\dfrac{\Delta}{2})\} + \Psi_{k-1}(s_1)) + \\ + \Delta \sum_{k=1}^{j-1} \Psi_{k-1}(s_1)\exp\{2\mu\tilde{K}(\dfrac{(j-i)\Delta}{2})\} & \text{, if } j \geq 2 \end{cases} \quad (5.17)$$

where $\tilde{K}(t) = K^*(t,w) - K^*(t,u)$. Finally, the n-th moment of $V(C), \mu_n$, is determined by the numerical approximation formula

$$\hat{\mu}_n = \lambda n! r^n \Delta\{\Psi_{n-1}(s_0)H(s_0) + \Psi_{n-1}(s_M)H(s_M))/2 + \sum_{j=1}^{M-1} \Psi_{n-1}(s_j)H(s_j)\}. \quad (5.18)$$

If the value of M is large one obtains very good approximation by applying formulae (5.17)-(5.18). In Table 5.1 we provide the first ten normalized moments of $V\{C\}$, i.e., $\mu_n^* = \hat{\mu}_n/(r(s''-s'))^n$, for the case of

Table 5.1 The Normalized Moments of V(C) and Their Beta-Mixture Approximations, for $s' = -\pi/18$, $s'' = -s'$, $M = 60$, $r = 1$, $u = 0.5$, $w = 0.75$, $a = 0.1$, $b = 0.3$.

Intensity μ	Normalized Moments										
	1	2	3	4	5	6	7	8	9	10	∞
1.	0.951	0.934	0.926	0.921	0.918	0.917	0.917	0.919	0.921	0.925	.901
	0.951	0.934	0.926	0.921	0.917	0.915	0.913	0.912	0.910	0.910	
3.	0.860	0.816	0.794	0.781	0.773	0.768	0.765	0.763	0.763	0.765	.730
	0.860	0.816	0.794	0.780	0.772	0.765	0.761	0.757	0.754	0.752	
5.	0.778	0.713	0.681	0.662	0.650	0.643	0.638	0.634	0.633	0.633	.592
	0.778	0.713	0.681	0.662	0.649	0.640	0.634	0.629	0.625	0.621	
7.	0.703	0.623	0.584	0.562	0.548	0.538	0.532	0.527	0.524	0.524	.480
	0.703	0.623	0.584	0.562	0.547	0.536	0.528	0.522	0.518	0.514	
9.	0.636	0.545	0.502	0.477	0.461	0.451	0.444	0.439	0.435	0.434	.389
	0.636	0.545	0.502	0.460	0.449	0.449	0.440	0.434	0.429	0.425	

Comments on Table 5.1

(1) Notice that $\mu_\infty^* = P_1$.

(2) Normalized moments μ_n^* are in upper line of each case.

(3) The moments of (5.2) are in lower line of each case.

(4) The slight increase of high order moments in some cases is due to accumulation of numerical errors.

Table 5.2 Parameters of the Mixed-Beta Approximating Distribution (Specifying parameters as in Table 5.1)

Intensity μ	σ	P_0	P_1	α	β
1	.1733	.0020	.9005	1.1161	1.0398
3	.2765	.0152	.7302	1.1405	1.0967
5	.3290	.0326	.5921	1.0956	1.1194
7	.3590	.0536	.4801	1.0450	1.1386
9	.3754	.0768	.3893	0.9937	1.0967

$s' = -\pi/18$, $s'' = -s'$, $M = 60$, $r = 1$, $u = .5$, $w = .75$, $a = .1$ and $b = .3$. The values of μ (the Poisson intensity) are 1 (2) 9.

In Table 5.2 we provide the values of $\sigma = (\mu_2^* - (\mu_1^*)^2)^{1/2}$, P_0, P_1, α and β corresponding to the cases of Table 5.1. We have also computed the first ten moments of (5.2) according to the parameters determined by the true moments, μ_n^*, and presented them in Table 5.1 (lower line of each case). We see that the actual moments and the ones of the approximating distribution are extremely close. This indicates that the mixed-beta distribution is apparently very close to the true one. We obtain in this manner also an estimate of the probability of complete coverage, P_0. An explicit formula for this parameter is not yet available.

REFERENCES

[1] Ailam, G. (1966). Moments of Coverage and Coverage Space, *J. Appl. Prob., 3:* 550-555.

[2] Chernoff, H. and J.F. Daly (1957). The distribution of shadows, *Jour. of Mathematics and Mechanics,* 6: 567-584.

[3] Feller, W. (1966). *An Introduction to Probability Theory and Its Applications,* Vol. II, 2nd ed. John Wiley, NY.

[4] Greenberg, I. (1980). The Moments of Coverage of a Linear Set, *J. Appl. Prob.* 17.

[5] Robbins, H.E. (1944). On the measure of random set, *Annals of Math. Statist., 15:* 70-74.

[6] Robbins, H.E. (1945). On the measure of random set, II. *Annals of Math. Statist.,* 16: 342-347.

[7] Siegel, A.F. (1978). Random space filling and moments of coverage in geometric probability, *Journal of Applied Probability,* 15: 340-355.

[8] Siegel, A.F. (1978). Random arcs on the circle, *Journal of Applied Probability,* 15: 774-789.

[9] Solomon, H. (1978). *Geometric Probability,* SIAM, Philadelphia.

[10] Yadin, M. and S. Zacks (1982). Random Coverage of a Circle with Applications to a Shadowing Problem, *Journal of Applied Probability,* 19: 562-577.

[11] Yadin, M. and Zacks, S. (1982). Visibility Probabilities and Moments of Measures of Visibility on Curves in the Plane for Poisson Shadowing Processes, Tech. Report No. 2, ONR Contract N00014-81-K-0407, Department of mathematical Sciences, SUNY-Binghamton.

[12] Yadin, M. and Zacks, S. (1982). The Distributions of Measures of Visibility on Line Segments in Three Dimensional Spaces under Poisson Shadowing Processes. Tech. Report No. 4, ONR Contract N00014-81-K-0407, Department of Mathematical Sciences, SUNY-Binghamton.

Some Factors Influencing Localization Accuracy*

Peter M. Schultheiss

Department of Electrical Engineering
Yale University
New Haven, CT

1 INTRODUCTION

This paper summarizes theoretical results obtained during the past several years in the general area of source localization using an array of receiving sensors. It attempts, in particular, to set absolute lower bounds on the mean square errors of bearing, range and other parameters important for localization, and to compare these bounds with the performance of actual instrumentations which have been built or at least proposed.

The general setting of the problem is sketched in Figure 1. The source radiates a random signal with the statistical properties of Gaussian noise (zero mean). An array of sensors receives this signal unaltered except for propagation delays. The precise values of these delays are determined by the source location relative to the array. Thus measurements of differential delays between sensor pairs provide the primary data concerning source position. The signal received at each sensor is contaminated with a zero mean Gaussian noise, statistically independent both of the signal and of the noise received at every other sensor. For the purposes of this paper we shall assume that spectral properties of signal and noise are known and that the observation time is long compared with the correlation time of both signal and noise. It is possible to relax both of these assumptions considerably. In particular it appears (from a less than exhaustive series of studies carried out to date) that lack of precise prior knowledge concerning spectral properties does not affect the mean square error expressions presented here. The practically most interesting case of "short" correlation time is that of a very narrowband signal. Its limiting version is a sinusoidal signal of random amplitude and at least the basic results concerning delay measurement are readily extended to that situation.

---

*This research was supported by the Office of Naval Research under contract N00014-80-C-0092.

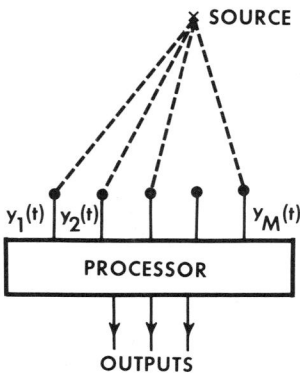

Figure 1 General geometry

The primary analytical tool used to derive the bulk of our results is the Cramer-Rao inequality. It asserts that any unbiased estimate $\hat{\theta}$ of a scalar parameter θ has a mean square error bounded by

$$D^2(\hat{\theta}) \geq - \frac{1}{E\left\{\dfrac{\partial^2 \ln p(x/\theta)}{\partial \theta^2}\right\}} \tag{1}$$

$p(\underline{x}/\theta)$ is the conditional probability density of the data vector $\underline{x}$ given the true parameter value θ. The right side of Equation (1) is not necessarily reachable, but a theorem asserts that it can be approached under very general conditions in the limit of very long observation times.

The Cramer-Rao inequality has a well known extension to the case of a vector parameter $\underline{\theta}$. The equivalent of Equation (1) is now

$$\text{Cov}(\hat{\underline{\theta}}) \geq J^{-1} \tag{2}$$

J is the Fisher matrix with elements

$$J_{ij} = - E\left\{\frac{\partial^2 \ln p(x/\theta)}{\partial \theta_i \, \partial \theta_j}\right\}. \tag{3}$$

The matrix inequality (2) should be read as asserting that $(\text{Cov}(\hat{\underline{\theta}}) - J^{-1})$ is nonnegative definite.

For our purposes we shall require a further extension of the Cramer-Rao theory to accommodate random parameters θ about whose statistics one has some prior information. One can then dispense with the requirement of an unbiased estimator and calculate a lower bound on the error correlation matrix averaged over the known prior distribution of $\underline{\theta}^1$. The relevant Fisher matrix is now composed of two components

$$J = J_x + J_\theta \tag{4}$$

where

$$J_x|_{ij} = - E_{\underline{x}, \underline{\theta}}\left\{\frac{\partial^2 \ln p(x/\theta)}{\partial \theta_i \, \partial \theta_j}\right\} \tag{5}$$

and

$$J_\theta|_{ij} = -E_\theta\left\{\frac{\partial^2 \ln p_\theta(\theta)}{\partial \theta_i \, \partial \theta_j}\right\}. \tag{6}$$

Thus J_x describes the average contribution of the data, J_θ that of the *a priori* distribution. The inverse of (4) is the desired lower bound on the average error correlation, entirely analogous to (3).

The data vector $\underline{x}$ can be any representation of the waveshapes received at the various sensors. When observation times are large compared with the correlation times of signal and noise, Fourier coefficients associated with different frequencies are essentially uncorrelated, so that the covariance matrix for $\underline{x}$ becomes block diagonal (dependence only over the spatial index at a common frequency). Even with this simplification the computational complexity is considerable and we omit almost all details of derivation.

We begin our summary of results with the simplest possible problem: Bearing and range estimation using a one dimensional (linear) array with equally spaced elements. The first generalization allows the array to be a member of a fairly large class of 2 dimensional configurations and addresses some problems related to optimal sensor placement. Next we allow the sensors to be randomly displaced from their nominal locations and examine the incremental error caused by such perturbations. Fairly complete results will be given for statically displaced sensors while only partial results are currently available for sensors in random motion. Finally we shall address one of the key problems associated with the realizability of the Cramer-Rao bound. The theorem quoted earlier assures asymptotic realizability in the limit of very large observation times. Just how large the observation time must be in practice depends critically on the signal to noise ratio and on the spectrum of the signal. In particular, narrowband signals generate ambiguity problems which make the Cramer-Rao bound realizable only at extremely high signal to noise ratios and/or very large observation times. Alternative bounds, which better describe attainable performance under realistic conditions, will therefore be discussed.

2 BEARING AND RANGE ESTIMATION WITH A LINEAR ARRAY

We consider a receiving array of M equally spaced sensors with total length L. Bearing α is measured relative to the array axis and range r is measured from the origin, placed at the center of the array as indicated in Figure 2. With this choice of coordinates one finds that the Fisher matrix for α and r is diagonal, so that the bearing and range errors are uncorrelated. Explicit expressions for the Cramer-Rao bounds on bearing and range errors assume the following form

$$D^2(\hat{\alpha}) \geq \frac{6c^2}{R} \frac{\frac{M-1}{M+1}}{L^2 \sin^2\alpha} \tag{7}$$

$$D^2(\hat{r}) \geq \frac{360\, c^2}{R} \frac{r^4}{L^4 \sin^4\alpha} \frac{(M-1)^3}{(M+1)(M^2-4)} \tag{8}$$

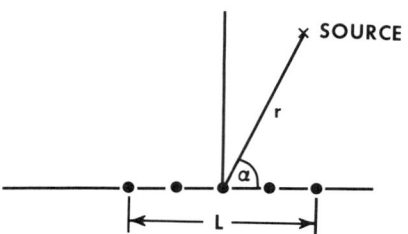

Figure 2 Uniformly spaced 1: near array

where

$$R = \frac{T}{2\pi} \int_o^\infty \frac{\omega^2 \left[M \frac{S(\omega)}{N(\omega)}\right]^2}{1 + M \frac{S(\omega)}{N(\omega)}} d\omega. \tag{9}$$

and c is the velocity of propagation. Signal and noise properties enter only into the gain factor R which, in turn, depends only on the post-beamforming signal to noise ratio $MS(\omega)/N(\omega)$. We note that L is $(M-1)$ times the sensor spacing, so that both (7) and (8) diverge for $M=1$, as they must, since bearing and range measurement is not possible with a single sensor (and a single propagation path). Equation (8) also diverges for $M=2$ because range estimation is based on a measurement of wavefront curvature and therefore requires at least three sensors. The r^4/L^4 dependence of (8) reflects the fact that wavefront curvature over the receiving array decreases rapidly with increasing range. We note further that for $M >> 1$ the only M dependence of (7) and (8) is due to the gain factor R.

A practical (though by no means simple) instrumentation for estimating bearing and range which has been discussed widely in the literature is the focused beamformer, sketched in Figure 3. For each value of bearing and range there is a set of delays τ_i which aligns the signal components of the received waveshapes and therefore maximizes the average output power. The focused beamformer searches through the delay combinations generated by all possible pairs (α, r) and reports as the estimate that pair which maximizes the output power. The performance of this device has been analyzed by Hahn[2]. Comparing his results for large TW products with Equation (7) and (8) one finds that the focused beamformer reaches the Cramer-Rao bound and is therefore optimal.

3 BEARING AND RANGE ESTIMATION WITH A SYMMETRICAL PLANAR ARRAY

The Cramer-Rao results of the last section can be generalized to an array with two spatial degrees of freedom. Relatively simple formulas for mean square error are obtained if the array geometry satisfies the following conditions:

(a) The origin is located at the centroid of the array
(b) Sensor locations are chosen symmetrically with respect to the origin (i.e., sensor pairs are symmetrically placed on lines through the origin).

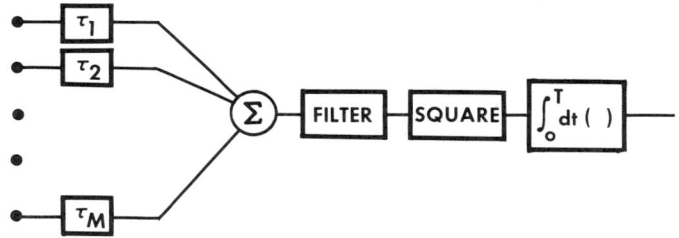

Figure 3 Focused beamformer

(a) is, of course, merely a convention concerning choice of the origin; only (b) imposes a real constraint on sensor placement. Figure 4 shows an arrangement satisfying this condition.

Under constraints (a) and (b) the Fisher matrix for bearing and range remains diagonal and the Cramer-Rao bounds for bearing and range are given by the following expressions.

$$D^2(\hat{\alpha}) \geq \frac{c^2}{2R} \frac{1}{E(u_i^2)} \tag{10}$$

and

$$D^2(\hat{r}) \geq \frac{2c^2}{R} r^4 \frac{1}{V(u_i^2)} \tag{11}$$

R remains the gain factor defined by Equation (9). The quantities $E(u_i^2)$ and $V(u_i^2)$ are defined by the relations

$$E(u_i^2) = \frac{1}{M} \sum_{i=1}^{M} u_i^2 \tag{12}$$

$$V(u_i^2) = \frac{1}{M} \sum_{i=1}^{M} \{u_i^4 - [E(u_i^2)]^2\} \tag{13}$$

u_i is the distance from the ith sensor to the line connecting the source to the origin (see Figure 4). $E(u_i^2)$ is the sample mean of the squared distances and

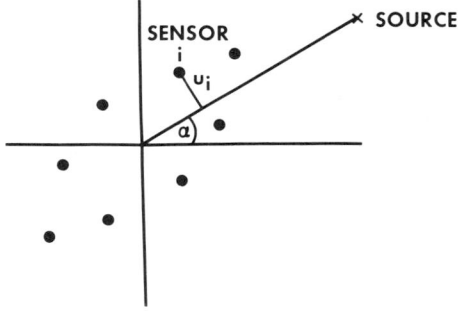

Figure 4 Symmetrical planar array

$V(u_i^2)$ is the sample variance. For effective bearing estimation one would therefore place sensors so as to maximize $E(u_i^2)$ while good range estimation requires sensor placement to maximize $V(u_i^2)$. The two conditions are clearly competitive, since the former calls for sensors placed as far away from the array center as possible (subject to space limitations and the requirement of noise independence from sensor to sensor) while the latter implies the use of at least three groups of sensors, one near the array centroid and two as remote from that point as permitted by space and independence constraints. It is interesting that the two parameters $E(u_i^2)$ and $V(u_i^2)$ summarize all information concerning array geometry which is relevant for bearing and range estimation under the stated assumptions.

4 STATICALLY PERTURBED ARRAYS

In practice the locations of various sensors in an array are often known with less than absolute precision. Observation of the sensor outputs can only yield estimates of the differential delays which happen to exist at the time of measurement. If these differ from the delays corresponding to the nominal sensor locations (on which translation to bearing and range is based) there will be an incremental error due to the uncertainty in sensor location. Thus

$$\text{Cov}(\hat{\alpha}, \hat{r}) = \text{Cov}(\hat{\alpha}, \hat{r})|_{\text{nominal}} + |\Delta \text{Cov}(\hat{\alpha}, \hat{r}). \tag{14}$$

The first term on the right side is the covariance matrix discussed in Sections 2 and 3. The second term is the increment averaged over the prior distribution of sensor displacements.

The most complete results have been obtained for a nominally linear, equispaced array (as in Figure 2), with sensors experiencing small independent displacements at right angles to the array axis, the displacements being characterized by a zero mean Gaussian distribution with variance σ^2. In that case one finds that $\Delta \text{Cov}(\hat{\alpha}, \hat{r})$ is diagonal and obtains

$$\Delta D^2(\hat{\alpha}) \geq \frac{\sigma^2}{L^2} \frac{1}{M} \frac{M+1}{M-1} \tag{15}$$

$$\Delta D^2(\hat{r}) \geq 720 \, \sigma^2 \frac{1}{M} \frac{r^4}{L^4 \sin^2\alpha} \frac{(M-1)^3}{(M+1)(M^2-4)} \tag{16}$$

The most interesting feature of Equations (15) and (16) is their independence of the signal and noise spectra. The incremental error is determined exclusively by array geometry and mean square sensor displacement. A second feature of interest is the factor $(1/M)$ in both equations. As the number of sensors increases, a spatial averaging effect reduces the incremental error.

The question of realizability of the Cramer-Rao bounds (15) and (16) also has a simple answer: The focused beamformer of Figure 3, using nominal sensor locations to construct appropriate delays for each (α, r) pair, realizes mean square errors equal to the bounds (for TW $>>$ 1). Its performance is clearly degraded by the uncertainty in sensor locations, but no other instrumentation can achieve a smaller total error[3].

No such absolute statement can be made when the sensor displacements are correlated. From a limited set of computations it appears, however, that significant error reductions below the bounds given by Equations (15) and (16) become possible only when the sensor displacements are so highly correlated as to make the situation of questionable practical interest.[3]

The Cramer-Rao bounds of Equations (15) and (16) can be generalized to two dimensional arrays of the form of Figure 4 whose sensors experience independent displacements of variance σ^2 in both the vertical and horizontal direction. Cov $(\hat{\alpha}, \hat{r})$ remains diagonal and the incremental variances assume the form

$$\Delta D^2(\hat{\alpha}) = \frac{4\sigma^2}{ME(u_i^2)} \tag{17}$$

$$\Delta D^2(\hat{r}) = \frac{16\sigma^2 r^4}{MV(u_i^2)} \tag{18}$$

Equations (17) and (18) retain the key features of Equations (15) and (16): The independence of signal and noise spectral properties (including power level) as well as the spatial averaging effect described by the factor 1/M. Furthermore, combining Equations (17) and (18) with (10) and (11) respectively, one observes that the *total* bearing and range errors satisfy

$$\frac{D^2(\hat{r})}{D^2(\hat{\alpha})} = 4r^4 \frac{E(u_i^2)}{V(u_i^2)}. \tag{19}$$

The trade off between bearing and range accuracy is therefore governed by the relative magnitude of the geometrical factors $E(u_i^2)$ and $V(u_i^2)$, regardless of whether or not the sensors experience displacement from their nominal locations.

5 DYNAMICALLY PERTURBED ARRAYS

In important practical problems the sensors are not only randomly displaced from their nominal locations, but they also move appreciably during the observation time. The effect of sensor motion on localization accuracy is a subject of active current study and only very partial results are available at the moment.

If there are only two sensors and the delay variation caused by sensor motion can be characterized as a zero-mean random Gaussian process with the spectrum

$$H(\omega) = \begin{cases} \dfrac{\sigma^2}{W} & 0 \leqslant \omega \leqslant W \\ 0 & \omega > W \end{cases} \tag{20}$$

the Cramer-Rao bound on the mean square error of differential delay is

$$D^2(\hat{\tau}) = D^2(\hat{\tau})|_{\sigma=0} + \Delta D^2(\hat{\tau}). \tag{21}$$

The component due to sensor motion is bounded by

$$\Delta D^2(\hat{\tau}) \leqslant \frac{\pi \sigma^2}{TW} \cdot \frac{2}{\pi} \int_0^{TW/2} \frac{\sin^2 x}{x^2} \, dx. \tag{22}$$

For TW/2 $\gg$ 1 the right side of (22) is very nearly equal to $\pi\sigma^2/(TW)$. For any finite signal-to-noise ratio one can show that $\Delta D^2(\hat{\tau})$ asymptotically approaches $\pi\sigma^2/(TW)$ as TW $\to \infty$. Hence for sufficiently large TW this quantity is not simply an upper bound on the incremental error, but a reasonable estimate of its magnitude. At the other extreme, when TW $\to 0$ one can easily establish that $\Delta D^2(\hat{\tau}) \to \sigma^2$. This is clearly the quasi-static situation: Since the measurement cannot differentiate between differential delays caused by source location and by sensor displacement, the incremental error is simply the mean square value of delay caused by sensor displacement.

The situation is more complicated for TW products which are neither very large nor very small. For any fixed TW one can show that the total error $D^2(\hat{\tau}) \to 0$ as the signal-to-noise ratio $\to \infty$. In physical terms one can now estimate the time function of delay during the observation interval with arbitrary accuracy. Because of the postulated strict band limitation of the sensor motion spectrum [Equation (20)], one can extrapolate the estimate to an arbitrarily large interval and then average to remove the (zero mean) fluctuation caused by sensor motion. When even small amounts of noise are present this extrapolation is no longer possible and it appears that the realizable incremental error will generally be very close to the right side of (22).

The above results have been extended to the problem of bearing and range estimation for the case of a nominally linear array consisting of M equally spaced sensors in independent motion at right angles to the array axis. If the rms sensor displacement σ_y is small compared with the sensor spacing d, one obtains the following expressions for the incremental errors

$$\Delta D^2(\hat{\alpha}) = \frac{12}{M(M^2-1)d^2} \gamma \frac{\pi\sigma_y^2}{TW} \qquad (23)$$

$$\Delta D^2(\hat{r}) = \frac{720\, r^4}{M(M^2-1)(M^2-4)d^4 \sin^2\alpha} \frac{\pi\sigma_y^2}{TW} \gamma \qquad (24)$$

For TW $\gg$ 1 and signal-to-noise ratios not so high as to permit the extrapolation discussed in the previous paragraph, the constant γ has a value extremely close to unity.

It does not appear to be a simple matter to relax the assumptions under which the above results were derived. Work is currently in progress to deal with sensor motions that are not characterized by flat spectra or are correlated from sensor to sensor.

6 AMBIGUITY PROBLEMS

All of the performance bounds discussed thus far were based on the Cramer-Rao inequality. As pointed out earlier, they can be approached only asymptotically, for sufficiently large observation times and sufficiently high signal to noise ratios. In qualitative terms, errors close to the bounds can be obtained when ambiguity problems have disappeared, so that the uncertainty concerning the measured parameter is confined to the immediate neighborhood of the true parameter value. If the signal spectrum has a small fractional bandwidth, as it does in many practical situations, this condition is easily violated. The realizable mean square error can then be very much larger than the Cramer-Rao

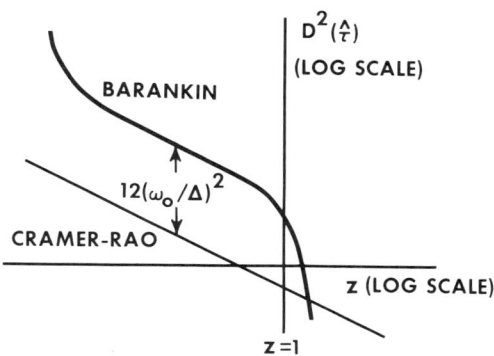

Figure 5 Bounds on mean square delay error

bound and it becomes desirable to set bounds that match attainable performance more closely. The Barankin bound has been used with some success to address this problem.

Relatively simple results can be obtained for the problem of two sensor delay estimation. Here, a simple version of the Barankin bound yields

$$D^2(\hat{\tau}) \geq \frac{(2\pi/\omega_0)^2}{e^z - 1} \tag{25}$$

where

$$z = \frac{\pi}{3} \frac{(S/N)^2}{1 + 2S/N} T\Delta \frac{\Delta^2}{\omega_0^2} \tag{26}$$

S/N is the signal-to-noise ratio, assumed to be constant over a narrow signal band Δ centered at frequency ω_0.

For $z \ll 1$ Equation (25) reduces to

$$D^2(\hat{\tau}) \geq \frac{(2\pi/\omega_0)^2}{z} = \frac{12\pi}{\frac{(S/N)^2}{1 + 2S/N} T\Delta \cdot \Delta^2} \tag{27}$$

$$= 12(\omega_0/\Delta)^2 \cdot \text{Cramer-Rao bound.}$$

The bound has therefore been increased by a factor proportional to $(\omega_0/\Delta)^2$, suggesting that the accuracy of the delay estimate is determined by the correlation time of the signal envelope rather than by the period of the center frequency. To approach this value of error, one must have a signal to noise ratio sufficiently high so that the envelope of the measured correlation function clearly rises above the noise background, but not so high that one can distinguish unambiguously between adjacent maxima of the quasi-sinusoidal correlation function.

For $z > 1$ Equation (25) falls off rapidly (See Figure 5). Once it has fallen below the Cramer-Rao bound it becomes uninteresting because the Cramer-Rao bound now imposes the tighter constraint. At very low values of z a more general version of the Barankin bound shows a second threshold, physically attributable to the fact that the correlation envelope is now no longer clearly distinguishable in the noise background. This threshold occurs in the neighborhood of the point $\frac{(S/N)^2}{1+2S/N} T\Delta = 1$, whereas the transition to the Cramer-Rao bound occurs near $z = 1$. The separation between the two thresholds is pronounced when (ω_0/Δ) is large.

Some extensions of the above results have been obtained for bearing and range estimation using an array. The effect of ambiguities on multipath ranging is currently under study.

REFERENCES

1. H.L. Van Trees, Detection and Estimation Theory, Part I, p. 84 (Wiley 1968).

2. W.R. Hahn, Optimal Signal Processing for Passive Sonar Range and Bearing Estimation, JASA, *58* 201 (1975).

3. P.M. Schultheiss, J.P. Ianniello, Optimum Range and Bearing Estimation with Randomly Perturbed Arrays, JASA, *68*, 167 (1980).

Adaptive Range Tracking of Underwater Maneuvering Targets Using Passive Measurements*

Richard L. Moose

Department of Electrical Engineering
Virginia Polytechnic Institute and State University
Blacksburg, VA

1 INTRODUCTION TO TARGET TRACKING

During the past several years much effort has been spent in the development of sophisticated digital filtering algorithms for tracking maneuvering targets. A common method has been to model the target dynamics in a rectangular coordinate system which results in a linear set of state equations, but forces the measurements to be nonlinear functions of the state variables. With this model an extended Kalman filtering algorithm is frequently used both to provide current state variable estimates and, by a one-step prediction process, to linearize the next measurement vector. This method works moderately well until the target makes an abrupt change in its trajectory in response to pilot or missile-guidance program commands. In this situation the velocity and position estimates can, and often do, diverge from the true unknown values. The inherent problems of this approach can lead to large bias errors and sometimes complete filter divergence.

Earlier work on the maneuvering target tracking problem includes Jazwinski's limited memory filtering [3], in which the filter gains are prevented from decaying to zero. Another technique, described by Thorp [4], involves switching between two Kalman filters in response to a detected maneuver. A third approach, due to Singer [5], models the target trajectory as a response of the target model to a time-correlated random acceleration. With this method additional state variables are used to generate the correlated forcing functions

---
*Sponsored by ONR under Contract No. N00014-77-C-0164.

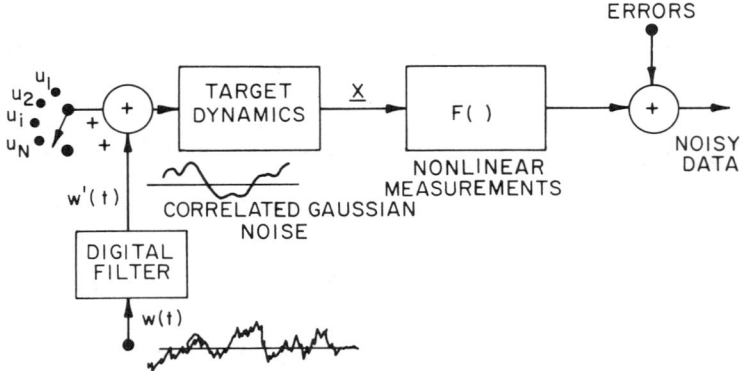

Figure 1 Maneuvering target motion model

which, in turn, increase the dimension of the Kalman filtering algorithm. In this manner the technique provides the filter with statistical information concerning target maneuvers based on an assumed range of possible accelerations. Singer's method was subsequently extended by many others.

Parallel to the effort was the method of modeling major changes in target trajectories by a semi-Markov process. An application of this approach to tracking maneuvering targets in two-dimensions by Moose [6] was successfully extended by Gholson and Moose [1] to three-dimensional tracking.

The general approach which uses the "adaptive semi-Markov maneuver model" of [6] and [7] implies a discretization of possible vehicle accelerations or velocities. The estimation algorithm then views the maneuvering vehicle as if it is responding to commands which are modeled by a semi-Markov process, i.e., a random process with a finite number of "states" (commands) which are selected according to the transition probabilities of a Markov process. A semi-Markov process differs from a Markov process in that the duration of time in one state prior to switching to another state is in itself a random variable. Incorporating the semi-Markov concept into a Bayesian estimator was done in [6] and [8]. This estimation algorithm provides a substantial improvement in filter stability, which means that large bias errors are prevented from being built up due to unmodeled target accelerations. An important aspect of this adaptive estimation algorithm is its elimination of a "growing memory" which is prevalent in many adaptive filters. It has been shown that by combining the concepts of Singer and Moose, et al., that the mean values required to prevent filter divergence is greatly reduced. This combination is illustrated in Figure 1.

The target trajectory is generated by the random selection of an input time-correlated Gaussian process whose mean value u_i is applied to the target plant dynamics for a random duration of time. This input disturbance process lasts until a new input u_j is randomly chosen from among a finite set of n possible inputs. With this model as a background and using an appropriate choice of state variable equations to represent target dynamics, either submarine or aircraft, it is possible to develop an "optimal" (in the minimum mean-square error sense) tracking filter that adaptively learns, then quickly adjusts itself for each major alteration of target trajectory.

Adaptive Range Tracking

2 LINEAR POLAR STATE VARIABLE MODEL

Referring to Figure 2, the polar coordinate system (ρ, z_{s0}) is attached to and moves with the observer. If (x_0, y_0, x_s, y_s) are the horizontal position coordinates of the observer and source, respectively, in a fixed (w.r.t. ocean floor) rectangular coordinate system, then

$$\rho = [(x_s - x_o)^2 + (y_s - y_o)^2]^{1/2}. \quad (1)$$

Equation (1) is linearly expanded as follows:

$$\rho_{k+1} = \rho_k + \frac{\partial \rho}{\partial x_s}\bigg|_k \left(x_{s_{k+1}} - x_{s_k}\right) + \frac{\partial \rho}{\partial y_s}\bigg|_k \left(y_{s_{k+1}} - y_{s_k}\right)$$

$$+ \frac{\partial \rho}{\partial x_0}\bigg|_k \left(x_{0_{k+1}} - x_{0_k}\right) + \frac{\partial \rho}{\partial y_0}\bigg|_k \left(y_{0_{k+1}} - y_{0_k}\right). \quad (2)$$

By modeling the source dynamics in the X and Z directions as a linear drag model (drag coefficient α) with a forcing function consisting of a Singer correlated Gaussian random process (time constant $= 1/a$), the following state variable model can be shown to result from (2) for a sample interval T.

$$\begin{bmatrix} \rho \\ \dot{\rho} \\ w'_{s_\rho} \end{bmatrix}_{k+1} = \begin{bmatrix} 1 & A & B \\ 0 & E & F \\ 0 & 0 & e^{-aT} \end{bmatrix} \begin{bmatrix} \rho \\ \dot{\rho} \\ w'_{s_\rho} \end{bmatrix}_k$$

$$+ \begin{bmatrix} C(A-T) \\ A(E-1) \\ 0 & 0 \end{bmatrix} \begin{bmatrix} u_{s_\rho} \\ V_0 \cos B_{s0} \end{bmatrix}_k + \begin{bmatrix} D \\ G \\ J \end{bmatrix} w_{s_{\rho k}} \quad (3)$$

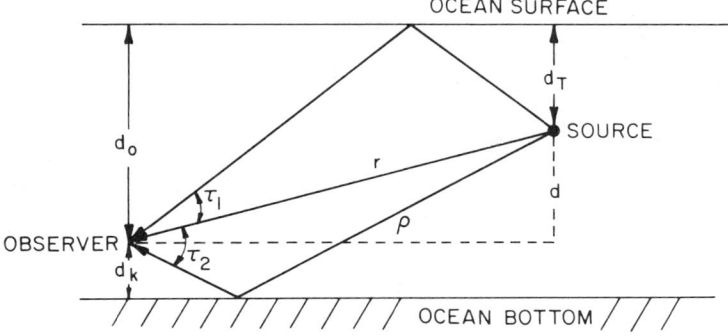

Figure 2 Two dimensional (range and depth) tracking geometry

where $A = (1 - e^{-\alpha T})/\alpha$, $B = [1 + (ae^{-\alpha T} - \alpha e^{-aT})/(\alpha - a)]/(\alpha a)$, $C = (\alpha T - 1 + e^{-\alpha T})/\alpha^2$, $D = [T + (aA - \alpha J)/(\alpha - a)]/(\alpha a)$, $E = e^{-\alpha T}$, $F = (e^{-aT} - e^{-\alpha T})/(\alpha - a)$, $G = (J - A)/(\alpha - a)$, and $J = (1 - e^{-aT})/a$.

V_0 is the observer's horizontal velocity and B_{s0} is the relative bearing of the source with respect to the observer. A similar state model exists for the vertical (Z) direction. Note that the state equation for $\ddot{\rho}$ in (3) is arrived at by applying the linear expansion (2) to the time derivative of (1).

Depth Channel Model

Using a discretized version of the basic linearized drag model of

$$\begin{bmatrix} \ddot{d}_T + \alpha \dot{d}_T = w_z' \\ \dot{w}_z' + aw_z' = w_z \end{bmatrix}$$

and letting d_T be the target depth the following state model is presented.

$$\begin{bmatrix} d_T \\ \dot{d}_T \\ w_z' \end{bmatrix}_{k+1} = \begin{bmatrix} 1 & A & B \\ 0 & E & F \\ 0 & 0 & e^{-aT} \end{bmatrix} \begin{bmatrix} d_T \\ \dot{d}_T \\ w_z' \end{bmatrix} + \begin{bmatrix} D \\ G \\ J \end{bmatrix} w_{z_k}. \quad (4)$$

The state model is also subject to the constraint of a constant velocity observer.

3 TRACKING GEOMETRY

The tracking geometry of Figure 2 has been investigated by Hassab [9] and shown to yield two different time delays τ_1 and τ_2. Here τ_1 is the difference in time between the direct path and the surface reflected path from source to observer, and τ_2 the direct and bottom bounce time difference. For *long* ranges i.e.,: $r^2 \gg d_0^2$, then $r = \rho$ the polar range, and

$$\tau_1 = \frac{2d_0 d_T}{\rho C}. \quad (5a)$$

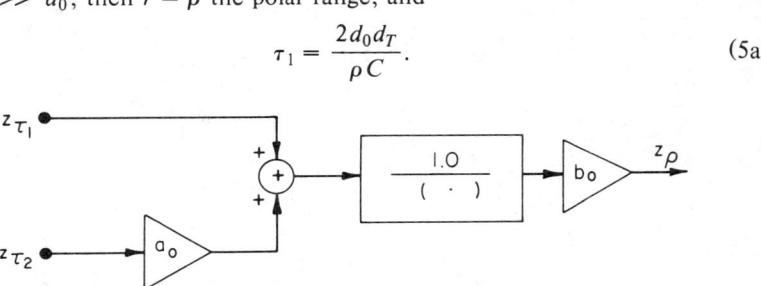

Figure 3 Nonlinear prefilter for the target range measurement

If, in addition, we have the "shallow water ranging situation," i.e., that (d_k^2/r^2) is $\ll 1$. For example, if $r \geq 2.25\, d_k$, our error of approximation is less than 1.4%, thus τ_2 becomes

$$\tau_2 = \frac{2d_k(d_w - d_T)}{\rho C}. \tag{5b}$$

Solving Equations (5a) and (5b) for target depth d_T then substituting into (5a) to get target polar range. ρ we find that:

$$\rho = \frac{b_0}{\tau_1 + a_0 \tau_2} \tag{6a}$$

$$d_T = \frac{\tau_1 d_w}{\tau_1 + a_0 \tau_2} \tag{6b}$$

where $a_0 = d_0/d_k$, $b_0 = 2d_0 d_w/C$, and C equals the sound velocity in sea water. In reality, we do not have τ_1 and τ_2 given to us but only the noisy set of measurements

$$z_{\tau_1} = \tau_1 + v_1$$

$$z_{\tau_2} = \tau_2 + v_2$$

where v_1 and v_2 are Gaussian random processes with zero mean and variances σ_1^2 and σ_2^2 respectively. This results in the noisy set of measurements

$$z_{d_T} = \frac{z_{\tau_1} d_w}{z_{\tau_1} + a_0 z_{\tau_2}}$$

$$z_\rho = \frac{b_0}{z_{\tau_1} + a_0 z_{\tau_2}}. \tag{7}$$

4 NONLINEAR PREFILTER DESIGN AND ANALYSIS

In Equation (7) we see that a noisy measurement (z_ρ) of polar range (ρ) is found by combining the weighted time delay measurements and dividing them into the geometry coefficient (b_0) previously defined. This process is shown in Figure 3. The output (z_ρ) consists of true range (ρ) plus nonzero mean and non-Gaussian additive noise (v_ρ).

After a statistical analysis the following probability density function for $v\rho$ was determined

$$p_2(v_\rho) = \frac{\rho \tau_T}{\sigma_T \sqrt{2\pi}\,(\rho + v_\rho)^2} \exp\left[\frac{-\tau_T^2 v_\rho^2}{2\sigma_T^2(\rho + v_\rho)^2}\right]$$

where $\sigma_T^2 = (\sigma_1^2 + a_0^2 \sigma_2^2)$. \hfill (8)

The nature of this density function is shown in the following figures for the target and observer at fixed depths, but closing in range.

It appears that the density function $p_2(v_\rho)$ given by Equation (8) is of a form that does not appear in complete sets of integral tables, nor does it seem

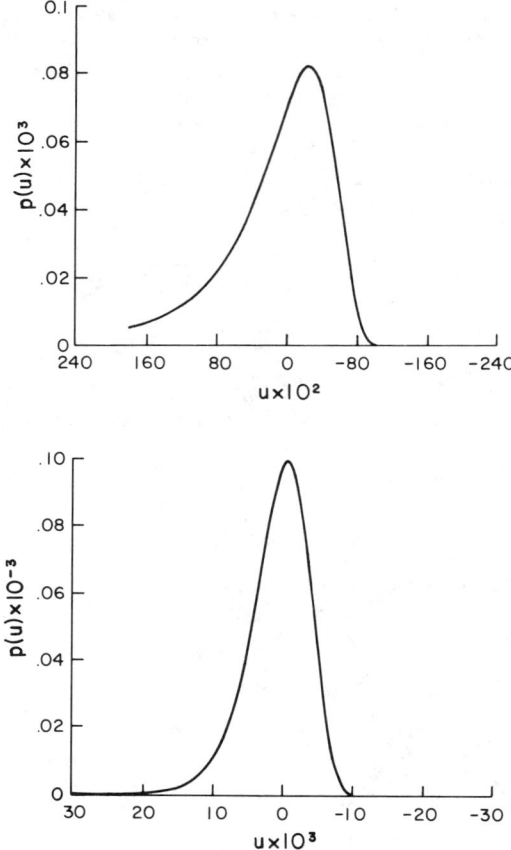

Figure 4 Measurement error density function for closing range

likely that there is a closed form for the expected value $(v_\rho = -ax/(x+k))$, or for the expected value of $v_\rho^2 = a^2x^2/(x+k)^2$ where x is $N(0,\sigma^2)$. Thus one must use *numerical integration or simulation techniques* in order to get an approximation for the mean and the variance of measurement error (v_ρ). Examining the density function, we see from Figure 4 that it is a smooth continuous curve. As target range closes the mean and variance becomes smaller and the density becomes more and more symmetrical approaching a near Gaussian like density.

Table I shows that v_ρ has a nonzero mean and a variance which are nonlinear functions of range and SNR. These statistics must be altered for they are not compatible with the Kalman filters that will process the range measurements. For example, v_ρ must be zero mean. This is achieved by a table look-up of the measurement noise mean based on the tracker's previous estimate of ρ and the SNR. This mean is then subtracted from the current measurement z_ρ producing a zero mean noise process. A similar table look-up method is

Adaptive Range Tracking

used to select the proper variance required by the Kalman filters. Simulations have shown that the values in Table I are sufficient to insure measurement compatibility with the Kalman filters.

Table I Linearized Measurement Error Means and Variances vs Range

Range	$\sigma_{1,2} = 1$ ms		$\sigma_{1,2} = 5$ ms	
	Mean	Var	Mean	Var
10,000	3.8	.209E + 05	62.	.549E + 06
20,000	23.	.336E + 06	487.	.102E + 08
30,000	72.	.172E + 07	1650.	.605E + 08
40,000	163.	.552E + 07	3518.	.192E + 09
50,000	310.	.137E + 08	5588.	.427E + 09
60,000	527.	.290E + 08	6841.	.780E + 09
70,000	831.	.555E + 08	7068.	.127E + 10
80,000	1238.	.972E + 08	6204.	.190E + 10

A limiter is required at the output of the nonlinear prefilter. This is due to the fact that a density function of the form given by Eq. (8) has a large variance and an occasional bad data point is observed. The limits are quite loose and were set at $\hat{\rho}_k \pm 10^5$.

5 STATE ESTIMATION AND ADAPTIVE TRACKING SYSTEM STRUCTURE

Introduction

In this section, we will discuss the basic estimator system structure. We will make use of the linearized measurements containing the nonstationary, non-Gaussian, statistics that were previously discussed in Section (4) of the paper.

Referring to Figure 5, the nonlinear time delay measurements $z_{\tau 1}$ and $z_{\tau 2}$ are fed into the nonlinear prefilter. This unit develops a linearized measurement of target range (ρ). The errors in measuring these target parameters are both nonGaussian and nonstationary depending upon the geometry of the tracking situation. As target range closes, or opens the mean value, and vari-

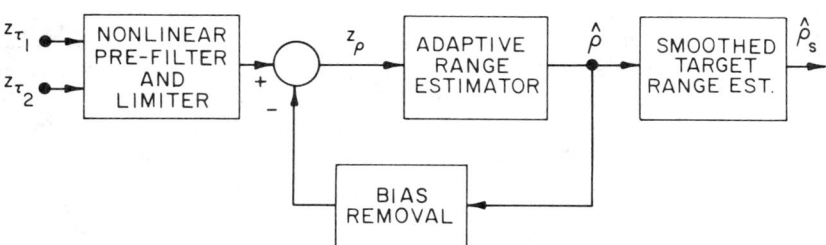

Figure 5 Basic range estimation structure

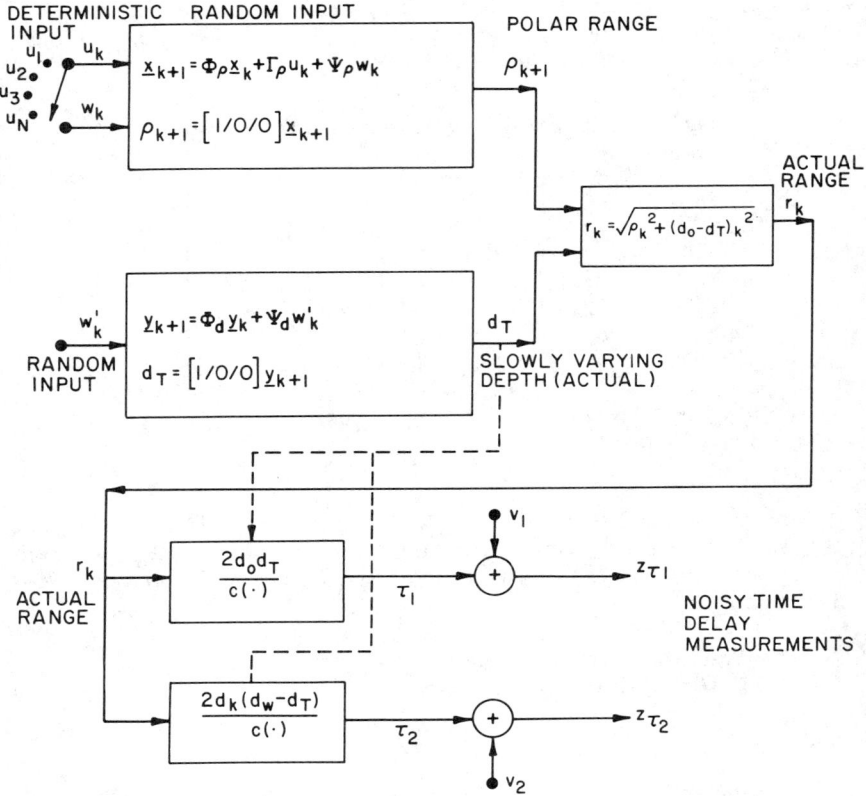

Figure 6 Data generation

ances of these errors change, and must be accounted for by the stored means and variances of Table I.

In Figure 6 a discrete time system model is presented showing the development of noisy time delay measurements $z_{\tau_1}(k)$ and $z_{\tau_2}(k)$. The upper portion shows the generation of $\underline{x}_{k+1} = [\Phi \underline{x}_k + \Gamma \underline{u}_k + \underline{\psi} w_k]$ where Φ, Γ, and $\underline{\psi}$ are discussed in Equation (3). The deterministic input u_k *is unknown to the tracking filter* and serves to generate large scale target maneuvers in velocity. A measure of randomness in target trajectory ($x_1 = \rho_k$ and $x_2 = \dot{\rho}_k$) is generated by applying a Gaussian random input $\underline{w}_k$ to the simulated target. This forcing function is exponentially correlated and is denoted by state variable $x_3 = w_\rho(k)$.

Once ρ_{k+1} is generated, it is acted upon in a nonlinear manner to generate τ_1 and τ_2, which when added with the Gaussian random measurement errors v_1, v_2, produce the noisy time delay measurements z_{τ_1} and z_{τ_2}.

The Polar Range Adaptive State Estimator

The heart of the adaptive filter summarized in this paper is in the forming of the total estimate of the target states, from a weighted sum of state estimates conditioned on the N possible discrete input levels $u_\rho^{(i)}$. Consider the state model of Equation (3). This state model views the target input acting in the polar direction as being derived from a time correlated Gaussian density having a mean value u_ρ. Next consider a series of N such Gaussian curves with displaced mean values $u_\rho^{(i)}$, $i = 1, 2, \ldots, N$ and partially overlapping "tails" outlined in Figure 1. If a bank of N Kalman filters is formed, each filter based on the state equations of Equation (3) with the deterministic input $u_\rho^{(i)}$ being a different one of these N mean values, then a series of N estimates is obtained, each conditioned on a different Gaussian curve. Next a weighted sum of these estimates is found to be the total unconditioned estimate of the target states given by the optimal mean-squared estimate

$$\hat{\underline{x}}(k+1) = \sum_{i=1}^{N} \hat{\underline{x}}^{(i)}(k+1) \, W_i(k+1). \tag{9}$$

Now as the target executes a series of evasive maneuvers in the polar channel, the changing input necessary to produce these maneuvers is viewed as randomly switching among the N Gaussian curves. By applying semi-Markov statistics to this switching process a series of N probabilities W_i, $i = 1, 2, \ldots, N$ is generated where

$$W_i \equiv Pr \, \{\text{target input is being derived from}$$
$$\text{the Gaussian curve whose mean value is } u_\rho^{(i)}\}$$

or,

$$W_i(k+1) = Pr \, \{u_\rho(k) = u^{(i)} | Z(k+1)\} \tag{10}$$

and

$$\hat{\underline{x}}^{(i)}(k+1) = E\{\underline{x}(k+1) | u(k) = u^{(i)}, Z(k+1)\}$$

represents the ith Kalman filter conditioned on $u_\rho = u^{(i)}$. For example, filter (i) is given by

$$\hat{\underline{x}}^{(i)}(k+1) = \Phi \hat{\underline{x}}^{(i)}(k) + \Gamma u^{(i)} + K(k+1) \, [z(k+1)$$
$$- H\Phi \hat{\underline{x}}^{(i)}(k) - H\Gamma u^{(i)}]. \tag{11}$$

where

$$K(k+1) = M(k+1)H^T[HM(k+1)H^T + R]^{-1},$$
$$M(k+1) = \Phi P(k)\Phi P(k)\Phi^T + \psi Q \psi^T$$
$$P(k+1) = [I - K(k+1)H]M(k+1).$$

The matrices Φ, Γ and ψ are used to denote the respective coefficient matrices of Equation (3).

The weighting probabilities can be expressed in a vector recursive form as

$$\underline{W}_{k+1} = C_k \underline{P}_k \underline{\theta}^T \underline{W}_k \tag{12}$$

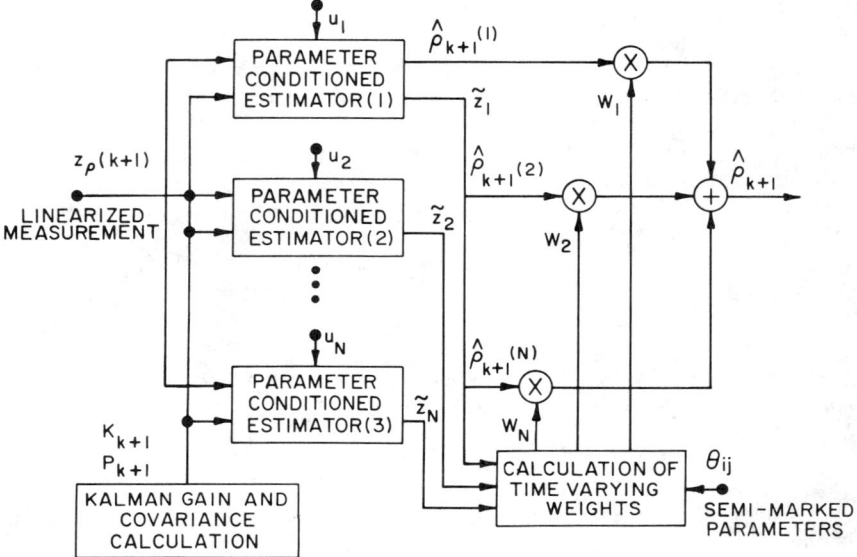

Figure 7 Adaptive polar range estimator

where $\underline{P}_k$ is a time varying diagonal matrix whose elements have been previously computed in each of the (N) Kalman filters (the measurement residual), i.e.,:

$$P_{ii} = N[m_i(k+1), V(k+1)].$$

The term C_k is a normalizing constant computed at each iteration to ensure that the sum of probabilities equals unity. The matrix $\underline{\theta}$ is a precomputed matrix whose elements contain statistical knowledge of the randomly switching plant inputs. In practice, the diagonal elements are nearly unity and the off diagonal terms are set equal.

Although it might appear from Equation (9) that an entire Kalman filter algorithm is being executed N times (for each of the possible inputs) at each time iteration, *such is not the case*, since the process and measurement covariances Q and R are the same for each filter.

What differentiates the different target "states" is the discrete levels u_i; however, the target dynamics remain unchanged! *The entire covariance, and gain analysis of the Kalman filter algorithm becomes identical for each state in a given channel and, consequently, need be executed just once rather than N times.* The adaptive filter structure is shown in Figure 7.

6 RANGE ESTIMATION RESULTS

In order to have a *benchmark* with which to compare the adaptive estimator's performance the following scenario was devised. A target at the extreme range of 100,000 was generated on a closing trajectory of 25 ft/s at a depth of 600.

Adaptive Range Tracking

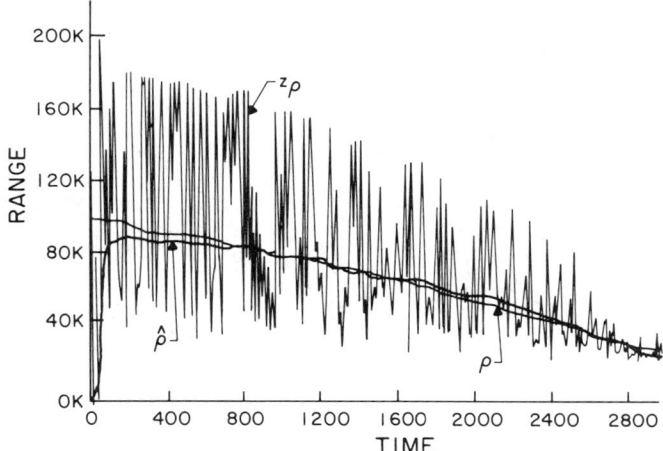

Figure 8 Range estimate of matched (unrealizable) Kalman filter (low SNR case)

The target's plant model was matched exactly with that of the filter. This was done to simulate and test the performance of the optimal (nonrealizable) tracking filter. In addition, it also provided a test of the nonlinear prefilter and the adaptive state estimator whose weights were all set at *zero* for the unmatched filters and *unity* for the single matched filter.

Target tracking results are shown in Figure 8 for additive noise of 5 ms. Examining the figure we see the performance that one would expect out of the unrealizable filter. Note that the variance of measurement error decreases with decreasing range as one might expect.

The data simulation of Figure 6 made use of the following parameters: the maneuver time constant $(1/\alpha)$ was chosen as 25 s, the Singer correlated acceleration time constant $(1/a)$ in Equation (3) was selected to be 40 s, and the data rate was chosen to be one sample every 10 s. Although studies were made indicating a marginal increase in tracking performance, for sample intervals of 2 and 5 s, the sample interval was retained at 10 s in order to reduce the computation burden for real world applications.

In the adaptive estimator of Figure 7, 6 levels of input u_i were chosen to span the expected target velocity range of $+30$ (ft/s) for an opening target and -45 (ft/s) for a closing target. If the velocity ranges are greater than this number, N could be increased as needed. The u_i's were chosen to model $+30$, $+15$, 0, -15, -30, -45 (ft/s).

The next Figure 9 shows the realizable adaptive filter tracking in the presence of low noise (1 ms). The target makes a random step change in velocity at time $k = 150$, which corresponds to 1500 s into the scenario. The range estimate $\hat{\rho}$ tracks very well converging very quickly after the target maneuver.

In Figure 10 the weighting coefficients are shown to switch appropriately from 0 to near unity as required.

In Figure 11, we see the results of tracking a medium range (0-45 k) target as it undergoes a maneuver at $k = 140$ and $k = 340$. The target is on a

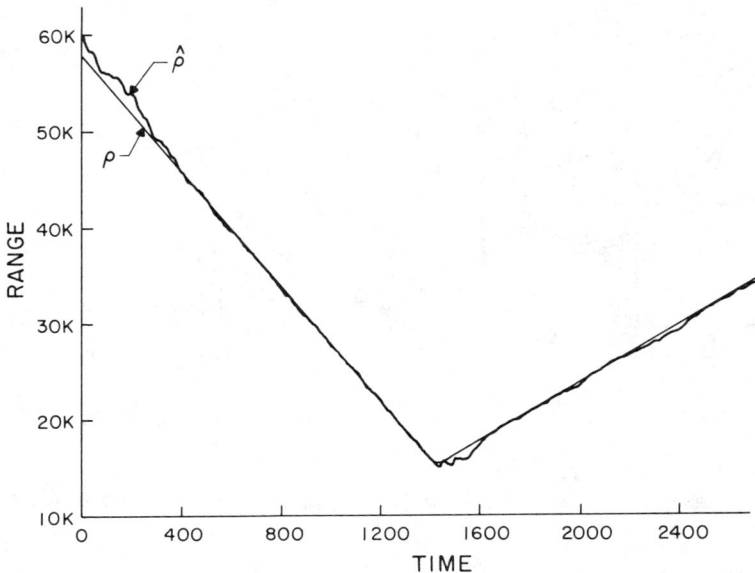

Figure 9 Target tracking for the low noise case realizable filter

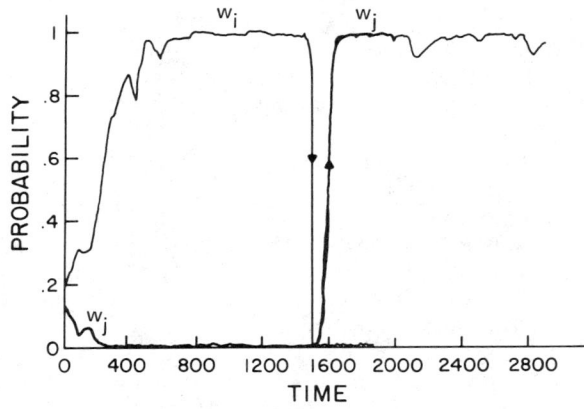

Figure 10 Plot of weighting coefficients vs time for target maneuver

closing trajectory then reverses its velocity at 140 and 340. The estimate $\hat{\rho}$ tracks very well at close ranges and progressively gives noiser estimates as the targets as the targets range increases. This is due to the variance of the measurements, which are not shown, increasing with range. The additive time delay noise is $\sigma_1 = \sigma_2 = 3$ m s, a considerable increase. Several conclusions can be drawn from Figure 11. The first is that we need to limit ourselves to close proximity targets, or smooth the estimate and weighting coefficients $\hat{w}_1(k)$, or

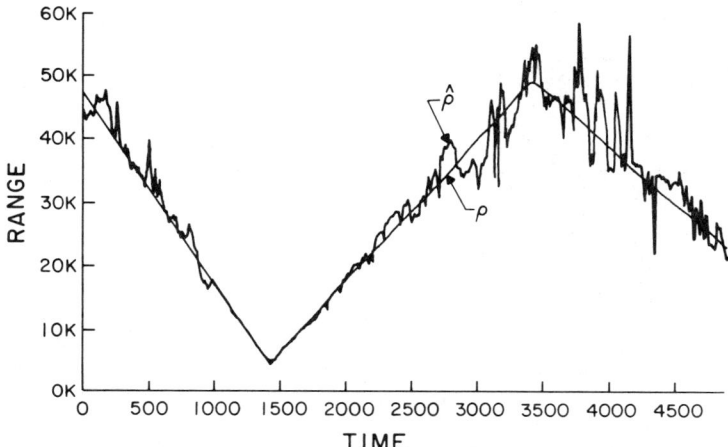

Figure 11 Target tracking of multiple maneuvering target under increased measurement error

operate in an environment with smaller additive noises which is totally unrealistic.

Modification to the Adaptive State Estimator

In order to operate in a noisy environment where the additive measurement errors attached to τ_1 and τ_2 are greater than or equal to 3 ms, it becomes necessary to smooth both the weighting coefficients and/or the output estimate $\hat{\rho}_{k+1}$. It was found that a simple first order digital filter of the form

$$\tilde{\rho}_{k+1} = a\tilde{\rho}_k + b\hat{\rho}_{k+1} \tag{13}$$

was successful for good smoothing and small lag in tracking maneuvers. The term $\tilde{\rho}_{k+1}$ represents the smoothed output range estimate at t_{k+1} and $\hat{\rho}_{k+1}$ the "rough" filter input. Coefficients a and b were set at 0.8 and 0.2 respectively.

In addition, to output smoothing, the weighting coefficients $w_i(k) i = 1, 2, \ldots, 6$ were averaged as follows:

$$\tilde{w}_i(k) = \frac{1}{N} \sum_{j=0}^{N-1} w_i(k - j). \tag{14}$$

Here $\tilde{w}_i(k)$ represents the time averaged output of the moving window averager Equation (14). Numerous simulations were performed as N was changed from 10 to 30. Some results follow.

Figures 12 and 13 illustrate the performance of the filter's range and velocity estimates and the adaptive nature of the weighting functions. The figures also show how the number of averaged data points $\tilde{w}_i(k)$ effects all of the above mentioned parameters. The scenario used to generate Figures 12 and 13 starts the target at a range of 50,000 ft and a velocity of −30 ft/s (-means toward observer) and changes the targets velocity to +30 ft/s after 2,500 s.

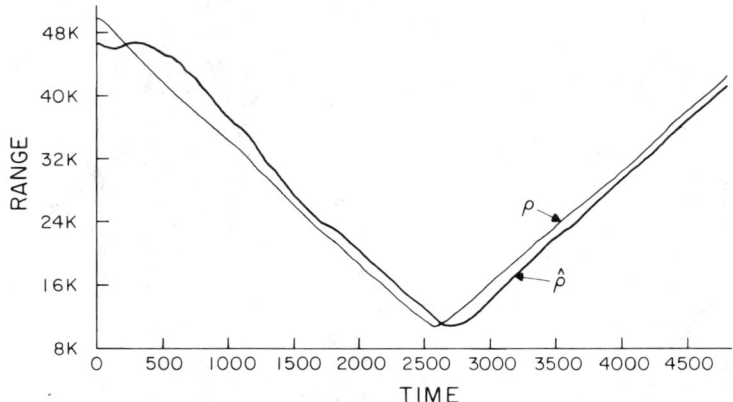

Figure 12 Estimate of range for approaching/opening target (20 points) averaged

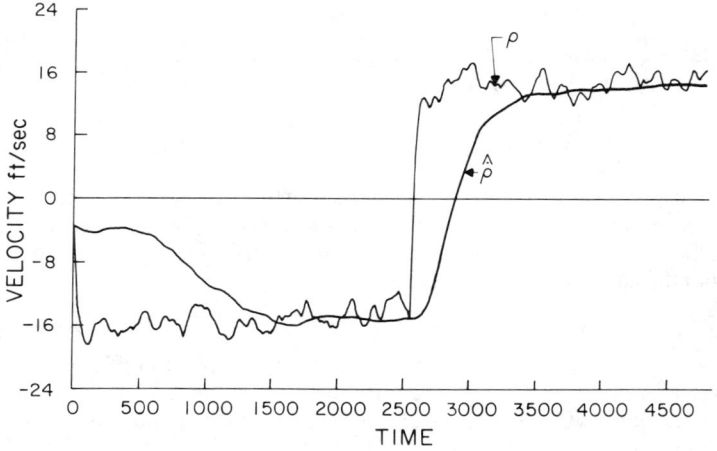

Figure 13 Estimate of velocity for maneuvering target (20 points) averaged

7 CONCLUSION

An adaptive state estimator has been developed and extensively tested to track a target making random large scale maneuvers in velocity. The target/observer scenario is constrained to the vertical plane in the ocean environment. This was intentionally done so as not to compete with well established bearing tracking programs.

The adaptive estimator made use of a nonlinear prefilter to *uncouple* the state variables that model target motion in both depth and range. An additional benefit was the elimination of all extended Kalman filters in the tracking system. This results in a more *robust* tracker and significantly fewer computations. The cost of doing this, is that the linearized measurements contain nonstationary and non-Gaussian measurement errors.

System inputs to the tracking system consists of noisy time difference measurements of bottom/direct, and surface/direct multipath time delays. The adaptive tracker prefilters the noisy multipath measurements in a nonlinear operation and then transmits the new linearized range measurements into its filtering channel. The range channel is fairly complex. Since the target is free to make major random velocity changes it required six Kalman filters and an adaptive weighting technique to span the expected range of all target velocities. Computationally this was quite *easily done* since each filter was only third order, and had the same Kalman gain and covariance matrix which only required one basic computation common to all six filters. The filters differed only in the deterministic input $u_i i = 1, 2, \ldots, 6$ built into each. As the target changed range in 5k increments new means and variances were programmed into the filter bank thus compensating for the effect of nonstationary linearized measurement errors.

Overall tracking results seem quite good, especially in the low signal to noise ratio cases. An on-going effort is underway to study techniques of averaging data to Gaussianize the measurement errors.

REFERENCES

1. N.H. Gholson and R.L. Moose, "Maneuvering target tracking using adaptive state estimation," *IEEE Trans. Aerosp. Electron. Syst.*, May 1977.

2. R.L. Moose, H.F. VanLandingham, D.H. McCabe, "Modeling and Estimation for Tracking Maneuvering targets," *IEEE Trans. Aerosp. Electron. Syst.*, vol. AES-15-No. 3., pp. 448-456, May 1979.

3. A.H. Jazwinski, "Limited memory optimal filtering," *IEEE Trans. Automat. Contr.*, vol. AC-13, Oct. 1968.

4. J.S. Thorp, "Optimal tracking of maneuvering targets," *IEEE Trans. Aerosp. Electron. Syst.*, vol. AES-9, July 1973.

5. R.A. Singer, "Estimating optimal tracking filter performance for manned maneuvering targets," *IEEE Trans. Aerosp. Electron. Syst.*, July 1970.

6. R.L. Moose, "Adaptive estimator for passive range and depth determination of a maneuvering target (U)," *U.S. Naval J. Underwater Acoustics*, July 1973.

7. R.A. Howard, "System analysis of semi-Markov processes," *IEEE Trans. Mil Electron.*, vol. MIL-8, pp. 114-124, April 1964.

8. D.H. McCabe and R.L. Moose, "Passive Source Tracking using Sonar Time Delay Data," *IEEE Trans. on Acoustics Speech, Signal Processing*, June 1981.

9. J.C. Hassab, "Passive Tracking of a Moving Source by a Single Observer in Shallow Water," *Journal of Sound and Vibration* (1976), 44 (1).

Passive Sonar Delay Estimate Improvement Using *A Priori* Knowledge and Increased Number of Sensors*

R. Lynn Kirlin

Department of Electrical Engineering
University of Wyoming
Laramie, WY

1 INTRODUCTION

This paper intends to show the improvement in noisy measurements of passive sonar delays by using *a priori* range information and increasing the number of sensors. The problem is cast into state variable form with the sonar delays as states. Kalman filtering is used to improve, through *a priori* information and multiple measurements, the delays to be estimated. The improvement can be examined by noting the decrease in the sonar delay estimate variances after one set of measurements is taken, and *a priori* information taken into account.

The work in this paper represents contracted work for the Office of Naval Research. Kirlin [2, 5] has derived analytical results for delay estimation improvement with three sensors. The research here has extended the results with numerical solutions for up to 24 sensors in three clusters with eight sensors each. Ideally, more measurements (more sensors) of the delays will make the delay estimates better, but the amount of improvement depends upon three major parameters: (1) measurement signal-to-noise ratio (SNR), (2) *a priori* range, and (3) the ratio (Q) of measurement error variance to *a priori* delay estimate variance. These three parameters are varied along with the sensor count to generate data. The data are then displayed in contour plots of improvement for fixed Q and sensor number and in improvement-vs-sensor number plots for fixed SNR and *a priori* range. Interpretation of the plots is given in Section IV. All the numerical results, plots, and the parent FORTRAN program are included in the Appendices of [6].

---

*Supported under ONR Contract Number N00014-82-K-0048

The estimation problem is defined first for three sensors and later expanded to the multiple sensor case. For all cases, the sensors are considered to be perfectly in line, as in an ideal, towed array. Other assumptions are that the signal spectrum is the same at every sensor and the noise spectra are also the same, but independent; that is, the noise is independent and identically distributed at each sensor.

2 DELAY ESTIMATION WITH THREE SENSORS

Figure 1 shows the state and measurement definitions for the three sensor delay estimate problem. These particular delays are chosen as states because d_1 and $(d_2 - d_1)$ are important in range estimation, and d_2 is important in bearing estimation. L is the distance between sensors one and two, and between two and three.

Using Kalman filtering, the delays can be considered as states to be estimated [1].

System Model

For subscripts k, referring to the kth periodic time sample, and letting the delays be states of a linear system, a general system model is

$$\underline{D}_k = \Phi_{k-1} \underline{D}_{k-1} + \underline{w}_{k-1}.$$

Assuming a nonmoving target and no system noise, $\Phi_{k-1} = I$ and $\underline{w}_k = 0$, then the system model becomes

$$\underline{D}_k = \underline{D}_{k-1}.$$

Measurement Model

The measurements $\underline{z}_k$ of the delays are given by

$$\underline{z}_k = H\underline{D}_k + \underline{e}_k,$$

where $\underline{e}$ is a zero-mean Gaussian white noise. The error vector $\underline{e}$ is not the noise at the sensors, but the error in the measurements in the delays. The members of $\underline{e}$ are not statistically independent. For the three sensor model, when the measurements are defined as in Figure 1, the measurement matrix H is

$$H = \begin{bmatrix} 1 & 0 \\ 0 & 1 \\ -1 & 1 \end{bmatrix}.$$

Also for the three sensor case, the *error* covariance matrix is

$$R = E[ee^T] = r_0^2 \begin{bmatrix} 1 & \gamma & -\gamma \\ \gamma & 1 & \gamma \\ -\gamma & \gamma & 1 \end{bmatrix},$$

where r_0^2 is the measurement error variance of the delay measurements. γ is due to cross correlations (of measurements) from sensor pairs, which have a

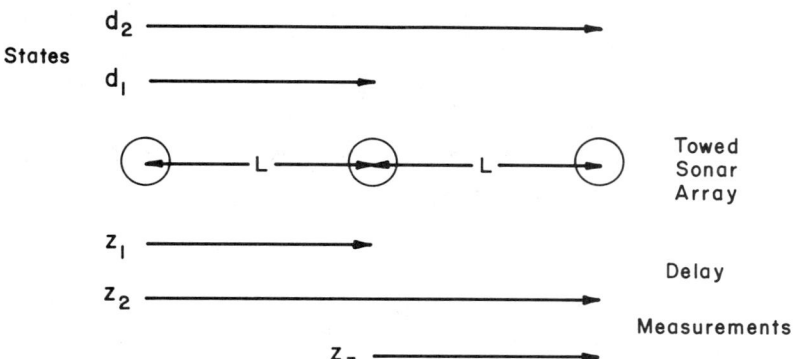

Figure 1 States and measurements for three sensor estimation

common sensor [2, 4, 5]. γ has the (+) sign if the common sensor is used either first or last in delay measurements, and has the (-) sign if the common sensor is used first in one delay and second in another. For example,

$$r_{13} = E[e_1 e_3] = E[e_{12} e_{23}] = -\gamma r_0^2$$

because sensor two is used first in measurement z_3, but second in z_1. If two measurements have no sensor in common, their covariance is zero.

For the situation under examination in this thesis, the signal spectra at every sensor are the same and flat; the noise spectra are the same and flat, but independent. Kirlin [2, 5], using Hahn's work [4], has shown γ is directly related to SNR by

$$\gamma = SNR/(1 + 2SNR).$$

γ therefore ranges in the interval $0 \leq \gamma \leq 0.5$ for SNR ranging between zero and infinity. In fact, for SNR greater than ten, γ is approximately 0.5, and γ is nearly zero only for SNR less than 0.01.

The error covariance matrix R is singular for $\gamma = 0.5$ and causes numerical difficulty, so for the numerical results $\gamma = 0.499$ is used for SNR approaching infinity ($\gamma = 0.499$ implies SNR = 250).

H and R matrices for multiple sensor cases are defined later.

Delay Estimate Update and Covariance Matrix

The new delay estimate after the kth measurement is

$$\hat{\underline{D}}_k = \hat{\underline{D}}_{k-1} + K_{k-1} [\underline{z}_k - H\hat{\underline{D}}_{k-1}],$$

where the Kalman gain is

$$K_k = P_k H^T R^{-1}$$

and the estimate covariance is

$$P_k = E[(\hat{\underline{D}}_k - \underline{D})(\hat{\underline{D}}_k - \underline{D})^T].$$

P_k contains the uncertainty in the delay estimate and is called the delay estimate covariance matrix. P_k for $\Phi_k = I$ and $\underline{w}_k = \underline{0}$ is [2]

$$P_k = (P_{k-1}^{-1} + H^T R^{-1} H)^{-1}.$$

For the one measurement vector to be taken and the *a priori* information available in P_0 (before the first measurement), the covariance equation update after one measurement ($k = 1$) is [2]

$$P_1 = (P_0^{-1} + H^T R^{-1} H)^{-1},$$

where

$$P_0 = \sigma_0^2 \begin{bmatrix} 1 & \beta \\ \beta & 4 \end{bmatrix}.$$

Let Θ be the *a priori* uniformly distributed counterclockwise bearing angle from broadside, and let v be the speed of sound in the medium. Then $\sigma_0^2 = (L/v)^2/2$ is the mean squared value of $d_1 \approx = (L/v)\sin\theta$, for ranges much greater than the sensor spacings L. $p_{22}(0) = 4\sigma_0^2$ because any *a priori* error in $\hat{d}_2$ is approximately twice the error of $\hat{d}_1$. β is a weight corresponding to knowledge of target range. For infinite range, $\beta = 2$ because any *a priori* error in $\hat{d}_1(0)$ must be twice as large in $\hat{d}_2(0)$.

For ranges much larger than the sensor spacing L, it may be shown $\beta \approx = 2 - L^2/R^2$, where R is the range of the target. For any reasonable range, β approaches two. It can also be demonstrated that $\beta = 0$ for ranges much less than sensor spacing L. Unfortunately, P_0 is singular for $\beta = 2$, so the numerical results are for $\beta = 1.9999$, corresponding to a range much greater than the sensor spacing L.

Since P_1 is the updated estimate covariance matrix, its diagonal values should be smaller (improved) than those of P_0, because a measurement has been used. The estimation variances of P_1 must also be smaller than or equal to the measurement error variances (r_0^2), because multiple measurements have been taken and *a priori* information used in the update process; that is, our estimates must be at least as good as our measurements, hopefully better. The amount of improvement of the diagonals of P_1 over r_0^2 due to *a priori* knowledge and multiple measurements is determined herein.

Following the earlier approach [2], the ratios

$$\eta_1 = p_{11}(1)/r_0^2 \text{ and } \eta_2 = p_{22}(1)/r_0^2$$

are the improvement factors of interest. These factors are scaled by r_0^2 to show the improvement over the measurement variance. For example, if η_1 equals one, then $p_{11}(1) = r_0^2$, and there was no improvement in the estimate of d_1 from *a priori* knowledge or the multiple measurements. On the other hand, if η_1 equals zero, then the estimate of d_1 is exact (no error). Similarly, η_2 shows the improvement in the variance of $\hat{d}_2$. So $0 < \eta_1 \leq 1$ and $0 \leq \eta_2 \leq 1$, where η equals one for no improvement over the measurements, and η approaches zero for exact estimates.

In general, η_1 and η_2 are functions of γ, β, $Q = r_0^2/\sigma_0^2$, and the number of sensors, M. These are the parameters varied so the computer generated data can be examined in plots of η_1 and η_2.

Table I Measurement variance improvement factors for extreme cases on apriori range (β) and SNR (γ)

β	γ	η_1	η_2
0	0 (SNR = 0)	$\dfrac{r_o^2/\sigma_o^2 + 8}{(r_o^2/\sigma_o^2)^2 + 10(r_o^2/\sigma_o^2) + 12}$	$\dfrac{r_o^2/4\sigma_o^2 + 1/2}{(r_o^2/4\sigma_o^2)^2 + 5/2(r_o^2/4\sigma_o^2) + 3/4}$
2	0 (SNR = 0)	$\dfrac{1}{r_o^2/\sigma_o^2 + 6}$	$\dfrac{1}{r_o^2/(4\sigma_o^2) + 3/2}$
0	1/2 (SNR = ∞)	$\dfrac{r_o^2/\sigma_o^2 + 16/3}{(r_o^2/\sigma_o^2)^2 + 20/3(r_o^2/\sigma_o^2) + 16/3}$	$\dfrac{r_o^2/4\sigma_o^2 + 1/3}{(r_o^2/4\sigma_o^2)^2 + 5/3(r_o^2/4\sigma_o^2) + 1/3}$
2	1/2 (SNR = ∞)	$\dfrac{1}{r_o^2/\sigma_o^2 + 4}$	$\dfrac{1}{r_o^2/4\sigma_o^2 + 1}$

Analytical results for η_1 and η_2 for the three sensor problem are reproduced in Table 1. Now the multiple sensor estimation problem is considered.

3 MULTI-SENSOR DELAY ESTIMATION

For optimal range and bearing estimation, the best array configuration for M sensors is three clusters of M/3 sensors each and equal spacing L between groups, as shown in **Figure 2** [3].

All sensors in a "pod" are assumed to be in the same location, i.e., there is no delay between sensors in the same group. The noise spectrum for each sensor is assumed to be the same, but independent from sensor to sensor. With more sensors, more delay measurements are possible, and the increased number of measurements of d_1 and d_2 should improve the Kalman estimate. The improvement can be examined using the updated estimation covariance matrix for different numbers of sensors after one iteration (one set of delay measurements):

$$P_1 = (P_0^{-1} + H^T R^{-1} H)^{-1}.$$

Multi-sensor Measurement Matrix

The H matrix for multiple sensors has more rows than that for three sensors. In general, there are $M^2/3$ possible useful pairwise measurements for M sensors grouped as in Figure 2. Figure 3 shows the measurement definitions for nine sensors assuming no delay or measurement between sensors in the same group.

Sensors in group I are paired with possible combinations of sensors in groups II and III. The sensors in II are then paired with those in III for the $M^2/3$ possibilities. The general measurement vector $\underline{z}_k$ is

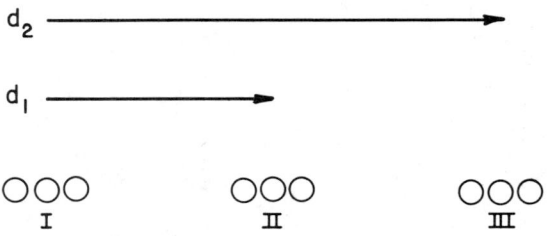

Equally Spaced Clusters

Figure 2 Multi-sensor sonar array configuration

Sonar Delay Estimate

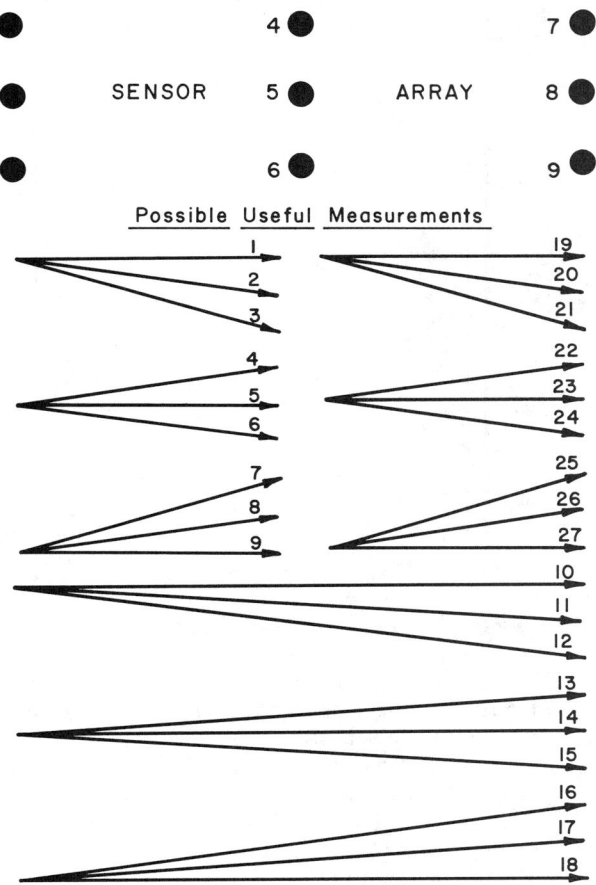

Figure 3 Measurement definitions for nine sensors

$$z_k = \begin{array}{c} \left.\begin{array}{c}\\ \\ \\ \\ \\ \end{array}\right\}\dfrac{M^2}{9} \\ \\ \left.\begin{array}{c}\\ \\ \\ \\ \\ \end{array}\right\}\dfrac{M^2}{9} \\ \\ \left.\begin{array}{c}\\ \\ \\ \\ \\ \end{array}\right\}\dfrac{M^2}{9} \end{array} \left| \begin{array}{cc} 1 & 0 \\ 1 & 0 \\ 1 & 0 \\ \cdot & \cdot \\ \cdot & \cdot \\ 1 & 0 \\ \hline 0 & 1 \\ 0 & 1 \\ 0 & 1 \\ \cdot & \cdot \\ \cdot & \cdot \\ 0 & 1 \\ \hline -1 & 1 \\ -1 & 1 \\ -1 & 1 \\ \cdot & \cdot \\ \cdot & \cdot \\ -1 & 1 \end{array} \right| \begin{bmatrix} d_1 \\ \\ \\ d_2 \end{bmatrix} + \underline{e}_k$$

Multi-sensor Measurement Error Covariance Matrix

For M sensors in clusters as in Figure 2, the R matrix is $M^2/3$ by $M^2/3$. An example of the R matrix for nine sensors and 27 measurements is in Figure 4.

The R matrix for multiple sensor cases is built just as in the three sensor case. The computer program decides if any two of the $M^2/3$ possible measurements have sensors in common and then adds the appropriate sign to γ. The diagonal members are initialized to r_0^2. The error covariance matrix is symmetric so only the components above the diagonal need be calculated. See [6] for more details on the initialization of the R matrix.

4 IMPROVEMENT FACTOR ANALYSIS

Improvement in $\hat{d}_1$:η_1

The parameter η_1 shows improvement of $\hat{d}_1$ over a delay measurement. The effect of range information can be examined in contour plots for fixed sensor count and Q. A typical contour plot is available in Figure 5. This particular plot is of η_1, for nine sensors with equal measurement error variance and estimation variance ($Q=1$). The contours are nearly linear for a given γ, except near $\beta=2$. Then a dramatic decrease of η_1 occurs for all γ's. This trend is particularly noticeable when several sensors ($M > 9$) are used in the model. The

Sonar Delay Estimate

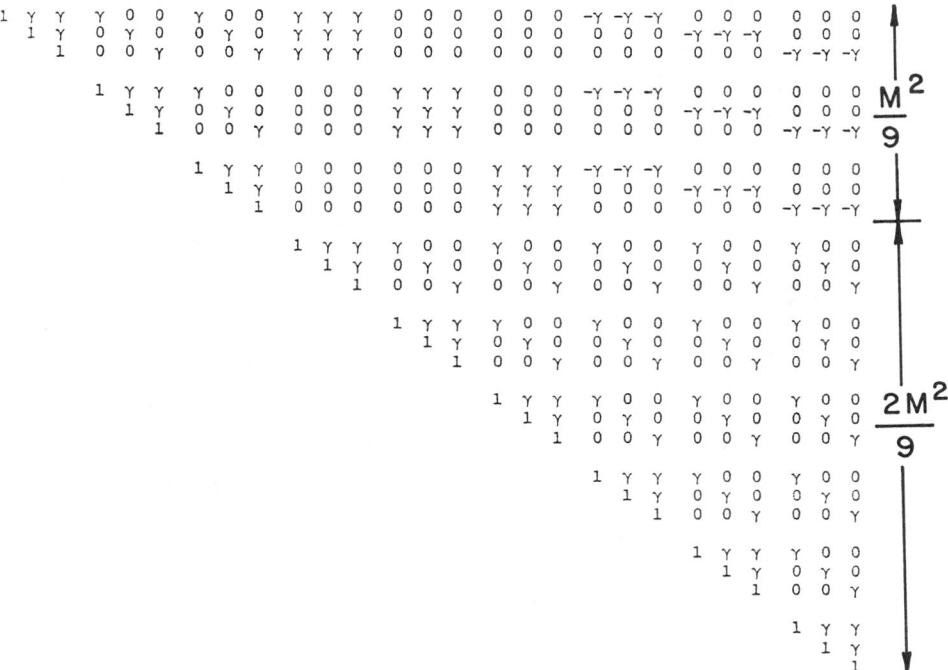

Figure 4 Error covariance matrix R for nine sensors, 27 measurements

dramatic improvement for β approaching two is particularly convenient because β is approximately two for all reasonable ranges ($R > > L$). From the data, it is apparent there is less improvement (η_1 increasing) as γ increases. This is logical, because as SNR increases (γ increases), the measurements must be better, and any improvement due to *a priori* range information makes less difference, so the ratio $p_{11}(1)/r_0^2$ must increase with SNR. In situations where r_0^2 is larger than $\sigma_0^2(Q=10)$, which implies poor measurements, *a priori* range information (β) has a significant effect. Figure 6 shows a case with $Q=10$ and the contour lines are not linear for a given γ. Since $Q=10$ implies poor measurements, then only the data near $\gamma=0$ (SNR small) is of interest, and these contours are much like the ones for the $Q=1$ case. Also for $Q=10$, the estimator depends heavily on *a priori* information because the measurements are poor; one would therefore expect β or *a priori* range information to have a greater influence.

The η_1 vs M plots for fixed γ and β give excellent insight as to the improvement of $\hat{d}_1$ by increasing the number of delay measurements taken. In general, there is always improvement in the estimate (decrease of η_1) as M increases; however, the curves for all cases have a knee at about $M=6$ or $M=9$. So after $M=9$, the addition of more sensors has much less influence than the change from three to six sensors. (See Figure 8).

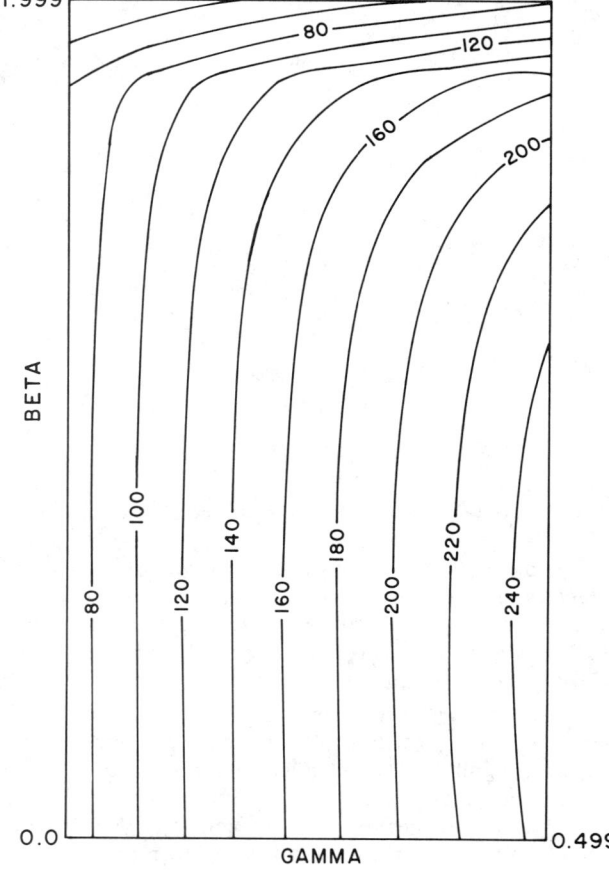

Figure 5 Contour of n_1 for nine sensors and Q = 1.0

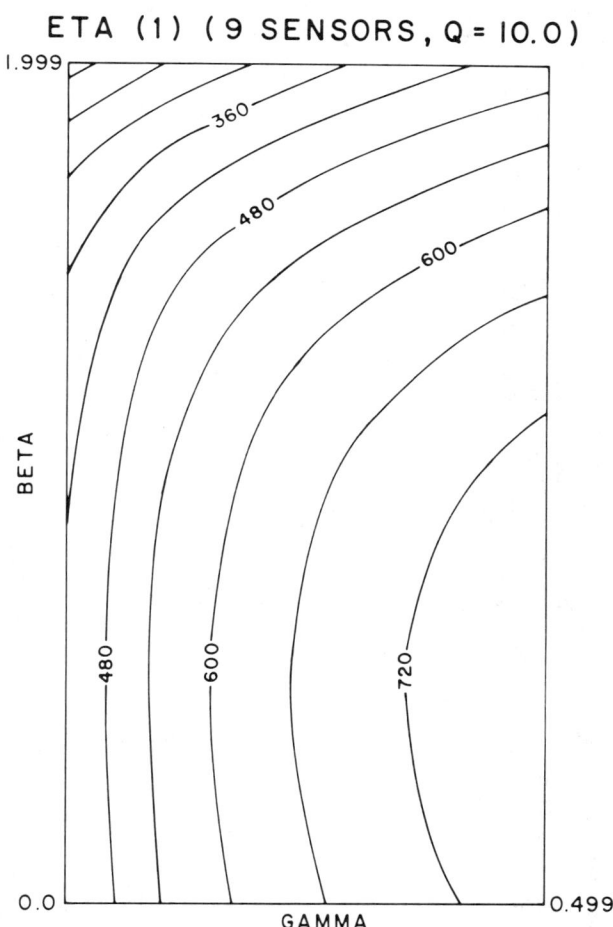

Figure 6 Contour of n_1 for nine sensors and $Q = 10.0$

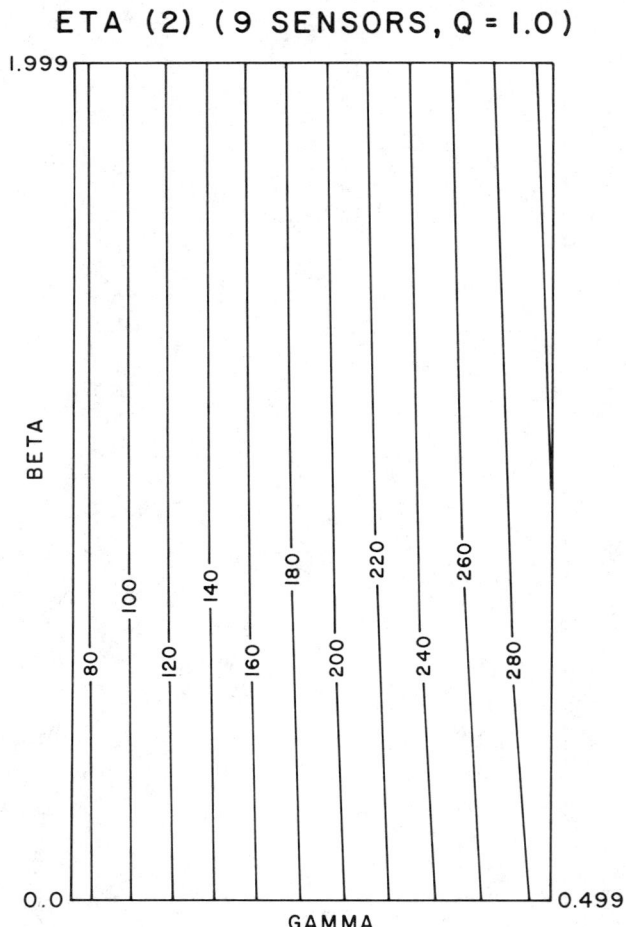

Figure 7 Contour of n_2 for nine sensors and $Q = 1.0$

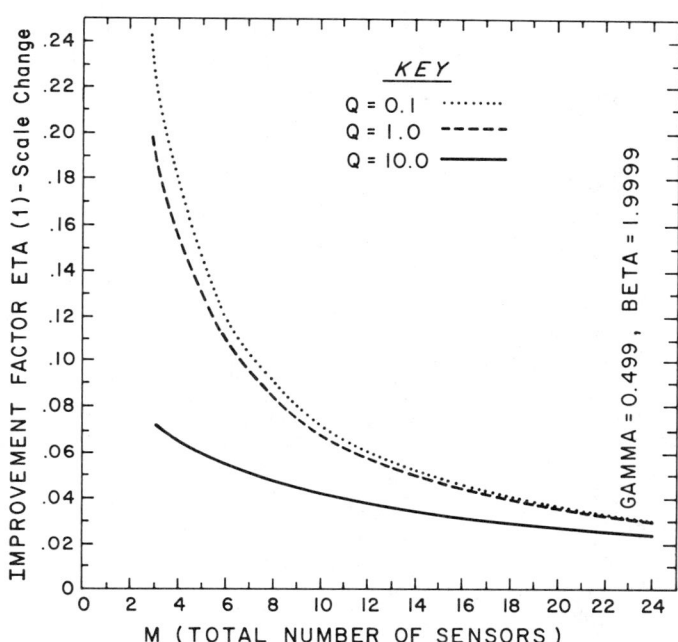

Figure 8 Effect of sensor count on n_1, $\gamma = 0.499$, $\beta = 1.9999$

Improvement in $\hat{d}_2$:η_2

Examination of the η_2 contours shows the improvement in $\hat{d}_2$ by multiple sensors is exactly as for η_1, so it will not be discussed further, although the results are summarized in Table II. The improvement from *a priori* range knowledge is entirely different.

Figure 7 shows a typical contour plot for the improvement factor η_2. Comparison of the contour plots of Figures 5 and 7, which are of similar situations, exhibits one major difference: the values for η_2 are larger in all instances. Why is $\hat{d}_1$ improved more than $\hat{d}_2$? The assumption that all measurements have been made with equal accuracy causes this peculiarity and is easiest to explain for the three sensor case. Here, three measurements have been taken: z_1, z_2, and z_3, all with variance r_0^2. z_2 can be used as a measurement of $2d_1$, yielding $\hat{d}_1$ with a variance of $r_0^2/4$, and is very helpful in the estimate of d_1. On the other hand, z_1 and z_3, also with variance r_0^2, are measurements of $d_2/2$, and if used for $\hat{d}_2$ do not improve the estimate significantly $((r_0^2 + r_0^2)/2 = r_0^2)$; therefore, η_2 greater than η_1 is expected.

Table II Improvement factors n_1 and n_2, tabulated qualitative results

		SNR → 0 ($\gamma=0$)		SNR → ∞ ($\gamma=0.499$)		
		apriori range $0<\beta<2$	# of sensors $3<M<24$	apriori range $0<\beta<2$	# of sensors $3<M<24$	
$Q=0.1$ $\sigma_0^2 >> r_0^2$ good measurements or poor apriori range information	n_1	dramatic improvement as $\beta \to 2$	SNR→0 implies very poor measurements. $Q=0.1$ and SNR→0 are contradictory.	dramatic improvement as $\beta \to 2$	knee at M=9 .0835 at $\beta \to 2$	n_1
	n_2	no effect		no effect	knee at M=9 .330 at $\beta \to 2$	n_2
$Q=1.0$ $\sigma_0^2 = r_0^2$ equal measurement and apriori range variances	n_1	increasing improvement as β increases (small M)	knee at M=6 .0401 at $\beta \to 2$	steadily increasing improvement as β increases	knee at M=9 .0769 at $\beta \to 2$	n_1
	n_2	less improvement as β increases (small M)	knee at M=6 .160 at $\beta \to 2$	less improvement as β increases (small M)	knee at M=9 .307 at $\beta \to 2$	n_2
$Q=10.0$ $r_0^2 >> \sigma_0^2$ poor measurements or very good apriori range info.	n_1		knee at M=6 .0294 at $\beta \to 2$	SNR→∞ implies very good measurements. $Q=10.0$ and SNR→∞ are contradictory.		
	n_2		knee at M=6 .117 at $\beta \to 2$			

326   Kirlin

5 CONCLUSION

This paper has examined the effect of using *a priori* information and multiple measurements on delay estimate variances. The estimate variances after one set of measurements is taken are scaled by the measurement variance to show improvement over using a simple measurement of the delays in question. Following is a quick summary of the major findings in this report.

1. The estimate of d_1 improved greatly due to knowledge that range is much greater than the sensor group spacing; however, $\hat{d}_2$ was not much improved by range information.
2. Plots of the improvement factors versus the number of sensors in the array demonstrate nine or twelve sensors is the critical number. Increased number of sensors provide very little extra improvement in the delay estimates for the conditions examined.

REFERENCES

[1]. A. Gelb, editor, *Applied Optimal Estimation*, M.I.T. Press, Cambridge, Mass., 1974.

[2]. R.L. Kirlin, D.F. Moore, and R.F. Kubichek, "Improvement of delay measurements from sonar arrays via sequential state estimation," *IEEE Trans. ASSP*, Vol. ASSP-29, No. 3; June 1981, pp. 804-806.

[3]. G.C. Carter, "Variance bounds for passively locating an acoustic source with symmetric line array," *Journal of the Acoustical Society of America*, Vol. 62, No. 4; Oct. 1977, pp. 922-926.

[4]. W.R. Hahn, "Optimum signal processing for passive sonar range and bearing estimation," *Journal of the Acoustical Society of America*, Vol. 58, No. 1; July 1975, pp. 201-207.

[5]. R.L. Kirlin, "Delay estimation among three sensors given measurement and *a priori* covariances," Dept. of Electrical Engineering, University of Wyoming, April 1982.

[6]. E.S. Gale, "Passive sonar delay estimate improvement using *a priori* knowledge and increased number of sensors," M.S./E.E. Thesis, Electrical Engineering Department, University of Wyoming, Laramie, Wyoming, 1982.

Capability of Array Processing Algorithms to Resolve Source Bearings*

Stuart R. DeGraaf and Don H. Johnson

*Department of Electrical Engineering
Rice University
Houston, Texas*

1 INTRODUCTION

The determination of the bearings of distant sources of acoustic energy in a noisy ocean environment is a primary objective of a passive sonar system. By processing data obtained from an array in various ways, one can evaluate the bearings from which acoustic energy is incident. A useful measure of the performance of an array and processing algorithm is the signal-to-noise ratio required to resolve two closely-spaced, equal-energy sources.

Classical or Bartlett beamforming, minimum energy (ME) adaptive beamforming,[1] and linear predictive (LP) processing [2] algorithms are commonly employed to estimate source bearing. Each algorithm involves the evaluation of a functional:

$$P_{\text{BART}}(\underline{k}) = \underline{W}'\mathbf{R}\underline{W} \tag{1}$$

$$P_{\text{ME}}(\underline{k}) = [\underline{W}'\mathbf{R}^{-1}\underline{W}]^{-1} \tag{2}$$

$$P_{\text{LP}_q}(\underline{k}) = |\underline{U}_q'\mathbf{R}^{-1}\underline{W}|^{-1} \tag{3}$$

---

*Work supported under Contract N00014-81-K-0565, Task NR 042-461 with the Office of Naval Research.
[1]This beamforming algorithm, due to Capon [1], is often referred to as the maximum likelihood method in the literature. Actually, the likelihood function is not maximized; the customary terminology is therefore misleading.

R is the cross spectral correlation matrix [2], and $\underline{W}$ is the direction of look vector with elements $W_m = \exp(-j2\pi f \underline{k} \cdot \underline{z}_m)$, where $\underline{z}_m$ is the location of the mth sensor and $\underline{k}$ is the bearing of look. $\underline{U}_q$ is the prediction element vector containing a one in the qth element and zeros elsewhere. The usual value of q for a linear array of equally-spaced sensors is 0; here we consider the variation of using other prediction elements. Formula (3) bears comment. Typically, a power of -2 rather than -1 is used for the LP algorithm. Use of the -1 power ensures that the units of (1,2,3) agree and allows direct comparison of the resolving capabilities of the algorithms. Plots of (1,2,3) vs bearing of look are known as **beam patterns** (see Fig. 1). When a single propagating plane wave is present, a global maximum occurs **on-target**, i.e., at the bearing of incidence; when more than one plane wave is present, the maxima in the beam pattern are no longer necessarily located exactly on-target. Consider the case in which two equal-energy plane waves propagate along nearly the same bearing. These bearings may be **resolved**, in which case the beam-patterns exhibit two distinct maxima as in Fig. 1. On the other hand, the beam-patterns may fail to resolve the source bearings. Instead of exhibiting two distinct maxima, they may display a broad maximum located at some intermediate bearing. The **resolution problem** is to determine the conditions under which each algorithm is capable of resolving source bearings. The **detection problem** is to determine accurately the number of targets present. The presence of sidelobes, local maxima located off target, complicates the problem.

Cox [3] analyzed the capability of the Bartlett and ME array processing algorithms to resolve two equal-energy, incoherent signals. He demonstrated that the ME algorithm can resolve closely-spaced source bearings at lower array signal-to-noise ratios[2] than can the Bartlett algorithm. Studying the single signal case, Seligson [4] observed that the ME algorithm's performance is quite sensitive to deviations of the signal wavefront from its assumed planar shape. Superposition of temporally correlated plane waves produces a net wavefront which is not planar. Since multipath propagation is prevalent in the ocean [5], there is a need to understand how signal correlation affects the resolving capabilities of array processing algorithms. This paper presents the results of an analysis [6] similar to Cox's to study the capabilities of Bartlett, ME, and LP algorithms to resolve **coherent** (i.e., correlated, narrowband) signals using linear arrays of equally-spaced sensors; detection capabilities of the algorithms are also discussed. In addition, guidelines are presented which indicate the amount of averaging necessary to suppress variability of the beam-patterns due to finite averaging in the empirical estimate of the cross spectral correlation matrix. The algorithms are compared on the basis of their capability to simultaneously detect and resolve coherent signals and their sensitivity to finite averaging.

2 DETECTION AND RESOLUTION OF SOURCE BEARING

We consider the case where the noise is spatially white, has zero mean, and is independent of the signals. To study the effects of signal bearing, signal-to-noise ratio, and signal coherence on the detection and resolution capabilities of

---

[2]The **array signal-to-noise ratio** (ASNR) is the product of M, the number of sensors in the array, with the signal-to-noise ratio seen at a single sensor.

Array Processing Algorithms

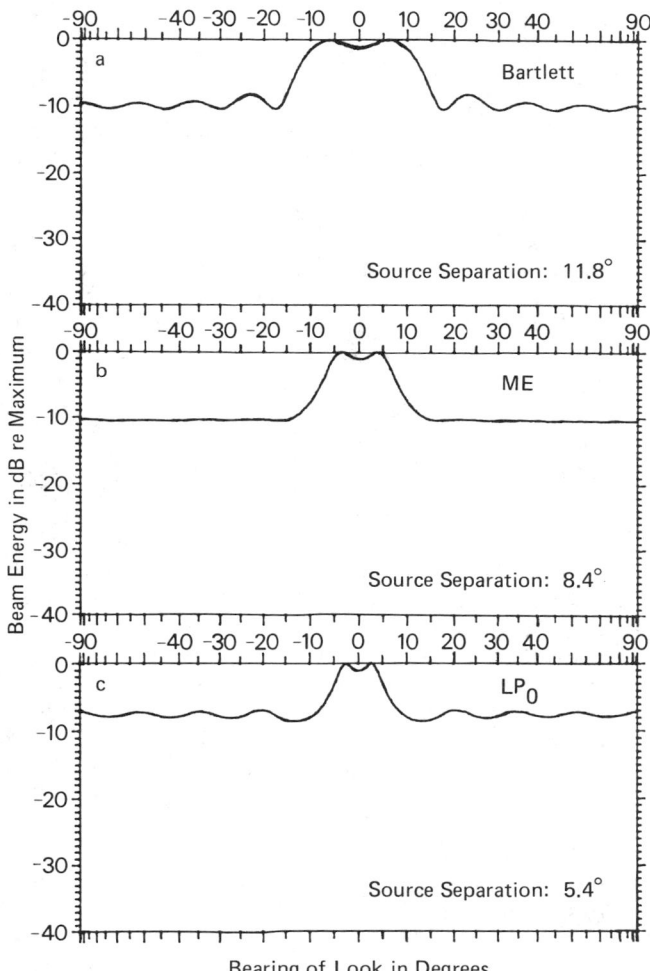

Figure 1 Beam patterns

the various array processing algorithms, we restrict our attention to the case where two equal-energy signals impinge on the array. The criterion used for detection of the signals is that the ratio of on-target to largest-sidelobe beam energy exceed a threshold value of two. The criterion used for resolution of the source bearings is that the ratio of on-target to between-target beam energy exceed a threshold value of $\frac{\pi^2}{8}$. This value is obtained by classical beamforming when the signal-to-noise ratio is large and the signals are separated by the Rayleigh resolution limit. Analytic expressions for these ratios were derived [6] for each of the array processing algorithms.

2.1 Resolution

The **resolvent** array signal-to-noise ratio, the minimum ASNR for which the ratio of on-to-between-target beam energies exceeds the threshold value, was computed as a function of source bearing separation. The resolvent ASNR is a function of the coherence magnitude and phase as well as source bearing separation.

Figure 2a compares the minimum ASNR necessary for a ten element array to resolve the bearings of incoherent sources using the Bartlett, ME, and LP_0 processing algorithms. Of the three, the LP_0 algorithm requires the least ASNR to resolve source bearings. The ME and LP_0 algorithms are both capable of resolving arbitrarily closely spaced source bearings if the ASNR is large enough. In contrast, the Bartlett algorithm is incapable of resolving bearings spaced more closely than the Rayleigh limit regardless of the ASNR. Figures 1a, 1b, and 1c illustrate Bartlett, ME, and LP_0 beam-patterns, respectively, which just resolve the indicated incoherent signals. In each case the ASNR is 10 dB and the bearing separation is the smallest resolvable (from Fig. 2a). Figures 2b and 2c compare the worst-case[3] resolvent ASNR for the three algorithms when the coherence magnitude is 0.7 and 0.99, respectively. As the magnitude of the coherence increases, the resolvent ASNR for the ME and LP_0 algorithms increases. The LP_0 algorithm retains its advantage over ME and Bartlett processing with regard to bearing resolution. The LP_0 algorithm consistently requires roughly 15 dB less ASNR than the ME algorithm to resolve closely-spaced source bearings. The Bartlett algorithm exhibits a minimum signal separation below which signals cannot necessarily be resolved; this limit increases as the magnitude of the coherence increases.

As discussed by Johnson [2], the linear predictive algorithm offers flexibility in the choice of prediction element q. Figure 3 compares the minimum ASNR necessary for a ten element array to resolve the bearings of incoherent sources[4] using the linear predictive algorithm with prediction elements[4] 0 to 4. For closely spaced signal bearings ($\theta \leq 4°$) the center prediction elements ($q = 4, 5$) require the least ASNR to resolve the source bearings. More widely separated bearings ($4° < \theta \leq 22°$) are most easily resolved by prediction elements at the ends of the array. The sharp peaks in the plots for prediction elements 3-6 correspond to situations in which these LP algorithms produce spurious peaks in their beam-patterns at bearings between the actual target bearings. As signal coherence increases in magnitude from zero to one, the resolvent ASNR rises quite uniformly as a function of bearing separation. For a ten element array, the resolvent ASNR for signals having 0.7 coherence magnitude is 6 dB greater than for incoherent signals; the resolvent ASNR for signals having 0.99 coherence magnitude is 15 dB greater than for incoherent signals.

---

[3] A sensible way to compare the resolving capabilities of the algorithms is to select the coherence phase which maximizes the resolvent ASNR for each source bearing separation. Thus the worst-case resolvent ASNR is sufficient to **guarantee** bearing resolution regardless of coherence phase; sources with particular coherence phases can usually be resolved at a somewhat lower ASNR.

[4] As a function of prediction element q, the resolvent ASNR is symmetric about the center of the array.

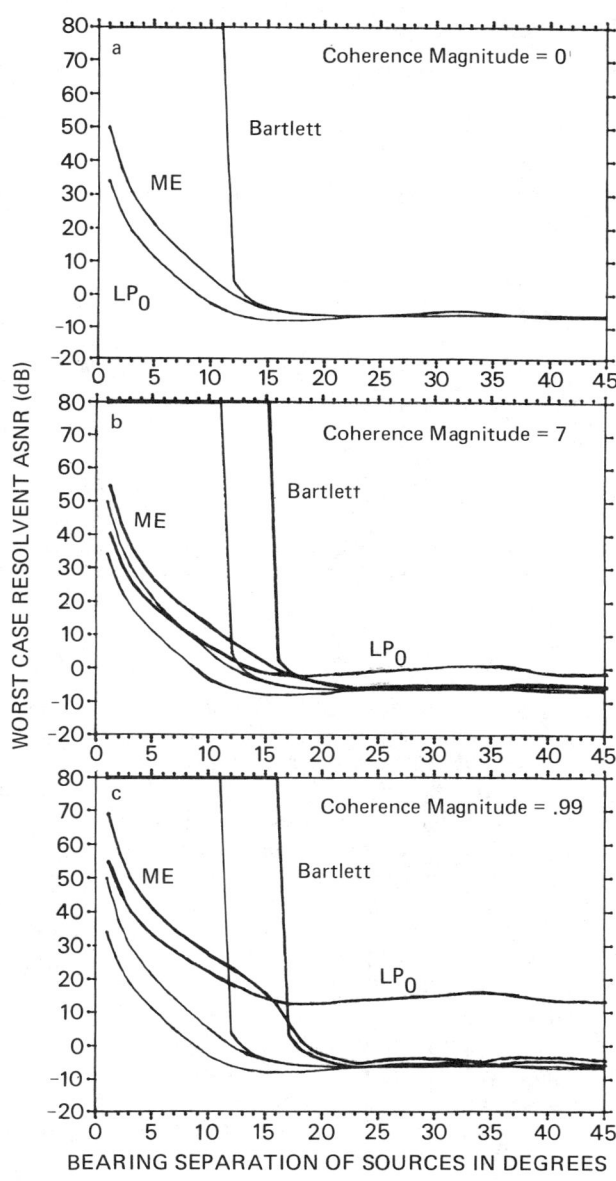

Figure 2 Minimum ASNR necessary for a 10 element array to resolve the bearings of incoherent sources using the Bartlett, MG, and LP_0 processing algorithms

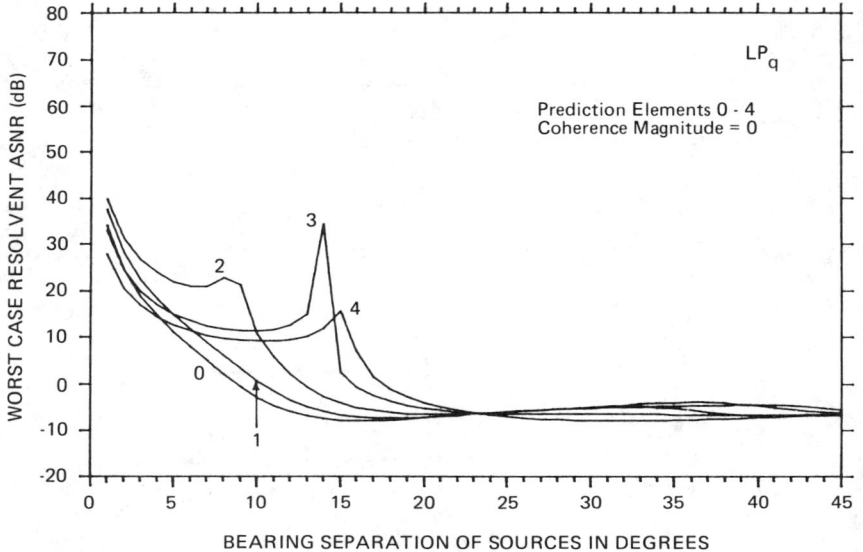

Figure 3 Comparison of minimum ASNR necessary for a ten element array to resolve the bearings of incoherent sources using the linear predictive algorithm with prediction elements 0 to 4

2.2 Detection

In addition to resolving source bearings, it is necessary for a processing algorithm to detect the presence of sources by exhibiting peaks in the beam-pattern which stand out against off-target ripple. For a large array ($M \gg 5$) in the single signal case, the off-target ripple in the ME beam-pattern asymptotically approaches 5 percent as the ASNR increases; the LP ripple approaches 27 percent. The ripple in the Bartlett beam-pattern increases proportionally to the ASNR. Generally, the sidelobes present in the multiple signal case are smaller than those in the single signal case. The **detectable** ASNR, the minimum ASNR for which the ratio of on-target to highest-sidelobe beam energies exceeds a value of two, was computed as a function of source bearing separation. As with the resolvent ASNR, the detectable ASNR is a function of the coherence magnitude and phase as well as source bearing separation. As coherence magnitude increases, the worst-case[5] detectable ASNR increases. The Bartlett algorithm requires the least ASNR to guarantee signal detection, followed by the ME algorithm. The LP algorithms require the highest ASNR to guarantee signal detection. Clearly a tradeoff exists between resolution and detection capabilities of the algorithms.

---

[5]The worst-case coherence phase maximizes the detectable ASNR. Thus the worst-case detectable ASNR is sufficient to **guarantee** source detection regardless of coherence phase; sources with particular coherence phases can usually be detected at a somewhat lower ASNR.

2.3 Combined Detection and Resolution

All of the algorithms require a higher ASNR to detect widely-separated sources than to resolve them. Conversely, all of the algorithms require a higher array SNR to resolve closely spaced sources than to detect them. Figures 4a, 4b, and 4c show the minimum ASNR required by the Bartlett, ME, and LP_0 processing algorithms to guarantee both detection and resolution of equal-energy signals with varying degrees of coherence.

3 EFFECTS OF FINITE AVERAGING

Results presented in the previous section were based on use of the true correlation matrix. In realistic situations, this matrix must be estimated from data. We now consider the effect of empirical computation of the correlation matrix on the beam-patterns. Typically, the correlation matrix is estimated by averaging the outer products of K data vectors by breaking the available sensor data into K segments [2]. K is known as the **time-bandwidth product**. By analyzing the variability of on-target and off-target beam energies with respect to the nonzero signal-noise cross terms in the single signal case [6], the following guidelines were established which dictate the amount of averaging necessary for each algorithm to produce beam-patterns with a small random component.

$$\text{Bartlett:} \quad K \gg \frac{M}{\left[\frac{\sigma_0}{\sigma_1} + M \frac{\sigma_1}{\sigma_0}\right]^2}$$

$$\text{ME:} \quad K \gg M$$

$$\text{LP:} \quad K \gg M \text{ and } K \gg M^2 \frac{\sigma_1^2}{\sigma_0^2}$$

σ_1^2 is the signal energy and σ_0^2 is the noise energy. Empirically, the ripple in the LP beam-patterns increases rapidly as the amount of averaging decreases; the ripple in the ME beam-pattern increases much more slowly. The ripple in the Bartlett beam pattern is the least sensitive to finite averaging.

4 CONCLUSIONS

The resolution and detection performances of the beamforming and linear predictive array processing algorithms discussed here are adversely affected by signal coherence and by limited averaging in the computation of the correlation matrix. Furthermore, the performance of each algorithm is affected to a different degree by these conditions. **No one algorithm is superior to the others in all situations.** Consequently, the coherence and amount of averaging possible in particular applications greatly influence the choice of an "optimal" array processing algorithm. Table 1 summarizes the results of Sections 2 and 3.

As the magnitude of the coherence increases from zero to one, the resolving capability of all of the algorithms diminishes. The incremental effect of a change in coherence magnitude on the resolving capabilities of the ME and LP algorithms is greatest when the coherence magnitude is large (near one);

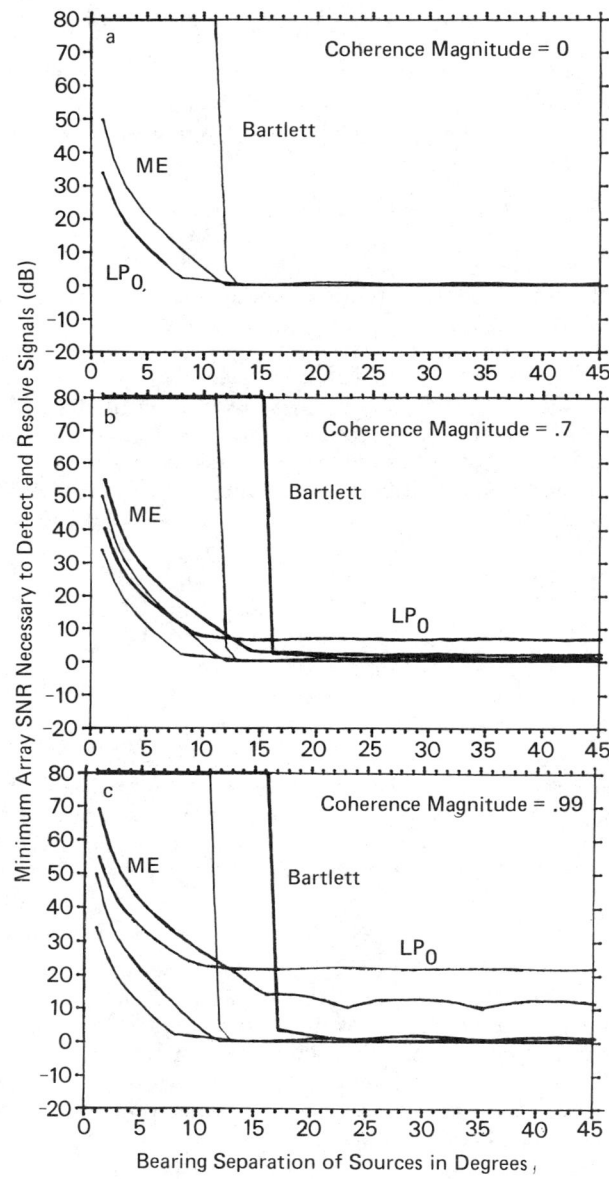

Figure 4 Minimum ASNR required by the Bartlett, MG and LP_0 processing algorithms to guarantee both detection and resolution on equal-energy signals with varying degrees on coherence

Table 1

Array processing algorithm	$\|c\|$	Minimum array SNR in dB guaranteed to resolve and detect signals			Averaging necessary to Minimize variability
		Closely spaced signals		Widely separated signals	
		$\theta = 1°$	$\theta = 10°$		
Bartlett	0	†	†	0	$K \gg \dfrac{M}{\left[\dfrac{\sigma_0}{\sigma_1} + M \dfrac{\sigma_1}{\sigma_0}\right]^2}$
	0.7	‡	‡	1	
	0.99	ˆ	ˆ	1	
ME	0	50	5	0	$K \gg M$
	0.7	54	12	2	
	0.99	69	27	11	
LP_{end}	0	34	2	1	
	0.7	40	8	7	
	0.99	55	22	22	
					$K \gg M$ and $K \gg M^2 \dfrac{\sigma_1^2}{\sigma_0^2}$
L_{center}	0	28	*	1	
	0.7	34	*	7	
	0.99	49	*	22	

†Rayleigh resolution limit = 11°
‡Rayleigh resolution limit = 15°
ˆRayleigh resolution limit = 16°
*Use of the center prediction elements for signal separations greater than 4° yields spurious peaks (see text).

for the Bartlett algorithm, the incremental effect is greatest when the coherence magnitude is small. Except for perfectly coherent signals, the LP processing algorithm is uniformly the most capable of resolving closely-spaced signals, followed by ME and Bartlett processing.

The capability of the three algorithms to detect incoherent signals is basically the same. Coherence has negligible effect on the detection capabilities of Bartlett processing. In contrast, as signal coherence increases, the detection capabilities of ME processing decrease, as Seligson [4] observed. The detection capabilities of LP processing are inferior to those of ME processing.

In Section 3 constraints on the time-bandwidth product were presented which ensure minimal sensitivity of the beam-pattern to imperfections in the correlation matrix that result from finite averaging. The Bartlett beam-pattern is least sensitive to finite averaging while LP beam-patterns are most sensitive. The sensitivity of the ME beam-pattern to finite averaging lies between these extremes. A tradeoff exists between resolving capability and sensitivity to finite averaging. The LP algorithms are most capable of resolving closely-spaced signals; however, their high sensitivity to finite averaging restricts their application to environments where large amounts of averaging are possible. In addition,

Table 2

Time-bandwidth product	Bearing separation of signals	
	Closely spaced	Widely separated
Small	ME for all coherences	Bartlett for all coherences
Large	LP_{end} for all coherences	Bartlett for all coherences
	LP_{center} for all coherences when signals are very close	LP or ME for small coherences

the LP algorithms require the highest SNR to detect widely-separated signals when the coherence is large. The Bartlett algorithm is least sensitive to finite averaging and requires the least SNR to detect widely-separated signals; however, its resolving capability is very poor.

The ME algorithm lies between the LP and Bartlett algorithms in terms of resolution and detection capabilities as well as sensitivity to noise. The ME algorithm seems best suited to applications where the amount of averaging possible is small and the capability to resolve closely-spaced signal bearings is requisite. The LP algorithms are of little value in these situations because of their extreme sensitivity to noise. The Bartlett algorithm is also of little value here because of its extremely poor resolving capability. Table 2 crudely suggests which array processing technique is best suited to each of four combinations of signal separation and time-bandwidth product.

The variation to the LP algorithm of utilizing different prediction elements proves to be of some value. Moving the prediction element from the end to the center of a linear array of equally-spaced sensors enhances the capability of the linear predictive algorithm to resolve extremely closely-spaced source bearings. However, the beam-patterns produced by the center prediction elements are more likely to exhibit spurious peaks when the sources are less closely spaced. Choice of prediction element has no effect on the capability of the LP algorithm to detect widely-separated sources. The effect of increasing signal coherence is the same for all prediction elements.

REFERENCES

[1] J. Capon, "High-Resolution Frequency-Wavenumber Spectrum Analysis," *Proc. IEEE,* Vol. 57, pp. 1408-1418, Aug. 1969.

[2] D.H. Johnson, "The Application of Spectral Estimation Methods to Bearing Estimation Problems," *Proc. IEEE,* Vol. 70, No. 10, Sept. 1982.

[3] H. Cox, "Resolving power and sensitivity to mismatch of optimum array processors," J. Acoust. Soc. Am., Vol. 54(3), pp. 771-785, March 1973.

[4] C.D. Seligson, "Comments on 'High-Resolution Frequency-Wavenumber Analysis,'" *Proc. IEEE,* Vol. 58, pp. 947-949, June 1970.

[5] R.J. Urick, *Principles of Underwater Sound,* pp. 93-180, McGraw-Hill, New York, NY, 1975.

[6] S.R. DeGraaf, "The Effect of Coherent Signals on the Capability of Array Processing Algorithms to Resolve Source Bearings," M.S. Thesis, Dept. of Elec. Eng., Rice University, Houston, TX, 1982.

Detection Thresholds for Multitarget Tracking in Clutter*

Thomas E. Fortmann

Automated Systems Department
Bolt, Beranek and Newman, Inc
Cambridge, MA

Yaakov Bar-Shalom

Department of Electrical Engineering and Computer Science
University of Connecticut
Storrs, CT

1 INTRODUCTION

Garden-variety tracking problems involve processing measurements (e.g., range and azimuth observed by a sensor) from a target of interest and producing, at each time step, an estimate of the target's current position and velocity vectors. Uncertainties in the target motion and in the measured values, usually characterized as random noise, lead to corresponding uncertainties in the target state.

A common and versatile approach to such problems involves assuming that the state dynamics and the measurements are both corrupted by additive, white, possibly Gaussian noise; the solution is then the celebrated Kalman-Bucy filter [1-5], which is the conditional mean state estimator, best linear estimator, maximum *a posteriori* estimator, maximum likelihood estimator, or least-squares estimator,[1] depending upon one's point of view. The parameters that determine tracking performance in such a filter are the system matrices in the equations describing target state dynamics and measurements, which will be considered fixed for the purposes of this discussion, and the covariance

---

*This research was supported by the Office of Naval Research under Contract N00014-80-C-0270.

[1] In the least-squares case, the assumptions about noise are replaced by assumptions about error weightings.

matrices of the process and measurement noise, which specify the uncertainties in target motion and measured values, respectively.

In many tracking problems, particularly those arising in surveillance, there is additional uncertainty regarding the *origin* of the received data, which may (or may not) include measurements from the target(s) of interest, interfering targets, or random clutter (false alarms). This leads to the problem of *data association* or *data correlation*, which has been attacked on a number of fronts [6-14] and surveyed in [15-17]. In this situation, tracking performance depends not only upon the noise covariances, but upon the amount of uncertainty in measurement origin. In some of the approaches cited above [6-10], this dependence is explicit and is characterized in terms of the *detection probability* P_D and *false alarm probability* P_F (which is proportional to clutter density).

In typical applications, measurement data are provided to a tracker by upstream signal processing and detection algorithms, as indicated in Figure 1. The process noise covariances are normally selected on the basis of experience and intuition (i.e., they are guessed). The measurement noise covariances are either provided by the signal processing algorithm, as shown in the figure, or they are selected in the same manner as the process noise. In any case, the true noise levels are usually fixed by target dynamics and sensor configuration and cannot be adjusted on line.

Detection and false alarm probabilities, on the other hand, are highly interdependent and adjustable via a *detection threshold:* raising the threshold lowers both probabilities, and vice-versa. This relationship, which also depends parametrically on the signal-to-noise ratio (SNR), is usually characterized by means of a set of receiver operating characteristic (ROC) curves, as discussed below in Section 3. The threshold is typically set by choosing a design point on the most applicable ROC curve, based on the perceived tradeoffs between false alarms and missed detections. However, to the best of our knowledge, these tradeoffs have never included any systematic or quantitative consideration of the effects downstream on data association and tracking performance.

In this paper we shall describe such a quantitative relationship. The dependence of a tracker's error covariance upon detection and false alarm probability is explicitly (but approximately) characterized by a scalar parameter q_2 in the covariance equation (modified Riccati equation). The scalar parameter depends upon the probabilities of detection and false alarm, and also upon the volume of the data association gate, which in turn depends on the state error covariance matrix $\underline{P}$. The modified Riccati equation can be iterated to convergence, yielding a steady-state $\bar{\underline{P}}$, and tracking performance can be characterized by a scalar metric such as determinant$(\bar{\underline{P}})$, trace$(\bar{\underline{P}})$, or (in surveillance applications) root-mean-square position error. This result is important for the following reasons:

1. Contour plots of the scalar tracking performance metric as a function of detection probability and false alarm probability form a set of *tracker operating characteristic* (TOC) curves, which can be superimposed on ROC curves for the detector or receiver of interest in order to determine graphically the operating points that optimize tracker performance.

Detection Thresholds

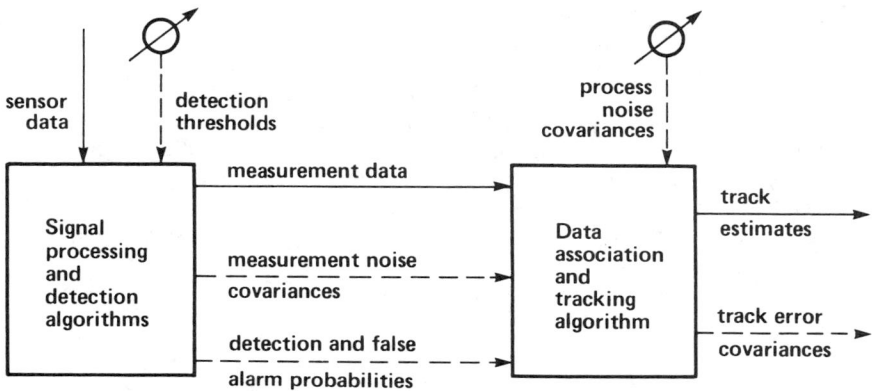

Figure 1 Upstream signal processing and detection algorithms

2. The stability of the tracking process depends critically on the detection and false alarm probabilities; indeed, a region of apparent instability of the modified Riccati equation exists in the P_D-P_F plane of the TOC curves. The implication of this for detector/receiver design is that there are settings of the detection parameters that render the output useless for downstream tracking.

3. Allocation of tracking resources (both computation and communication) requires prediction of future state error covariances under various resource configurations, i.e., as a function of detection and false alarm probability and of process and measurement noise covariance.

4. The same derivations provide a solution to the related problem of determining the statistical properties of the modified likelihood function [18], used for decision making (e.g. maneuver detection) when measurement origins are uncertain.

The key TOC results are summarized in the next section, followed by examples in Section 3. Conclusions and suggestions for further research may be found in Section 4.

2 APPROXIMATE COVARIANCE EQUATION

This section provides a summary of the key TOC results; a detailed derivation may be found in [20].

We assume a linear, state-space target model with the usual white, Gaussian noise, and a Kalman-Bucy filter tracker [2-5]. We further assume candidate measurements are associated with targets by means of a *probabilistic data association* (PDA) scheme [6-8, 15], in which the filter update uses a weighted sum of the candidate measurements, where the weights are the posterior probabilities of the respective measurements being correct. This leads to a covariance update equation (modified Riccati equation) that includes stochastic terms.

The posterior probabilities, and hence the stochastic Riccati equation, depend upon detection probability P_D and clutter density C, and the latter is proportional to the probability of false alarm P_F. The principal purpose of this research is to quantify the dependence of tracking performance on P_D and P_F, using a deterministic approximation to the stochastic Riccati equation.

The approximation consists of replacing certain stochastic quantities with their expected values. This yields a deterministic covariance equation, which is reduced via a very complex derivation to

$$\underline{P}_{k|k-1} = \underline{F}\underline{P}_{k-1|k-1}\underline{F}' + \underline{G}\underline{Q}\underline{G}' \qquad (1)$$

$$\underline{P}_{k|k} = \underline{P}_{k|k-1} - q_2(\underline{S}_k; P_D, P_F)\underline{W}_k\underline{S}_k\underline{W}'_k$$

where $\underline{P}_{k|j}$ is the filter covariance at time k given measurements up to time j, $\underline{W}_k$ is the gain matrix, and $\underline{S}_k$ is the innovations covariance matrix. The scalar quantity q_2 reduces the effect of the new information; note that when $q_2 = 1$, (1) is the standard Kalman filter prediction/upgrade equation. We shall now use (1) to characterize the dependence of tracking performance on P_D and P_F.

For most values of P_D and P_F, (1) can be iterated until it converges to a steady-state covariance matrix $\underline{\bar{P}}(P_D, P_F)$ (the stability issue is discussed below). In order to obtain a scalar tracking performance metric, one can then extract the steady-state root-mean-square (RMS) position error

$$e(P_D, P_F) \triangleq \sqrt{\bar{p}_{11}(P_D, P_F) + \bar{p}_{22}(P_D, P_F)} \qquad (2)$$

where $\bar{p}_{11}$ and $\bar{p}_{22}$ are the diagonal elements of $\underline{\bar{P}}$ that correspond to target position.

We shall refer to a contour plot of (2) as a *tracker operating characteristic* (TOC). This name is chosen because the well-known *receiver operating characteristic* (ROC) curve in the same P_D-P_F plane is the locus of possible operating points for a detector/receiver, where a particular operating point on the curve is determined by the *detection threshold* level. Thus, if the ROC curve is superimposed on the TOC contours, the dependence of tracking performance on detection threshold can be determined directly. This will be illustrated in the next section with an example.

There are various other performance metrics that can be used, of course, such as the determinant or trace of $\underline{\bar{P}}$. In many applications, the steady-state covariance may not be appropriate: one can instead use the value of $\underline{P}_{k|k}$ obtained by iterating (1) over a fixed period of time from a standard $\underline{P}_{0|0}$.

3 EXAMPLES

The target/sensor geometry shown in Figure 2 was used in the multitarget tracking examples of [8]. Taking the (linearized) values of $\underline{F}$, $\underline{G}$, $\underline{H}$, $\underline{Q}$, and $\underline{R}$ from the initial time in that example, we have iterated (1) to obtain the steady-state RMS position error (2) for various values of P_D and P_F. Evaluation of $q_2(\underline{S}_k; P_D, P_F)$ was carried out using a look-up procedure from tables generated off-line.

Detection Thresholds

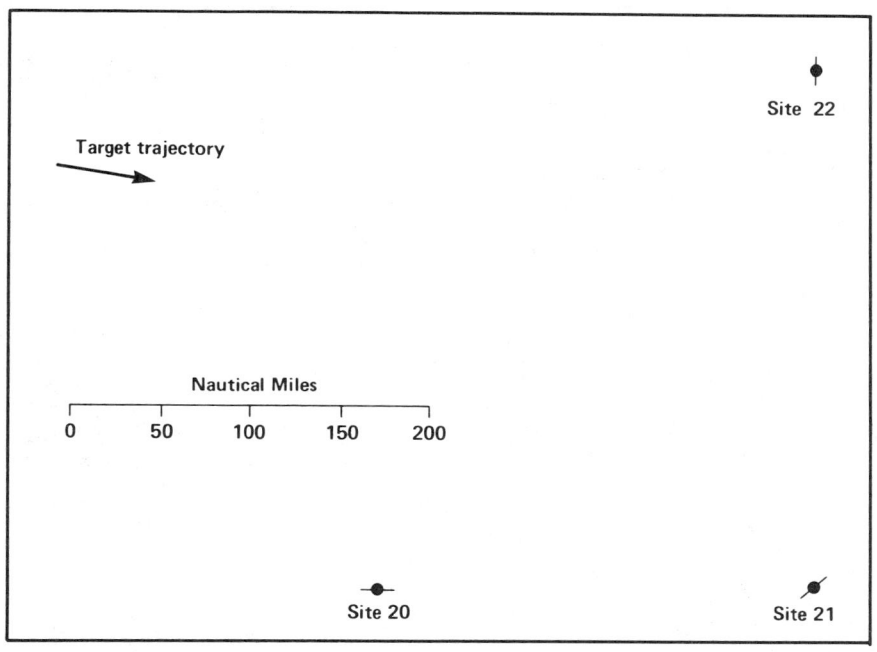

Figure 2 Target/sensor geometry

Tracker operating characteristics will be shown for two different measurement types. In the first example, the target is tracked using measurements of bearing (azimuth) and frequency from sensors 20 and 22 at 5-minute intervals. The process noise matrix ($\underline{GQG'}$) is diagonal, with standard deviations of .2° in course, .2 knots in speed, and .01 Hz in source frequency.

The measurement noise matrix, also diagonal, has a standard deviation of 5° in bearing and .08 Hz in frequency. We further assume that the sensor signal processing is able to resolve signals separated by about 4° and .15 Hz. We may thus view the space of bearing/frequency measurements as a collection of resolution cells, with the tracker's validation gate encompassing some subset of these cells. In practice, the detector/receiver will have an *ad hoc* rule prohibiting detections in adjacent cells, so that the *effective cell volume* is about $V_c = .3°$-Hz. We assume further that false alarms occur independently in each cell with probability P_F, so that the clutter density (expected number of false alarms per unit volume) is

$$C = P_F/.3 \text{ deg-Hz.} \qquad (3)$$

With these assumptions, the TOC contours shown in Figure 3 have been computed (note that P_F ranges only from 0 to 0.1). This is a contour plot of (2), with performance improving (i.e., position error decreasing) toward the upper left-hand corner. Performance degrades in the other direction, as P_D decreases and/or P_F increases, and there is a region in which the modified Riccati equation (1) does not appear to converge to a finite steady-state covariance $\underline{\bar{P}}$.

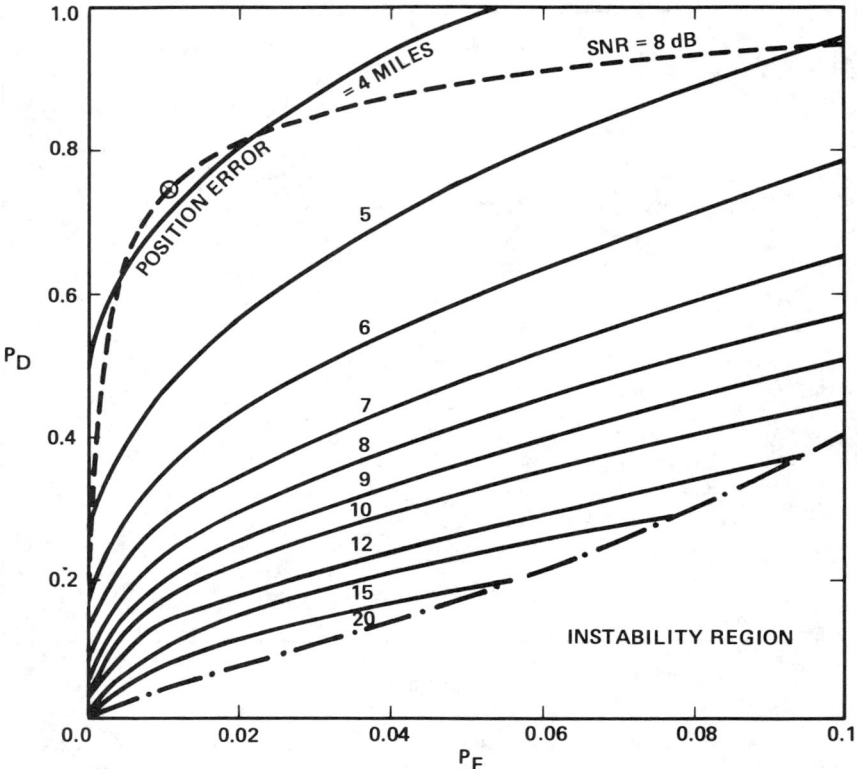

Figure 3 TOC contours

Figure 3 specifies tracking performance as a function of P_D and P_F. In order to determine what values of these probabilities are achievable, we need *receiver operating characteristic* (ROC) curves for the detection system that provides measurements to the tracker. To this end, we shall assume that the detection algorithm is equivalent to a set of classic *quadrature receivers* or *incoherent matched filters* [19], one operating on each resolution cell in bearing/frequency space. The quadrature receiver assumes a sinusoidal signal of unknown phase. Under the signal-plus-noise hypothesis, the test statistic has a Rician distribution, which reduces to a Rayleigh distribution in the noise-only case. Expressions for P_D and P_F may be derived [19] and used to compute the ROC curves shown in Figure 4.

For a given signal-to-noise ratio (SNR), the corresponding ROC curve is the locus of possible *operating points* that the detector can assume, depending on where one sets the *detection threshold*. In Figure 3 one such curve (SNR = 8 dB) is superimposed as a dashed line on the TOC contours. This shows graphically how tracking performance depends on the operating point of the detector (i.e., on the detection threshold). In particular, performance is optimal at the operating point indicated by ⊗. There is a relatively broad region about this

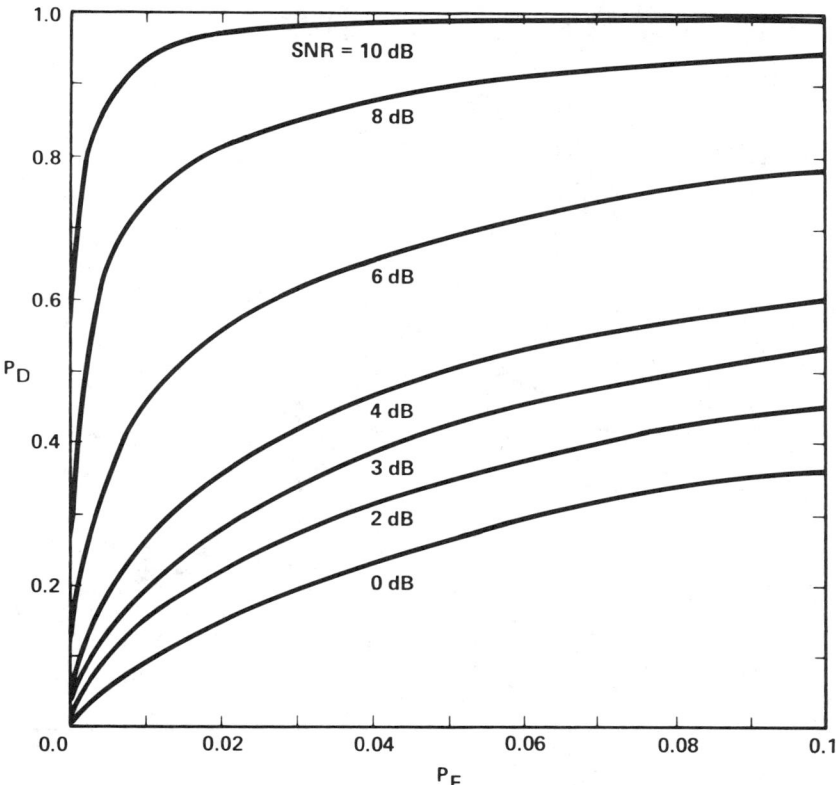

Figure 4 ROC curves

point where performance is near-optimal, but performance degrades significantly thereafter.

4 COHERENCE MEASUREMENTS

In the second example, the target is tracked by cross-correlating signals between pairs of sensors to obtain measurements of time delay difference and Doppler difference from sensor pairs 20/21 and 21/22 (see Figure 2) at 5-minute intervals. Standard deviations of 4 sec in time difference and .004 Hz in Doppler difference are assumed, and the effective resolution cell volume is .008 sec-Hz, so that

$$C = P_F/.008 \text{ sec - Hz.} \tag{4}$$

This leads to the TOC contours shown in Figure 5.

Analysis of the cross-correlation algorithm is somewhat more difficult than that of the quadrature receiver. Nevertheless, if we assume that both signal and noise are sample functions from white, gaussian, random processes and that the time-bandwidth product is 500 sec × .25 Hz, we can obtain the ROC

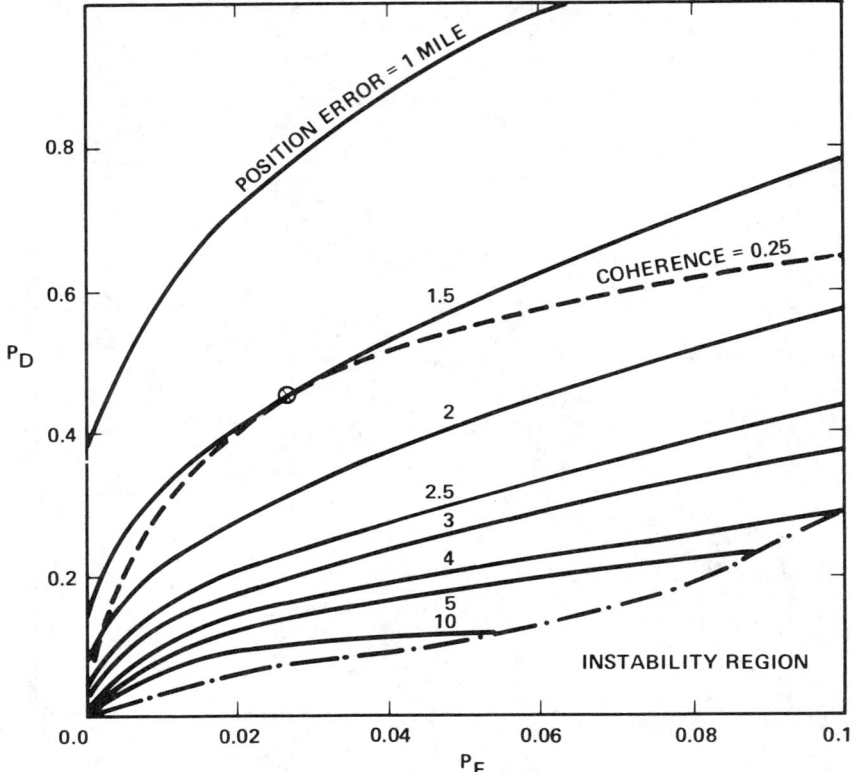

Figure 5 TOC contours

curves shown in Figure 6. The parameter is now coherence between the two channels, rather than SNR.

Again, for a given coherence, the corresponding ROC curve is the locus of possible operating points for the coherence detector. In Figure 5, the curve corresponding to a coherence of .025 is superimposed as a dashed line on the TOC contours to show graphically how tracking performance depends on the operating point of the coherence detector (i.e., on the detection threshold), and the optimal point is indicated by ⊗.

5 CONCLUSION

We have established, for the first time, an important relationship between thresholds in detector/receivers and performance in downstream trackers. More specifically, a modified Riccati equation determines the approximate state error covariance of a probabilistic data association (PDA) tracking filter as a function of the threshold-dependent probabilties of detection and false alarm. By plotting contours of tracking performance (in this case, steady-state RMS

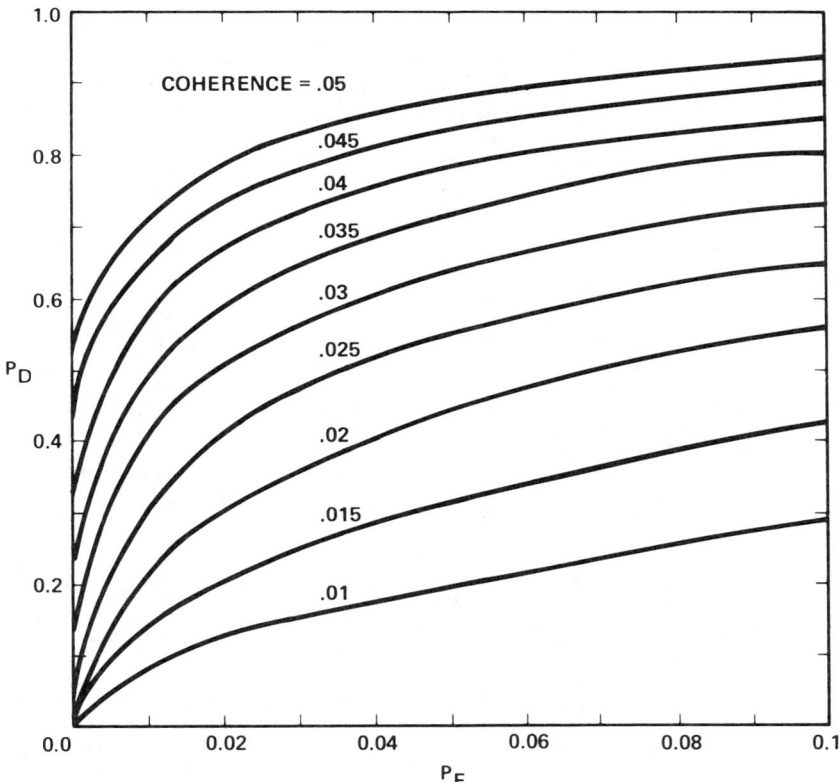

Figure 6 ROC curves

position error) in the P_D-P_F plane and then superimposing a ROC curve for a particular SNR, one can determine the optimal detection threshold graphically.

Several extensions of this concept are of interest. The graphical method for selecting an operating point can be replaced by a mathematical optimization: an obvious necessary condition is that the ROC and TOC curves be tangent. However, the practical difficulty of computing the required differentials to solve the necessary conditions is substantial. An approximate (e.g., table look-up) procedure for optimization would be useful for dealing with multi-dimensional TOCs, such as will occur if different receivers are allowed to have different thresholds or if bearing/frequency and time/Doppler measurements are used simultaneously.

Another issue of major importance is the optimization of tracking performance when the signal's SNR is not known. In this case, several ROC curves are involved and one must select a threshold that is best (in some sense) for a whole range of SNRs. Alternatively, an *adaptive thresholding* scheme can be devised, whereby the SNR is monitored and the threshold adjusted so as to maximize performance along the current ROC curve.

REFERENCES

1. R.E. Kalman and R.S. Bucy, "New Results in Filtering and Prediction Theory," *Trans. ASME: J. Basic Eng.*, Vol. 83, March 1961, pp. 95-108.

2. I.B. Rhodes, "A Tutorial Introduction to Estimation and Filtering," *IEEE Trans. Auto. Control*, Vol. AC-16 December 1971, pp. 688-706.

3. A.H. Jazwinski, *Stochastic Processes and Filtering Theory*, Academic Press, 1970.

4. P.S. Maybeck, *Stochastic Models, Estimation, and Control—Volume 1*, Academic Press, 1979.

5. B.D.O. Anderson, *Optimal Filtering*, Prentice-Hall, 1979.

6. Y. Bar-Shalom and E. Tse, "Tracking in a Cluttered Environment with Probabilistic Data Association," *Automatica*, Vol. 11, September 1975, pp. 451-460.

7. T.E. Fortmann and S. Baron, "Problems in Multi-Target Sonar Tracking," *Proc. 1978 IEEE Conf. on Decision and Control*, San Diego, California, January 1979.

8. T.E. Fortmann, Y. Bar-Shalom, and M. Scheffe, "Multi-Target Tracking Using Joint Probabilistic Data Association," *Proc. 1980 IEEE Conference on Decision and Control*, Albuquerque, New Mexico, December 1980.

9. R. Singer, R. Sea, and K. Housewright, "Derivation and Evaluation of Improved Tracking Filters for use in Dense Multitarget Environments," *IEEE Trans. Info. Theory*, Vol. IT-20, July 1974, pp. 423-432.

10. D.B. Reid, "An Algorithm for Tracking Multiple Targets," *IEEE Trans. Auto. Control*, Vol. AC-24, December 1979, pp. 843-854.

11. E. Taenzer, "Tracking Multiple Targets Simultaneously with a Phased Array Radar," *IEEE Trans. Aerospace and Electronic Systems*, Vol. AES-16, September 1980, pp. 604-614.

12. C.L. Morefield, "Application of 0-1 Integer Programming to Multitarget Tracking Problems," *IEEE Trans. Auto. Control*, Vol. AC-22, June 1977, pp. 302-312.

13. D.L. Alspach, "A Gaussian Sum Approximation to the Multitarget Identification— Tracking Problem," *Automatica*, Vol. 11, May 1975, pp. 285-296.

14. R.W. Sittler, "An Optimal Data Association Problem in Surveillance Theory," *IEEE Trans. Mil. Electron.*, Vol. MIL-8, April 1964, pp. 125-139.

15. Y. Bar-Shalom, "Tracking Methods in a Multi-Target Environment," *IEEE Trans. Auto. Control*, Vol. AC-23, August 1978, pp. 618-626.

16. H.L. Wiener, W.W. Willman, I.R. Goodman, and J.H. Kullback, "Naval Ocean-Surveillance Correlation Handbook, 1978," NRL Report 8340, Naval Research Laboratory, October 1979.

17. I.R. Goodman, H.L. Wiener, and W.W. Willman, "Naval Ocean-Surveillance Correlation Handbook, 1979," NRL Report 8402, Naval Research Laboratory, September 1980.

18. T.E. Fortmann and Y. Bar-Shalom, "Modification of the Likelihood Function to Account for Probabilistic Data Association," BBN Report 3964A (revised), Bolt Beranek and Newman Inc., November 1979, Contract N00039-78-C-0296.

19. A.D. Whalen, *Detection of Signals in Noise*, Academic Press, 1971.

20. T.E. Fortmann, Y. Bar-Shalom, M. Scheffe, and S. Gelfand, "Detection Thresholds for Multi-Target Tracking in Clutter," *Proc. 1981 IEEE Conference on Decision and Control*, San Diego, California, December 1981.

Multitarget Tracking Using Joint Probabilistic Data Association[*]

Yaakov Bar-Shalom

Department of Electrical Engineering and Computer Science
University of Connecticut
Storrs, CT

Thomas E. Fortmann

Automated Systems Department
Bolt, Beranek & Newman
Cambridge, MA

1 INTRODUCTION

The problems and issues involved in multitarget ocean tracking using a heterogeneous set of passive acoustic measurements were outlined in [1]; chief among these are data association and maneuver detection. An approach to solving these problems was also described. The resulting experimental algorithm involves an extended Kalman-Bucy filter (EKF) with both geographic and acoustic states, and handles measurement vectors such as bearing/frequency and delay/Doppler difference. Fundamental to the tracking algorithm is a *Probabilistic Data Association* (PDA) scheme based on [2], in which posterior association probabilities are computed for all current candidate measurements in a validation gate and used to form a weighted sum of innovations for updating the target's state in a suitably modified version of the EKF. A correction term in the Riccati equation accounts for the uncertainty in measurement origin, and the resulting state estimate is the conditional expectation, subject to the assumption that the state estimate is gaussian prior to each update. This assumption is questionable, but it leads to quite acceptable results in practice.

---

[*]This research was supported by the Office of Naval Research under Contracts N00014-78-C-0529 and N00014-80-C-0270.

The basic PDA algorithm assumes that each target is isolated from all other targets: false measurements in a validation gate are modeled as independent clutter points drawn from a Poisson distribution with spatial density C, and detection of a target is an independent event at each sample time with probability P_D.

The Poisson assumption breaks down and performance degrades significantly in the presence of *persistent interfering sources*, i.e., when interfering targets are detected consistently by one or more sensors. In this case, the interfering sources are usually trackable and the PDA algorithm can be extended to correct for them by computing the posterior probabilities jointly across clusters of targets; only the clutter is then modeled as Poisson. In an initial derivation [3], some inappropriate assumptions led to erroneous prior probabilities, and the combinatorics of maintaining separate validation gates for each target made it very difficult to implement. The main result of this paper is a new *Joint Probabilistic Data Association* (JPDA) algorithm for multiple targets in Poisson clutter.

This is a *target-oriented* approach, in the sense that a set of established targets is used to form gates in the measurement space and to compute posterior probabilities, in contrast to the *measurement-oriented* algorithms of Reid [4] and others, where each measurement is considered in turn and hypothesized to have come from some established track, a new target, or clutter.

This is also a *non-backscan* (or zero-scan) approach, meaning that all hypotheses are combined after computation of the probabilities, for each target at each time step. While it can be extended to retain n scans as in [5], this would lead in the passive sonar environment to enormous memory requirements because of the relatively high clutter densities (0.2-2.0 false detections per gate) and the large number of separately measured variables (2-12 validation gates per target) that are typically encountered. Moreover, because of the spurious nature of many of the false detections, we believe that the JPDA is *more cost-effective* than a brute-force n-scan algorithm in this environment. Retaining multiple scans would also cause enormous complications for the maneuver-detection and hypothesis-testing machinery that is built on top of the JPDA tracker.

The multitarget data association problem is formulated in the next section, followed by a derivation of the new JPDA algorithm in Section 3. The performance improvements gained with the JPDA algorithm in a passive sonar application are illustrated in Section 4 using simulated data from two heavily interfering targets in clutter.

2 PROBLEM FORMULATION

Consider a dynamic system (target model) of the familiar form

$$\underline{x}_{k+1} = \underline{F}\underline{x}_k + \underline{G}\underline{w}_k \tag{2.1}$$

$$\underline{y}_k = \underline{H}\underline{x}_k + \underline{v}_k \tag{2.2}$$

where $\underline{x}$ is the target state vector, $\underline{y}$ is the measurement vector, $\underline{w}$ and $\underline{v}$ are zero-mean, mutually independent, white, gaussian noise vectors with covariance matrices $\underline{Q}$ and $\underline{R}$, respectively, and k is a discrete time index. The matrices $\underline{F}$, $\underline{G}$, $\underline{H}$, $\underline{Q}$, and $\underline{R}$ are assumed known and their dependence on k is suppressed here for notational convenience. The initial state is assumed Gaussian with mean $\hat{\underline{x}}_{0|0}$ and covariance $\hat{\underline{P}}_{0|0}$. A specific target model is described in[1].

The tracker's estimate of the target state $\underline{x}_k$ at time k, given data up to time i, is denoted $\hat{\underline{x}}_{k|i}$. The error in this estimate is $\tilde{\underline{x}}_{k|i} \triangleq \underline{x}_k - \hat{\underline{x}}_{k|i}$, with error covariance matrix $\underline{P}_{k|i} \triangleq E\{\tilde{\underline{x}}_{k|i} \tilde{\underline{x}}'_{k|i}\}$, where E denotes expectation. In the absence of measurement origin uncertainty, the discrete-time Kalman-Bucy filter [6,7,8,9] yields the state estimate and covariance via the recursions

$$\hat{\underline{x}}_{k|k} = \hat{\underline{x}}_{k|k-1} + \underline{W}_k \tilde{\underline{y}}_k = \underline{F}\hat{\underline{x}}_{k-1|k-1} + \underline{W}_k \tilde{\underline{y}}_k \qquad (2.3)$$

$$\underline{P}_{k|k} = \underline{P}_{k|k-1} - \underline{W}_k \underline{S}_k \underline{W}'_k = \underline{F}\underline{P}_{k-1|k-1}\underline{F}' + \underline{G}\underline{Q}\underline{G}' - \underline{W}_k \underline{S}_k \underline{W}'_k \qquad (2.4)$$

where the innovation vector

$$\tilde{\underline{y}}_k \triangleq \underline{y}_k - \hat{\underline{y}}_{k|k-1} \qquad (2.5)$$

has the covariance matrix

$$\underline{S}_k \triangleq E\{\tilde{\underline{y}}_k \tilde{\underline{y}}'_k\} = \underline{H}\underline{P}_{k|k-1}\underline{H}' + \underline{R} \qquad (2.6)$$

and the filter gain matrix is

$$\underline{W}_k = \underline{P}_{k|k-1}\underline{H}'\underline{S}_k^{-1}. \qquad (2.7)$$

The resulting state estimate, under the above assumptions, is the *conditional mean*

$$\hat{\underline{x}}_{k|k} = E\{\underline{x}_k | Y^k\} \qquad (2.8)$$

where Y^k denotes the set of all data vectors $\underline{y}_i$ for $i \leq k$.

In order to avoid cluttering the discussion to follow, this brief summary ignores a number of complications that arise in practice. If the system is nonlinear, for example, then it can usually be linearized and the same basic equations can be applied to deviations from the nominal trajectory [7,9]. If the target occasionally deviates from the assumed motion model, e.g., by maneuvering, then some decision-making or other machinery must be provided to deal with these instances.

A fundamental characteristic of this class of problems is that the size and composition of the measurement vector are unpredictable from one time to the next; in other words, $\underline{y}_k$ comprises a time-varying set of independent subvectors, as discussed below in Section 4. We shall avoid the resulting notational morass by restricting Equations (2.3)-(2.6) to apply to a single measurement subvector $\underline{y}_k$ from a single sensor. In addition, *we will suppress the time index k* from all variables except $\underline{P}$ and Y, unless it is required for clarity. Without any loss of generality, the *data association problem* may now be formulated as follows.

At each time step, the sensor provides a set of candidate measurements to be associated with targets (or rejected). In most approaches, this is done by forming a "validation gate" around the predicted measurement from each target and retaining only those detections that lie within the gate. There are many different approaches to establishing a correspondence between candidate measurements and targets; in this paper we shall focus on the Probabilistic Data

Association (PDA) method [1, 2, 11]. The m candidate measurements at time k are denoted $\underline{y}_j$, $j=1, \ldots m$, i.e.,

$$Y^k = \{\underline{y}_1, \ldots \underline{y}_m\} \cup Y^{k-1}, \tag{2.9}$$

and the corresponding innovations are

$$\underline{\tilde{y}}_j \triangleq \underline{y}_j - \underline{\hat{y}}, \quad j=1, \ldots m. \tag{2.10}$$

The term measurement will be used interchangeably for $\underline{y}_j$ and $\underline{\tilde{y}}_j$, since they contain equivalent information [9].

Considering a single target independently of any others, X_j denotes the event that the jth measurement belongs to that target and X_0 the event that none of the measurements belongs to it (no detection). The PDA approach builds upon the assumptions that the estimation errors $\underline{\tilde{x}}$ and $\underline{\tilde{y}}$ have gaussian densities at each time step (this is approximate, since there is an exponentially growing tree of possible measurement sequence hypotheses and the true densities are weighted sums of gaussians). It is also assumed that the correct measurement is detected with probability P_D (independently at each time) and that all other measurements are Poisson-distributed* with parameter CV, where C is the expected number of false measurements per unit volume and V is the volume of the validation gate.

The gate is normally a "g-sigma" ellipsoid $\{\underline{\tilde{y}} : \underline{\tilde{y}}' \underline{S}^{-1} \underline{\tilde{y}} \leq g^2\}$ and P_G is the probability that the correct measurement, if detected, lies within the gate. The gate volume is thus $V = c_M g^M |\underline{S}|^{1/2}$, where M is the dimension of $\underline{\tilde{y}}$ and $c_M = \pi^{M/2}/\Gamma(M/2 + 1)$ is the volume of the M-dimensional unit sphere ($c_1 = 2$, $c_2 = \pi$, $c_3 = 4\pi/3$, etc.).

In the PDA filtering approach, the conditional mean estimate $\underline{\hat{x}}$ is obtained from (2.3) by using the combined (weighted) innovation

$$\underline{\tilde{y}} \triangleq \sum_{j=1}^{M} \beta_j \underline{\tilde{y}}_j \tag{2.11}$$

where $\beta_j = P\{X_j | Y^k\}, j=0, \ldots m$, is the posterior probability that the j-tH measurement (or no measurement, for $j=0$) is the correct one. The update part of the covariance equation (2.4) becomes [2,11]

$$\underline{P}_{k|k} = \underline{P}_{k|k-1} - (1-\beta_0) \, \underline{W}_k \underline{S}_k \underline{W}'_k + \underline{\tilde{P}}_k \tag{2.12}$$

where the positive semidefinite matrix

$$\underline{\tilde{P}}_k = \underline{W}_k \left[\sum_{j=1}^{m} \beta_j \underline{\tilde{y}}_j \underline{\tilde{y}}_j' - \underline{\tilde{y}} \underline{\tilde{y}}' \right] \underline{W}'_k \tag{2.13}$$

accounts for the measurement origin uncertainty. Note that the data-dependent factors β_0 and $\underline{\tilde{P}}_k$ transform the original deterministic Riccati equation into a stochastic one.

---

*Equivalently, the number n of false measurements has probability mass function $P(n) = e^{-CV}(CV)^n/n!$ and the location of each false measurement is uniformly distributed in the gate.

Multiple Interfering Targets and JPDA

In order to deal with multiple interfering targets, consider a cluster† of targets (established tracks) numbered $t = 1, \ldots, T$ at a given time k. The set of m candidate measurements associated with this cluster (i.e., found in the validation gates for targets $1, \ldots, T$) is denoted $\underline{y}_j$, $j = 1, \ldots m$, as above. Each measurement belongs either to one of the T targets or to the set of false measurements (clutter), which is denoted by target number $t = 0$.

Denoting the predicted measurement for target t by $\hat{\underline{y}}^t$, the innovation (2.10) corresponding to measurement j becomes

$$\tilde{\underline{y}}_j^t \triangleq \underline{y}_j - \hat{\underline{y}}^t \quad (2.14)$$

and the combined (weighted) innovation (2.11) becomes

$$\tilde{\underline{y}}^t = \sum_{j=1}^{m} \beta_j^t \tilde{\underline{y}}_j^t \quad (2.15)$$

where β_j^t is the posterior probability that measurement j originated from target t and β_0^t is the probability that none of the measurements originated from target t (i.e., it was not detected). This is used in targets t's copy of (2.3) to update the state estimate $\hat{\underline{x}}^t$.

In other words, the Joint Probabilistic Data Association (JPDA) and PDA approaches utilize the same estimation equations; the difference is in the way the association probabilities are computed. Whereas the PDA algorithm computes β_j^t, $j = 0, 1, \ldots m$, separately for each t, under the assumption that *all* measurements not associated with target t are false (i.e., Poisson-distributed clutter), the JPDA algorithm computes β_j^t *jointly* across the set of T targets and clutter. From the point of view of any target, this accounts for false measurements from both discrete interfering sources (other targets) and random clutter. Details are given in the next section.

3 JOINT PROBABILITIES

The key to the JPDA algorithm is evaluation of the conditional probabilities of the following joint events,

$$\underline{X} = \bigcap_{j=1}^{m} X_{jt_j} \quad (3.1)$$

where X_{jt_j} is the event that measurement j originated from target t_j, $0 \leq t_j \leq T$.

Using Bayes' rule, the probability of a joint event conditioned on all measurements up to the present time is obtained as in [10]

---

†A cluster is a set of targets whose validation gates are "connected" by measurements lying in their intersections [3,4]. Note that a different measurement subvector (e.g., from another sensor) will lead to a different target cluster.

$$P\{\underline{X}|Y^k\} = \frac{C^\phi}{c} \prod_{j:\tau_j=1} \frac{\exp[-\frac{1}{2}(\tilde{y}_j^{t_j})'\underline{S}_{t_j}^{-1}(\tilde{y}_j^{t_j})]}{(2\pi)^{M/2}|\underline{S}_{t_j}|^{1/2}}$$

$$\prod_{t:\delta_t=1} P_D^t \prod_{t:\delta_t=0} (1 - P_D^t) \qquad (3.2)$$

where c is the normalization constant, C is the spatial clutter density; $\tau_j = 1$ indicates that measurement J is associated with a target; ϕ is the number of false measurements; δ_t is the detection indicator for target t.

The probability β_j^t that measurement j belongs to target t may now be obtained by summing over all feasible events $\underline{X}$ for which this condition is true:

$$\beta_j^t = \sum_{\underline{X}} P\{\underline{X}|Y^k\}\hat{\omega}_{jt}(X) \qquad j = 1, \ldots m; \quad t = 0, 1, \ldots T \qquad (3.3)$$

$$\beta_j^t = 1 - \sum_{j=1}^{M} \beta_j^t t = 0, 1, \ldots T \qquad (3.4)$$

where $\hat{\omega}_{jt} = 1$ if measurement j is associated with target T. These probabilities are used to form the combined innovation (2.15) for each target.

4 TRACKING RESULTS

A simulation program was used to create realistic passive sonar data on which the JPDA algorithm could be tested. One such data set contains measurement files of bearing/frequency lines and time/Doppler differences for two hypothetical targets, with a common 12 Hz source frequency and courses that result in **severe interference**. The target-sensor geometry is indicated in Figure 1. Targets 1 and 2 travel at 6 knots on courses of 100° and 80° respectively, and cross midway through the 6-hour period shown.

Measurement data were created by dead reckoning target motion (no process noise), adding noise to the computed true measurements, and then adding clutter measurements (false detections). The standard deviation of the measurement noise was 5° for bearings, 80 mHz for frequency, 3.6 sec for time difference, and 4 mHz for Doppler difference. The true measurement was detected at any given time with probability $P_D = 0.7$. The number of clutter points was Poisson-distributed and their locations in the measurement space were uniformly distributed over a very broad region about the actual track. The clutter density was $C = 0.25$/Hz-deg for bearing/frequency and $C = 0.25$/hz-sec for time/Doppler difference; with varying gate sizes, this ranged from 0.2 to 2.0 false detections per gate.

Several tracks were made using this data; the initial 2-sigma confidence ellipse areas were 25π mi^2, with course and speed confidence of $\pm 20°$ and ± 3 knots, respectively. In practice, one would use somewhat looser limits to initialize uncertain targets, but the object here was to simulate a situation in which the tracks are already well-established before they intersect. Although the data had no process noise, the filter was given process noise standard deviations of to avoid divergence.

Figure 2 shows tracks made using the well-known "nearest-neighbor" data association scheme [13], in which the candidate measurement closest to the

Probabilistic Data Association

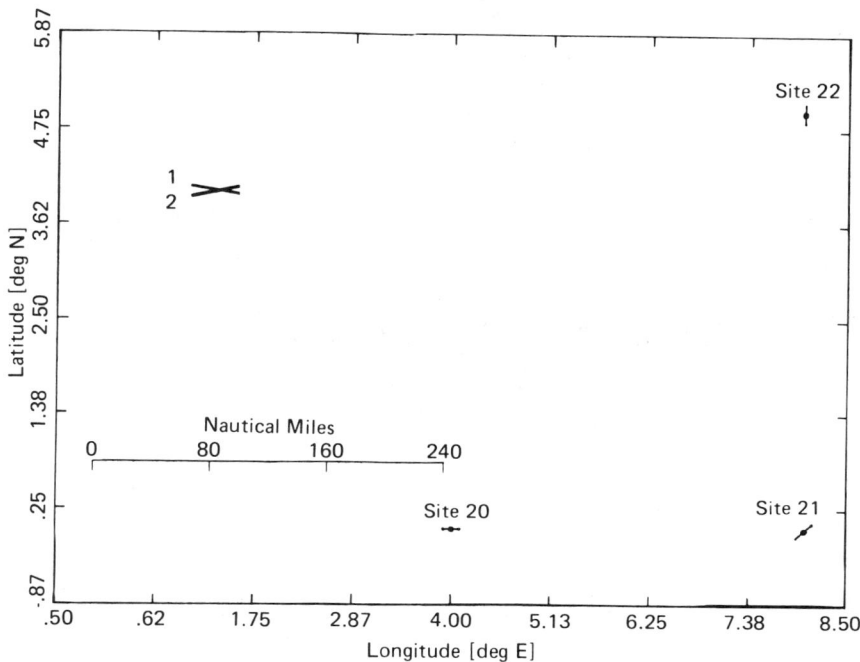

Figure 1 Target/sensor geometry

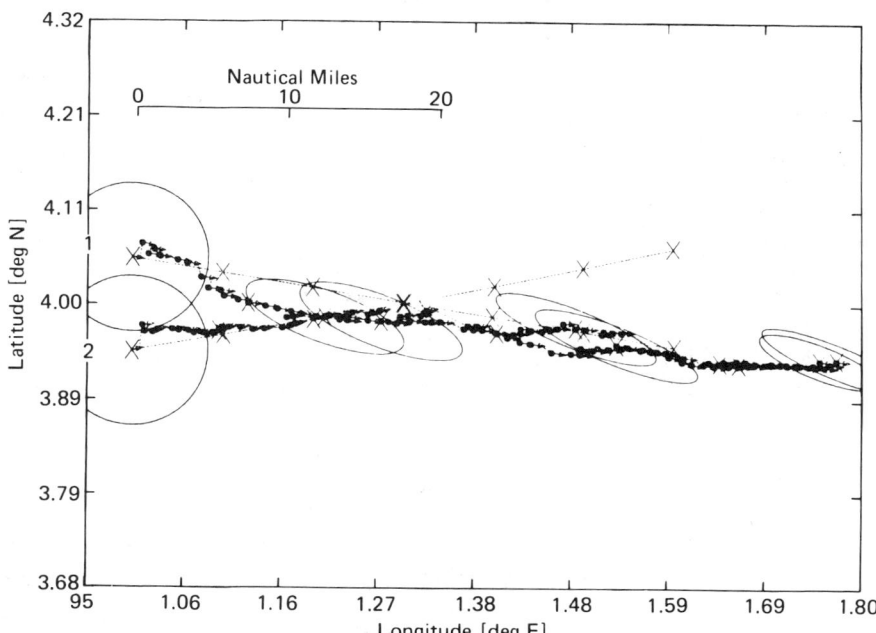

Figure 2 Nearest neighbor data association

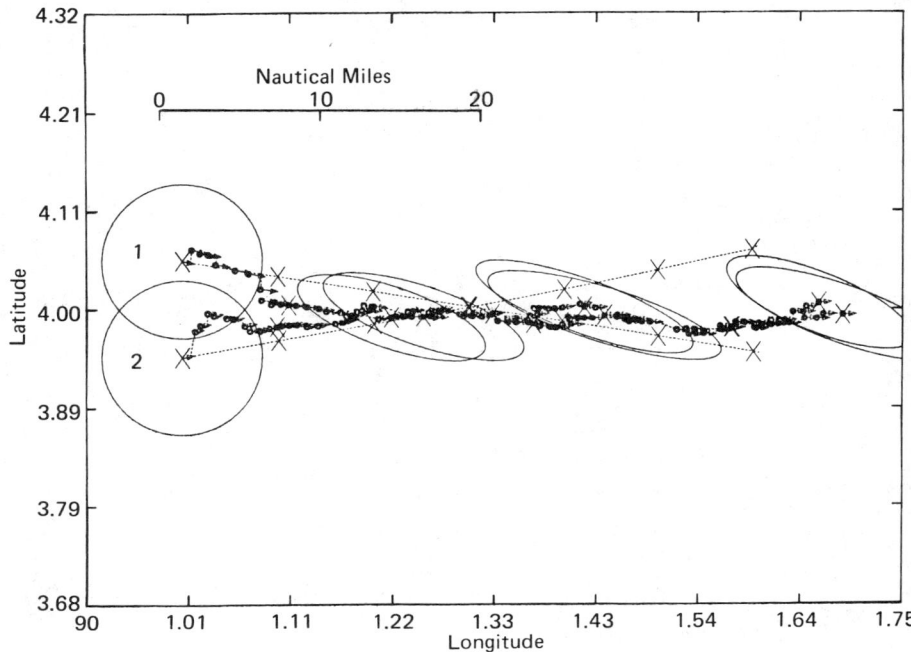

Figure 3 Ordinary probabilistic data association (multi target logic OFF)

center of each validation gate is accepted as correct, all others are ignored, and the covariance equation is not augmented to account for association errors. As one might anticipate, the tracker becomes hopelessly confused and loses one of the targets altogether.

The ordinarily PDA filter (without multi-target logic) fares only slightly better in this situation. As shown in Figure 3, both tracks lock onto a sort of "compromise" and end up lost midway between the two actual targets. This behavior is to be expected, since a basic assumption in the standard PDA algorithm (applied to each target in turn) is that measurements not originating from the target under consideration are random clutter points with a Poisson/uniform distribution.

The joint PDA method, described above in Section 3, corrects this situation by allowing the probabilistic weights used in data association to be computed jointly across all known targets. As shown in Figure 4, this improves the tracks dramatically, and the 2-sigma confidence ellipses contain the true position in all cases. Note that the ellipses are larger than those of Figure 3, particularly near the point of intersection, reflecting the fact that the algorithm "hedges" its decisions and relies more on dead reckoning when the targets are too close to determine with high confidence which measurement belongs to which target.

Finally, tracks made using perfect data association are shown in Figure 5. Comparing these with Figure 4, it is clear that the JPDA tracker performs

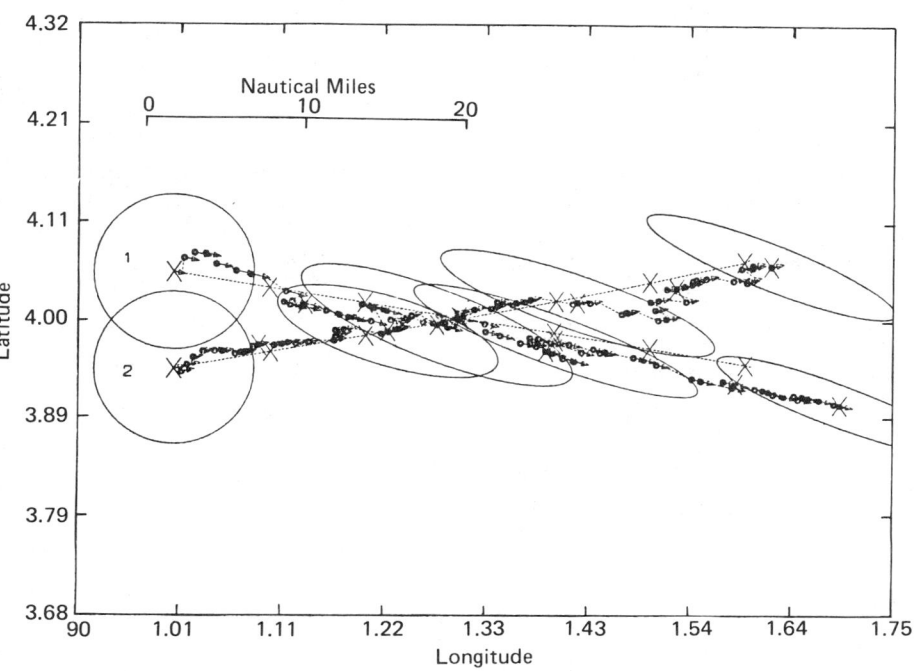

Figure 4 Joint probabilistic data association (multi target logic ON)

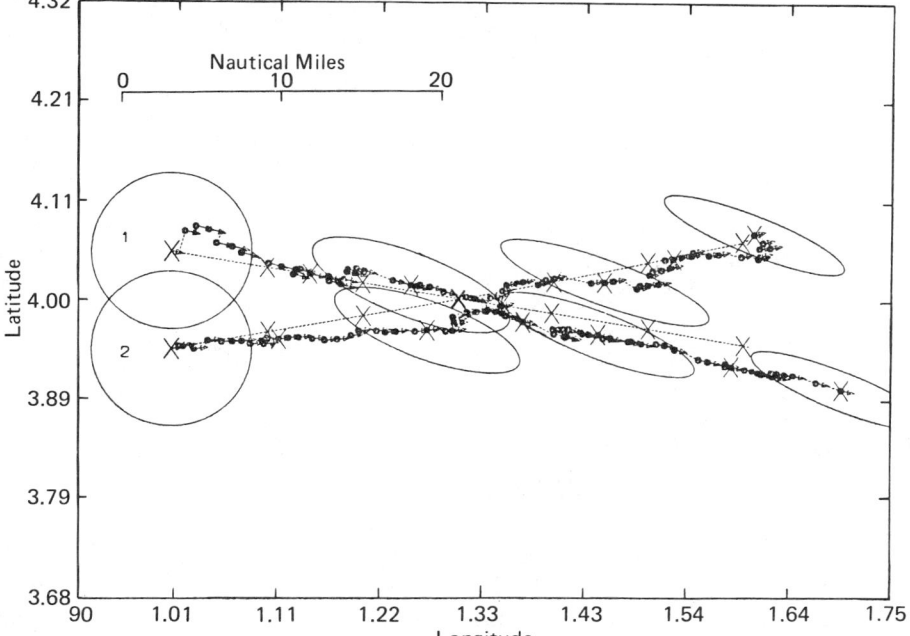

Figure 5 Perfect data association

remarkably well despite the severe interference between targets. Indeed, by the end of the 6-hour period, the two sets of tracks are nearly identical.

5 CONCLUSION

Report-to-track correlation has been identified as one of the major problems in ocean surveillance [14,15], and the results of this paper demonstrate the importance of correlating (i.e., associating) each report with all tracks simultaneously, rather than separately. The joint probabilistic data association framework is particularly well-suited to this task in the passive sonar environment, where additional measurements arise from both interfering targets and random clutter, detection probability is significantly less than 1, the measurement space is large and heterogeneous, a target's state is not fully observable from any single measurement, and maneuvers are common.

Continuing research in this and related areas includes assessment of tracking performance as a function of detection probability and clutter density (this has important implications for the setting of gains and thresholds in the signal processing algorithms that provide measurements to the tracker), extension of the likelihood function and other hypothesis-testing machinery (e.g., for maneuver detection) to the JPDA situation, the use of more complex models for the detection and clutter processes, and automated schemes for track initiation.

REFERENCES

1. T.E. Fortmann and S. Baron, "Problems in Multi-Target Sonar Tracking," *Proc. 1978 IEEE Conf. on Decision and Control*, San Diego, California, January 1979.

2. Y. Bar-Shalom and E. Tse, "Tracking in a Cluttered Environment with Probabilistic Data Association," *Automatica*, vol. 11, September 1975, pp. 451-460.

3. Y. Bar-Shalom, "Extension of the Probabilistic Data Association Filter to Multitarget Environment," *Proc. Fifth Symp. on Nonlinear Estimation*, San Diego, California, September 1974.

4. D. B. Reid "An Algorithm for Tracking Multiple Targets," *IEEE Trans. Auto. Control*, vol. AC-24, December 1979, pp. 843-854.

5. R. Singer, R. Sea, and K. Housewright, "Derivation and Evaluation of Improved Tracking Filters for use in Dense Multitarget Environments," *IEEE Trans. Info. Theory*, vol. IT-20, July 1974, pp. 423-432.

6. I. B. Rhodes, "A Tutorial Introduction to Estimation and Filtering," *IEEE Trans. Auto. Control*, vol. AC-16, December 1971, pp. 688-706.

7. A.H. Jazwinski, *Stochastic Processes and Filtering Theory*, Academic Press, 1970.

8. P.S. Maybeck, *Stochastic Models, Estimation and Control—Volume 1*, Academic Press, 1979.

9. B.D.O. Anderson, *Optimal Filtering*, Prentice-Hall, 1979.

10. T.E. Fortmann, Y. Bar-Shalom, and M. Scheffe, "Multi-Target Tracking Using Joint Probabilistic Data Association," *Proc 1980 IEEE Conference on Decision and Control*, Albuquerque, New Mexico, December 1980.

11. Y. Bar-Shalom, "Tracking Methods in a Multi-Target Environment," *IEEE Trans. Auto. Control*, vol. AC-23, August 1978, pp. 618-626.

12. C.L. Morefield, "Application of 0-1 Integer Programming to Multitarget Tracking Problems," *IEEE Trans. Auto. Control*, vol. AC-22, June 1977, pp. 302-312.

13. E. Taenzer, "Tracking Multiple Targets Simultaneously with a Phased Array Radar," *IEEE Trans. Aerospace and Electronic Systems*, vol. AES-16, September 1980, pp. 604-614.

14. H.L. Weiner, W.W. Willman, I.R. Goodman, and J.H. Kullback, "Naval Ocean-Surveillance Correlation Handbook, 1978" NRL Report 8340, Naval Research Lab, October 1979.

15. I.R. Goodman, H.L. Wiener, and W.W. Willman, "Naval Ocean-Surveillance Correlation Handbook, 1979," NRL Report 8402, Naval Research Laboratory, September 1980.

Selection of Processing Parameters for Generating Ambiguity Surfaces*

Joseph LaPointe, Jr.

Analytical Technology Applications Corporation
Mountain View, CA

1 INTRODUCTION

The detection performance of the linearly thresholded sample magnitude-squared correlation coefficient has been documented for the case of equal signal and noise bandwidths (herein called matched containment) (1,2). The statistics of the sample magnitude-squared correlation coefficient (MSCC) are also well known for the matched containment case (3,4). It is necessary to filter the data prior to computing the sample MSCC with the filter bandwidth called the processing bandwidth. In actual practice it is not always possible to compute the sample MSCC with a processing bandwidth, W_P, that is equal to the signal bandwidth, W_S, because the signal bandwidth and the signal center frequency are only known approximately. When this is the case, it is the usual practice to select a sufficiently large W_P so that the signal is fully contained within the processing bandwidth. The signal will overcontained whenever $W_P > W_S$.

The detection performance of the sample MSCC can be controlled through the selection of the processing bandwidth and the integration time. Overcontaining the signal decouples the signal and noise "time-bandwidth" products (N_S and N, respectively). The detection performance of the sample MSCC is presented for the overcontained case. It will be shown that the input signal-to-noise ratio for a given Probability of Detection (P_D) and Probability of False Alarm (P_{FA}) can be reduced by overcontaining the signal whenever N_S is small.

---

*Research supported by the Office of Naval Research under Contract N00014-80-C-0698.

2 STATISTICS

Let $Z(l)$ be a two-dimensional zero mean complex Gaussian random column vector with elements $z_1(l)$ and $z_2(l)$ denoting the frequency coefficients from channels 1 and 2, respectively, at frequency l/T Hz for $l = $, 2, ..., N where $TW_P = N$ and T is the integration time. Typically the frequency coefficients are generated from the time series data by a discrete Fourier transform. The two-dimensional positive Hermitian sample autocorrelation matrix is

$$A = \sum_{l=1}^{N} Z(l) \, Z'(l) \tag{1}$$

where ' denotes complex conjugate of the transpose. Let $A = [a_{lk}]$ for $l, k = 1, 2$. Then the sample MSCC can be computed from the sample autocorrelation matrix by

$$\rho^2 = \frac{|a_{12}|^2}{a_{11} \, a_{22}}. \tag{2}$$

The definition of the sample MSCC given in Equation (2) corresponds to computing the sample MSCC according to

$$\rho^2 = \frac{\left|\sum_{l=1}^{N} z_1(l) \, z_2^*(l)\right|^2}{\sum_{l=1}^{N} |z_1(l)|^2 \sum_{l=1}^{N} |z_2(l)|^2} \tag{3}$$

where * denotes complex conjugation.

It is easily shown that $Z(l)$ and $Z(k)$ are independent when $l \neq k$ for strictly bandlimited spectra when the frequency coefficients are spaced at intervals of $1/T$ Hz (5). In the signal overcontainment case, let there be $N_S = TW_S$ frequency coefficients containing signal and $M = T(W_P - W_S)$ frequency coefficients containing only noise, where $N = N_S + M$. Define the overcontainment ratio

$$OVC = W_P/W_S$$
$$= (N_S + M)/N_S. \tag{4}$$

Note that matched containment occurs when $W_P = W_S$ or $M = 0$. Finally, define the cross-spectral density matrix of $Z(k)$ to be

$$R_z(k) = E\{Z(k) \, Z'(k)\}$$
$$= \begin{cases} R_{z_1} = R_S + R_N, & k = 1, 2, \ldots, N_S \\ R_{z_0} = R_N, & k = N_S + 1, \ldots, N \end{cases} \tag{5}$$

where R_N is the cross-spectral density matrix for the noise, R_S is the cross-spectral density matrix for the signal, and $E\{\cdot\}$ denotes statistical expectation.

The probability density function (PDF) and the cumulative density function (CDF) of the sample MSCC can be derived from the PDF of the sample autocorrelation matrix A by the change of variables indicated in Equation (2) and by integrating out the auxiliary variables a_{11}, a_{22}, and the phase of a_{12}. The PDF of A is easily derived by first deriving the characteristic function of A and then computing the inverse Fourier transform of the characteristic function. The characteristic function of A is

$$M_A(\theta) = E\{e^{jTR(A\theta)}\}$$

$$= \frac{1}{|I - jR_{z_1}\theta|^{N_S} |I - jR_{z_0}\theta|^M} \quad (6)$$

where $TR(\cdot)$ denotes trace, θ is a two-dimensional positive definite Hermitian matrix, and $|\cdot|$ denotes determinant (5). By computing the inverse Fourier transform of $M_A(\theta)$, the PDF of A is

$$f(A) = \frac{|A|^{N-2} e^{-TR(R_{z_1}^{-1}A)}}{\pi \Gamma(N)\Gamma(N-1)|R_{z_0}|^M |R_{z_1}|^{N_S}}$$

$$\cdot {}_1\tilde{F}_1(M, N, \Delta R A) \quad (7)$$

where ${}_1\tilde{F}_1(\cdot, \cdot, \cdot)$ is the confluent hypergeometric function of matrix argument; $\Gamma(\cdot)$ is the Gamma function; and $\Delta R = R_{z_1}^{-1} - R_{z_0}^{-1}$ (5). Note that for matched containment, $M = 0$ and

$$f(A) = \frac{|A|^{N_S-2} e^{-TR(R_{z_1}^{-1}A)}}{\pi \Gamma(N_S)\Gamma(N_S-1)|R_{z_1}|^{N_S}} \quad (8)$$

which is the complex Wishart Density from which the statistics of the sample MSCC are obtained for matched containment (6).

The PDF for the sample MSCC is obtained by making the change of variables indicated above. The CDF of the sample MSCC is

$$F(\rho_t^2|\rho_T^2, M, N_S) = \int_0^{\rho_t^2} f(\rho^2|\rho_T^2, M\ N)\ d^2 \quad (9)$$

where $\rho_t^2 \in [0, 1]$ is the threshold on the sample MSCC, $f(\cdot|\cdot, \cdot, \cdot)$ is the PDF of the sample MSCC, and ρ_T^2 is the true MSCC in the signal band. The equation for the PDF and the CDF of the sample MSCC are rather involved. Therefore, only the equation for the CDF will be displayed. The reader is referred to (5) for the equation for the PDF. By performing the indicated change of variables and evaluating Equation (9), the CDF of the sample MSCC for equal channel signal-to-noise ratios (SNR) in the signal band and for perfectly correlated signal components is

$$F(\rho_t^2|\rho_T^2, M, N_S) = (1 - \rho_T^2)^N (SNR + 1)$$

$$\sum_{l=0}^{\infty} \left[\left(\frac{SNR}{SNR + 1}\right)^{2l} \frac{\Gamma(N + l)\Gamma(N_S + 2l)}{\Gamma(N_S(\Gamma(N + 2l)l!)} \right.$$

$$\left. I_{\rho_t^2}(l + 1, N - 1)\ {}_2F_1\!\left(M, -N; N + 2l; \frac{SNR}{2SNR+1}\right) \right] \quad (10)$$

where $I_x(a, k)$ is the incomplete Beta function (Equation 26.5.1 of (7)). The true MSCC in the signal band for equal channel SNR's and for perfectly correlated signal components is

$$\rho_T^2 = \frac{SNR^2}{(SNR + 1)^2}. \tag{11}$$

It is easily shown that when signal is absent, $\rho_T^2 = 0$ and

$$F(\rho_t^2|0, M, N_S) = 1 - (1 - \rho_t^2)^{M + N_S - 1}. \tag{12}$$

The CDF of the sample MSCC given in Equation (10) reduces to the CDF of the sample MSCC for matched containment given in (1) when $M = 0$.

3 PERFORMANCE

The detection performance is defined as the SNR in the signal band needed to achieve a desired P_D and P_{FA} for a specified overcontainment ratio and degrees of freedom in the signal band. The H_0 hypothesis occurs when only uncorrelated noise is present. The P_{FA} and P_D are "one minus the CDF" under the appropriate conditions. From Equations (10) and (12),

$$P_{FA} = (1 - \rho_t^2)^{N-1} \tag{13}$$

and

$$P_D = 1 - F(\rho_t^2|\rho_T^2, M, N_S). \tag{14}$$

The procedure used to quantify the detection performance is to (a) select a

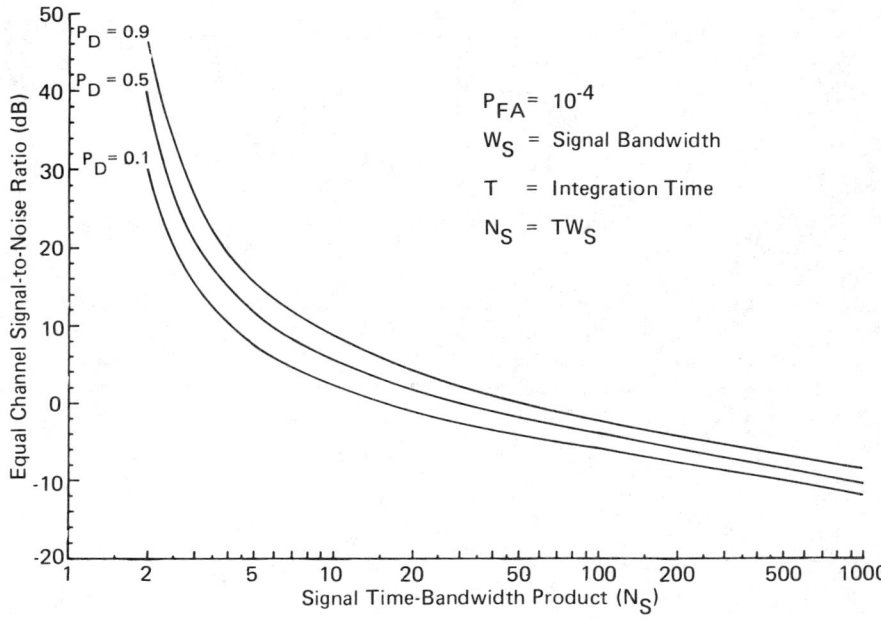

Figure 1 Matched containment detection performance

Generating Ambiguity Surfaces

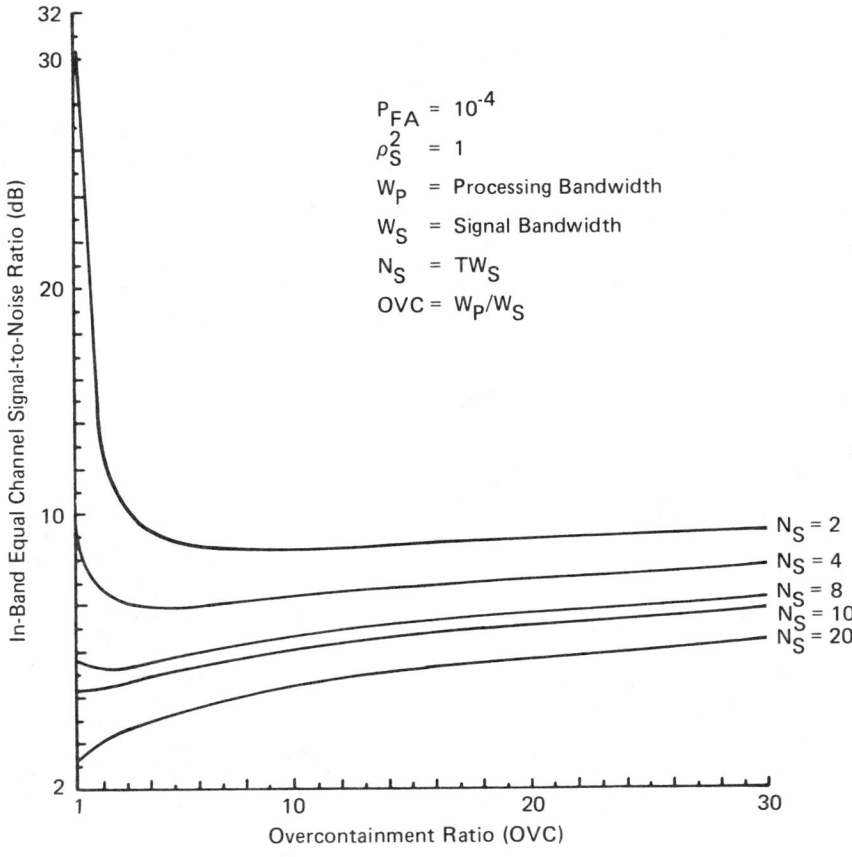

Figure 2 Effects of overcontainment for $P_D = 0.1$

P_D, P_{FA}, N_S, and OVC, (b) find the threshold, ρ_t^2, from Equation (13) and (c) numerically solve Equation (14) for the SNR.

The matched containment SNR's need to achieve a $P_{FA} = 10^{-4}$ and P_D's of 0.1, 0.5, and 0.9 are shown in Figure 1 as a function of N_S. This figure is provided as a point of reference for the matched containment detection performance. As is well known, the SNR decreases rapidly with increasing N_S for small N_S. Once N_S is greater than 64, the SNR decreases 2 dB per doubling of N_S.

The SNR required to obtain a $P_{FA} = 10^{-4}$ and P_D's of 0.1, 0.5, and 0.9 are shown in Figures 2, 3, and 4, respectively, as a function of OVC. It is immediately obvious that the SNR can be reduced by overcontaining the signal for $N_S < N_{S_0}$ where $N_{S_0} = 10$, 12, and 14 for $P_D = 0.1$, 0.5, and 0.9, respectively. When $N_S < N_{S_0}$, there is a value of OVC, OVC_0, that minimizes the SNR. For OVC's larger than OVC_0, the SNR increases at a rate of about 1 dB per doubling of OVC. For $N_S > N_{S_0}$ the SNR increases with OVC and also

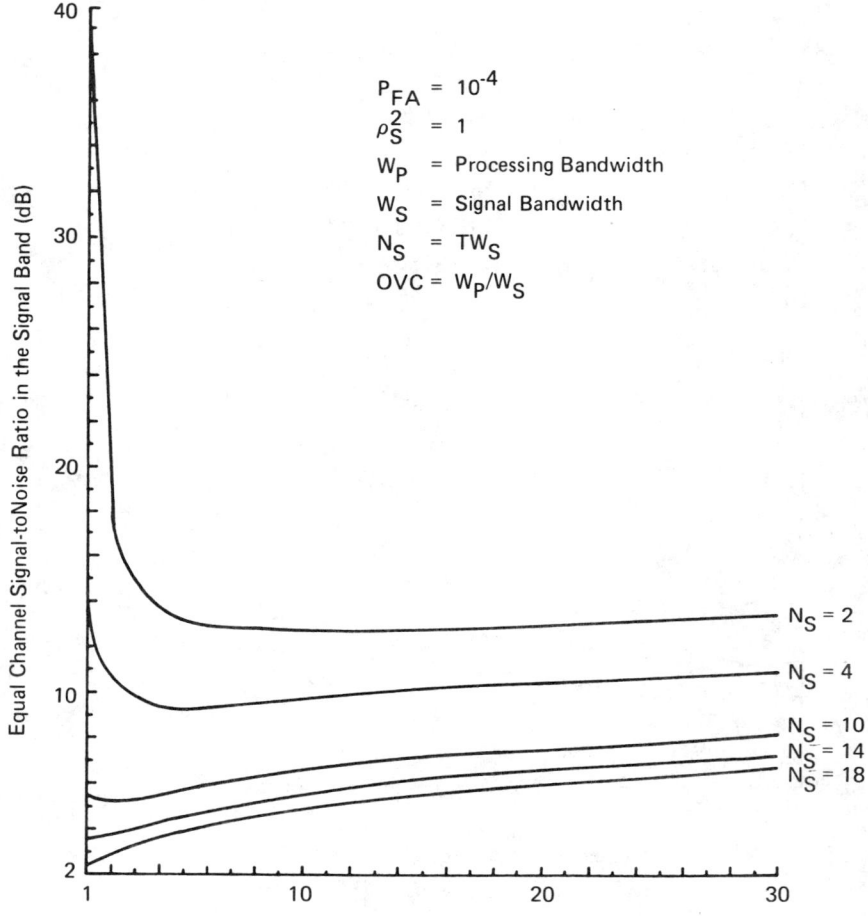

Figure 3 Effects of overcontainment for $P_D = 0.5$

reaches an asymptote of 1 dB per doubling of OVC for the range of OVC's considered.

Reducing the SNR by increasing the signal overcontainment is counterintuitive because increasing the noise power is not expected to improve the performance. This gain is caused by the effect overcontainment has on the noise. The noise degrees of freedom are increased with respect to N_S, and the noise power is increased with respect to the matched containment noise power. The increase in both cases is equal to OVC. For small N_S, the increase in N causes a reduction in the threshold that more than offsets the increase in the noise power, thereby producing the observed gains. However, for moderate to large N_S, the increase in noise power offsets the reduction in the threshold. It can be said that the observed gains for small N_S are caused by the improvement in the knowledge of the background noise.

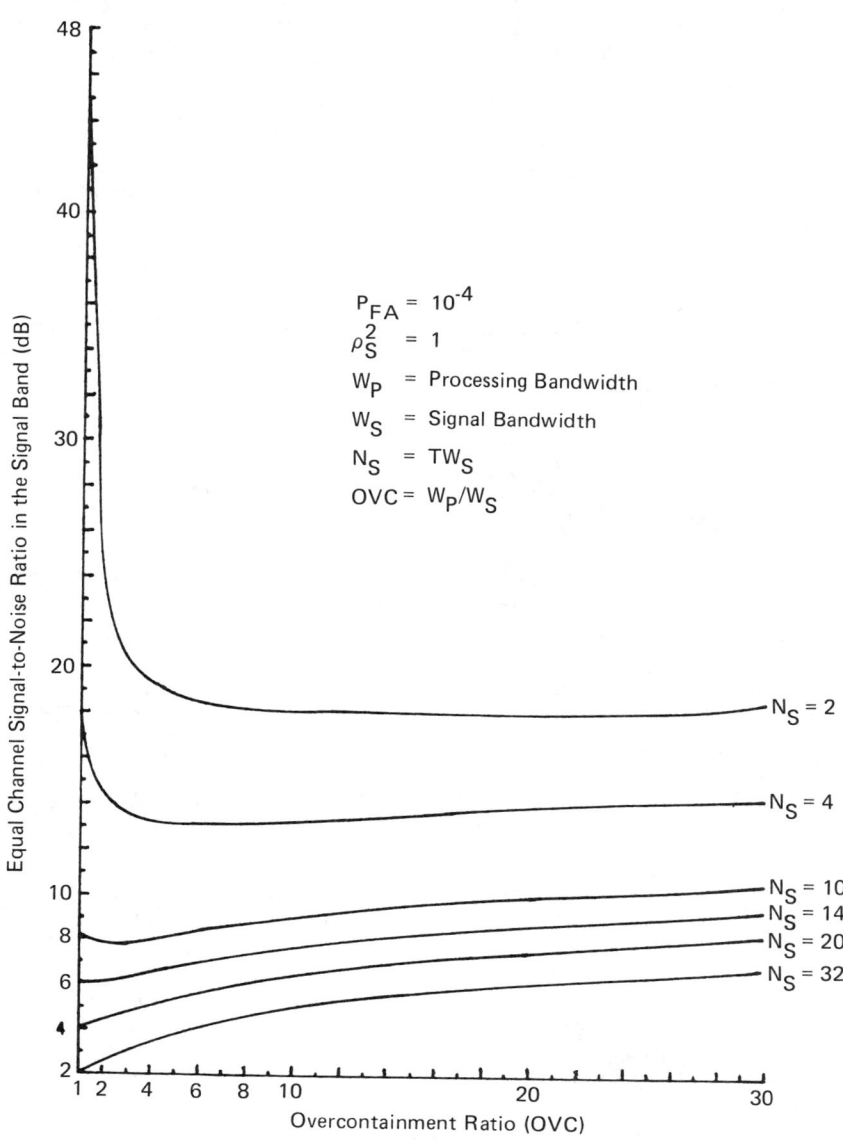

Figure 4 Effects of overcontainment for $P_D = 0.9$

4 CONCLUSION

The SNR needed to detect a signal can be controlled with overcontainment. For small N_S, the SNR can be reduced by increasing the overcontainment because overcontainment improves the estimate of the background noise at a faster rate than the increase in noise power. On the other hand, for moderate to large N_S, the SNR slowly increases with increasing overcontainment because the noise power increases at a faster rate than the improvement in the estimate of the background noise. It can be concluded that (a) gains are obtainable by signal overcontainment for small N_S and (b) when gains are not obtainable, the losses are insensitive to the amount of overcontainment.

REFERENCES

Carter, G.C., "Receiver Operating Characteristics for a Linearly Thresholded Coherence Estimation Detector," *IEEE Trans. Acoust., Speech, Signal Processing*, ASSP-25, February 1977, pp. 90-92.

Gosselin, J.J., "Comparative Study of Two Sensor (Magnitude Squared Coherence) and Single Sensor (Square Law) Receiver Operating Characteristics," *IEEE Int. Acoust., Speech, Signal Processing Conf. Rec.*, Hartford, CT, May 9-11, 1977, pp. 311-314.

Carter, G.C., Knapp, C.H. and Nuttall, A.H., "Estimation of the Magnitude-Squared Coherence Function via Overlapped Fast Fourier Transform Processing," *IEEE Trans. Audio Electroacoust.*, Vol. AU-21, August 1973, pp. 337-344.

Carter, G.C. and Nuttall, A.H., "Statistics of the Estimate of Coherence," *Proc. IEEE*, Vol. 60, April 1972, pp. 465-466.

LaPointe, J. *Ambiguity Surface Statistics and Overcontainment*, ATAC Final Report SV8007-1, ATAC, Sunnyvale, CA, 30 September 1981.

Goodman, N.R., "Statistical Analysis Based on a Certain Multivariate Complex Gaussian Distribution (An Introduction)," *Annals. of Math. Stat.*, Vol. 34, 1963, pp. 152-177.

Abramowitz, M. and Stegun, I.A. (eds.), *Handbook of Mathematical Functions with Formulas, Graphs, and Mathematical Tables*, U.S. Government Printing Office, Washington, D.C., 1964.

Part V: Statistical Image Processing

Syntactic Approach to Signal and Image Analysis*

K. S. Fu

School of Electrical Engineering
Purdue University
W. Lafayette, IN

1 INTRODUCTION

The many different mathematical techniques used to solve pattern analysis problems may be grouped into two general approaches [6,11]. They are the decision-theoretic (or discriminant) approach and the syntatic (or structural) approach. In the decision-theoretic approach, a set of characteristic measurements, called features, are extracted from the patterns. Each pattern is represented by a feature vector, and the classification of each pattern is usually made by partitioning the feature space. Many results from discriminant analysis and statistical decision theory have been applied to pattern classification problems [6]. On the other hand, in the syntactic approach, each pattern is expressed as a composition of its components, called subpatterns and pattern primitives [5,8]. This approach draws an analogy between the structure of patterns and the syntax of a language. The recognition of each pattern is usually made by analyzing the pattern structure according to a given set of structure or syntax rules. In this paper, we briefly review the recent progress in syntactic approach to signal and image analysis.

In syntactic methods, a pattern is represented by a sentence in a language which is specified by a grammar. The language which provides the structural description of patterns, in terms of a set of pattern primitives and their composition relations, is sometimes called the "pattern description language." The rules governing the composition of primitives into patterns are specified by the so-called "pattern grammar." An alternative representation of the structural information of a pattern is to use a "relational graph," of which the nodes

---
*This work was supported by the ONR Contract N00014-79-C-0574.

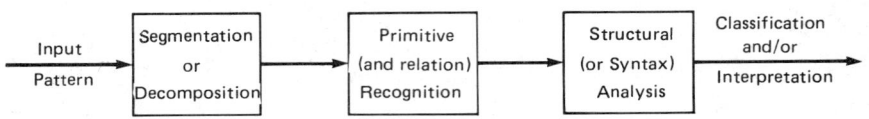

Figure 1 Block diagram of syntactic pattern recognition system

respresent the subpatterns and the branches represent the relations between subpatterns. A block diagram of syntactic pattern recognition system is given in Figure 1.

2 PRIMITIVE SELECTION AND PATTERN GRAMMARS

Since pattern primitives are the basic components of a pattern, presumably they are easy to recognize. Unfortunately, this is not necessarily the case in some practical applications. For example, strokes are considered good primitives for script handwriting, and so are phonemes for continuous speech, however, neither strokes nor phonemes can easily be extracted by machine. A compromise between its use as a basic part of the pattern and its easiness for recognition is often required in the process of selecting pattern primitives.

There is no general solution for the primitive selection problem at this time. For line patterns or patterns described by boundaries or skeletons, line segments are often suggested as primitives. A straight line segment could be characterized by the locations of its beginning (tail) and end (head), its length, and/or slope. Similarly, a curve segment might be described in terms of its head and tail and its curvature. The information characterizing the primitives can be considered as their associated semantic information or as features used for primitive recognition. Through the structural description and the semantic specification of a pattern, the semantic information associated with its subpatterns or the pattern itself can then be determined. For pattern description in terms of regions, half-planes have been proposed as primitives. Shape and texture measurements are often used for the description of regions.

After pattern primitives are selected, the next step is the construction of a grammar (or grammars) which will generate a language (or languages) to describe the patterns under study. It is known that increased descriptive power of a language is paid for in terms of increased complexity of the syntax analysis system (recognizer or acceptor). Finite-state automata are capable of recognizing finite-state languages although the descriptive power of finite-state languages is also known to be weaker than that of context-free and context-sensitive languages. On the other hand, nonfinite, nondeterministic procedures are required, in general, to recognize languages generated by context-free and context-sensitive grammars. The selection of a particular grammar for pattern description is affected by the primitives selected, and by the tradeoff between the grammar's descriptive power and analysis efficiency.

A number of special languages have been proposed for the description of patterns such as English and Chinese characters, chromosome images, spark chamber pictures, two-dimensional mathematics, chemical structures, spoken words, and fingerprint patterns [3,7,8]. For the purpose of effectively describing high dimensional patterns, high dimensional grammars such as web grammars, array grammars, graph grammars and tree grammars have been used for syntactic pattern recognition [8,15,19,26,27].

Ideally speaking, it would be nice to have a grammatical (or structural) inference machine which would infer a grammar from a given set of patterns. Unfortunately, not many convenient grammatical inference algorithms are presently available for this purpose. Nevertheless, recent literatures have indicated that some simple grammatical inference algorithms have already been applied to syntactic pattern recognition, particularly through man-machine interaction [8].

3 SYNTACTIC PATTERN ANALYSIS USING STOCHASTIC LANGUAGES

In some practical applications, a certain amount of uncertainty exists in the process under study. For example, due to the presence of noisy and variation in the pattern measurements, segmentation error and primitive recognition error may occur, causing ambiguities in the pattern description languages. In order to describe noisy and distorted patterns under ambiguous situations, the use of stochastic languages has been suggested [5,8]. With probabilities associated with grammar rules, a stochastic grammar generates sentences with a probability distribution. The probability distribution of the sentences can be used to model the noisy situations.

By associating probabilities with the sentences, we can impose a probabilistic structure on the language to describe noisy patterns. The probability distribution characterizing the patterns in a class can be interpreted as the probability distribution associated with the sentences in a language. Thus, statistical decision rules can be applied to the classification of a pattern under ambiguous situations (for example, use the maximum-likelihood or Bayes decision rule). Furthermore, because of the availability of the information about production probabilities, the speed of syntactic analysis can be improved through the use of this information [5]. Of course, in practice, the production probabilities will have to be inferred from the observation of relatively large numbers of pattern samples [8]. When the imprecision and uncertainty involving in the pattern description can be modeled by using the fuzzy set theory, the use of fuzzy languages for syntactic pattern recognition has recently been suggested [8,11].

4 SYNTACTIC RECOGNITION

Conceptually, the simplest form of recognition is probably "template-matching." The sentence describing an input pattern is matched against sentences representing each prototype or reference pattern. Based on a selected "matching" or "similarity" criterion, the input pattern is classified in the same class as the prototype pattern which is the "best" to match the input. The structural information is not recovered. If a complete pattern description is required for recognition, a parsing or syntax analysis is necessary. In between the two extreme situations, there are a number of intermediate approaches. For example, a series of tests can be designed to test the occurrence or nonoccurrence of certain subpatterns (or primitives) or certain combinations of them. The result of the tests, through a table lookup, a decision tree, or a logical operation, is used for a classification decision. Recently, the use of discriminant grammars has been proposed for the classification of syntactic patterns [22].

There are many parsing algorithms proposed for context-free languages [8]. A parsing procedure for recognition is, in general, nondeterministic and, hence, is regarded as computationally inefficient. Efficient parsing could be achieved by using special classes of languages such as finite state and deterministic languages for pattern description. The tradeoff here between the descriptive power of the pattern grammar and its parsing efficiency is very much like that between the feature space selected and the classifier's discrimination power in a decision-theoretic recognition system. Special parsers using sequential procedures or other heuristic means for efficiency improvement in syntactic pattern recognition have recently been constructed [25,30,33].

In practical applications, pattern distortion and measurement noise often exist. Pattern segmentation errors and misrecognitions of primitives (and relations) and/or subpatterns will lead to erroneous or noisy sentence rejection by the grammar characterizing its class. Recently, the use of an error-correcting parser as a recognizer of noisy and distorted patterns has been proposed [8,31,34]. In the use of an error-correcting parser as a recognizer, the pattern grammar is first expanded to include all the possible errors into its productions. The original grammar is transformed into a covering grammar that generates not only the correct sentences, but also all the possible erroneous sentences. For string grammars, three types of error— substitution, deletion and insertion—are considered. Misrecognition of primitives (and relations) are regarded as substitution errors, and segmentation errors as deletion and insertion errors.

A distance between two strings is defined in terms of the minimum number of error transformations used to derive one from the other by Aho and Peterson [1]. When the error transformations are defined in terms of substitution, deletion and insertion errors, the distance measurement coincides with the definition of Levenshtein metric [18]. A weighted Levenshtein distance can be defined by assigning nonnegative weights to the three transformations respectively. In addition, a weighted metric that would reflect the difference of the same type of error made on different terminals has also been proposed [8].

For a given input string y and a given grammar G, a minimum-distance error-correcting parser† (MDECP) is an algorithm that searches for a sentence z in $L(G)$ such that the distance between z and $y, d(z,y)$ is the minimum among the distances between all the sentences in $L(G)$ and y. The algorithm also generates the value of $d(z,y)$. We simply define this value to be the distance between $L(G)$ and y and denote it as $d_1(L(G),y)$. We note that the minimum-distance correction of y is y itself if $y \epsilon L(G)$. The sequential parsing procedure suggested by Persoon and Fu [25] has been applied to error-correcting parser to reduce the parsing time [8]. In addition, error-correcting parsing for transition network grammars and tree grammars has also been studied [8]. Another approach to reduce the parsing time is the use of parallel processing [2].

---

†If the pattern grammar is stochastic, the maximum-likelihood and Bayes criteria can be applied [5,8].

As the distance between a string (a syntactic pattern) and a language (a set of syntactic patterns) is defined, a minimum-distance decision rule can be stated as follows: suppose that there are two classes of patterns, C_1 and C_2 characterized by grammar G_1 and G_2 respectively. For a given syntactic pattern y with unknown classification, decide $y \in \genfrac{}{}{0pt}{}{C_1}{C_2}$ if $d(L(G_1),y) \lessgtr d(L(G_2),y)$.

In statistical pattern recognition, a pattern is represented by a vector, called a feature vector. The similarity between two patterns can often be expressed by a distance, or more generally speaking, a metric in the feature space. Cluster analysis can be performed on a set of patterns on the basis of a selected similarity measure [6]. In syntactic pattern recognition a similarity measure between two syntactic patterns must include the similarity of both their structures and primitives. In this section, we have discussed distance measures for strings, which leads to the study of clustering analysis for syntactic patterns. The conventional clustering methods, such as, the minimum spanning tree, the nearest (or K-nearest) neighbor classification rule and the method of clustering centers can be extended to syntactic patterns [8].

5 SYNTACTIC APPROACH TO SHAPE AND TEXTURE ANALYSIS

Recently, syntactic methods have been applied to both shape description and recognition. Pavlidis and Ali [24] have proposed a general model of syntactic shape analyzer. The first major component of the model is a curve-fitting algorithm which achieves the noise elimination and data reduction. The split-and-merge algorithm is used to obtain a polygonal approximation of the boundary of the original picture or object. It is assumed that the boundaries of the objects of interest consist of concatenations of the following subpatterns or nonterminals: QUAD (arcs approximated by a quadric curve), TRUS (sharp protrusions or intrusions), LINE (long line segments), and BREAK (short segments with no regular shape). Each of the nonterminals has a set of attributes as its semantic information. The production rules of the proposed general shape grammar consist of both syntactic and semantic rules. Stochastic finite automata are used as parsers for shape recognition.

Another method recently proposed for syntactic shape description and recognition is the use of attributed grammars [9,33]. Two types of primitive with attributes are proposed. The first type is a curve segment with its direction (the vector from the starting point to the end point), total length, total angular change, and a measure of its symmetry as the four attributes. The second type of primitive is an angle primitive with its attribute specified by the angular change at the concatenating point of two consecutive curve segments. Finite-state and context-free attributed grammars are used for shape description and recognition. Each production rule of the attributed grammar has a symbolic part like the conventional grammar rule and a semantic part for processing the attributes of the terminals and nonterminals in the symbolic part. The primitive extraction process is embedded in the parsing of the strings describing the boundaries of objects. Modified Earley parser and finite automata are used as shape recognizers.

One important property of pictures is texture. Recently, a syntactic approach to texture analysis and discrimination has been proposed [8]. A texture pattern is divided into fixed-size subpictures or windows. Using the gray level of a pixel or of a small array of pixels as primitives, we can represent each window by a tree with a prespecified tree structure. A tree grammar is used to characterize windowed patterns of the same texture. Since the windowed patterns are also a part of the global structure of the texture, one or more higher level tree structures can be employed to describe the arrangement of windowed patterns. Error-correcting tree automata (SPECTRA) constructed according to the texture (tree) grammars can be applied for texture discrimination. Stochastic tree grammars have been suggested for the modelling of noisy and distorted texture patterns.

6 APPLICATIONS TO WAVEFORM AND SIGNAL PROCESSING

Syntactic pattern recognition has been applied to waveform analysis, ECG interpretation, speech recognition and understanding, character recognition, fingerprint classification, recognition of two-dimensional mathematical notation, modelling of Earth Resources Satellite data, machine parts recognition and automatic visual inspection [3,7,8,12-15,19,21]. We briefly review some of the recent applications of syntactic methods to waveform and signal processing.

Waveforms are basically one-dimensional signals, which appear to be naturally suitable for the application of syntactic methods. A waveform could be represented by a concatenation of waveform segments. However, the selection of primitives (basic waveform segments) and subpatterns could be quite different from different application points of view. Linear and/or quadric segments through functional approximation have proposed as waveform primitives [8]. Ehrich and Faith [4] have proposed the use of a relational tree to describe a waveform in terms of its peaks and valleys. Sankar and Rosenfeld [28] have recently proposed an alternative method of using a peak relational tree in terms of fuzzy connectivity to describe waveforms.

Horowitz [6] has proposed a deterministic context-free grammar that can be used to recognize positive and negative peaks in a waveform represented by a string of "positive slope," "negative slope" and "zero slope" primitives. Stockman et al., have suggested the use of a waveform parsing system to analyze carotid pulse waves [30]. Le Chevalier et al., [17] have proposed the use of syntactic decoding method (error-correcting string parser with substitution error only) for syntactic signal processing. Waveform segments (e.g., segments of sinusoids) are selected as primitives. A signal is described as a sequence (concatenation) of primitives. In addition to the distance suggested in error-correcting parsing, the correlation is also used as a similarity measure between two strings. Practical applications include signal detection, radar target identification and adaptive antenna processing.

Giese, et al., [12] have applied a syntactic method to the analysis of electroencephalogram (EEG). A typical EEG pattern consists of 100-sec 4-channel waveform, and each 100-sec waveform is segmented into 1-sec segments. Each 1-sec segment is considered a primitive, and seven different classes of primitives are identified. The recognition of primitives is accomplished from seventeen features by a linear classifier. A context-free grammar is used to describe various EEG patterns and a simple bottom-up parser has been constructed for normal vs abnormal recognition. Recently, Liu and Fu [20] have applied the syntactic approach to seismic pattern classification. 321 seismic records recorded at LASA in Montana were used for the discrimination between nuclear explosion and earthquake. Each record is decomposed into 20 segments and each segment is classified into one of the ten classes using zero-crossing count and log energy as features. Classification of a complete record is performed using a nearest-neighbor decision rule on the basis of the string-to-string distance defined in Section 4.

7 CONCLUDING REMARKS

We have briefly reviewed some recent advances in the area of syntactic pattern recognition. Due to noise and distortions in real world patterns, syntactic approach to pattern recognition was regarded earlier as only effective in handling abstract and artificial patterns. However, with the recent development of distance or similarity measures between syntactic patterns and error-correcting parsing procedures, the flexibility of syntactic methods has been greatly expanded. Errors occurring at the lower-level processing of a pattern (segmentation and primitive recognition) could be compensated at the higher level using structural information. Using a distance or similarity measure, nearest-neighbor and k-nearest-neighbor classification rules can be easily applied to syntactic patterns. Furthermore, with a distance or similarity measure, a clustering procedure can be applied to syntactic patterns. Such a nonsupervised learning procedure can also be very useful for grammatical inference in syntactic pattern recognition [8].

It has been noticed from the recent advances that semantic information has been used more and more with the syntax rules in characterizing patterns. Quite often, semantic information involving spatial information can be expressed syntactically such as attributed grammars, and relational trees and graphs [9,24,33]. Parsing efficiency has become a concern in syntactic recognition. Special grammars and parallel parsing algorithms have been suggested for speeding up the parsing time. Structural information of an image can also be used as a guide in the segmentation process through the syntactic approach [15,32]. On the other hand, simple fixed-size segmentation procedures are often used in syntactic pattern recognition in spite of the fact that the application of these extremely simple procedures may result in unnatural subpatterns and primitives [12,20]. Syntactic representation of patterns such as hierarchical trees and relational graphs should also be very useful for database organization. Several recent publications have already shown such a trend [10,16].

REFERENCES

[1] A.V. Aho and T.G. Peterson, "A Minimum Distance Error-Correcting Parser for Context-Free Languages," *SIAM Journal on Computing,* Vol. 4, Dec. 1972.

[2] N.S. Chang and K.S. Fu, "Parallel Parsing of Tree Languages," Proc. 1978 IEEE Computer Society Conference on Pattern Recognition and Image Processing, May 31-June 2, Chicago, Ill.

[3] R. DeMori, "On Speech Recognition and Understanding" in *Pattern Recognition: Theory and Application* ed. by K.S. Fu and A.B. Whinston, Noordhoff Publ. Co., Leyden, Netherlands, 1977.

[4] R.W. Ehrich and J.P. Foith, "Representation of Random Waveforms by Relational Trees," *IEEE Trans. on Computers*, Vol. C-25, July 1976, pp. 725-736.

[5] K.S. Fu, *Syntactic Methods in Pattern Recognition*, Academic Press, 1974.

[6] K.S. Fu, *Digital Pattern Recognition*, Springer-Verlag, 1976.

[7] K.S. Fu, *Syntactic Pattern Recognition Applications*, Springer-Verlag, 1977.

[8] K.S. Fu, *Syntactic Pattern Recognition and Applications*, Prentice-Hall, 1982.

[9] K.S. Fu, "Attributed Grammars for Pattern Recognition—A General (Syntactic-Semantic) Approach," Proc. 1982 IEEE Computer Society Conference on Pattern Recognition and Image Processing, June 14-17, Las Vegas.

[10] K.S. Fu and N.S. Chang, "An Integrated Image Analysis and Image Database Management System," Proc. 1981 COMPCON Fall, September, Washington, D.C.

[11] K.S. Fu and A. Rosenfeld, "Pattern Recognition and Image Processing," *IEEE Trans. on Computers*, Vol. C-25, No. 12, 1976.

[12] D.A. Giese, J.R. Bourne, and J.W. Ward, "Syntax Analysis of Electroencephalogram," *IEEE Trans. on Systems, Man and Cybernetics*, Vol. SMC-9, August 1979.

[13] R. Jakubowski and A. Kasprzak, "A Syntactic Description and Recognition of Rotary Machine Elements," *IEEE Trans. on Computers*, Vol. C-26, No. 10, Oct. 1977, pp. 1039-1042.

[14] J.F. Jarvis, "Regular Expressions as a Feature Selection Language for Pattern Recognition," Proc. Third International Joint Conference on Pattern Recognition, Nov. 8-11, 1976, Coronado, Calif., pp. 189-192.

[15] J. Keng and K.S. Fu, "A Syntax-Directed Method for Land-Use Classification of LANDSAT Images," Proc. Symposium on Current Mathematical Problems in Image Science, Nov. 10-12, 1976, Monterey, Calif.

[16] T. Kunii, S. Weyle and J.M. Tenenbaum, "A Relational Data Base Scheme for Describing Complex Pictures with Color and Texture," Proc. Second International Joint Conference on Pattern Recognition, August 13-15, 1974, Copenhagen, Denmark.

[17] F. LeChevalier, G. Bobillot and C. Fugier-Garrel, "Syntactic Signal Processing," 1978 International Symposium on Information Theory, Oct. 10-14, Ithaca, N.Y.

[18] V.I. Levenshtein, "Binary Codes Capable of Correcting Deletions, Insertions and Reversals," *Sov. Phys. Dokl.,* Vol. 10, Feb. 1966.

[19] R.Y. Li and K.S. Fu, "Tree System Approach to LANDSAT Data Interpretation," Proc. Symposium on Machine Processing of Remotely Sensed Data, June 29—July 1, 1976, Lafayette Indiana.

[20] H.H. Liu and K.S. Fu, "A Syntactic Approach to Seismic Pattern Recognition," *IEEE Trans. Pattern Analysis and Machine Intelligence,* Vol. PAMI-4, March 1982.

[21] J.L. Mundy and R.E. Joynson, "Automatic Visual Inspection Using Syntactic Analysis," Proc. 1977 IEEE Computer Society Conference on Pattern Recognition and Image Processing, June 6-8, Troy, N.Y.

[22] C. Page and A. Filipski, "Discriminant Grammars, An Alternative to Parsing for Pattern Classification," Proc. 1977 IEEE Workshop on Picture Data Description and Management, April 20-22, Chicago, Ill.

[23] T. Pavlidis, "Linear and Context-Free Graph Grammars," *J. ACM,* Vol. 19, 1972, pp. 11-22.

[24] T. Pavlidis and F. Ali, "A Hierarchical Syntactic Shape Analysis," *IEEE Trans. Pattern Analysis and Machine Intelligence,* Vol. PAMI-1, January 1979.

[25] E. Persoon and K.S. Fu, "Sequential Classification of Strings Generated by SCFG's," *International Journal of Computers and Information Sciences,* Vol. 4, Sept. 1975.

[26] J.L. Pfaltz and A. Rosenfeld, "Web Grammars," Proc. First International Joint Conference on Artificial Intelligence, Washington, D.C., 1969.

[27] A. Rosenfeld, *Picture Languages,* Academic Press, 1979.

[28] P.V. Sankar and A. Rosenfeld, "Hierarchical Representation of Waveforms, " TR-615, Computer Science Center, University of Maryland, College Park, Md. 20742, Dec. 1977.

[29] A.C. Shaw, "Picture Graphs, Grammars and Parsing, " in *Frontview of Pattern Recognition* ed. by S. Watanabe, Academic press, 1972.

[30] G. Stockman, L.N. Kanal and M.C. Kyle, "Structural Pattern Recognition of Carotid Pulse Waves Using a General Waveform Parsing System," *Comm. ACM,* Vol. 19, No. 12, Dec. 1976, pp. 688-695.

[31] M.G. Thomason and R.C. Gonzalez, "Error Detection and Classification in Syntactic Pattern Structures," *IEEE Trans. on Computers,* Vol. C-24, 1975.

[32] S. Tsuji and R. Fujiwara, "Linguistic Segmentation of Scenes into Regions," Proc. Second International Joint Conference on Pattern Recognition, August 13-15, 1974, Copenhagen, Denmark.

[33] K.C. You and K.S. Fu, "A Syntactic Approach to Shape Recognition Using Attributed Grammars," *IEEE Trans. on Systems, Man and Cybernetics,* Vol. SMC-9, June 1979.

[34] K.C. You and K.S. Fu, "Distorted Shape Recognition Using Attributed Grammars and Error-Correcting Techniques," *Computer Graphics and Image Processing,* Vol. 13, 1980, pp. 1-16.

Application of Map Estimation Techniques to Image Segmentation*

Howard Elliott and M. F. Tenorio

Department of Electrical and Computer Engineering
University of Massachusetts
Amherst, MA

Fred R. Hansen

System Studies Department
Sandia National Laboratories
Livermore, CA

Lalita Srinivasan

Department of Electrical Engineering
Colorado State University
Fort Collins, CO

1 INTRODUCTION

Systems for computer vision, that is systems which use computers for automatically extracting information from digitized images, have many applications. They can be used to analyze medical data, for nondestructive testing, and for control of robots and manipulators. Another important application involves analysis of images from space based sensors. This latter application poses problems which do not arise in many other applications. For example environmental conditions, e.g., lighting, object contrast, etc., can often be controlled in industrial applications. Furthermore, there is usually a single sensor generating relatively small amounts of data for analysis. On the other hand, remote space based systems can generate enormous amounts of multi-sensor data, and there is little or no control of environmental conditions. In this paper we present

---

*This work was supported by the Office of Naval Research under Contract N00014-82-K-0076.

some algorithms for image segmentation, a basic component process of computer vision systems. Simply, image segmentation algorithms partition an image into regions with similar properties of features. The algorithms discussed are geared toward segmentation of remotely sensed data. In particular, they can work with multi-sensor data, and with high levels of noise corruption.

In order to work at low signal to noise ratios, spatial dependence of pixel intensity has been exploited by modelling the true scene as a Markov field. In the case of one dimensional digital signals the Markov field model reduces to a Markov chain. Markov chain models for stochastic processes have received considerable attention in the statistical literature [1], [2], and found extensive application in the control, communication, and signal processing fields [3]-[6]. Estimation and detection problems associated with random signals modelled by Markov chains can be formulated as likelihood maximization problems where one wants to maximize the joint likelihood of the data and the Markov state sequence. Maximization of this joint likelihood is completely equivalent to generating what is known as the the maximum *a posteriori* probability (MAP) estimate of the Markov chain [18]. In the one dimensional case, this leads to elegant and reasonably efficient dynamic programming algorithms for computing the estimate which maximizes the joint maximum *a posteriori* probability of the entire data string (digital signal) in a sequential manner. Unfortunately, this approach does not generalize in a natural way to the case of two dimensional signals such as digital images. As a result, this has somewhat hampered the use of MAP formulations in image processing. The few exceptions [7]-[12] are discussed more carefully below.

Recent work by Kaufman and co-workers, [7], and Therrien [8], [9] have also made use of Markov field models and MAP formulation. In [7] this approach was combined with reduced update Kalman filtering techniques for image enhancement, while in [8] and [9] it was combined with two-dimensional autoregressive texture models for texture based segmentation. However, the algorithm described in [7] and one of the algorithms described in [8] and [9] fail to exploit the true spatial dependence imposed by the model. In particular they don't attempt to maximize a joint likelihood of all the data but rather the individual likelihoods at points or in small regions within the image. This problem is partially overcome by a second multi-pass or iterative algorithm proposed in [8] and [9]. However, the relation between the estimate obtained using this approach and a true joint MAP estimate is unclear.

Cooper and Elliott [10]-[12] have applied MAP techniques to boundary estimation in noisy images as well. Although boundary estimation can be formulated as a one-dimensional signal estimation problem where the independent variable is arc length along the boundary, it is interesting to note that even in this case the two dimensional image data necessitated the use of a suboptimal algorithm.

We have taken a very simple Markov field model, and have developed an algorithm which approximates the behavior of an optimal sequential estimation algorithm. It makes use of two stages of dynamic programming. In the first stage a "generalized" dynamic programming algorithm is applied to each row of the image. This yields a set of candidate segmentations for each row. In the second stage, a final segmentation is "pieced" together from the candidate row segmentations using dynamic programming as well. The algorithm requires

Image Segmentation

only a single pass over the image data, and the version for moderate signal to noise ratios can be performed in a highly parallel fashion. The version which works best at low signal to noise ratios requires a sequential raster processing of the image. This paper presents a generalization of the results presented in [19]. That paper was geared completely toward two region or binary scenes. Here we present a very general formulation and show results of applying the algorithm to four region scenes. We also briefly discuss the problem of parameter estimation for algorithm initialization.

Section II presents an image model and formulates the segmentation problem as a MAP estimation problem. The algorithms are presented in Section 3, and examples of its performance are given in Section 4. Finally Section 5 contains concluding remarks.

2 MODELLING ASSUMPTIONS AND PROBLEM FORMULATION

Let a set of digitized observations of a scene be characterized by an $N_1 \times N_2$ matrix of pixel intensities $G_k = [g_{ij}^k]$. In general we will assume K data measurements for each pixel. These can correspond to measurements taken directly from different sensors or correspond to features obtained by preprocessing the raw data. The latter approach can be helpful when dealing with textured images. Thus the complete set of observations will be contained in the sequence of matrices $G = \{G_k\}_{k=1}^K$.

Assume the scene to be composed of M region types, and the observations corrupted by region dependent additive noise. If pixel (i,j) is in region type m, then g_{ij}^k will be modelled as

$$g_{ij}^k = b_{ij}^k + n_{ij}^k$$

where $\{b_{ij}^k\}$ and $\{n_{ij}^k\}$ are stochastic fields characterizing the underlying scene, and the observation noise, respectively, in data set k. In this report we make the simplifying assumption that each region type in each data set is characterized by a constant intensity, r_{km}, i.e., $b_{ij}^k = r_{km}$ if pixel (ij) is in region type m. Furthermore we assume the additive noise field to be spatially uncorrelated, and Gaussian so that the vector of observation noise

$$n_{ij} = [n_{ij}^1, n_{ij}^2, \ldots, n_{ij}^K]^T \quad (1)$$

is multivariate normal with mean zero and covariance matrix C_m in region type m. This implies that the observation vector

$$g_{ij} = [g_{ij}^1, g_{ij}^2, \ldots g_{ij}^K]^T \quad (2)$$

will be multivariate normal with mean

$$r_m = [r_{1m}, r_{2m}, \ldots, r_{Km}]^T \quad (3)$$

and covariance C_m if pixel (i,j) is in region type m.

For segmentation one wants to generate an estimate $\hat{S}$ of the sets $S = \{S_m\}_{m=1}^M$ where

$$S_m = \{(i,j) : b_{ij}^k = r_{km}\}. \quad (4)$$

Thus the output of the algorithm would be an M-level image matrix $\hat{B} = [\hat{b}_{ij}]$ where $\hat{b}_{ij} = m$ if S_m contains pixel (i,j). Define

$$p_m(x) = (2\pi)^{-K/2}|C_m|^{-\frac{1}{2}} \exp((x-r_m)^T C_m^{-1}(x-r_m)). \tag{5}$$

Classical maximum likelihood segmentation methods would assign pixel (i,j) to region set S_m if

$$p_m(g_{ij}) \geq P_l(g_{ij}) \; \forall \; l, \; 1 \leq \ell \leq M. \tag{6}$$

This approach is adequate if the observations are relatively noise free.

In the noisy case, to develop robust segmentation procedures one needs to incorporate spatial continuity of pixel intensities into the data model. One approach might involve preprocessing to average or smooth the data. We've chosen to incorporate spatial continuity by modelling the underlying scene as Markov field with M states. Analogous Markov chain models for one dimensional data have received considerable attention in the control, communication and signal processing literature [4]-[6], [15]. For a scene, the Markov field is defined by the relation

$$p(b_{ij}^k = r_{km}, 1 \leq k \leq K \mid b_{rs}^k, 1 \leq r \leq N_1, 1 \leq s \leq N_2, (r,s) \neq (i,j), 1 \leq k \leq K)$$
$$= p(b_{ij}^k = r_{km}, 1 \leq k \leq K \mid b_{rs}^k, (r,s)\epsilon\eta_{ij}, 1 \leq k \leq K) \tag{7}$$
$$= P_{ijm}$$

where the support η_{ij} is a local neighborhood of pixel (i,j). In this presentation we will limit ourself to the simple causal support

$$\eta_{ij} = ((i-1,j), (i,j-1)). \tag{8}$$

This will allow development of an algorithm which processes the data in a raster scan fashion from left to right and top to bottom.

With this model one can formulate the segmentation problem as a maximum a posteriori probability (MAP) estimation problem. In particular, let $l(\;)$ represent a log-likelihood function. One would then like to find the estimate $\hat{S}$ which maximizes the conditional likelihood,

$$l(\hat{S}/G) = l(G/\hat{S}) + l(\hat{S}) - l(G). \tag{9}$$

Or since $l(G)$ is independent of $\hat{S}$, more simply one can maximize

$$l(\hat{S},G) = L(G/\hat{S}) + l(\hat{S}). \tag{10}$$

In this case,

$$l(G/\hat{S}) = \sum_{m=1}^{M} \sum_{(i,j)\epsilon S_m} \ln p_m(g_{ij}) \tag{11}$$

and

$$l(\hat{S}) = \sum_{m=1}^{M} \sum_{(i,j)\epsilon S_m} \ln P_{ijm} \tag{12}$$

where p_m and P_{ijm} are defined in (5) and (7) respectively.

In the next section a sequential, suboptimal algorithm is presented for generating an estimate which approximates the true MAP estimate obtained by maximizing (10)-(12). The algorithm makes use of dynamic programming techniques.

3 SUBOPTIMAL SEQUENTIAL SEGMENTATION

Dynamic programming can be applied in a straightforward manner to obtain true MAP estimates of one dimensional or singly indexed Markov chains. In order to apply it to doubly indexed Markov fields one must make approximations so that the data can be processed in a one dimensional manner. On the other hand, one also wants to make use of the two dimensional nature of the data to obtain reasonable performance at low or moderate signal to noise ratios. The algorithm described below represents a compromise. It has two stages. During the first stage, the data in each image row is processed using dynamic programming techniques to obtain Q likely candidate segmentations for each row. These are then pieced together into a single segmentation of the entire image by use of a second stage of dynamic programming.

To develop the algorithm we make the following key approximation. Recall that the Markov transition probabilities were defined as

$$P_{ijm} = P(b_{ijm}^k = r_{km}, 1 \leq k \leq K | b_{i,j-1}^k, b_{i-1,j}^k, 1 \leq k \leq K).$$

Let us approximate P_{ijm} by $\tilde{P}_{ijm}$ defined by the relation

$$\tilde{P}_{ijm} = R_{ijm} C_{ijm} \tag{13a}$$

$$C_{ijm} = P(b_{ij}^k = r_{km}, 1 \leq k \leq K | b_{i-1,j}^k, 1 \leq k \leq K) \tag{13b}$$

$$R_{ijm} = P(b_{ij}^k = r_{km}, 1 \leq k \leq K | b_{i,j-1}^k, 1 \leq k \leq K). \tag{13c}$$

R_{ijm} represents a transition probability from the preceding pixel in the same row, while C_{ijm} represents the transition probability from the preceding pixel in the same column.

Using (13), (10) can be approximated by

$$\tilde{l}(\hat{S}, G) = \tilde{l}_R(\hat{S}, G) + \tilde{l}_C(\hat{S}) \tag{14}$$

$$\tilde{l}_R(\hat{S}, G) = \sum_{m=1}^{M} \sum_{(i,j) \in S_m} (\ln p_m(g_{ij}) + \ln R_{ijm}) \tag{15}$$

$$\tilde{l}_C(\hat{S}) = \sum_{m=1}^{M} \sum_{(i,j) \in S_m} (\ln C_{ijm}) \tag{16}$$

Let us next introduce the notation

$G_R(i,J) \triangleq$ Observations of first J pixels in row i

$\hat{S}_R(i,J) \triangleq$ Segmentation of first J pixels in row i

$G(I) \triangleq$ Observations of first I rows of scene

$\hat{S}(I) \triangleq$ Segmentation of first I rows of scene.

For row $i, l_R(\cdot)$ can then be recursively written as

$$\tilde{l}_R(\hat{S}_R(i,J), G_R(i,J)) = l_R(\hat{S}_R(i,J-1), G_R(i,J-1)) + \ln R_{iJm} + \ln p_m(g_{iJ}) \tag{17}$$

assuming pixel (i,J) is assigned to set S_m.

Similarly we can see that the total likelihood can be recursively written as

$$\tilde{l}(\hat{S}(I), G(I)) = \tilde{l}(\hat{S}(I-1), G(I-1)) + \tilde{l}_c(\hat{S}_R(I,N_2), G_R(I,N_2)) \quad (18)$$

An Algorithm for Moderate Signal to Noise Ratios

Recursions (17) and (18) are of the form necessary for maximization via dynamic programming. For moderate signal to noise ratio, which will be defined below, one can implement the following algorithm.

Stage 1 Apply a generalization of dynamic programming to each row to generate the Q most likely segmentations using the recursive likelihood (17).

Stage 2 Apply standard dynamic programming to the Q candidate segmentations of each row using recursion (18) to obtain the final segmentation.

More details regarding this algorithm, and in particular the dynamic programming can be found in [19].

One advantage of this algorithm is its potential for parallel implementation. In particular, during the first stage each row can be processed in parallel. However, as will be demonstrated in the examples section to follow, to improve performance at low signal to noise ratios modifications are necessary which require processing of the rows sequentially.

Modified Algorithm for Low Signal to Noise Ratios

For lower signal to noise ratio images two modifications are necessary. The first involves incorporating information from the previous row during the first stage row segmentation. The second involves post filtering the segmented image to remove directional burst type errors.

First, to incorporate previous row information into the algorithm stages 1 and 2 are interweaved. That is after processing rows one and two of the image a single step of the stage 2 algorithm is performed. Then after processing row 3 another step of the stage 2 algorithm is performed, etc. Furthermore in processing all but the first row, transition probabilities from the most likely segmentation of the previous row, as determined by the last step of the stage 2 algorithm, are incorporated into the row likelihood expression. After determining the Q most likely segmentations for a row in this manner, these transition probability terms are then subtracted out before performing the next stage 2 step. This ensures that column transitions are never counted twice during likelihood calculations. The column transition probabilities used during stage 1 are not necessarily the same as those used during stage 2. More will be said regarding choice of transition probabilities in the next section.

The second modification involves incorporation of a third stage into the algorithm. After obtaining an m-level segmented image, $\hat{B}$, using the above procedure, a majority filter is passed across this image. This filter places a (5×1) template over each pixel in the image $\hat{B}$, and reassigns the value of that pixel to the value occuring most frequently in that template. This tends to remove burst type errors which occur during the stage 1 row segmentations. Burst errors are a well known phenomena in MAP estimation.

Calculating Transition Probabilities

In order to implement this algorithm one must determine two ($M \times M$) transition probability matrices giving the probabilities of successive pixels changing state along a row, R_{ijm}, and along a column C_{ijm}. In [19], results of an analytical study were presented for two region images ($M = 2$). It showed how appropriate values can be calculated as a function of object sizes and signal to noise ratio. The main tradeoffs involve detecting small objects versus introducing "ghost" objects. In the next section we will show results for the four region case ($M = 4$). Although we are presently looking into analytical methods for calculating probabilities in this case, for these examples we tried to use our intuition in extending the results presented in [19] for the two level case. First, we assumed no preference between level changes. Thus a change between two very different intensity levels had the same probability as a change between two more similar intensity levels. In practice one would attempt to use a priori information in weighting such transitions. It was shown in [19] that at low signal to noise ratios one needed to choose the probability of no state change to be much larger than the probability of a change. It was this basic notion that was extended to the four region case. It might be noted that this algorithm reduces to simple maximum likelihood, or threshold segmentation if all transitions are weighted equally.

Estimating Image Statistics

For implementation of the algorithm the mean vectors, r_m, and covariance matrices C_m, $1 \leq m \leq M$ must be estimated. Thus far we have considered the special case of independent data sets so that

$$C_m = \text{diag}\,(\sigma_{km}^2).$$

We have then attempted to consider each data matrix G_k as a set of samples from a population mixture and applied the well known method of moments for estimating the mixture parameters. This works well for images of just two or three region types down to a signal to noise ratio of 0 dB. Although still in the experimental stage, our approach for 4 or more region types involves use of a hierarchical procedure which first partitions the data into subsets each containing data which is a mixture of three or fewer populations. The method of moments is then applied to each subset.

4 EXAMPLES

In this section some examples are given to show the performance of the algorithm in segmenting images with $M = 4$ region types. Figures 1a, b and 2a, b compare the use of the modified MAP algorithm to classical maximum likelihood or threshold segmentation. Two geometric patterns were used, one containing overlapping disc patterns and the other containing disjoint diamond patterns. Each picture shows the clean scene (upper left), the noise corrupted version (upper right), the segmented scene (lower left), and the result of post filtering the segmented scene through a majority filter (lower right). For the

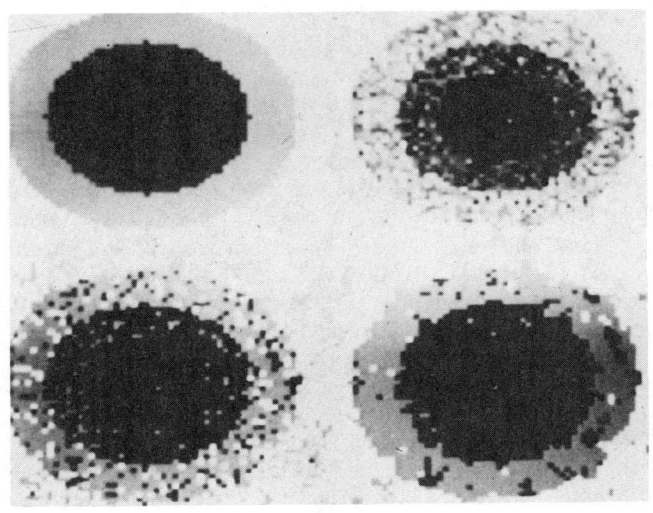

Figure 1 Application of thresholding and post filtering to geometric patterns at S/N=2

Image Segmentation

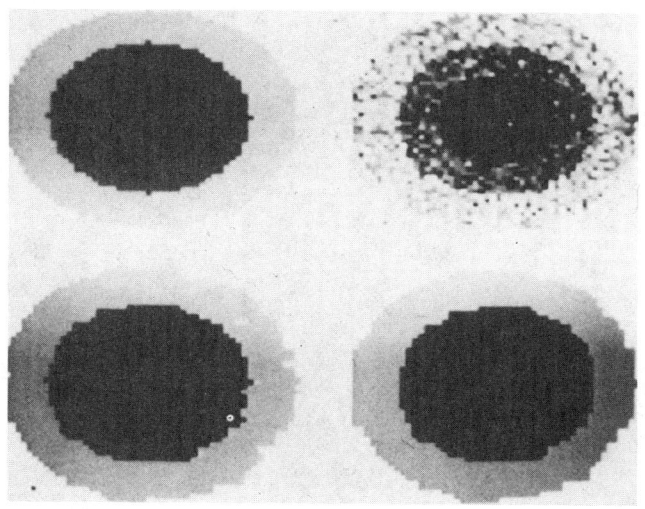

Figure 2 Application of MAP algorithm and post filtering to geometric patterns at S/N=2

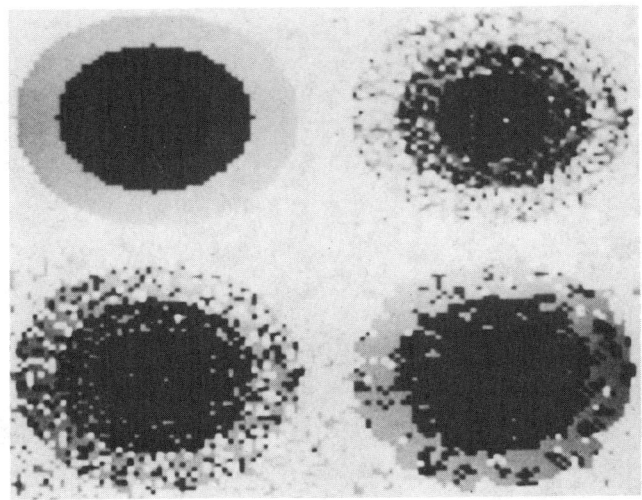

Figure 3 Application of thresholding and post filtering to circular pattern at S/N=1.5

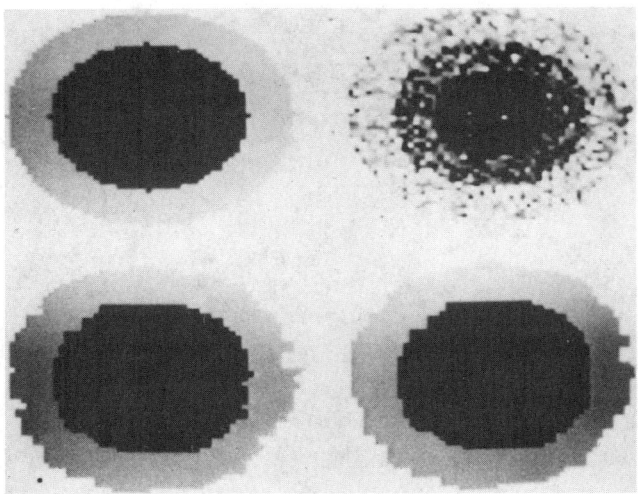

Figure 4 Application of MAP algorithm and post filtering to circular pattern at S/N=1.5

Image Segmentation 395

Figure 5 Application of MAP algorithm and post filtering to diamond pattern at S/M=1.5

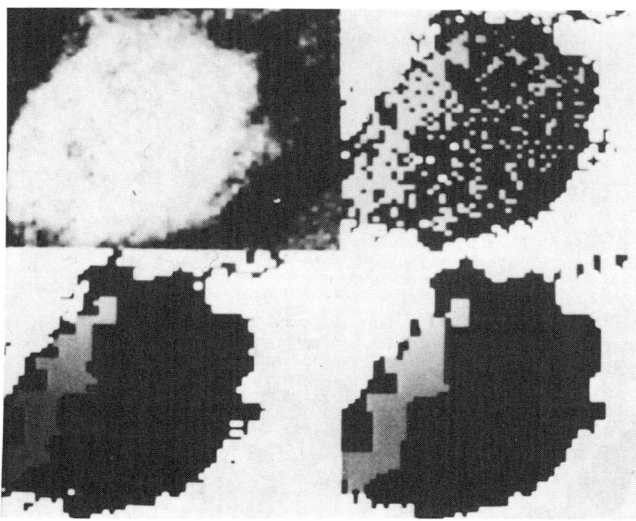

Figure 6 Application of thresholding and MAP algorithm to satellite image of a cloud

thresholded image a symmetric, 5-pixel, + shaped, filter template was used, while for the MAP segmented image a 5 × 1 template was used. For these examples, a single data set was used so that $K = 1$, and the noise variance, σ^2, was the same in each of the four region types. In this case the signal to noise ratio can be defined as

$$S/N = \min_{\substack{1 \leq l, m \leq 4 \\ l \neq m}} \frac{|r_l - r_m|}{\sigma^2}$$

For Figures 1 and 2 $S/N=2$. For the MAP algorithm all transition probabilities were chosen so that the probability of changing states was .13 and probability of staying in the same state was .61. Figures 3, 4 and 5 show similar results for these patterns at $S/N=1.5$. Finally Fig. 6 shows the result of applying this algorithm to a noisy satellite image of a cloud. For this example four regions were obtained by defining two region types within the cloud corresponding to two different brightnesses, and two background regions, one corresponding to the cloud shadow on the Earth, and the other to the remainder of the background. For these examples, region means and noise variances were precalculated and intput to the program.

5 CONCLUDING REMARKS

This paper described a method for segmenting multi-region, multi-sensor images. Thus far we have applied to images with four or less region types. Both the time and storage requirements of the algorithm grow rapidly as the number of region types increases and a practical limitation for implementation appears to be six regions, although our software is presently only able to handle four regions. On the other hand, time and storage requirements increase relatively little as the number of data sets increases. Our future work will thus be directed at extending this approach to textured images by preprocessing and incorporating textural information as additional data sets.

REFERENCES

[1] J.L. Doob, *Stochastic Processes*, John Wiley, New York, p. 153.

[2] K.L. Chung, *A Course in Probability Theory*, Academic Press, New York, 1974 (Second Edition).

[3] H. Kushner, *Introduction to Stochastic Control*, Holt, Rinehart and Winston, New York, 1971.

[4] G.D. Forney, "Maximum Likelihood Sequence Estimation of Digital Sequences in the Presence of Intersymbol Interference," IEEE Trans. on Info. Theory, Vol. IT-18, May 1973, pp. 363-377.

[5] A.J. Viterbi, "Error Bounds for Convolutional Codes and an Asymptotically Optimum Decoding Algorithm," IEEE Trans. on Info. Theory, Vol. IT-13, April 1967, pp. 363-377.

[6] L.L. Scharf, D.D. Cox, C.J. Masreliez, "Modulo-2 Phase Sequence Estimation," IEEE Trans. on Info. Theory, Vol. IT-26, Sept. 1980, pp. 615-620.

[7] H. Kaufman, J.W. Woods, V.K. Ingle, R. Mediqvilla, and A. Radpour, "Recursive Estimation: A Multiple Model Approach," Proc. 18th Conf. on Dec. and Control, Fort Lauderdale, Dec. 1979.

[8] C.W. Therrien, "Linear Filtering Models for Texture Classification and Segmentation," Proc. of 5th Int. Conf. on Pat. Rec., Miami, Dec. 1980.

[9] C.W. Therrien, "Linear Filtering Models for Terrain Image Segmentation," MIT Lincoln Laboratory Tech. Rep. #552, February 1981.

[10] H. Elliott, D.B. Cooper, F. Cohen, P. Symosek, "Implementation, Interpretation and Analysis of a Suboptimal Boundary Finding Algirithm," IEEE Trans. on PAMI, March, 1982.

[11] H. Elliott, L. Srinivasan, "An Application of Dynamic Programming to Sequential Boundary Estimation," Computer Graphics and Image Processing, December, 1981.

[12] D.B. Cooper, H. Elliott, F. Cohen, L. Reiss, P. Symosek, "Stochastic Boundary Estimation and Object Recognition," Computer Graphics and Image Processing, April 1980, pp. 326-355.

[13] R. Bellman, *Dynamic Programming*, Princeton University Press, Princeton, 1957.

[14] J. Ravin, "Decision Making in Markov Chains Applied to the Problem of Pattern Recognition," IEEE Trans. on Info. Theory, Vol. IT-13, Oct. 1967, pp. 536-551.

[15] L.L. Scharf and H. Elliott, "Aspects of Dynamic Programming in Signal and Image Processing," to appear IEEE Trans. on Aut. Control, Bellman Issue, October, 1981.

[16] H. Elliott and F.R. Hanse, "An Application of Adaptive Algorithms to Image Processing," Proceedings of 19th Conference on Decision and Control, Albuquerque, 1980.

[17] F.R. Hansen, "Application of Markov Field Models to the Design and Analysis of Image Segmentation Algorithms," M.S. Thesis, Colorado State University, July 1981.

[18] A.P. Sage, J.L. Melsa, *Estimation Theory with Application to Communications and Control*, McGraw-Hill, New York, 1971.

[19] F.R. Hansen and H. Elliott, "Image Segmentation Using Simple Markov Field Models," Computer Graphics and Image Processing, October 1982.

A Cluster Analysis Program for Image Segmentation*

Melvin F. Janowitz

*Department of Mathematics and Statistics
University of Massachusetts
Amherst, MA*

1 INTRODUCTION

A more detailed description of the contents of this paper occurs in [7]. The present version is intended only to be a survey, and for that reason will omit proofs of results. The thrust of the work is to relate the image segmentation problem to a rather general model for cluster analysis that was introduced in [3]. Some techniques suggested by the model are described, and their implementation illustrated. Section 2 is devoted to some background material from the theory of partially ordered sets, with Section 3 containing the image segmentation model. Section 4 has an elementary discussion of some of the underlying statistical considerations, with Sections 5 and 6 devoted to the description of certain segmentation techniques suggested by the model. Finally, in Section 7 the techniques are implemented on real data, and the results are presented.

2 BACKGROUND MATERIAL FROM THE THEORY OF PARTIALLY ORDERED SETS

A basic familiarity with the theory of partially ordered sets and lattices will be assumed. Despite this, it will be convenient to specifically develop certain terminology here. Unless otherwise specified, all partial orders will be denoted $\leqslant$, with $\subseteq$ reserved for set inclusion. The symbol $P(X)$ represents the Boolean algebra of all subsets of the set X, ordered by set inclusion. Unions and inter-

---

*Research supported by ONR Contract N00014-79-C-0629.

sections have the standard symbols of ∪ and ∩, and it will be useful to let R denote the nonnegative real numbers, ordered in the usual manner.

Let P, Q be partially ordered sets. A mapping $\phi: P \to Q$ is called *isotone* if $p_1 \leqslant p_2$ in P implies that $\phi(p_1) \leqslant \phi(p_2)$ in Q. It is *residuated* if it is isotone and there is an isotone mapping $\phi^+: Q \to P$ such that $p \leqslant \phi^+\phi(p)$ and $q \geqslant \phi\phi^+(q)$ for every $p \in P$, $q \in Q$. The mapping ϕ^+ is the *residual mapping associated with* ϕ, and it is completely determined by ϕ; likewise, ϕ is completely determined when ϕ^+ is specified, so for a given residual mapping θ, it will often be convenient to write θ^* for the residuated mapping with which θ is associated. The notation Res (P,Q) or Res (P) when $P = Q$ will be used to denote the set of all residuated mappings from P into Q, with Res$^+(Q,P)$ and Res$^+(P)$ for the corresponding sets of residual mappings. It turns out that residuated and residual mappings play a natural role in the Jardine-Sibson model [8] for hierarchical clustering. This is because ([3], Lemma 4.1, p. 60) $C: R \to P(X)$, where X is a finite nonempty set, is residual if and only if it satisfies the following three conditions:

(1) C is isotone.
(2) $C(h) = X$ for some $h \in R$.
(3) Corresponding to each $h \in R$, there is a positive real number $\delta = \delta(h)$ such that $C(h) = C(h + \delta)$.

If X is the set of all 2 element subsets of the set P, this turns out to be equivalent to C being a *numerically stratified clustering* in the sense of Jardine and Sibson [8], p. 61.

The theory of residuated mappings is rather extensively developed in [2], and the reader is referred to that source for further information.

3 THE IMAGE SEGMENTATION PROBLEM

The underlying input data is a function $F: P \to R$, where P represents a closed bounded rectangle. One thinks of P as representing a picture having certain natural regions that are characterized in some known manner by the values of F. This might involve regions on which F is constant, or some sort of texture measure, or some combination of attributes of F. Unfortunately, no direct knowledge of the values of F is available. Rather, the actual input is a function $G: X \to R$, where $X = \{1, 2, \ldots, m\} \times \{1, 2, \ldots, n\}$ for suitable positive integers m and n. One can think of G as being some sort of sample of F or as somehow summarizing the values of F, possibly by averaging F over some small subregions of P. Often there is some noise involved in the passage from F to G, so G can only be regarded as an estimate of F. The idea is to use G to somehow recapture the principal regions of P as they are defined by F. A more precise and detailed description of these underlying assumptions occurs in [7], but the present version is sufficient for the current discussion.

A rather broad model for hierarchical clustering was presented in [3]. It included both the Jardine-Sibson model and a graph-theoretic model due to Matula [9]. It will now be shown that it also includes the image segmentation problem as we have stated it. Recall that the actual input data is a mapping $G: X \to R$, where $X = \{1, 2, \ldots, m\} \times \{1, 2, \ldots, n\}$. The idea is to transform G into a function $H: X \to R$, where H has fewer distinct levels of range values than does G. The values of H should be thought of as an estimate of the regions represented by F. There is a natural bijection between mappings $G: X \to R$ and elements of $\text{Res}^+(R,P(X))$. It is given by associating with G the unique residuated mapping $G^*: P(X) \to R$ to which it extends, the extension being given by the rule $G^*(M) = v\{G(m): m \in M\}$. One then takes the residual mapping G^+ associated with G^*. Thus the image segmentation problem may be viewed as the study of transformations of $\text{Res}^+(R,P(X))$ into $\text{Res}^+(R,P(Y))$, where $Y \subseteq X$. Such transformations will be called *segmentation methods*. The reason for $Y \subseteq X$ will be apparent from the concrete examples that will be presented. This places the segmentation problem squarely within the framework of the model contained in [3].

No assumption has thus far been made as to whether the input data has ordinal or numerical significance. Suppose that the input data has only ordinal significance in that the actual numerical values of G have no significance, but one might still be able to attach some significance to an assertion of the form $G(i,j) < G(k,l)$. Is it possible to describe the class of segmentation methods that are suitable for this type of data? Indeed it is, but before the description can be presented, some additional terminology is required. Suppose one is given a segmentation method and a residual mapping θ on R. To say that F is θ-*compatible* ([3], p. 68) is to say that for every $C \in \text{Res}^+(R,P(X))$, it is true that $F(C \cdot \theta) = F(C) \cdot \theta$. A minimal requirement for F is that it be θ-compatible for all order automorphisms θ of R. Such techniques are called *monotone equivariant* by Jardine and Sibson. A precise mathematical proof is given in [5] of the fact that every segmentation technique suitable for use with ordinal data is in a sense equivalent to a monotone equivariant technique. The equivalence is the fact that if the outputs were rank ordered, they would be identical. This now makes one inquire into the nature of monotone equivariant techniques. This question was posed and answered in [4] (Theorem 1, p. 149). For $C \in \text{Res}(R,P(X))$, one says that $h(h \in R)$ is a *splitting level* of C in case h is the image of some subset of X under C^*, the residuated mapping with which C is associated. If $0 < h_1 < \ldots < h_t$ denotes the sequence of splitting levels of C, then the segmentation method F is monotone equivariant if and only if it is true that:

(i) every splitting level of $F(C)$ is a splitting level of C,
(ii) the sequence $FC(0) \leqslant FC(h_1) \leqslant \ldots \leqslant FC(h_t) = Y$ depends only upon the sequence $C(0) < C(h_1) < \ldots < C(h_t) = X$ and is independent of the actual values of the h_i's;
(iii) conditions (i) and (ii) hold for every C in $\text{Res}^+(R,P(X))$.

One can associate with each subset T of $\text{Res}^+(R)$ the collection $\alpha(T)$ of all segmentation methods that are θ-compatible for every $\theta \in T$. If the nature of the input data makes compatibility with a certain set T of residual mappings desirable, then the determination of $\alpha(T)$ will produce a description of the most general type of segmentation technique that one should properly use. We have just noted that if T denotes the set of all order automorphisms of R, then $\alpha(T)$ is the set of all monotone equivariant mappings. An extremely important result is provided by F. Baulieu [1]. It yields an explicit description of *all* classes $\alpha(T)$ that can arise for T containing the set of all order automorphisms of R. Here is a description of these classes:

(S1) *Flat methods.* $FC(h)$ depends only on $C(h)$.
(S2) *Semiflat methods.* $FC(h)$ depends upon $C(h)$ and $C(0)$.
(S3) *Divisive methods.* $FC(h_i)$ depends upon $C(h_i)$, $C(h_{i+1})$, ..., $C(h_t)$, where h_t is the highest splitting level of C.
(S4) *SF(1) methods.* $FC(0)$ depends only on $C(0)$; for $h_i \neq 0$, $C(h_i)$ depends upon $C(h_i)$, $C(h_{i+1})$, ..., $C(h_t)$.
(S5) *Monotone equivariant methods.*

The simplest of these classes of segmentation methods is of course the class of flat methods, and it is to this class that we now direct our attention. Such methods are easy to describe and easy to implement. Given a fixed nonempty finite set X and an increasing sequence

$$M_1 \subset M_2 \subset \ldots \subset M_t = X$$

of subsets of X, where $M_i = C(h_i)$, the goal is to produce a sequence

$$N_1 \subseteq N_2 \subseteq \ldots \subseteq N_t = Y$$

of subsets of X that somehow summarize or represent the underlying picture that led to the original input data. In that N_i depends only upon M_i, this amounts to defining an isotone mapping γ on $P(X)$.

Not every isotone mapping on $P(X)$ is appropriate for defining a flat segmentation method. One wants to attach some spatial significance to the decision as to whether a point x belongs to $\gamma(M)$. One way of doing this is to insist that γ be *point-based* in that for each $x \in \gamma(X)$ there is a subset $N(x)$ of X containing x such that:

(PB1) $\gamma(\emptyset) = \emptyset$, and $\gamma(N(x)) \neq \emptyset$.
(PB2) For $A \subseteq N(x)$, if $\gamma(A) \neq \emptyset$, then $x \in \gamma(A)$.
(PB3) For $M \subseteq X$, $x \in \gamma(M)$ if and only if $x \in \gamma(M \cap N(x))$.

It is immediate that for every subset M of X, $\gamma(M) = \bigcup_{x \in \gamma(X)} \gamma(M \cap N(x))$. If one wants the output of γ to be independent of the polarity of the image, it is useful to also insist that γ *preserve complements* in that it also satisfy:

(PB4) For $M \subseteq X$, $\gamma(X \setminus M) = \gamma(X) \setminus \gamma(M)$.

The local version of this is the content of Theorem 1.

Theorem 1. For a point-based isotone mapping γ on $P(X)$, axiom (PB4) is equivalent to the assertion that for each subset A of $N(x)$, exactly one of $\gamma(A)$ and $\gamma(N(x) \setminus A)$ shall be nonempty.

If $\{N(x)\}$ and $\{M(x)\}$ are each families of neighborhoods satisfying axioms (PB1) through (PB3), then so is $\{M(x) \cap N(x)\}$. In view of this there is no harm in assuming;

(PB5) If $M(x)$ satisfies (PB1) through (PB3), then $N(x) \subseteq M(x)$.

Such a minimal family of subsets of X will be called the *system of neighborhoods* associated with γ. Henceforth, when we speak of a point-based isotone mapping γ, it will always be assumed that $\{N(x)\}_{x \in \gamma(X)}$ denotes the system of neighborhoods associated with γ.

Definition. The point-based isotone mapping γ is said to be *frequency-defined* if there exist positive integers j and k such that for every $x \in \gamma(X)$, (i) $k = \#N(x)$, and (ii) $x \in \gamma(M)$ if and only if $j \leq \#(M \cap N(x))$. Here $\#A$ denotes the *cardinality* of the set A.

Theorem 2. Let γ be a point-based, complement-preserving isotone mapping on $P(X)$. Necessary and sufficient conditions for γ to be frequency-defined are that:

(i) the neighborhoods $N(x)$ all have the same cardinality k, where k is odd, and

(ii) if A_x has minimal cardinality among those subsets A of $N(x)$ for which $x \in \gamma(A)$, then $\#A_x = (1+k)/2$.

To say that the mapping γ is a join homomorphism is to say that $\gamma(M \cup N) = \gamma(M) \cup \gamma(N)$. There is a dual notion of *meet homomorphism* and to say that γ is a *homomorphism* is to say that it is both a join and a meet homomorphism. The next theorem shows that these conditions are rather powerful, and will only be met in trivial situations.

Theorem 3. Let γ be a point-based isotone mapping on $P(X)$. Then (1a) $\Leftrightarrow$ (1b), (2a) $\Leftrightarrow$ (2b) and (3a) $\Leftrightarrow$ (3b).

(1a) γ is a join homomorphism.

(1b) For each $x \in \gamma(X)$ there is a subset A_x of $N(x)$ such that $x \in \gamma(M)$ if and only if $M \cap A_x \neq \emptyset$

(2a) γ is a meet homomorphism.

(2b) Corresponding to each $x \in \gamma(X)$ there is a subset B_x of $N(x)$ such that $x \in \gamma(M)$ if and only if $B_x \subseteq M$.

(3a) γ is a homomorphism.

(3b) For each $x \in \gamma(X)$ there corresponds an element y_x of $N(x)$ such that $x \in \gamma(M)$ if and only if $y_x \in M$.

Corollary 4. If γ is also complement-preserving, then all six conditions of the theorem are mutually equivalent.

A *translation* on X is a mapping of the form $T_{p,q}$ where p,q are fixed integers and $T_{p,q}(i,j) = (p+i, q+j)$. Unless $p = 0 = q$, the domain and image of $T_{p,q}$ will be proper subsets of X. We agree to call the point-based isotone mapping *translation-invariant* provided it satisfies:

(PB6) If $N(x)$ is contained in the domain of the translation T, and if $y = T(x)$, then $N(y) = T(N(x))$, and for $A \subseteq N(x)$, $x \in \gamma(A)$ if and only if $y \in \gamma(T(A))$.

Remark 5. If M is acted upon by a translation to produce N, it is desirable that $\gamma(M)$ produce $\gamma(N)$ when it is acted on by that same translation. As long as γ is translation-invariant, $M \subseteq$ domain T, and $M \cup T(M) \subseteq \gamma(X)$, it is easy to see that this is true. However, it is equally easy to show that this can fail even if γ is translation-invariant.

Remark 6. The actual construction of a point-based isotone mapping on $P(X)$ may now clearly be understood. A system of neighborhoods $\{N(x)\}$ for points x in some subset Y of X is first chosen. One then defines a family of mappings $(\gamma_x)_{x \in Y}$, where γ_x is an isotone mapping on $N(x)$ such that:

(1) $\gamma_x(\emptyset) = \emptyset$, and $\gamma_x(N(x)) = \{x\}$.

To produce a complement-preserving mapping, one also wants

(2) For $A \subseteq N(x)$, exactly one of $\gamma_x(A)$ and $\gamma_x(N(x) \setminus A)$ is nonempty.

The mapping γ is now defined by the rule $x \in \gamma(M)$ if and only if $\gamma_x(M \cap N(x)) = \{x\}$. This is the technique that will be used for the remainder of the paper.

4 UNDERLYING STATISTICAL CONSIDERATIONS

The construction of a flat segmentation method involves the definition of an isotone mapping γ on $P(X)$. The input data represents a supset M of X that is an estimate of the set M^* of X that is the true input data. The idea is to try to define γ so that $\gamma(M)$ is in some sense a better estimate of M^* than is M. A crude statistical model may be constructed by assuming that membership in M has probability $p(p > 0.5)$ of providing a correct estimate of membership in M, that membership in $X \setminus M^*$ has a probability $q(q > 0.5)$ of providing a correct estimate of membership in $X \setminus M^*$, and that these probabilities are independently distributed over the members of X. An *interior point* of M^* is defined to be a point x for which $N(x) \subseteq M^*$; dually, x is an *exterior point* of M^* if $N(x) \subseteq X \setminus M^*$. The *gain* of the isotone mapping γ is defined to be the sum of the probability of correct classification for an interior point of M^* and that of an exterior point. Needless to say, these are idealized concepts, as M^* is precisely the unknown set that we are trying to estimate. If $G(\gamma)$ denotes the gain of γ, we then have

Lemma 7. *Let γ be complement-preserving with $k = \#N(x)$ odd, and let γ' be the frequency-defined, point-based isotone mapping having the same neighborhood system as γ. Then $G(\gamma) \leq G(\gamma')$.*

Theorem 8. *If $p = q$, and if γ and γ' have the same system $N(x)$ of neighborhoods with γ' frequency-defined, then $G(\gamma) \leq G(\gamma')$.*

When $p = q$, the above theorem shows that the "gain" from a flat segmentation method can be maximized by using a point-based isotone mapping that is complement-preserving and frequency-defined. With this thought in mind, we shall concentrate on such mappings, defining them using k by k neighborhoods of points with k an odd integer. The j/k^2 rule will be the unique such mapping defined on a k by k neighborhood, and the 3/5 rule will refer to the γ that is based on a point together with its 4 immediate direct neighbors (North, South, East and West). These rules were discussed in some detail in [7], and examples were given there to show that near the boundary of M^*, the use of these rules can actually decrease the probability of correct classification. This suggests using an isotone mapping γ that is based on a weighted mean or on $M \cap N(x)$ containing certain desirable subsets. For example, one would be more likely to conclude that $x \in \gamma(A)$ if A were the left-hand subset than if A were the right-hand set:

```
0 0 0 0 0        1 0 0 1 0
0 1 1 1 0        0 0 0 0 1
0 1 1 1 0        1 0 0 1 0
0 1 1 1 0        0 1 0 0 1
0 0 0 0 0        1 0 0 0 1
```

5 THE SEGMENT PROGRAM

This was described in some detail in [6] and [7] For the reader's convenience, a brief description is also included here.

A. *Input.* An m by n matrix A having nonnegative integers as entries.

B. *Prefiltering.* A k by k mean or median filter is applied to smooth the data of A, and the output is rounded to produce B. Spot noise is removed by deleting the highest and lowest i values when computing the k by k mean.

C. *Thinout.* A frequency count is made of the values appearing in B. Those values that occur with frequency less than some threshhold are deleted, and a nearest value rule used to reassign them. Alternately, the k highest occuring values can be retained. The resulting matrix is denoted C.

D. Suppose matrix C has data values $j_1, j_2, \ldots, j_k$. The nondirected 3 by 3 *dispersion* of value j_1 is defined as follows: For each point in C having value j_1, look at a 3 by 3 neighborhood centered on that point. Calculate the number of points in that neighborhood that do *not* have value j_1, and average this figure over all points having value j_1. This is the dispersion for cluster j_1. One can do a similar calculation for $j_2, \ldots, j_k$. An analogous definition can be formulated for larger neighborhoods, and certain options are included that compute directional versions of the dispersion.

E. Various options exist for deleting one or more of the data values having high dispersion levels. The simplest is to just delete the highest value dispersion. Other options might include deleting all data values whose dispersion exceeds 0.9 on the first pass, and then dropping this cutoff down by increments of 0.1 until a stopping criterion is reached. Various choices also exist for the reassignment of points whose value has been deleted. They proceed on both a global and a local basis. Cluster means can be computed for each remaining cluster, and pixels reassigned to the cluster to which their input value is closest, or an entire cluster can be reassigned to the cluster to which its mean is closest. A second technique involves the reassignment of points to the next higher or lower cluster that is still valid, either on a local or a global basis. As these techniques have been described, they are not monotone equivariant, but if the data are first rank ordered, and the output is labeled according to which value corresponds to which rank, the resulting techniques do become monotone equivariant.

6 THE SLICER PROGRAM

A crude segmentation program can easily be devised using the material of Sections III and IV. It has two phases:

Phase 1. Find levels at which to slice the data.

Phase 2. Make appropriate slices and use a j/k^2 rule to increase the probability of correct classification.

The Phase 2 portion of the program is a flat segmentation method. The program itself can either be used as a cleaning algorithm, or as a tool for obtaining a first approximation to the principal regions of a digital picture.

Description of Phase 1. A total of six methods of determining the slice levels are built into the program.

Method 1. Compute the mean M and standard deviation S of the input data, and take slices at $M - 2S$, $M - S$, M, $M + S$, and $M + 2S$. If the extreme values are outside the data range, they are modified to bring them within the minimum and maximum values of the data.

Method 2. Slice at relative maxima of the histogram of the data.

Method 3. Slice at relative minima of the nondirected 3 by 3 dispersion.

Method 4. This is similar to Method 3, except that if A is the vector of distinct data values, then at level $A[i]$, one is interested in those values that lie outside the range from $A[i-1]$ to $A[i+1]$.

Method 5. Slice at relative minima of the average variance in t by t neighborhoods.

Method 6. User inputs slice values.

Description of Phase 2. Having arrived at a list of slice levels, a j/k^2 rule is chosen. The actual slices are constructed so that they are halfway between the slice input levels. Thus if one wanted slices at 17, 20, 22, 26, and 30, then one would look at the sets $M_1 \subset M_2 \subset M_3 \subset M_4 \subset M_5$, where $M_i = \{x: x < k_i\}$ with $k_1 = 18.5$, $k_2 = 21$, $k_3 = 24$, $k_4 = 28$ and $k_5 = $ the maximum value of X. This forms 5 clusters, and they are assigned the values of 17, 20, 22, 26 and 30. There are also occasions when one would want the actual slice values to coincide with the input slice values. If γ represents the isotone mapping specified by the j/k^2 rule, the output would then be the sequence;

$$\gamma(M_1) \subseteq \gamma(M_2) \subseteq \gamma(M_3) \subseteq \gamma(M_4) \subseteq \gamma(M_5).$$

If the actual slices are taken at levels $k_1, k_2, \ldots, k_t$ the implementation of this process is very simple. One starts by forming a Boolean matrix with 1 denoting a value $\leqslant k_1$ and 0 otherwise. One then looks at k by k regions. If there are j or more 1's in such a region, the output is 1; otherwise it is 0. The output for level k_2 is added to this, and the process continues until the supply of slice levels is exhausted.

The actual implementation of the SLICER program can be done more efficiently by first performing a k by k median filter, and then applying the algorithm described at the beginning of Phase 2. That way, if one wanted to examine the effect of various types of slicing, one would only have to perform the j/k^2 rule a single time, as this decision rule is in reality implemented by the median filter.

7 SOME EXAMPLES

Figures 1-4 contain a comparison of the output of the SEGMENT program with that of program SLICER. The figures all follow the same format, so a single explanation of them should suffice. There are 6 pictures per figure. The upper left picture is the original data set. The middle picture on the upper level is the SEGMENT output using a nondirected 3 by 3 dispersion, a 3 by 3 mean filter, and the version of the program that removes 1 cluster at a time, reassigning its members locally to the next higher or lower available cluster. The parameters were set to produce 5 clusters. The remaining pictures relate to the SLICER program, though it should be noted that the program was modified to make it iterate with a stopping criterion set for when an iteration did not change the cluster means. At each stage, cluster means were computed, and these values used as slice level input for the next iteration. The upper right picture was the result of the first, third and fifth levels of option 1 of SLICER, the bottom left was the result of option 1, the bottom middle the result of 3 slices equally placed between the minimum and maximum values of the input data, and the bottom right the result of 5 slice levels equally placed. Thus for 3 levels, the slices are taken 1/4, 2/4 and 3/4 of the way from the minimum to the max-

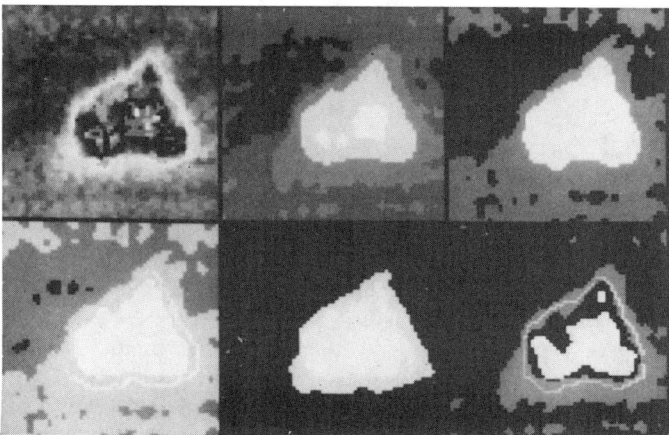

Figure 1 Portion of Record 21 of Westinghouse FLIR data tape

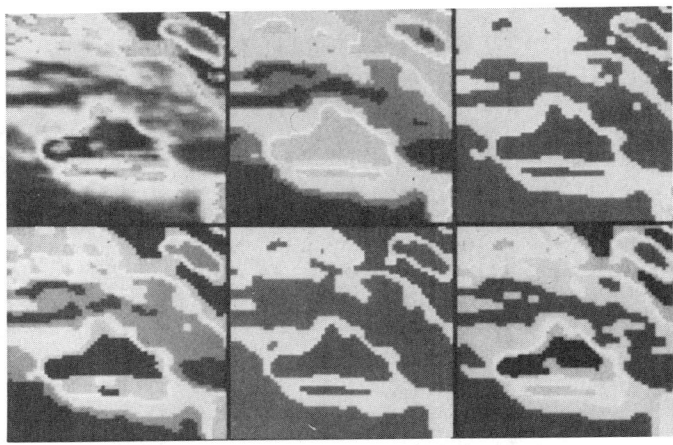

Figure 2 Portion of Record 33 of Westinghouse FLIR data type

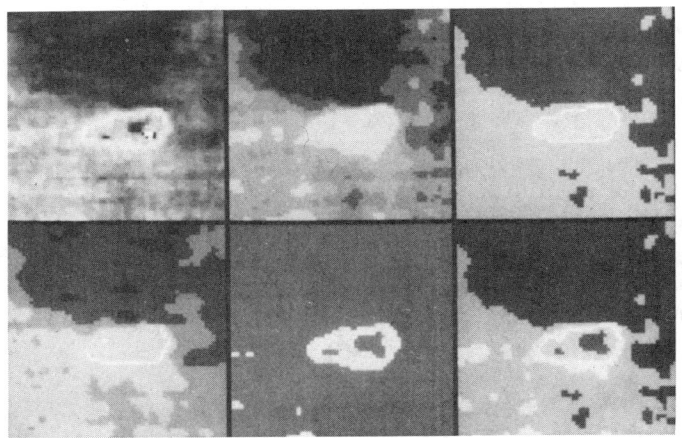

Figure 3 Portion of Record 12 of Westinghouse FLIR data tape

imum data value; for 5 levels, the slices are 1/6, 2/6, 3/6, 4/6, and 5/6 of the way. In each case a 5/9 rule was applied to produce the final output. No further cleaning algorithm was applied. It should be noted that the various SLICER outputs produced reasonable crude segmentations of the data. Further work is clearly needed on the initial selection of slice levels. The outputs are especially useful when one resorts to line printer representations, as one can intelligently reduce a picture to a desired number of grey levels.

Figure 3 is worthy of note. It illustrates one of the pitfalls of the SEGMENT program. This program operates essentially by merging clusters; no provision is made for a subsequent splitting of a cluster, unless the dispersion cri-

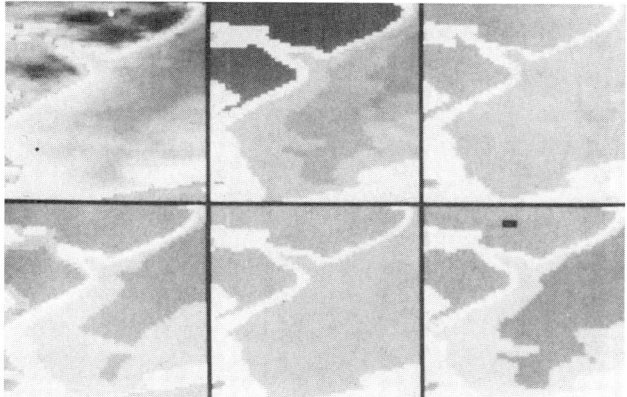

Figure 4 Portion of weather tape furnished by Air Force Geophysical Laboratory. The picture shows the region off of the Chesapeake Bay with the edge of the Gulf Stream visible in the lower right corner.

terion deems that cluster to be too noisy to be maintained. The vehicle in the center of the picture is too large because of some early mergers that were not undone. In the lower right-hand corner of Figure 4, a portion of the Gulf Stream is visible. In all but 1 of the outputs, this cluster was merged with warm coastal water. This was caused largely by the iterations using cluster means as outputs. The original iteration more correctly merged the Gulf Stream into the same cluster as the land masses, thus correctly identifying it as being much warmer than the coastal water.

REFERENCES

1. F. Baulieu, *Order theoretic classification of cluster methods*, University of Massachusetts doctoral dissertation, 1981.

2. T.S. Blyth and M.F. Janowitz, *Residuation theory*, Pergamon Press, Oxfort, 1972.

3. M.F. Janowitz, *An order theoretic model for cluster analysis*, SIAM J. Appl. Math. 34 (1978), 55-72.

4. M.F. Janowitz, *Monotone equivariant cluster methods*, ibid. 37 (1979), 148-165.

5. M.F. Janowitz, *Preservation of global order equivalence*, J. Math. Psych. 20, (1979), 78-88.

6. M.F. Janowitz, *Cluster analysis algorithms for image segmentation*, University of Massachusetts Technical Report J8102, 1981.

7. M.F. Janowitz, *Monotone equivariant segmentation techniques*, University of Massachusetts Technical Report J8201, 1982.

8. N. Jardine and R. Sibson, *Mathematical taxonomy*, Wiley, New York, 1971.

9. D.W. Matula, *Graph theoretic techniques for cluster analysis algorithms*, in "Classification and Clustering" (J. van Ryzin, ed.), Academic Press, New York, 1977.

DCT Image Compression Over Noisy Channels*

Jerry D. Gibson

Department of Electrical Engineering
Texas A&M University
College Station, TX

1 INTRODUCTION

The efficient digital representation of images is important for transmission, storage, and encryption purposes. A popular technique for digital image compression at 1 bit/pel is the two-dimensional discrete cosine transform (DCT). Unfortunately, this image compression technique, like most other data compression methods, is extremely sensitive to bit errors. This paper discusses two methods for reducing the effects of bit errors, soft decision demodulation (SDD) and forward error correction (FEC) coding. The soft decision demodulation system monitors the reliability of the most significant bits of the highest energy DCT coefficients. If a bit of an important coefficient is deemed unreliable, the entire coefficient is replaced by an estimate derived from adjacent analysis blocks. Theoretical, simulation, and subjective viewing results indicate that a noticeable improvement in image quality is possible by monitoring only six bits out of each 16 by 16 block of 256 bits. The advantage of SDD is that no modifications are required at the transmitter. If it is possible to modify the system transmitter, FEC block coding techniques prove useful. In particular, the use of (7,4), (15,11) and (31,26) Hamming codes for protecting the most significant bits of the highest energy coefficients can provide a substantial improvement in image quality at a bit error rate of 10^{-2}. Design criteria and performance results are presented. Reconstructed images representing SDD and FEC system performance at 1 bit/pel are shown and discussed.

---

*Supported by the Office of Naval Research under Contract No. N00014-81-K-0299.

2 TWO-DIMENSIONAL DCT

The monochrome images used for this work consist of 256 by 256 pixels with each pixel represented by an 8-bit word. The two-dimensional DCT (2D-DCT) is a popular transform for image compression at 1 bit/pixel [1], and it is considered exclusively in this work. The 2D-DCT is defined by

$$F(u,v) = \frac{2}{N} c(u) c(v) \sum_{j=0}^{N-1} \sum_{k=0}^{N-1} f(j,k).$$
$$\cos\left[\frac{(2j+1)\pi u}{2N}\right] \cos\left[\frac{(2k+1)\pi v}{2N}\right]; \quad (1)$$

for $u, v = 0, 1, \ldots, N - 1$, $c(0) = 1/\sqrt{2}$, and $c(u) = 1$ for $u = 1, 2, \ldots, N - 1$. The inverse 2D-DCT is

$$f(j,k) = \left(\frac{2}{N}\right) \sum_{u=0}^{N-1} \sum_{v=0}^{N-1} c(u) c(v) F(u,v) \cos\left[\frac{(2j+1)\pi u}{2N}\right].$$
$$\cos\left[\frac{(2k+1)\pi v}{2N}\right], \quad (2)$$

for $j, k = 0, 1, \ldots, N - 1$. One advantage of the 2D-DCT is that it can be computed using "fast" algorithms.

To use the 2D-DCT in a data compression system, $F(u,v)$ in Eq. (1) is calculated over an N by N block ($N = 16$ for this paper), the lowest energy coefficients are discarded, and the highest energy coefficients are quantized and coded. The scheme used for bit allocation is due to Wintz and Kurtenbach [2]. The coefficients were quantized using minimum mean squared error (MMSE) nonuniform Laplacian quantizers for the soft decision demodulation (SDD) work and MMSE nonuniform Gaussian-assumption quantizers for the forward error correction (FEC) coding studies. The quantizer output levels are represented digitally by the folded binary code (FBC). In the absence of channel errors, this method produces good quality reconstructed images at a rate of 1 bit/pixel.

3 SOFT DECISION DEMODULATION

The soft decision demodulation (SDD) system (shown in Fig. 1) differs from the standard 2D-DCT compression system in that at the receiver, not only is the codeword representing a coefficient decoded to the corresponding reconstructed coefficient $\hat{c}_i$, but in addition, the most significant bits (MSB) of the highest energy coefficients are monitored for reliability [3]. Specifically, assuming binary antipodal signaling, if the received signal for bit j of c_i falls in the erasure zone $(-T_{ij}\sqrt{E}, T_{ij}\sqrt{E})$, $0 < T_{ij} < 1$, then that bit is considered unreliable, and the entire codeword is rejected and replaced by an estimate. The reconstructed image, $\hat{x}$, is then obtained by the inverse DCT operation, H^{-1}.

In this system, there are three types of error events that can occur:

1. A channel error occurs, but the receiver does not detect it,

DCT Image Compression

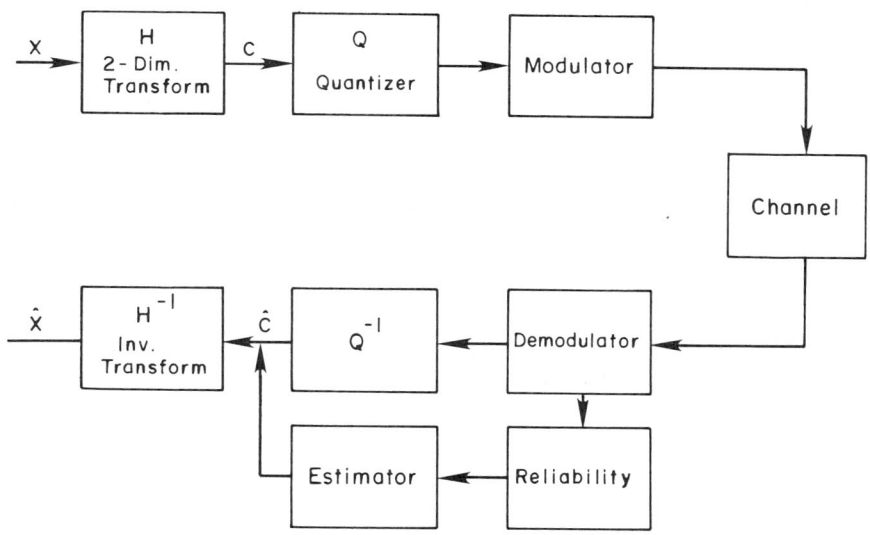

Figure 1 Soft decision demodulation system

2. A channel error occurs, and the receiver detects it and replaces the affected coefficient with an estimate, or
3. No channel error occurs, but the receiver thinks one has, and it replaces the affected coefficient with an estimate.

In designing a soft decision system it is desirable to minimize the Type 1 events, (equivalently, maximize Type 2 events) and to minimize the Type 3 events. Since increasing the probability of a Type 2 event also increases the probability that Type 3 event will occur, the goals stated above are conflicting, and some trade off must be made between the different types of errors. The criterion used here in optimizing the system is the familiar minimum mean squared error between the original and reproduced image. Although minimizing mean squared error does not necessarily coincide with the best perceptual image, it is useful in that it constitutes a basic framework for the overall system design.

The number of MSB that must be monitored and their associated thresholds T_{ij}, are determined by quantities called A-factors and the bit energy to noise spectral density ratio $2E/N_0$ (assuming an additive white Gaussian noise channel). The A-factors, denoted $A(j, NB_i)$, represent the mean squared error in reproducing a quantized codeword with a single bit error in bit j of an NB_i bit quantizer. The A-factors and quantization noise power for folded binary codes and Laplacian quantizers of size one to eight bits are shown in Table 1.

The total mean squared error for a block of coefficients is given by

$$\epsilon_T^2 = \sum_{i=1}^{N^2} [\epsilon_q^2(NB_i) + \epsilon_a^2(i)] \cdot \sigma_i^2 \qquad (3)$$

where coefficient c_i with variance σ_i^2 has been allocated NB_i bits, ϵ_q^2 represents

Table 1 Quantization Noise and Single Bit A-Factors for 1-8 Bit Laplacian Quantizers

No. of bits	ϵ_q^2	Single Bit A-Factors							
		1	2	3	4	5	6	7	8
1	.499	2.00							
2	.176	3.29	2.00						
3	.0543	3.78	3.08	.638					
4	.0153	3.94	4.16	.819	.182				
5	.00405	3.98	5.09	.970	.209	.0498			
6	.00285	3.99	7.01	1.75	.438	.109	.0274		
7	.000908	4.00	9.41	2.35	.588	.147	.0367	.00919	
8	.000277	4.00	12.1	3.04	.759	.190	.0475	.0119	.00296

quantization noise, and ϵ_a^2 represents the error power introduced by the channel. For a system with no error correction/detection, the channel noise can be expressed as

$$\epsilon_a^2(i) \doteq P \cdot \sum_{j=1}^{NB_i} A(j, NB_i), \quad (4)$$

while for a soft decision receiver,

$$\epsilon_a^2(i) \doteq \sum_{j=1}^{M_i} A(j, NB_i) \cdot P_{u_{ij}} + P \cdot \sum_{j=M_i+1}^{NB_i} A(j, NB_i) + \Delta\sigma_i^2 P_{r_i}. \quad (5)$$

In (5) M_i is the number of most significant bits that are monitored, $P_{u_{ij}}$ is the probability that bit j is in error but the received signal is not in the erasure zone, P is the channel bit error probability, and P_{r_i} is the probability that at least one of the M_i most significant bits were designated unreliable, and $\Delta\sigma_i^2$ is the normalized mean square estimation error for coefficient i. Minimizing (5) with respect to T_{ij} yields

$$T_{ij} = \frac{1}{2(2E/N_0)} \log_e \left[\frac{A(j, NB_i)}{\Delta\sigma_i^2} - 1 \right] \quad (6)$$

and M_i is chosen to be the largest j in (6) such that T_{ij} is defined and greater than zero.

The proposed estimate of c_i, given that a soft decision error has occurred, is simply the average of the corresponding coefficients in neighboring blocks. For example, assume that coefficient c_1 in block (k,m) has been determined unreliable, then it is rejected and replaced by

$$\hat{c}_1(k,m) = (1/3)[\hat{c}_1(k-1,m) + \hat{c}_1(k-1,m-1) + \hat{c}_1(k,m-1)]. \quad (7)$$

These three neighboring blocks will have already been decoded, and it is assumed that they have been decoded correctly.

Coefficients 1, 2, and 17 were chosen for soft decision monitoring. Coefficient 1 is the dc coefficient, coefficient 2 depends on vertical edges, and

Table 2 Parameters for the Soft Decision Receiver

Design $2E/N_0$	Coefficient	$\Delta\sigma_i^2$	Bit	Threshold
7.34 dB	1	.3	1	.231
			2	.388
			3	.203
			4	.039
	2	.5	1	.178
			2	.236
			3	.084
	17	.5	1	.178
			2	.236
			3	.084

coefficient 17 depends on horizontal edges. The mean squared estimation error, $\Delta\sigma_i^2$, depends on the estimation rule and the input image, but due to a lack of a good model for the coefficients, these terms were chosen empirically. Table 2 summarizes the soft decision design parameters.

Theoretical and Monte Carlo simulation results are in close agreement. Images illustrating system performance are shown in Fig. 2. Figure 2(a) is the input image to the system and Fig. 2(b) is the output image at a rate of 1 bit/pixel and with a noiseless channel. Figure 2(c) is the output image for a hard decision receiver and a bit error rate (BER) of 10^{-2}. Figure 2(d) is the corresponding output for the soft decision receiver. In Fig. 2(c), there are many blocks which have been whited out due to errors in the MSB of the dc coefficient. By using SDD, these errors have been detected, thus resulting in the image in Fig. 2(d). A type 3 error is also evident in Fig. 2(d) where there is a light block in the girl's hair, which is not evident in Fig. 2(c).

4 HAMMING CODING

An alternative to SDD is to use forward error correction (FEC) to protect the transmitted bits. In this section, the use of (7,4), (15,11), and (31,26) Hamming codes to protect the quantized DCT coefficients is investigated [4]. In order to apply these codes, it is necessary to determine how many bits to protect and which bits to protect. The latter question is answered first by calculating the mean squared reconstruction error contributed by each of the 256 bits in a block and then ranking these bits from largest to smallest error.

Without FEC, the mean squared error (MSE) due to the channel is given by Eq. (4). For those bits protected by channel coding, the probability of bit error is denoted P_c so the channel MSE expression becomes

$$\epsilon_a^2(i) = P_c \sum_{j=1}^{r} A(j, NB_i) + P \sum_{j=r+1}^{NB_i} A(j, NB_i) \qquad (8)$$

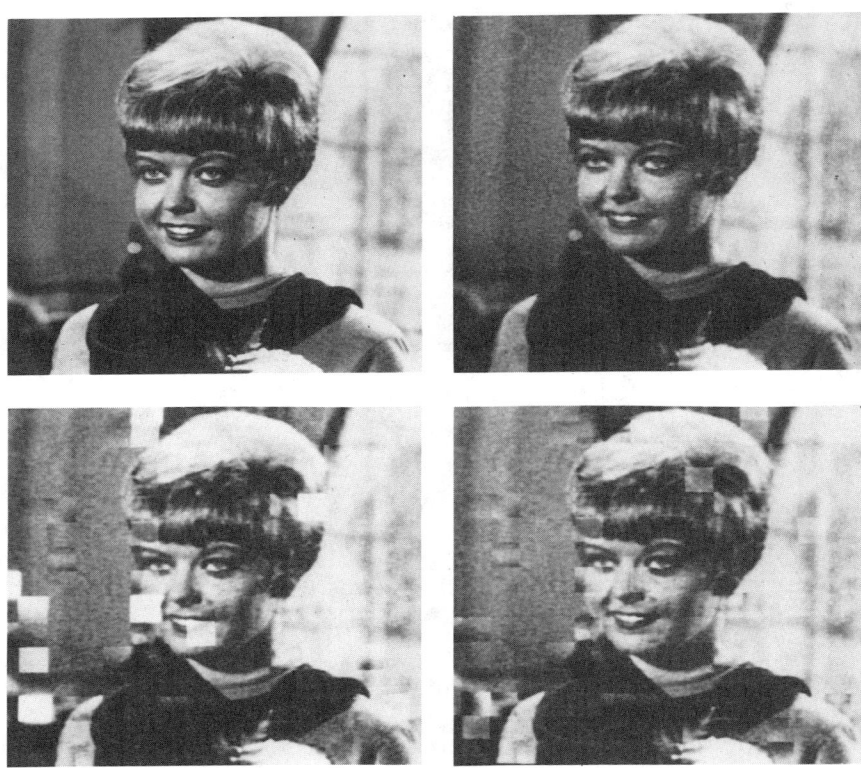

Figure 2 Simulation results for channel errors and soft decision

where r bits are assumed protected and errors with coding are independent and equally likely. The normalized mean squared reproduction error (NMSE) is the total error in (3) divided by the sum of the variances of the coefficients. For the FEC work, the coefficients are assumed to be Gaussian distributed, and Table 3 lists the appropriate A-factors and ϵ_q^2.

The question of how many bits to protect involves a tradeoff between bits allocated to source coding and bits allocated to channel coding with overall rate constrained to 1 bit/pixel. By using Eqs. (3) and (8), the NMSE as the number of channel coding bits is increased can be computed for each Hamming code at a BER of 10^{-2}. The minimum NMSE for each of the three codes is shown in Table 4. As can be seen from this table, the (7,4) code yields the best performance and the optimal number of bits to code for the girl image is 44.

Since the total transmitted data rate is fixed at 1 bit/pixel, using 33 bits per block for channel coding reduces the number of bits available for source coding to 223. It is important to ascertain how much degradation in image quality is imposed by this reduction of source coding bits when the channel is error-free. Figure 3(a) is the reconstructed image at 1 bit/pixel with no bits

Table 3 Quantization Plus Clipping Error and Single Bit A-Factors for 1-8 Bit Gaussian Quantizers

No. of bits	$\epsilon_q^2 + \epsilon_c^2$	Single Bit A-Factors							
		1	2	3	4	5	6	7	8
1	3.63×10^{-1}	2.547							
2	1.17×10^{-1}	3.532	1.117						
3	3.45×10^{-2}	3.858	1.506	.3763					
4	9.50×10^{-3}	3.961	1.828	.4525	.1095				
5	2.50×10^{-3}	3.997	2.072	.5096	.1216	.02968			
6	1.04×10^{-3}	4.001	2.772	.6933	.1733	.04333	.01083		
7	3.04×10^{-4}	4.007	3.307	.8267	.2067	.05168	.01292	.003227	
8	8.88×10^{-5}	4.009	3.860	.9650	.2413	.06031	.01508	.003774	.0009436

Table 4 No. of Bits coded that achieves minimum NMSE.

Girl Image			
Code rate	Error rate	No. bits coded	NMSE
4/7	10^{-2}	44	2.273×10^{-2}
11/15	10^{-2}	66	3.036×10^{-2}
26/31	10^{-2}	78	4.544×10^{-2}

allocated to channel coding and a zero BER, while Fig. 3(b) is the compressed image at 1 bit/pixel with 33 bits per block allocated to channel coding. As is evident, Figs. 3(a) and 3(b) do not differ substantially, and hence, allocating bits to channel coding does not seriously reduce error-free system performance at 1 bit/pixel.

Monte Carlo simulation results were obtained for each BER of interest by processing each image 25 times with a different random error sequence for each run. The average output SNR over each set of 25 runs is then computed for system evaluation. Simulations were performed both for systems with channel coding and without channel coding at BER's of 10^{-4}, 10^{-3}, 10^{-2}, and 10^{-1}. At the design error rate of 10^{-2}, channel coding provides a 3.2 dB advantage in output SNR for this image.

Of course, the final important question is how much channel coding improves the reconstructed image visual quality. To provide an indication of the full range of possible reconstructed images over the many Monte Carlo runs, a subjective selection of the worst and best images was made. Figures 4(a) and (b) show the worst and best girl images without channel coding at a BER of 10^{-2}, and Figs. 4(c) and (d) show the worst and best images, respectively, for channel coding at a BER of 10^{-2}. Clearly, the channel coding scheme proposed here provides a substantial, noticeable improvement in reconstructed image quality.

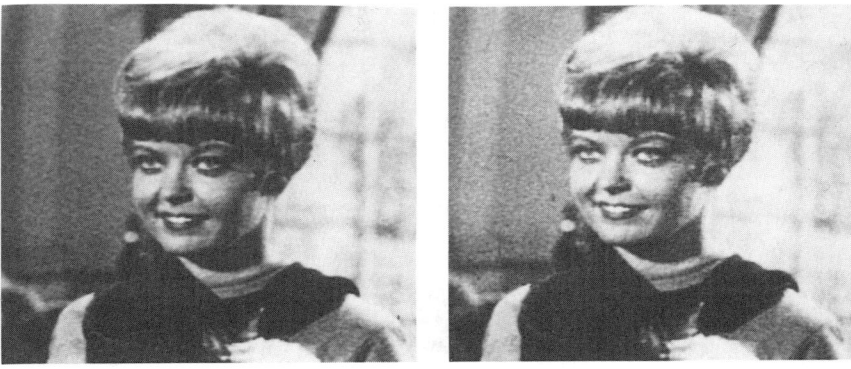

Figure 3 Original girl image and data compressed reconstructed images with and without channel coding (BER=0)

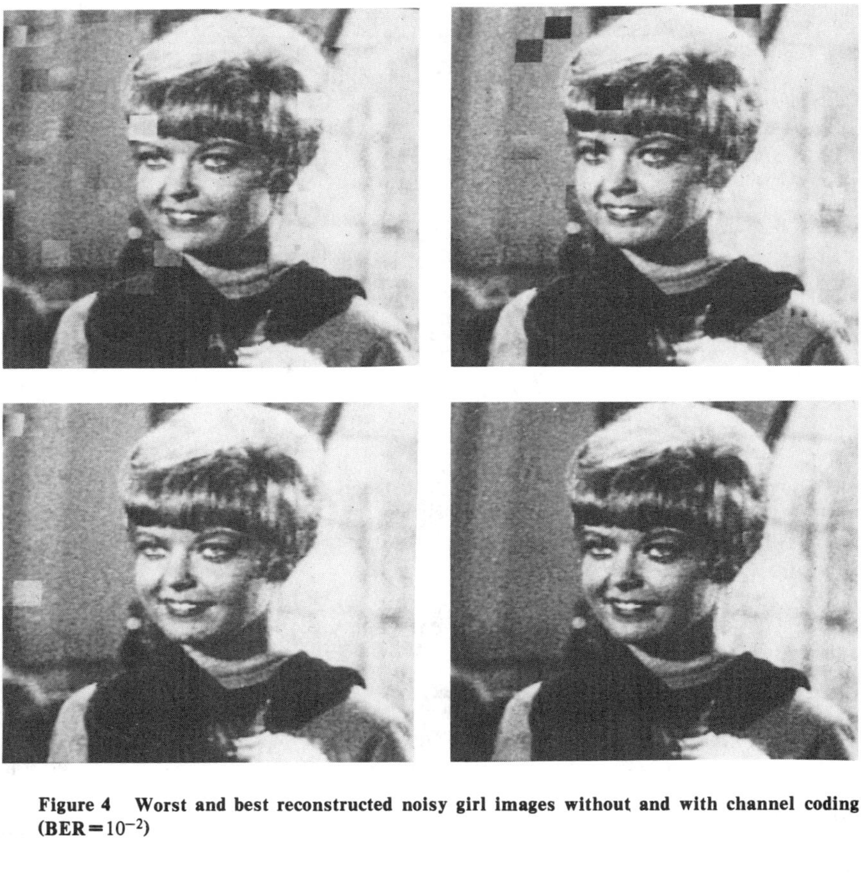

Figure 4 Worst and best reconstructed noisy girl images without and with channel coding (BER=10^{-2})

5 CONCLUSIONS

In soft decision demodulation, if certain bits of a received codeword are unreliable, the codeword is replaced by an estimate. By monitoring only the most significant bits of the three highest energy DCT coefficients, the reconstructed image can be improved for a system using binary antipodal signals over an additive white Gaussian noise channel. Only an extremely simple averaging scheme was used to obtain the estimate which replaces an unreliable coefficient, and more sophisticated methods may yield a greater advantage for soft decision.

For forward error correction, simulation results indicate that using the (7,4) Hamming code to protect the most significant bits of the 2D-DCT quantized coefficients can substantially improve reconstructed image quality at a BER of 10^{-2} and a rate of 1 bit/pixel. A somewhat surprising result is that the allocation of 33 out of 256 bits per block to channel coding does not noticeably degrade reconstructed image quality in the absence of channel errors. If the transmitter can be modified, it thus seems preferable to use FEC as opposed to SDD.

REFERENCES

1. N. Ahmed and K.R. Rao, *Orthogonal Transforms for Digital Signal Processing*, New York: Springer-Verlag, 1975.

2. P.A. Wintz and A.J. Kurtenbach, "Waveform error control in PCM telemetry," *IEEE Trans. Inform. Theory*, Vol. IT-14, pp. 650-661, September 1968.

3. R.C. Reininger and J.D. Gibson, "Soft decision demodulation and transform coding of images," *Conf. Rec.*, 1982 Int. Conf. Commun., Phila., PA, June 13-17, pp. 4H.3.1-4H.3.6.

4. D.R. Comstock, "Hamming Coding of DCT-Compressed Images Over Noisy Channels," Master's Thesis, Dept. of Elec. Eng., Texas A & M Univ., May 1982.

On Segmentation of Time Series and Images in the Signal Detection and Remote Sensing Contexts*

Stanley L. Sclove

Departments of Mathematics and Quantitative Methods
University of Illinois at Chicago Circle
Chicago, IL

1 INTRODUCTION

Problems of segmenting time series or digital images are considered. In both cases the observations fall into classes, and this assignment of observations to classes, or "labeling," is unknown. Thus each point gives rise to a pair, the observation itself, together with its unknown label. In the context of this model, segmentation is estimation of the labels. In time series the points are time points; in digital images, the points are points of the image (picture elements, or pixels). An image is two-dimensional, while time is one-dimensional, so time series are treated first.

2 SEGMENTATION OF TIME SERIES

The problem of segmentation considered here is: Given a time series

$$\{x_t, t=1, 2, \ldots, n\},$$

partition the set of values of t into subseries (segments, regimes) for which the values x_t are relatively homogeneous. The segments are assumed to fall into several classes. In cyclic processes the classes are phases of the cycle.

Examples.
   (i) Segment a received signal into background, target, background again, another target, etc.

---
*Work supported under Contract N00014-80-C-0408, Task NR 042-443 with the Office of Naval Research.

(ii) Segment an EEG of a sleeping person into periods of deep sleep and restless or fitful sleep (two classes of segment).

(iii) Segment an ECG into rhythmic and arhythmic periods (two classes of segment).

(iv) Segment an economic time series into periods of recession, recovery, and expansion. Here there are three classes of segment.

In some applications the observation X is a vector of several measurements. For example, for blood pressure, X is a vector of the two measurements, systolic and diastolic. The discussion of segmentation of digital images, later in the paper, will be in terms of vector measurements. Time series will be discussed in terms of scalar (single) measurements, although the ideas and methods readily generalize to multiple time series.

2.1 The Model

One can imagine a series which is usually relatively smooth but occasionally rather jumpy as being composed of subseries which are first-order autoregressive, the autocorrelation coefficient being positive for the smooth segments and negative for the jumpy ones. One might try fitting such data with a segmentation of two classes, one corresponding to a positive autocorrelation, the other, to a negative autocorrelation.

The mechanism generating the process changes from time to time, and these changes manifest themselves at some unknown time points (epochs, or change-points)

$$t_1, t_2, \ldots t_{m-1}$$

i.e., the number of segments is m. The integer m and the epochs are unknown. Generally there will be fewer than m generating mechanisms. The number of mechanisms (classes) will be denoted by k; it will be assumed that k is at most m. In some situations, k is specified; in others, it is not. With the cth class is associated a stochastic process, P_c, say. For example, above we spoke of a situation with $k=2$ classes, where, for $c=1,2$, the process P_c is first-order autoregressive with coefficient, say, ϕ_c.

Now with the tth observation ($t=1, 2, \ldots, n$) associate the *label* γ_t, which is equal to c if and only if x_t arose from class c, $c = 1, 2, \ldots, k$. Each time-point t gives rise to a pair

$$(x_t, \gamma_t),$$

where x_t is observable and γ_t is not. The process $\{x_t\}$ is the observed time series; $\{\gamma_t\}$ will be called the *label process*.

Define a *segmentation*, then, as a partition of the time index set $\{t : t = 1, 2, \ldots, n\}$ into subsets

$$S_1 = \{1, 2, \ldots, t_1\}, S_2 = \{t_1 + 1, \ldots, t_2\}, \ldots, S_m = \{t_{m-1}+1, \ldots, n\},$$

where the t's are subscripted in ascending order. Each subset $S_g, g = 1, 2, \ldots, m$, is a *segment*. The integer m is not specified. In the context of this model, to segment the series is merely to estimate the γ's.

The idea underlying the development here is that of *transitions* between classes. The labels γ_t will be treated as random variables Γ_t with transition probabilities

$$Pr(\Gamma_t = d | \Gamma_{t-1} = c) = p_{cd},$$

taken as stationary, i.e., independent of t. The k - by - k matrix of transition probabilities will be denoted by $\underline{P}$, i.e.,

$$\underline{P} = [p_{cd}].$$

If a process is strictly cyclic, like intake, compression, combustion, intake, etc., for a combustion engine, or recession to recovery to expansion to recession, etc., in the business cycle, then this condition can be imposed by using a transition probability matrix with zeros in the appropriate places. We shall consider a matrix like this in Section 2.3.2.

Segmentation will involve the simultaneous estimation of several sets of parameters, the distributional parameters of the within-class stochastic processes, the transition probabilities, and the labels.

A joint probability density function (p.d.f.) for $\{(X_t, \Gamma_t), t = 1, 2, \ldots, n\}$ can be obtained by conditioning each variable on all the preceding ones. The working assumptions behind the method of this paper are as follows.

(A.1) The labels are a first-order stationary Markov chain, independent of the observations; i.e., the probability of being in state d at time $t+1$ given state c at time t, is p_{cd}, which does not involve t or the values of the observations.

(A.2) The distribution of the random variable X_t depends only on its own label and previous X's, not previous labels.

Further details and a mathematical formulation corresponding to these assumptions are given in Sclove (1981).

In regard to (A.2), in the simplest case the X's are (conditionally) independent, given the labels. That is, the distribution of X_t depends only on its label, and not previous X's. In the examples in the present paper this assumption is made. In this case the p.d.f.'s $f(x|\gamma_t = c), c = 1, 2, \ldots, k$, are called *class-conditional* densities. In the parametric case they take the form

$$f(x_t | \gamma_t = c) = g(x_t; \beta_c), \tag{1}$$

where β is a parameter indexing a family of p.d.f.'s of form given by the function g. For example, in the case of Gaussian class-conditional distributions, β_c consists of the mean and variance for the cth class.

This model, with transition probabilities, has certain advantages over a model formulated in terms of the epochs. The epochs behave as discrete parameters, and, even if the corresponding generalized likelihood ratio was asymptotically chi-square, the number of degrees of freedom would not be clear. On the other hand, the transition probabilities vary in an interval and it is clear that they constitute a set of $k(k-1)$ free parameters.

2.2 An Algorithm

2.2.1 Development of the Algorithm

It follows from the assumptions that the likelihood L, i.e., the joint p.d.f., considered as a function of the parameters, can be written in the form

$$L = A(\{p_{cd}\}, \{\gamma_t\}) B(\{\gamma_t\}, \{\beta_c\}). \tag{2}$$

Hence, for fixed values of the γ's and β's, L is maximized with respect to the p's by maximizing the factor A. Now let n_{cd} be the number of transitions from class c to d. (These n's are functions of the labels.) The factor A is merely the point multinomial probability function, the parameters being the n's and p's. It follows that the maximum likelihood estimates of the p's, for fixed values of the other parameters, are given by

$$p_{cd} = n_{cd}/n_c, \tag{3}$$

where

$$n_c = n_{c1} + n_{c2} + \ldots + n_{ck}.$$

Further, given the p's and γ's, the estimates of the distributional parameters—the β's—are easy to obtain. This suggests the following algorithm.

Step 0. Set the β's at initial trial values, suggested, e.g., by previous knowledge of the phenomenon under study. Set the p's at initial trial values, e.g., $1/k$. Set $f(\gamma_1)$ at initial trial values, e.g., $f(\gamma_1) = 1/k$, for $\gamma_1 = 1, 2, \ldots, k$.

Step 1. Estimate γ_1 by maximizing $f(\gamma_1)f(x_1|\gamma_1)$.

Step 2. For $t = 2, 3, \ldots, n$, estimate γ_t by maximizing

$$\gamma_{t-1}\gamma_t f(x_t|\gamma_t, x_{t-1}, \ldots, x_1),$$

as, under (A.1) and (A.2), the likelihood can be expressed as a product of such factors.

Step 3. Now, having labeled the observations, estimate the distributional parameters, and estimate the transition probabilities according to (3).

Step 4. If no observation has changed labels from the previous iteration, stop. Otherwise, repeat the procedure from Step 1.

Step 2 is Bayesian classification of x_t. Suppose the $(t-1)$st observation was tentatively classified into class c. Then the prior probability that the tth observation belongs to class d is p_{cd}, $d = 1, 2, \ldots, k$. Hence all the techniques for classification in particular models are available (e.g., use of linear discriminant functions when the observations are multivariate normal with common covariance matrix).

2.2.2 The First Iteration

When the k class-conditional processes consist of independent, identically distributed normally distributed random variables with common variance, one can start choosing initial means and labelling the observations by a minimum-distance clustering procedure. [This is one iteration of ISODATA (Ball and Hall, 1967). One could iterate further at this stage.] From this clustering initial estimates of transition probabilities and the variance are obtained.

2.2.3 Restrictions On The Transitions

As mentioned above, one might wish to place restrictions on the transitions, e.g., to allow transitions only to adjacent states. (For example, "recovery" is adjacent to "recession," "expansion" is adjacent to "recovery," but "expansion" is not adjacent to "recession.") The model does permit restrictions on the transitions. The maximization is conducted, subject to the condition that the corresponding transition probabilities are zero. This is easily implemented in the algorithm. Once one sets a given transition probability at zero, the algorithm will fit no such transitions, and the corresponding transition probability will remain set at zero at every iteration.

2.3 An Example

Here, in the context of a specific numerical example, the problem of fitting the model for a fixed k and the problem of choice of k will be illustrated. (The additional problem of prediction is considered in Sclove 1981.)

Quarterly gross national product (GNP) in current (nonconstant) dollars for the twenty years 1947 to 1966 was considered. (This makes a good size dataset for expository purposes here.) Parameters were estimated from the first 19 years, the last four observations (1966) being saved to test the accuracy of predictions. (See Sclove 1981.) The data and first difference are given in Table 1. The raw series is nonstationary, so the first differences (increases in quarterly GNP) were analyzed. (Study of more recent data suggested use of first differences of logs; this will be discussed in another context in a later report.) The notation is

$$x_t = \text{GNP}_{t+1} - \text{GNP}_t, \ t = 1, 2, \ldots, 79;$$

e.g., GNP_1 is the GNP at the end of the quarter 1947-1, GNP_2 is that at the end of 1947-2, and $x_1 = \text{GNP}_2 - \text{GNP}_1$ is the increase in GNP during the second quarter of 1947. (A negative value of an x indicates a decrease in GNP for the corresponding quarter.) A Gaussian model was used.

2.3.1 Fitting the Model

In this section we discuss the fitting of a model with $k = 3$ classes, discussion of the choice of k being deferred to the next section. The three classes may be considered as corresponding to recession, recovery, and expansion, although some may prefer to think of the segments labeled as recovery as level periods corresponding to peaks and troughs. The approximate maximum likelihood

Table 1 GNP. Units: billions of current (nonconstant) dollars (from Nelson (1973), pp. 100-101)

Quarter	1	2	3	4	1	2	3	4
1947-48 GNP	224	228	232	242	248	256	263	264
change	4.0	4.2	10.3	5.9	7.6	6.9	1.4	−5.4
1949-50 GNP	259	255	257	255	266	27	293	305
change	−3.3	1.9	−2.1	11.0	9.4	17.7	11.4	13.5
1951-52 GNP	318	326	333	337	340	339	346	358
change	7.8	7.0	4.1	2.6	−0.4	6.5	12.1	6.5
1953-54 GNP	364	368	366	361	361	360	365	373
change	3.3	−1.7	−5.0	−0.1	−0.3	4.3	8.7	12.8
1955-56 GNP	386	394	403	409	411	416	421	430
change	8.2	8.1	6.3	1.8	5.6	4.4	8.9	7.4
1957-58 GNP	437	440	446	442	435	438	451	464
change	3.0	6.4	−4.8	−6.8	3.6	13.1	13.0	9.6
1959-60 GNP	474	487	484	491	503	505	504	503
change	12.9	−2.9	6.5	12.5	1.7	−0.5	−0.9	0.3
1961-62 GNP	504	515	524	538	548	557	564	572
change	11.3	9.3	13.5	10.1	9.4	7.2	7.6	5.4
1964-64 GNP	577	584	595	606	618	628	639	645
change	6.8	10.5	11.1	11.9	10.3	10.9	6.2	17.7
1965-66 GNP	663	676	691	710	730	743	756	771
change	12.9	15.4	18.9	19.5	13.8	12.6	14.8	13.5

solution found by the iterative procedure was (units are billions of current (nonconstant) dollars) −1.3, 6.2, and 12.3 for the means, 2.28 for the standard deviation, and

$$\begin{matrix} .625 & .250 & .125 \\ .156 & .625 & .219 \\ .039 & .269 & .692 \end{matrix}$$

for the transition probabilities. The estimated labels are given below; labels (r = recession, e = expansion) resulting from fitting $k = 2$ classes (see Section

Segmentation 427

2.3.2) are also given. The process was in state 1 for 21% of the time, in state 2 for 44% of the time, and in state 3 for 35% of the time.

An interesting feature of the model and the algorithm is that, as the iterations proceed, some isolated labels change to conform to their neighbors. This should be the case when p_{cc} is large relative to p_{cd}, for d different from c.

2.3.2 Choice of Number of Classes

Various values of k were tried, the results being scored by means of Akaike's information criterion (AIC). (See, e.g. Akaike 1981.) To estimate k one uses the minimum AIC estimate, where

$$AIC(k) = -2\log_e[\max L(k)] + 2m(k).$$

Here $L(k)$ is the likelihood when k classes are used, max denotes its maximum over the parameters, and $m(k)$ is the number of independent parameters when k classes were used. The statistic AIC (k) is a natural estimate of the "cross-entropy" between f and $g(k)$, where f is the (unknown) true density and $g(k)$ is the density corresponding to the model with k classes. (See, e.g., the paper by Parzen in the proceedings of this Workshop for a discussion of the cross-entropy, $H(f;g)$.) According to AIC, inclusion of an additional parameter is appropriate if $\log_e[\max L]$ increases by one unit or more, i.e., if max L increases by a factor of e or more.

The model was fit with several values of k and unrestricted transition probabilities. Also, since it seems reasonable to restrict the transitions to those between adjacent states, such models were evaluated as well. In the case of $k = 3$, where the states might be considered as recession, recovery, and expansion, this means setting equal to zero the transition probabilities corresponding to the transitions, recession-to-expansion and expansion-to-recession. The results are given in Table 2. The best segmentation model, as indicated by minimum AIC, is that with only two classes.

The results for $k = 2$ classes (which might be called recession, expansion) were 0.43 and 10.09 for the means, 3.306 for the standard deviation, and

$$\begin{matrix} .667 & .333 \\ .170 & .830 \end{matrix}$$

for the transition probabilities. The process was in state 1 for 37% of the time and state 2 the other 63% of the time. The labels were given above.

A model with only two classes enjoys advantages relating to its relative simplicity.

3 A MODEL AND METHOD FOR SEGMENTATION OF DIGITAL IMAGES

A digital (i.e., numerical) image may be considered as a rectangular array of picture elements (pixels). These will be indexed by (i,j). At each pixel the same p features are observed. We denote the features by

$$X_1, X_2, \ldots, X_p.$$

Table 2 Fitting models

Model	AIC
Segmentation, 2 classes	481.4[a]
Segmentation, 3 classes, full trans. prob. matrix	483.6
Segmentation, 3 classes, sparse trans. prob. matrix[b]	488.5⁻
Segmentation, 4 classes, full trans. prob. matrix	507.1
Segmentation, 4 classes, sparse trans. prob. matrix[b]	486.8
Segmentation, 5 classes, full trans. prob. matrix	506.5⁺
Segmentation, 5 classes, sparse trans. prob. matrix	stopped[c]
Segmentation, 6 classes, full trans. prob. matrix	stopped[c]

a. Optimum, among segmentation models considered.

b. Allows transitions only to adjacent states.

c. Stopped, i.e., the algorithm reached an iteration where it allocated no observations to one of the classes.

The vector of features is

$$\underline{X} = (X_1, X_2, \ldots, X_p).$$

The digital image is

$$\{\underline{X}_{ij}, i = 1, 2, \ldots, I, j = 1, 2, \ldots, J\},$$

where

$$\underline{X}_{ij} = (X_{1ij}, X_{2ij}, \ldots, X_{pij})$$

is the vector of numerical values of the p features at pixel (i,j).

Examples.

(i) In television, we have $p = 3$ colors,

x_{1ij} = red level at pixel (i,j),

x_{1ij} = green level at pixel (i,j),

and

x_{3ij} = blue level at pixel (i,j).

(ii) In LANDSAT data, $p = 4$ spectral channels, one in the green/yellow visible range, the second in the red visible range, and the other two in the near infrared range.

An *object* is a set of contiguous pixels which may be assumed to be members of a common class. One task of image processing is segmentation, grouping of pixels with a view toward identifying objects.

In this context the *conceptual model* is that the image is a set of pixels, and, also, the image consists of several segments. Each pixel belongs to one and only one segment. The segments fall into several classes. For example, in a picture of a house the classes might be brick, sky, grass, shadow and brush. Note that there might be several separate areas of, say, grass. Each of these areas is a segment, but they all belong to the class, "grass."

The *statistical model* accompanying this conceptual model is as follows:

— with each class of segment is associated a probability distribution for the feature vector $\underline{X}$;
— with each pixel is associated a label which, were it known to us, would tell us which class of segment the pixel belongs to.

Each pixel thus gives rise to a pair $(\underline{X}, \gamma)$, where $\underline{X}$ is observable and γ is not. This is the same as the time-series segmentation model. In the context of this statistical model segmentation is estimation of the set of labels.

Often one considers parametric models, in which the class-conditional probability functions $f_c(.)$ are assumed known, except for the values of distributional parameters. That is,

$$f_c(.) = f(.;\beta_c),$$

where β_c is the parameter. E.g., in the multivariate Gaussian case β_c consists of the mean and covariance matrix for class c.

In order to model the spatial correlation of images, one can assume that the labels Γ_{ij} form a stochastic process, say a Markov process. One reads through the array in a fixed order, say, first row, left to right, second row, left to right, and conditions a given pixel on pixels preceding it in this ordering. Here a first-order Markov process would be one where a given pixel is conditioned on the pixels to the north and west of it. Thus the transition probability matrix has the following form for $k=3$ classes of segment.

Pixel to north	west	1	2	3
1	1	$p_{11,1}$	$p_{11,2}$	$p_{11,3}$
1	2	$p_{12,1}$	$p_{12,2}$	$p_{12,3}$
1	3	$p_{13,1}$	$p_{13,2}$	$p_{13,3}$
2	1	$p_{21,1}$	$p_{21,2}$	$p_{21,3}$
2	2	$p_{22,1}$	$p_{22,2}$	$p_{22,3}$
2	3	$p_{23,1}$	$p_{23,2}$	$p_{23,3}$
3	1	$p_{31,1}$	$p_{31,2}$	$p_{31,3}$
3	2	$p_{32,1}$	$p_{32,2}$	$p_{32,3}$
3	3	$p_{33,1}$	$p_{33,2}$	$p_{33,3}$

The total number of parameters, distributional parameters and transition probabilities, is large. But, as mentioned by Jerome Friedman in this

Workshop, with very large datasets one ought not necessarily shy away from using models with many parameters.

The algorithm developed for segmenting images according to this model is similar to that for segmenting time series, except now the transition-probability matrix is more complicated.

As a sample "image" the Fisher iris data were used. This dataset consists of 4 features measured on 150 flowers, 50 in each of three species. To form an image the 150 flowers were arranged into a 15 × 10 rectangular array, rows 1-5 being species 1, rows 6-10 being species 2, rows 11-15 being species 3. This means that the true segmentation is as follows.

True Segmentation:

Row:	Column:									
	1	2	3	4	5	6	7	8	9	10
1	1	1	1	1	1	1	1	1	1	1
2	1	1	1	1	1	1	1	1	1	1
3	1	1	1	1	1	1	1	1	1	1
4	1	1	1	1	1	1	1	1	1	1
5	1	1	1	1	1	1	1	1	1	1
6	2	2	2	2	2	2	2	2	2	2
7	2	2	2	2	2	2	2	2	2	2
8	2	2	2	2	2	2	2	2	2	2
9	2	2	2	2	2	2	2	2	2	2
10	2	2	2	2	2	2	2	2	2	2
11	3	3	3	3	3	3	3	3	3	3
12	3	3	3	3	3	3	3	3	3	3
13	3	3	3	3	3	3	3	3	3	3
14	3	3	3	3	3	3	3	3	3	3
15	3	3	3	3	3	3	3	3	3	3

Below are given results obtained by starting with initial means equal to the measurements on flowers 50, 100 and 150. (These are easy for the algorithm in the sense that they are in fact from the three different species, but not so easy as, e.g., flowers 1, 51 and 101, which are further apart. Starting with means that are from correct classes is analogous to military applications where something is known about the physical characteristics of target, background, and clutter.) The results in successive iterations were as follows. Convergence was reached on the fourth iteration, i.e., on that iteration no pixel changed class.

Segmentation on Iteration 1:

Row: Column:

	1	2	3	4	5	6	7	8	9	10
1	1	1	1	1	1	1	1	1	1	1
2	1	1	1	1	1	1	1	1	1	1
3	1	1	1	1	1	1	1	1	1	1
4	1	1	1	1	1	1	1	1	1	1
5	1	1	1	1	1	1	1	1	1	1
6	3	3	3	2	3	2	3	2	3	2
7	2	2	2	3	2	2	2	2	2	2
8	3	2	3	2	2	2	3	3	2	2
9	2	2	2	3	2	3	3	2	2	2
10	2	3	2	2	2	2	2	2	2	2
11	3	3	3	3	3	3	2	3	3	3
12	3	3	3	3	3	3	3	3	3	3
13	3	3	3	3	3	3	3	3	3	3
14	3	3	3	3	3	3	3	3	3	3
15	3	3	3	3	3	3	3	3	3	3

Confusion Matrix:
True Class

		1	2	3	
	1	50	0	0	50
Label	2	0	36	1	37
	3	0	14	49	63
		50	50	50	150

Segmentation on Iteration 2:

Row: Column:

	1	2	3	4	5	6	7	8	9	10
1	1	1	1	1	1	1	1	1	1	1
2	1	1	1	1	1	1	1	1	1	1
3	1	1	1	1	1	1	1	1	1	1
4	1	1	1	1	1	1	1	1	1	1
5	1	1	1	1	1	1	1	1	1	1
6	2	3	3	2	3	2	3	2	3	2
7	2	2	2	2	2	2	2	2	2	2
8	3	2	2	2	2	2	2	3	2	2
9	2	2	2	3	2	2	2	2	2	2
10	2	2	2	2	2	2	2	2	2	2
11	3	3	3	3	3	3	2	3	3	3
12	3	3	3	3	3	3	3	3	3	3
13	3	3	3	3	3	3	3	3	3	3
14	3	3	3	3	3	3	3	3	3	3
15	3	3	3	3	3	3	3	3	3	3

Confusion Matrix:
True Class

		1	2	3	
	1	50	0	0	50
Label	2	0	42	1	43
	3	0	8	49	57
		50	50	50	150

Segmentation on Iteration 3:

Row: Column:

	1	2	3	4	5	6	7	8	9	10
1	1	1	1	1	1	1	1	1	1	1
2	1	1	1	1	1	1	1	1	1	1
3	1	1	1	1	1	1	1	1	1	1
4	1	1	1	1	1	1	1	1	1	1
5	1	1	1	1	1	1	1	1	1	1
6	2	3	2	3	2	3	2	3	2	3
7	2	2	2	2	2	2	2	2	2	2
8	3	2	2	2	2	2	2	2	2	2
9	2	2	2	2	2	2	2	2	2	2
10	2	2	2	2	2	2	2	2	2	2
11	3	3	3	3	3	3	3	3	3	3
12	3	3	3	3	3	3	3	3	3	3
13	3	3	3	3	3	3	3	3	3	3
14	3	3	3	3	3	3	3	3	3	3
15	3	3	3	3	3	3	3	3	3	3

Confusion Matrix:
True Class

		1	2	3	
	1	50	0	0	50
Label	2	0	44	0	44
	3	0	6	50	56
		50	50	50	150

All computations reported here were carried out using FORTRAN computer programs written by the author. These programs have been sent to the Statistics Program at ONR for deposit in NRL.

REFERENCES

Akaike, Hirotugu (1981). "Likelihood of a Model and Information Criteria." *Journal of Econometrics 16*, 1-14.

Ball, Geoffrey H., and Hall, David J. (1967). "A Clustering Technique for Summarizing Multivariate Data." *Behavioral Science 12*, 153-155.

Nelson, Charles R. (1973). *Applied Time Series Analysis for Managerial Forecasting*. Holden-Day, Inc., San Francisco.

Sclove, Stanley L. (1981). "On Segmentation of Time Series." Technical Report No. 81-1, ONR Contract N00014-80-C-0408, Task NR042-443, Math. Dept., University of Illinois at Chicago Circle (submitted for publication).

Efficient Algorithms for Digital Processing of Remotely Sensed Imagery*

C. H. Chen

*Department of Electrical and Computer Engineering
Southeastern Massachusetts University
N. Dartmouth, MA*

1 INTRODUCTION

With increased computer capability many optimal digital image processing algorithms that require very extensive computations can now be implemented. Such algorithms however do not guarantee better performance in practice than some robust and computationally less demanding algorithms. Thus effort was made to develop computationally efficient algorithms for segmentation and classification with remotely sensed imagery. Three such algorithms are presented in this paper. The first algorithm employs a two-dimensional (2-D) maximum entropy spectral estimation method. The second algorithm is on automatic spatial clustering. The third algorithm is for tracking a time varying image sequence. The algorithms are demonstrated for their effectiveness by computer results.

2 ALGORITHM FOR 2-D MAXIMUM ENTROPY SPECTRAL ANALYSIS

The 2-D fast Fourier transform suffers the same drawback as the one-dimensional fast Fourier transform with limited spectral resolution for short record length. Several 2-D maximum entropy spectral estimation methods have been developed with significantly improved spectral resolution. However

---

*Work supported under Contract N00014-79-C-0494, Task NR 042-422 with the Office of Naval Research.

so far only the method of Lim and Malik [1] is suitable for minicomputer implementation. They developed the method for estimating two-dimensional sinusoids with limited or incomplete correlation data. The method is adapted to the image analysis [2] [3] for texture feature computation. The method is based on the notion that the given correlation points in region A is consistent and the corresponding correlation coefficients should be zero outside region A after each iteration. By repeatedly correcting the correlation coefficients after each iteration with the known correlation points, the method will converge to the correct spectrum. The method is computationally simple due to the utilization of fast Fourier transform.

Figure 1a shows a test Seasat image along with 64 × 64 subimage. The contour plot of 2-D maximum entropy spectrum of the subimage is shown in Figure 1b with increment of 5dB. The corresponding three-dimensional display at (0.5, 0.5) point of view is shown in Figure 1c. Extensive further spectral estimation results are reported in Reference 3. The accuracy of spectrum estimation depends on the size of the region of known correlation points. The amount of computation increases with the region size. A typical choice of size 7 × 7 is required for moderate to large signal-to-noise ratio. Exact expression for computational requirement remains to be determined.

For texture image classification, two spectral features are often obtained from the estimated spectrum. The ring feature is the power of a ring-shaped region obtained by integrating the spectrum over two concentric circles. The wedge feature is the power of a wedge-shaped region obtained by integrating the spectrum over two angles. As the spectrum is computed at discrete frequencies, the integration must be replaced by summation. The ring feature represents the texture coarseness while the wedge feature represents texture directionality. Neither feature can represent both properties which are equally important in texture discrimination. It is determined [2] that equalized features which are the powers over various intersections of rings and wedges perform much better. Furthermore two or more equalized features must often be used to provide a desired classification accuracy.

An important parametric modeling of images is the AR (autoregressive) model. The 2-D maximum entropy spectral analysis is based on the assumption of autoregressive process for the data. Thus accurate AR coefficients of the image model can be calculated by the 2-D maximum entropy method. The probability density of the image can then be determined from which each pixel can be classified as belonging to one of several regions or segments. A maximum likelihood approach is to choose a region that maximizes the probability density for the image, given a specified set of regions. A maximum a posteriori approach is to maximize the probability of a given set of regions conditioned on our observation of the image [4]. These two approaches are employed for terrain image segmentation.

Figure 2a shows a digitized aerial photograph of a rural area containing some trees and fields, with the result of maximum likelihood approach shown in its right. The image is of size 128 × 128 pixels with gray levels 0-255. Two regions are assumed and the data used to determine the AR model coefficients are shown in the white boxes (32 × 32 pixels). The order of the AR model is 4 in this example. The maximum a posteriori approach is an iterative Bayes estimation procedure where the regions are treated not as unknown parameters

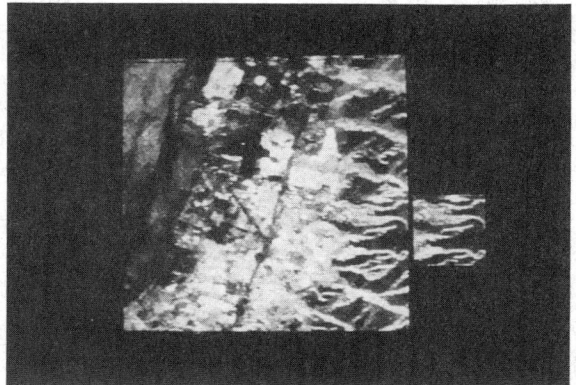

Figure 1a Test seasat image with 64×64 subimage

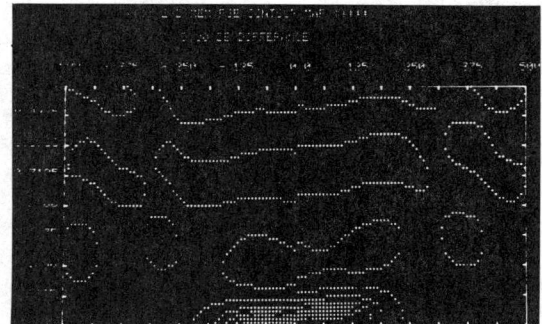

Figure 1b Increment of 5dB

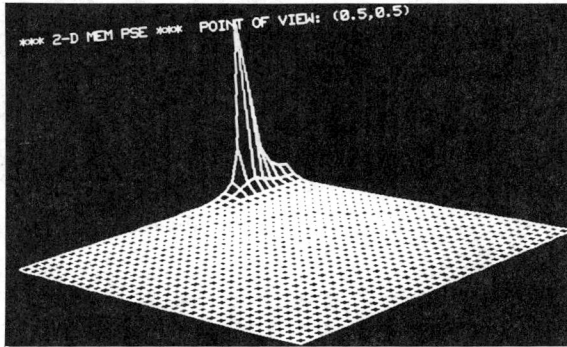

Figure 1c Three dimensional display

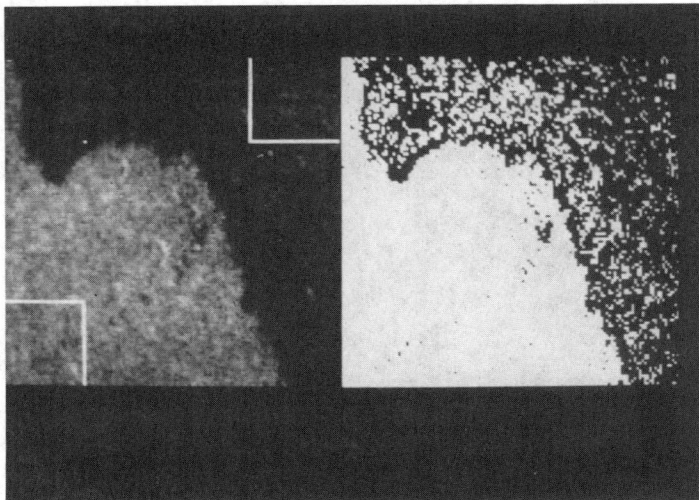

Figure 2a Digitized aerial photograph

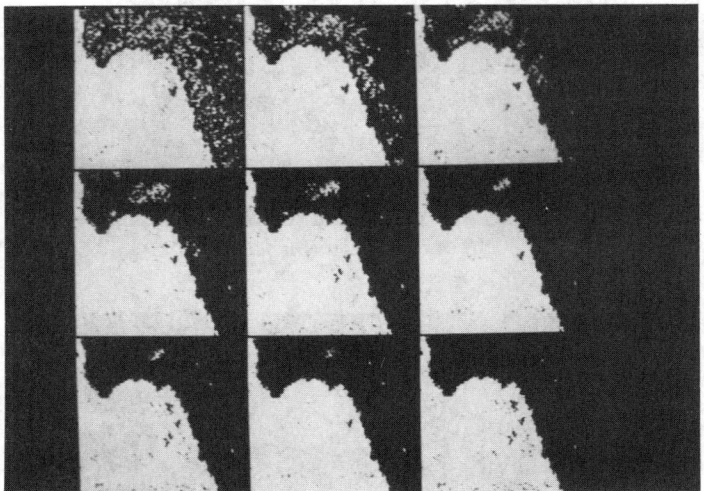

Figure 2b Convergence of segmentation

as in the maximum likelihood approach, but as random quantities whose statistical properties are described by Markov transition probabilities. The solution procedure bears some resemblance to relaxation labeling technique [5], in that the state assignments, viz, the classification of all pixels, are iteratively updated by considering the state assignments of neighboring pixels. Figure 2b shows the convergence of the segmentation by starting with zero iteration (same as maximum likelihood result) and proceeding with 1 to 8 iterations. The approach converges quite well after 8 iterations.

3 ALGORITHM FOR AUTOMATIC SPATIAL CLUSTERING

Conventional clustering methods do not preserve the spatial relations in an image. Spatial clustering for image analysis has been considered [6] [7]. However feature extraction was not taken into account. Furthermore the computation involved is quite extensive. A more efficient algorithm is developed that employs properly selected local features. The procedures are as follows:

(1) Form a feature set, for each pixel, consisting of local mean and gradient. Other features may also be used.
(2) For each 2 × 2 subarea, measure the mean vector and dispersion.
(3) Determine the critical dispersion, and calculate the merging distance d.
(4) Merge adjacent subareas with distance less than d to form subregions. Calculate the mean vectors of subregions.
(5) Group these mean vectors into clusters using K-mean algorithm which converges to several cluster centers representing the mean vectors of regions.

To have the algorithm automatically adjusted to an appropriate number of clusters (regions), an interregion threshold distance must be specified, which can determine the required number of regions. The algorithm can also work with a specified number of clusters.

(6) Classify every pixel into regions through the minimum distance between this pixel and the cluster centers.

Besides the clustering objective, the algorithm clearly provides image data compression and contrast enhancement. For another set of Seasat SAR image as shown in Figure 3a (Los Angeles area), the result of spatial clustering to 7 clusters is shown in Figure 3b. Because of the multiplicative noise, a simple filtering is performed before clustering. The filtering is based on the technique suggested by J. S. Lee [8]. For each 5 × 5 subarea the average gray level value is compared with the three-standard deviation (3σ) of the image histogram. If the value exceeds 3σ, then no averaging is used. If the value is within 3σ then the center pixel is replaced by the average value. The automatic spatial clustering of the filtered image is shown in Figure 3c. The "spotty" noises are considerably reduced. The filtering operation adds a slight computation to the algorithm.

4 ALGORITHM FOR TARGET TRACKING VIA IMAGE SEGMENTATION

In surveillance and tracking using imagery sensors, targets must be detected and tracked at various signal-to-noise ratios for a fixed or varying target size. Related work in this area includes the use of Bayes classifier for image segmentation [9] and the use of a semicasual recursive filter for image enhancement such that the target can be detected and tracked [10]. Our algorithm uses the Fisher's linear discriminant for pixel classification to segment simulated static

Figure 3a Set of seasat SAR image

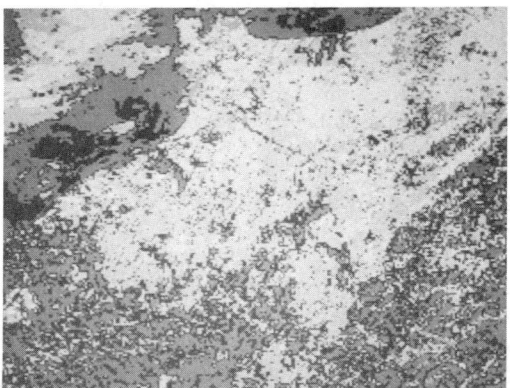

Figure 3b Spatial clustering to 7 clusters

Figure 3c Automatic spatial clustering of the filtered image

Figure 4a 32 × 32 artificial picture

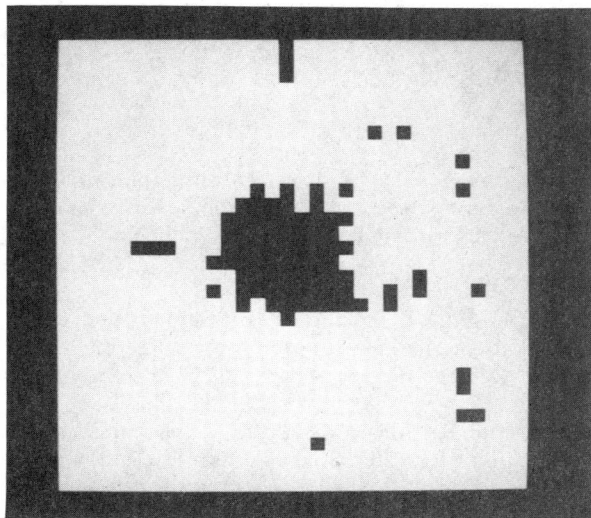

Figure 4b Pixel classification using 100 learning samples

and dynamic scenes. With estimated mean vectors and covariance matrices for target and background, the segmentation can be performed efficiently. The segmentation performance, which translates to detection performance, versus the learning sample size for dynamic scenes can be determined empirically [11]. Small target can still be detected if a sufficiently large learning sample size is available. Figure 4a is an original 32 × 32 artifical picture where the object or target box (8 by 8) is Gaussian distributed with mean 10 and variance 2 while the background is also Gaussian distributed with mean 8 and variance 2. The result of pixel classification using 100 learning samples is shown in Figure 4b. The object clearly can be extracted. The experiment performed has included various target sizes and learning sample sizes. For more complex scenes the performance of the algorithm is obviously limited as the interimage information is not utilized. The algorithm must be improved to incorporate such information.

5 CONCLUDING REMARKS

Three efficient image processing algorithms have been presented especially for minicomputer use under limited memory capacity. The algorithms perform well in image segmentation and classification. No comparison has been made with other algorithms in terms of the number of computations. Improvement in each algorithm is still desirable to reduce further the computation requirements. Parallel processing may be considered especially when local statistics is employed. The computer results presented have illustrated the effectiveness of the algorithms.

ACKNOWLEDGMENT

The digital Seasat images were kindly provided by Dr. Jong-Sen Lee of Naval Research Laboratory who also brought to our attention of his work on the removal of multiplicative noises.

REFERENCES

[1] J.S. Lim and N.A. Malik, "A new algorithm for two dimensional maximum entropy power spectral estimation," IEEE Trans. on Acoustic Speech and Signal Processing, Vol. ASSP - 29, No. 3, pp. 401-413, June 1981.

[2] C.H. Chen, "A study of texture classification using spectral features," Proc. of the 6th International Conference on Pattern Recognition, Munich, Germany, October 1982.

[3] C.H. Chen, "Nonlinear Maximum Entropy Spectral Analysis Methods for Signal Recognition," Research Studies Press, England, October 1982.

[4] C.H. Chen, et al., "Some Experimental Results on Linear Estimation for Image Analysis," Technical Report prepared for Contract N00014-79-C-0494, TR-81-4, January 28, 1981.

[5] A Rosenfeld, R.A. Hummel, and S.W. Zucker, "Scene labeling by relaxation operations," IEEE Trans. on Systems, Man and Cybernetics, Vol. SMC - 6, p. 420, 1976.

[6] R.M. Haralick and I. Dinstein, "A Spatial Clustering Procedure for Multi-Image Data," IEEE Trans. on Circuits and Systems, Vol. C-22, pp. 440-450, 1975.

[7] Y. Fukada, "Spatial Clustering Procedures for Region Analysis," Pattern Recognition, Vol. 12, No. 6, pp. 395-403, 1980.

[8] J.S. Lee, "The Sigma filter and its application to speckle smoothing of synthetic aperture radar image," *in this volume*.

[9] C. Skevington, G.M. Flachs, and B. Schaming, "A Statistical Approach to Image Segmentation," Proc of IEEE Pattern Recognition and Image Processing Conference, pp. 267-272, August 1981.

[10] J. Leite Pereira Filho, "Interframe Image Processing with Application to Target Detection and Tracking," Ph.D. Thesis, Naval Postgraduate School, March 1979.

[11] C.H. Chen and W.H. Yang, "An Experiment on Target Tracking Via Image Segmentation," Technical Report prepared for Contract N00014-79-C-0494, TR-82-2, Jan. 11, 1982.

The Sigma Filter and Its Application to Speckle Smoothing of Synthetic Aperture Radar Images[*]

Jong-Sen Lee

*Aerospace Systems Division
U.S. Naval Research Laboratory
Washington, DC*

1 INTRODUCTION

Generally, digital image smoothing techniques fall into two categories. In the first category, the noisy image is processed globally in the sense that the whole or a large section of a noisy image are correlated to obtain a smoothed image. Techniques in transform domain using Wiener or least square filtering [1-2] and techniques applying one-dimensional or two-dimensional Kalman filter are in this category. Statistical models for the signal (noise free image) and the noise are required for the implementation of these techniques. Unfortunately, the statistical model for most images is either unknown or impossible to describe adequately with a simple random process. The smoothed images display blurred edges and conceal subtle details. In addition, they are computationally costly. In the second category local operators are applied to noisy images. Only those pixels in a close neighborhood of the concerned pixel are involved in the computation. The immediate advantage of these techniques is their efficiency. They have great potential for real-time or near real-time implementation, because several pixels can be processed in parallel without waiting for their neighboring pixels to be processed. Recent researches in image smoothing and segmentation favor the local techniques.

There are many local smoothing methods. The well-known Median filter [4] of one or two dimensions attracted much attention. Edge preserving

---

[*]Work supported under Contract N00014-81-C-0534, Task NR 042-447 with the Office of Naval Research.

smoothing by Nagao and Matsugyama [5], gradient inverse weighting scheme by Wang, et al. [6], Box filtering algorithm [7], and Local statistics method by Lee [8, 9] are just a few other algorithms in this category. Obviously, it is nearly impossible to rank them, because an algorithm may be effective for a class of images, but ineffective for others. In this paper a new class of local smoothing schemes is introduced. It is motivated by the sigma probability of Gaussian distribution. The basic idea is to replace the pixel to be processed by the average of only those neighboring pixels having their intensity within a fixed sigma range of the center pixel. Replacing the center pixel by the average of selected neighboring pixels has been explored by many algorithms. The Nagao's filter [5] replaces the center pixel by the average of a subregion which has the minimum variance. Lee [9] in his refined local statistics method selected the region by using the gradient information. Graham [10] and Prewitt [11] replace a pixel by the average of surrounding area if the absolute value of their difference is smaller than some threshold. Rosenfeld [1] in his region growing and tracking algorithm excludes high contrast edges, lines, and points from the average by judging the gray level difference between the average of the region and the new pixel. The extended box-filtering algorithm [7] restricts the average to only neighboring pixels within a fixed intensity range. The main difference between the box filter and the sigma filter of this paper is that the former has the intensity range fixed throughout the whole image, while the latter lets the intensity range float with the intensity of the center pixel. The advantages are numerous: (1) noise near edge areas will be smoothed without blurring the edge because only pixels on one side of the edge are included in the average; (2) subtle details of several pixel clusters and linear features of one to three pixels in width will be preserved since only those pixels and not the background are included in the average; (3) it will not create artifacts and will retain shapes, because no directional masks are used, contrary to algorithms of Nagao [5] and Lee [9]; (4) it is computationally efficient, since only simple compare and fixed point add instructions are involved.

The sigma filter can be easily modified to smooth multiplicative noise corrupted images. Multiplicative noises or speckles occur in coherent optical images as well as synthetic aperture radar images. The peculiar granular pattern produced by the speckling effect is due to the interference of the coherent and dephased wavelets in the former case, and the coherent processing of radar signals in the latter [13]. Several SEASAT SAR images which were digitally correlated by Jet Propulsion Laboratory (JPL) are adapted to substantiate the effectiveness of the sigma filter. Comparisons are also made with the straight averaging filter and the median filter.

This paper is a combination of two recent studies ([15] and [16]). For a more detailed discussion please refer to them.

2 THE SIGMA FILTER

The noise in an image is generally considered as spatially uncorrelated and with continuous intensity spectrum. The white Gaussian noise is an example. We shall regard noise as any random clutter of the size of three or fewer pixels. It is well-known that the "straight" averaging filter will smooth noise at the expense of blurring edges and smearing subtle details. An indiscriminated average of pixels in a window is the cause of the problem. As mentioned in the introduction section many schemes were developed to overcome this problem. Merits of these algorithms will be explored in more detail in the next section. In this section, a conceptually simple algorithm is developed which easily excludes significantly different pixels from the average.

Most image noise is Gaussian in distribution. The two-sigma probability is defined as the probability of a random variable within two standard deviations from its mean. The two-sigma probability for one-dimensional Gaussian distribution is .955. It can also be interpreted that 95.5% of random samples lie within the range of two standard deviations. In image smoothing, any pixel outside the two-sigma range comes most likely from a different population and, therefore, should be excluded from the average. If we assume that the a priori mean is the gray level of the pixel to be smoothed, we can establish a two-sigma range from the gray level and include in the average only those pixels within the two-sigma intensity. Let $x_{i,j}$ be the intensity or gray level of pixel (i,j) and $\hat{x}_{i,j}$ be the smoothed pixel (i,j). Also we assume that the noise is additive with zero mean and standard deviation σ. The multiplicative noise case will be discussed later in Section IV. The sigma filter procedure is then described as follows:

(1) establish an intensity range $(x_{i,j} - \Delta, x_{i,j} + \Delta)$, where $\Delta = 2\sigma$.
(2) sum all pixels which lie within the intensity range in a $2n+1, 2m+1)$ window.
(3) compute the average by dividing the sum by the number of pixels in the sum.
(4) $\hat{x}_{i,k}$ = the average. (To reduce sharp spot noise, step (4) will be modified later in this section.)

Or, mathematically, let

$$\delta_{k,1} = \begin{cases} 1, & \text{if } (x_{i,j} - \Delta) \leq x_{k,1} \leq (x_{i,j} + \Delta) \\ 0, & \text{otherwise} \end{cases}$$

then

$$\hat{x}_{i,j} = \sum_{k=i-n}^{n+i} \sum_{l=j-m}^{m+j} \delta_{k,l} x_{k,l} \Big/ \sum_{k=i-n}^{n+i} \sum_{l=j-m}^{m+j} \delta_{k,l}. \quad (2)$$

The two-sigma range is generally large enough to include 95.5% of pixels from the same distribution in the window, yet in most cases it is small enough to exclude pixels representing high-contrast edges and subtle details. The main drawback is that sharp spot noise represented by clusters of one or two pixels will not be smoothed. This could be very annoying especially for a fairly noisy image. To remedy this, we shall replace the two-sigma average with the center pixel's immediate neighbor average, if M the total number of pixels within the intensity range is less than a prespecified value K. In other words, the step (4) is replaced by

$$\hat{x}_{i,j} = \begin{cases} \text{two-sigma average, if } M > K. \\ \text{immediate neighbor average, if } M \leq K. \end{cases} \quad (4)$$

The value of K should be carefully chosen to remove isolated spot noise without destroying thin features and subtle details. For a 7×7 window, K, should be less than 4, and should be less than 3 for a 5×5 window. It should be noted that subtle textures within the two-sigma range will be wiped out after a few iterations. If conservation of texture information is required, small Δ range and no more than one or two iterations should be applied.

For images with unknown noise characteristics, the intensity range Δ, can be determined either from a rough estimation of noise standard deviation in a flat area, or from the desirability of retaining the gray level difference between the desirable features and its background. The sigma filter can be applied repeatedly with reduced σ after each iteration. Two or three iterations are generally sufficient to reduce the noise level significantly.

For illustration, Figure 1 (A) shows a medical image of cell structure. The results of applying the 7×7 sigma filter once, twice and three times, are shown Figure 1 (B), 1(C), and 1(D) respectively. The result of applying the median filter twice is shown in Figure 1(F). It should be noted that 1(E) is the result of applying a derivative version of the sigma filter to be discussed in the next section.

3 A COMPARISON OF LOCALLY SMOOTHING ALGORITHMS

Numerous local image smoothing algorithms have been developed recently. It is practically impossible to compare all of them in detail. The straight local average method is known to blur edges and details. Lee [9, 13] using local statistics method produced good results for images corrupted by both additive and multiplicative noise. However, artifacts are observed in some cases, and the computation of the local variance makes this algorithm somewhat inefficient. It is excluded in the present comparison. The recently published gradient inverse method [6], the edge preserving smoothing scheme by Nagao and Matsuyama [5], and the well known median filter are chosen instead. The gradient inverse weighting scheme employs a 3×3 window and computes for each pixel its weighted average of inverse gradient with its neighboring pixels. The idea is to weight less those pixels having greater absolute differences with their center pixel. Nagao and Matsuyama [5] proposed an algorithm which selects the most homogeneous neighborhood and replaces the pixel by its neighborhood average. They created nine overlapped subregions in a 5×5 window. The means and variance of the nine subregions are computed, and the center pixel is replaced by the mean of the subregion having the minimum variance. The median filter is more flexible. It can be applied column wise, row wise and area wise. In our study, 3×3 window is used, and the median of the nine pixels in the window represents the smoothed pixel. The reason for not using a large window is that a large window will smear details and edges not to mention the higher computational load. Two test images shown in Figure 2 and Figure 3, of dimension 128×128 pixels are used in our comparison. A

Sigma Filter

Figure 1 This figure shows the results of the 7×7 sigma filter when applied once, twice, and three times to a medical image. The biased sigma filter when applied to (D) is shown in (E), and the result of 3×3 median filter is shown in (F) for comparison.

computer generated pattern of bars with increasing width from one pixel, three pixels to 15 pixels is created, and corrupted with noise to test the ability to preserve linear features, the ability to smooth noise along edges, and the effectiveness of noise reduction in general. The average intensity of the bar is 150 and the background is 50. The second image is a natural aerial scene but artificially corrupted with noise. The intensity levels in all images in this paper are between 0 and 255. Each algorithm is applied to the noisy image repeatedly 3 times. The sigma filter is applied in a 7 × 7 window with the intensity intervals diminishing from $2\sigma, \sigma,$ to $\sigma/2$, and $K=2$.

3.1 Effectiveness in Noise Smoothing

The efficiency of smoothing noise can be measured by the reduction in noise standard deviation or variance. For images of Figure 2 the standard deviations of each smoothed image are computed from a flat area in the lower left corner. The gradient inverse filter is apparently the least efficient smoothing algorithm due to its small mask and the nature of its weighting scheme. The sigma filter is significantly superior in smoothing noise with a reduction of standard deviation by approximately a factor of ten. Nagao's and the 3 × 3 median filters are comparable in their ability to reduce noise.

3.2 Preservation of Subtle Details and of Linear Features

In some images it is important to retain highly distinguishable subtle details and line features, such as piers and roads. In other application such as the image segmentation, it may be desirable to remove subtle details. The sigma filter is effective in preserving subtle details and line features as long as the intensity difference between them and their background is greater than the two-sigma intensity range. The gradient inverse method theoretically will smear any feature of any size if applied a sufficient number of times, since it includes all pixels in the average and only weighs them less if the difference is large. Similarly the Nagao's filter, will blur and eventually devour any feature with dimension of three pixels or less in any direction. As seen in Figure 2(D) and 2(I), the bars of one and three pixels are almost completely wiped out. The 3 × 3 median filter will wipe out single pixel line in one application, since in a 3 × 3 mask, among the nine pixels, six of them will be background pixels. Thus the median will approach the background pixel. For a 5 × 5 median filter, the three pixel wide bar will be wiped out in one application. Images in Figure 3 further substantiate the characteristics of these algorithms under study. Figure 3(C), 3(D), 3(E) and 3(F) are the results of applying the respective smoothing algorithm three times. The gradient inverse scheme did not do much about the noise and reduced slightly the contrast of the image. As shown in Figure 3(E), Nagao's filter smeared bridges and subtle detail and created artifacts. The 3 × 3 median filter smeared the bridge and generally blurred the image. The sigma filter performed fairly well except for the sharp spot noise problem.

3.3 Immunity From Shape Distortion

The gradient inverse method is not effective in smoothing noise, but it is relatively free from artifacts and shape distortion. The Nagao's filter on the other

Sigma Filter

Figure 2 Two noise corrupt images ((A) and (F)) of bars with increasing width from one pixel, three pixels to 15 pixels. Several noise smoothing algorithms are applied, and the results are shown in (B) - (E) and (G) - (J) respectively, to test their effectiveness in smoothing noise while retaining subtle details.

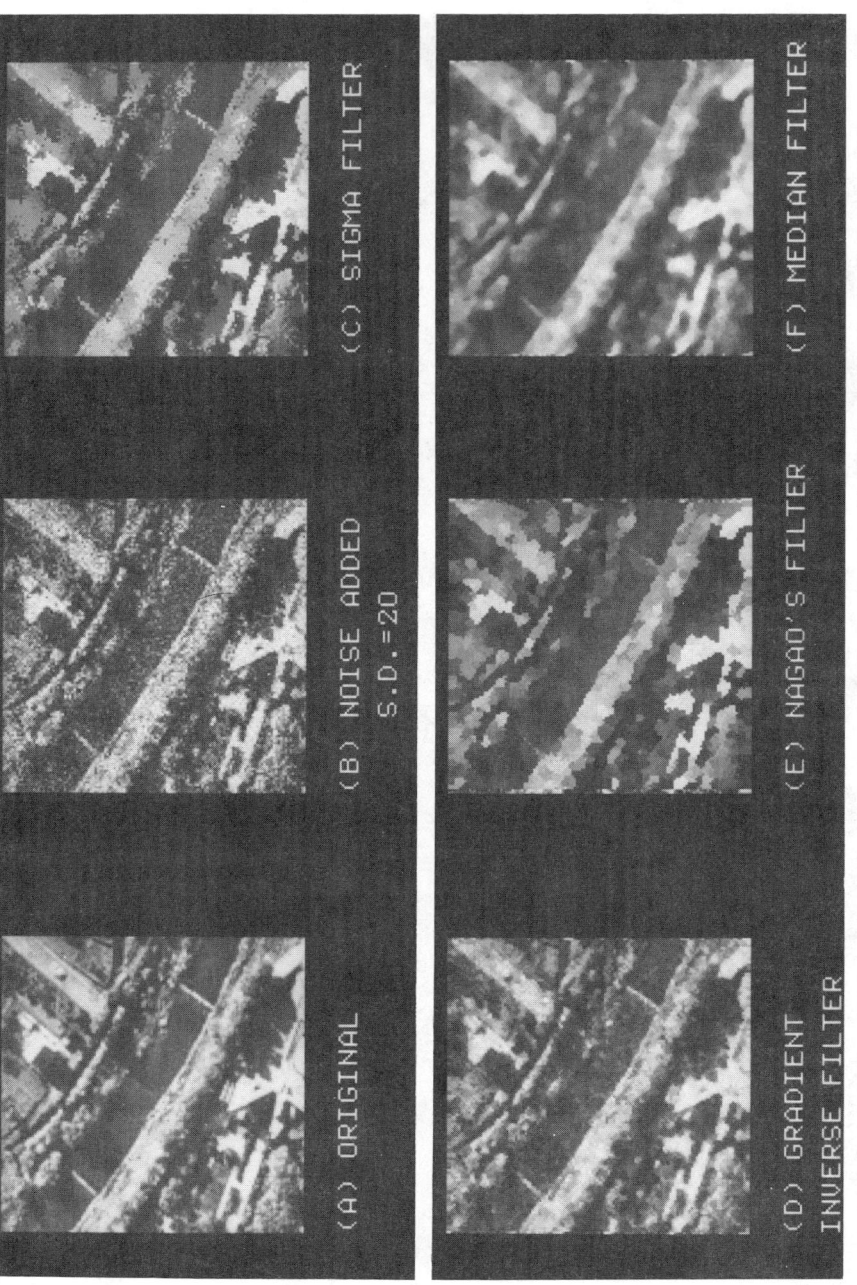

Figure 3 (A) An aerial image (courtesy of Image Processing Institute, USC. (B) Gaussian noise of standard deviation 20 is artificially added to (A); (C), (D), (E) and (F) are the results applied by the sigma filter, gradient inverse filter, Nagao's filter, and median filter respectively.

hand, as shown in Figure 3(E) does create significant distortion because of directional subregion average. It will round off corners less than 90 degrees. Median filter is known to create artifacts. A 3 × 3 median filter will round off corners and produce patterns of patches, the same as Nagao's filter. As shown in Figure 3(C), the sigma filter is practically free of shape distortion.

3.4 Retention of Step Edges and Sharpening Ramp Edges

The intensity variations in the direction perpendicular to a sharp edge form a step edge. Retaining the sharpness of a step edge is highly desirable in both image smoothing and segmentation. The gradient inverse filter will blur the step edge as it computes average on all pixels. The median filter will maintain a noise-free step edge; yet, it will smear a noisy step edge. Sharpening a ramp edge is generally of interest in studies of image segmentation by gray level difference. In this application Nagao's filter is excellent due to its directional subregion average. The other three algorithms will not sharpen a ramp edge but all will maintain a ramp edge fairly well.

3.5 Removing Spot Noise

The median filter is well known for its effectiveness in removing sparsely positioned sharp spot noise, since the spot noise has the intensity in either end of the intensity scale. Nagao's filter is also effective, but requires a few iterations. The gradient inverse filter weighs the spot noise much higher than its surrounding pixels. Consequently, it is not effective. The sigma filter with large window size is highly susceptible to spot noise, since no other pixel, but the spot noise itself is within the two sigma range. The modified version with threshold K, discussed in the last section will remove most isolated spot noise. However, spot noise near the edges remains because the 7 × 7 mask contains several edge pixels which will fall into the two sigma range. Increasing the value of K will further reduce the spot noise, but at the expense of blurring edges and subtle details.

3.6 Computational Efficiency

In our comparison, algorithms were coded in Fortran and no special efforts were devoted to accelerate their executions. The computations were carried out on a Data General NOVA 800 with Comtal 8000 image display. The ratio of computational time required for image of size 128 × 128 for each filter and each iteration is listed in increasing order as follows

1. The sigma filter (7 × 7), 1 unit of time
2. The median filter (3 × 3), 1.5 unit of time
3. The gradient inverse filter, 4.0 unit of time
4. The Nagao's filter, 11.0 unit of time

The sigma filter is the fastest algorithm in this group even with a 7 × 7 window. In our simulation, it took no more time than computing a straight

7 × 7 average. Nagao's filter is extremely slow, since it requires the computations of variance for nine subregions.

In conclusion, we found that the sigma filter performs better than other algorithms except the ability of removing spot noises, and sharpening ramp edges. A derivative of the sigma filter which will sharpen a ramp edge is given in Reference [15]. The result of applying it to Figure 1(A) is given in Figure 1(E).

4 MULTIPLICATIVE NOISE SMOOTHING BY THE SIGMA FILTER

It has been mentioned in the introduction section that speckles in synthetic aperture radar (SAR) images are multiplicative in nature. Let $x_{i,j}$ be the intensity of an observed image pixel and $y_{i,j}$ be the noise-free image pixel which we wish to recover. Then

$$x_{i,j} = y_{i,j} v_{i,j} \qquad (3)$$

in which $v_{i,j}$ represents the multiplicative noise with mean = 1 and variance = σ_v^2. The validity of this model for SEASAT SAR images has been tested and verified [14] with images obtained from JPL. The value of σ_v is found to be 0.28. For details, please refer to Reference[14]. The extention of the sigma filter for additive noise to that for multiplicative noise is straightforward. The two-sigma range is no longer constant but changes with $x_{i,j}$. From Equation (3), the standard deviations of $x_{i,j}$ is equal to $y_{i,j}\sigma_v$. Based on the same assumption of additive noise case, we let $x_{i,j}$ be the a priori mean of $y_{i,j}$. Thus, we have $\sigma = x_{i,j}\sigma_v$. Consequently, the two sigma range Δ in step (1) of Section 2 will be modified to $2\sigma_v x_{i,j}$. In other words, the average number of window pixels which falls in the intensity range of $(x_{i,j} - 2\sigma_v x_{i,j}, x_{i,j} + 2\sigma_v x_{i,j})$ replace the center pixel as the smoothed value of $x_{i,j}$.

5 EXPERIMENTS WITH SEASAT SAR IMAGES

Several SEASAT SAR images digitally processed by JPL are utilized to demonstrate the effectiveness of the sigma filters. All images contain 256 × 256 picture elements. They are shrunk by a factor of two from the original to cover a larger geographical area. The shrinking process generates a pixel from an average of four neighbors. Consequently, σ_v, the standard deviation of noise, is reduced by a factor of 2, or $\sigma_v = 0.14$.

In Figure 4, the sigma filter is compared with the local statistics method [14], the median filter [4], and the straight average filter. The sigma filters are applied four times, twice with the 7 × 7 window and another two times with the 5 × 5 biased sigma filter. The local statistics method is applied the same way as in the Reference [14]. The window size of the median filter is chosen to be 3 × 3, as large window size has the tendency of blurring the image. The result of applyng the median filter will blur the edges and smear fine features if applied several times. The straight 3 × 3 average filter shows the blurring effect in general, with noise reduction comparable with that of the median filter. The edge blurring effect is further illustrated in Figure 5 by comparing the intensity profiles of a scan line for the original, the sigma filter, the straight

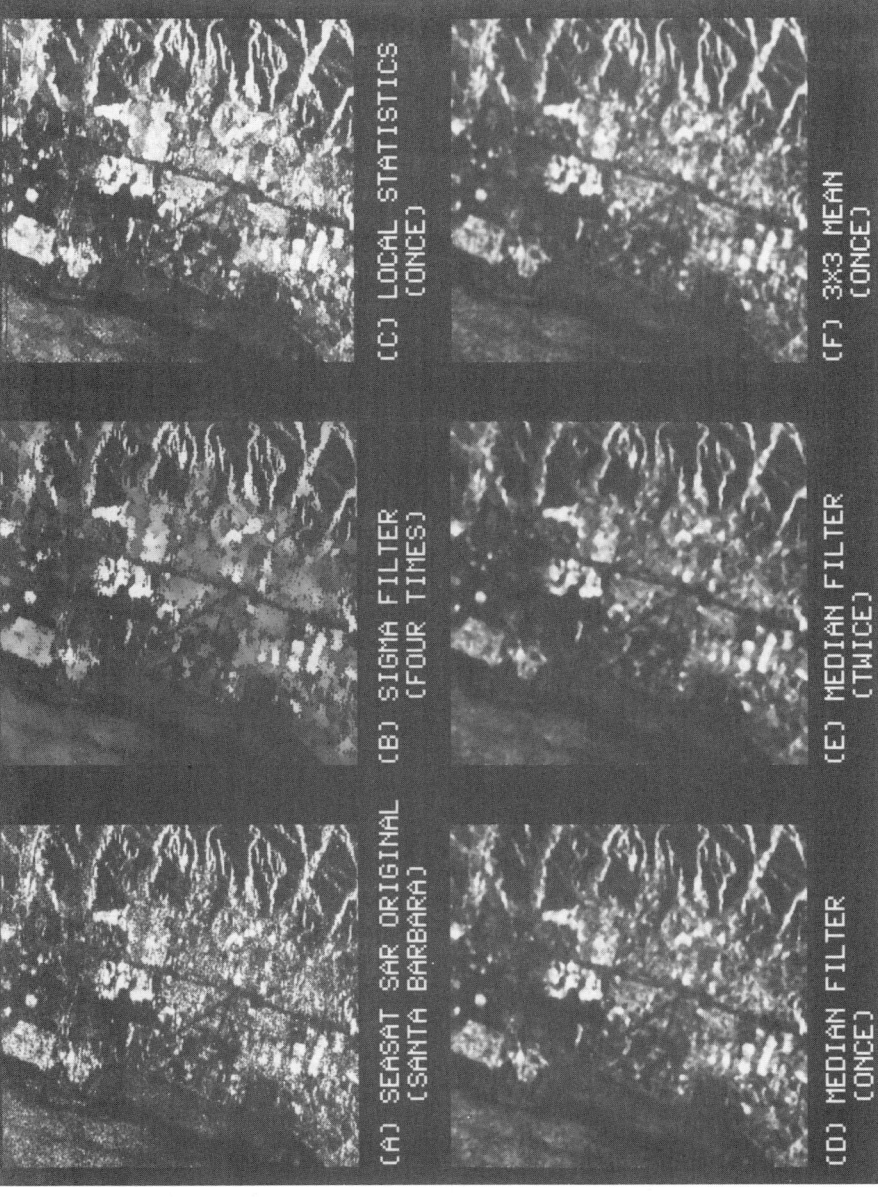

Figure 4 The result of applying the sigma filter (B) is compared with that of the local statistics method (C), the median filter (D) (E), and the straight 3×3 average (F)

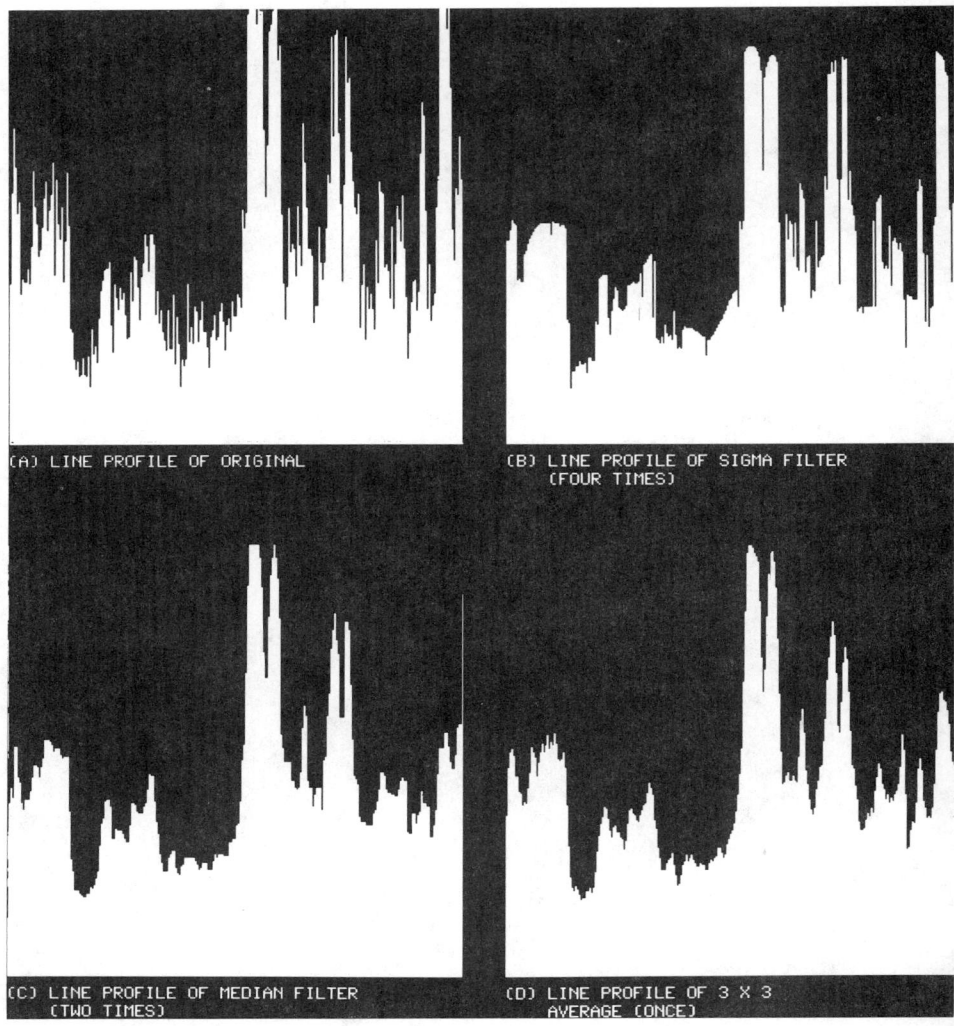

Figure 5 The intensity profiles of a scan line for the images in 4(A), 4(C), 4(E) and 4(F). The edge blurring effect of the median filter and the 3×3 straight average filter is clearly shown.

3 × 3 average, and the median filter. Figure 6 shows the Baltimore harbor area; the quality of the image has been improved significantly by suppressing speckles while preserving the sharpness of bridges and shore lines. The water detail of the harbor is also better defined. Another application is to use the sigma filter together with multilevel thresholding to segment the ocean into sea states. Loosely speaking, the intensity of an ocean area in a SAR image increases with the speed of surface wind. Figure 7(A) shows an ocean area of 50 km × 50 km, and Fig. 7(B) shows the results of segmentation for sea states.

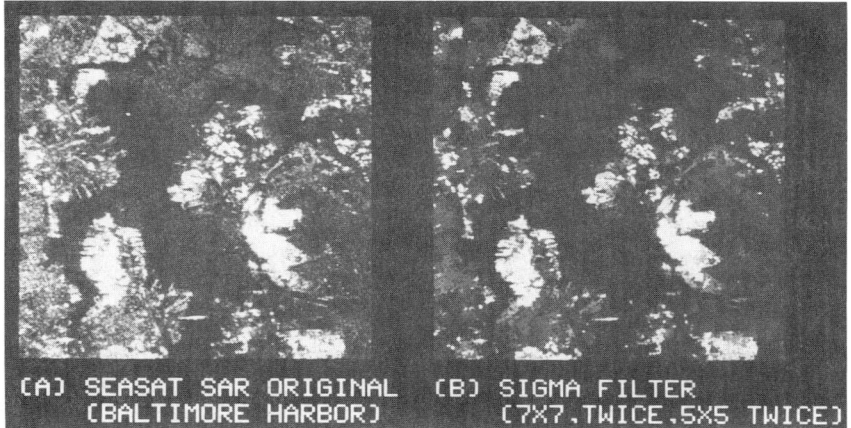

Figure 6 The result of applying sigma filter on the Baltimore harbor image (256×256). The quality of the image has been improved significantly and the water detail is also better defined.

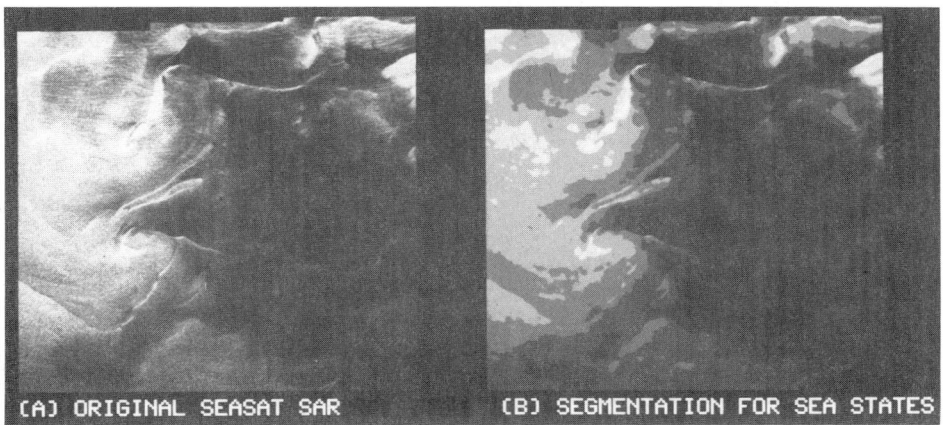

Figure 7 The sigma filter is used to segment the 50km×50 km ocean area (A). The segmented image is shown in (B).

6 CONCLUSIONS

A simple, effective and computationally efficient noise smoothing algorithm has been developed. Detailed comparisons with a few local smoothing algorithms are made to substantiate the basic characteristics of this filter. The procedure and strategy of utilizing this filter has been explored. The sigma filter has been extended to smooth images with multiplicative noise and sucussfully applied to smooth speckles in the SAR images. It is hoped that the sigma filter will be accepted as a basic digital image processing technique because of its simplicity and effectiveness.

ACKNOWLEDGMENT

The author wishes to thank Dr. I. Jurkevich and Dr. A. F. Petty for many helpful discussions.

REFERENCES

1. A. Rosenfeld and A.C. Kak, Digital Picture Processing, Academic Press, New York, N.Y. 1976.

2. H.C. Andrews and B.R. Hunt, Digital Image Restoration, Prentice-Hall, Englewood Cliffs, N.J. 1977.

3. A.K. Jain A Semicausal Model for Recursive Filtering of Two Dimensional Images," *IEEE Trans. Computer*, Vol. C-26, April 1977.

4. W.K. Pratt, Digital Image Processing, Wiley, New York, 1978.

5. M. Nagao and T. Matsuyama, "Edge Preservintg Smoothing," *Computer Graphics and Image Processing*, Vol. 9, 394-407, 1979.

6. D. Wang, A. Vagnucci and C. Li, "Image Enhancement by Gradient Inverse Weighted Smoothing Scheme," *Computer Graphics and Image Processing*, Vol. 15, 167-181, 1981.

7. M.J. McDonnall, "Box-Filtering Techniques," *Computer Graphics and Image Processing*, Vol. 17, 65-70, 1981.

8. J.S. Lee, "Digital Image Enhancement and Noise Filtering by Use of Local Statistics," *IEEE Trans. on Pattern Analysis and Machine Intelligence*, Vol. 2, No. 2, March 1980.

9. J.S. Lee, "Refined Filtering of Image Noise Using Local Statistics," *Computer Graphics and Image Processing*, Vol. 15, 380-389, 1981.

10. R.E. Graham, "Snow Removal—A Noise-Stripping Process for Picture Signal," *IRE Trans. Information Theory IT-8*, 129-144, 1966.

11. J.M.S. Prewitt, "Object Enhancement and Extraction in Picture Procession and Phycho-pictures," (B.S. Lipkin and A. Rosenfeld, Eds.) Academic Press, New York, 1970.

12. A. Leu, S. Zucker, and A. Rosenfeld, "Interactive Enhancement of Noisy Images," *IEEE Trans. Syst. Man. Cybern.* SMC - 7, No. 6, 435-442, 1977.

13. J.W. Goodman, "Some Fundamental Properties of Speckles," *J. Opt. Soc. Am.*, Vol. 66, No. 11, Nov. 1976.

14. J.S. Lee, "Speckle Analysis and Smoothing of Synthetic Aperture Radar Image," *Computer Graphic and Image Processing*, 17, 24-32, 1981.

15. J.S. Lee, "Digital Image Soothing and the Sigma Filter," *Computer Graphic and Image Processing*. (to appear).

16. J.S. Lee, "A Simple Speckly Smoothing Algorithm for Synthetic Aperture Radar Images," IEEE Trans. on System, Man, and Cybernatics, Vol. 13, No. 1, 85-89, Jan. - Feb., 1983.

Encoding Techniques for a Pictorial Database

Chung-Chun Yang*

*Systems Research Branch
Aerospace Systems Division
U.S. Naval Research Laboratory
Washington, DC*

Shi-Kuo Chang**,†

*Information Systems Research Laboratory
University of Illinois at Chicago
Chicago, IL*

1 INTRODUCTION

Computerized image processing is one of the main trends in technologies today and will undoubtedly grow in the future. More and more, industry, agriculture, medicine, commerce, science and military are collecting and making use of vast amounts of images gained from various sources such as radar, camera, remote sensing, x-rays, etc. All these two dimensional images are also called pictorial data. Actually, right now, too much data have been gathered through various kinds of instruments. The problems that are facing us now is how to handle or manipulate them and make use of them efficiently.

In picture processing, image understanding, and pictorial information retrieval, it is often desirable to partition a large picture into smaller pieces, sometimes called *pages* or *tiles*, so that the pieces can be stored economically (as far as secondary storage space requirement is concerned) and retrieved easily. The pieces of the picture are often organized hierarchically into a tree structure [CHANG82]. The concept of hierarchy is widely used in computer graphics

---
*The research was supported by Office of Naval Research under work request NR609-007
**The research was supported by National Science Foundation under Grant ECS-8005953 and by the Naval Research Laboratory under contract N00014-82-C-2156
†Present affiliation: Electric Engineering Department, Illinois Institute of Technology, Chicago, IL.

and image processing [CHANG79b, CHIEN80, KLING77, KLING79, McKEO77, MILGR79, OMOLA79, SHAPI79a, TANIM76]. The *picture encoding* problem can therefore be described as the partitioning of a large picture into smaller pieces, and the organization of such pieces into a hierarchical structure to facilitate pictorial information storage and retrieval.

In this report, we present a methodology of picture encoding for a pictorial database. In Sections 2, 3, and 4, we discuss hypercube encoding, minimal hypercube encoding, and the picture covering problem. The hypercube encoding technique generalizes tightly-closed-boundary encoding and picture paging techniques. In Section 5, the hierarchical hypercube encoding technique is introduced. In Section 6, we discuss how micro-pictures can be associated with a hypercube encoding. The micro-picture consists of logical picture objects, relational objects, and physical pictures. The picture reconstruction problem can be regarded as the materialization of a support picture from a micro-picture. In Section 7, the concepts of picture tree and picture query are introduced. We then describe a methodology for structured picture retrieval.

By combining pattern recognition techniques, a potential application of our research will be automatic recognition, (storage, and retrieval) of certain recurrent patterns (or features) in large amounts of images, which would make it possible, for instance, to extract some enemy's military installations or targets from satellite images of the earth.

2 PRELIMINARIES

A 2-dimensional picture function f is a mapping, $f: N \times N \to \{0, 1, \ldots, L-1\}$, where N represents the set of natural numbers $\{1, \ldots, N\}$, and $\{0, 1, \ldots, L-1\}$ is the *pixel value set* or *gray level set*. $f(x, y)$ then represents the pixel value at (x, y) [ROSEN78].

Similarly, an n-dimensional picture function f is a mapping,

$$f: N^n \to \{0, 1, \ldots, L-1\}$$

and $f(x_1, x_2, \ldots, x_n)$ represents the pixel value at $(x_1, \ldots, x_n)$.

It is often convenient to extract from a picture those points with pixel value greater than or equal to a certain threshold. A *picture point set* or simply a *point set* extracted from a picture f with threshold t is defined by,

$$S(f, t) = \{(x, y): f(x, y) \geq t\}$$

where t is the given threshold. Similarly, for an n-dimensional picture f, the picture point set is defined by,

$$S(f, t) = \{(x_1, x_2, \ldots, x_n): f(x_1, x_2, \ldots, x_n) \geq t\}.$$

As an example, the picture function f is,

```
y  0 1 0
   0 2 0
   1 0 1
   f        x
```

and $S(f, 1) = \{(1, 1), (3, 1), (2, 2), (2, 3)\}$. A picture f is often represented by its picture point set $S(f, t)$ with an appropriate threshold.

Given a picture point set S, we are often interested in finding picture point sets H_i, so that their union contains the original S. The point sets H_i usually have some nice properties to facilitate encoding. One family of point sets useful for such encoding purposes is the family of n-dimensional hypercubes, defined as follows:

$$H(x_{1a}, x_{1b}; x_{2a}, x_{2b}; \ldots; x_{na}, x_{nb})$$
$$= \{(z_1, \ldots, z_n): x_{ia} \leqslant z_i \leqslant x_{ib}, 1 \leqslant i \leqslant n\}.$$

In other words, hypercubes are also regarded as point sets: A hypercube *covers* a point set S, if every point in S is also in H. For example, $H(1,3;1,4)$ covers $S(f,1)$. We also say that $S(f,1)$ is covered by, or contained in $H(1,3;1,4)$. An example of a 3-dimensional hypercube is given by $H(1,3;1,3;1,3)$.

Given a picture function f and a point set S, f/S denotes the *restricted picture function* which is defined only for points in S, i.e., $f/S: (x_1, \ldots, x_n) = f(x_1, \ldots, x_n)$ if $x = (x_1, x_2, \ldots, x_n)$ is in S. f/S is called the *support picture function* or *support picture* for S.

3 MINIMAL HYPERCUBE ENCODING

We are interested in technique to find coverings of a point set $S(f,t)$ using hypercubes. The simplest approach, called *minimal hypercube (MH) encoding* [MERRI73], is to find the smallest hypercube H containing $S(f,t)$. The *minimal enclosing hypercube* $H(a_1, b_1; a_2, b_2; \ldots; a_n, b_n)$ of a point set S is given by,

$a_j = \min\{y_j: \text{for some } z_k, (z_1, \ldots, y_j, \ldots, z_n) \text{ is in } S\}$
$b_j = \max\{y_j: \text{for some } z_k, (z_1, \ldots, y_j, \ldots, z_n) \text{ is in } S\}.$

As a notational convenience, a GH hypercube can be written as

$$GH(x_1, \ldots, x_{m-1}; a_m, \ldots, a_n; b_m, \ldots, b_n)$$

where $(x_1, \ldots, x_{m-1})$ is the *handle vector*, $(a_m, \ldots, a_n)$ is the lower bound vector, and $(b_m, \ldots, b_n)$ is the *upper bound vector*. The index set $\{1, 2, \ldots, m-1\}$ is called the *handle set*.

It can be seen that the selection of the handle set determines the GH encoding. Therefore, the GH encoding problem is to select a handle set $\{i_1, i_2, \ldots, i_{m-1}\}$ from $\{1, 2, \ldots, n\}$ so that the least number of GH encoded tuples are generated.

The above heuristic algorithm was tested on random vectors with normal distributions. The results were compared to optimal solutions. To find the optimal GH_m encoding, exhaustive approach was taken and the GH_m codes with minimum number of GH_m tuples was taken as the optimal GH_m encoding. Table 1 shows the heuristic algorithm performs rather well [SINGH79].

4 ARBITRARY HYPERCUBE ENCODING AND PICTURE COVERING PROBLEM

The GH encoding technique essentially uses $(n - m + 1)$-dimensional hypercubes with handles to cover the original point set S. A further generalization is to use an arbitrary collection of hypercubes to cover S. This technique is called

Table 1 Dimension, $N = 5$. This table contains the No. of optimal GH_m encoded tuples

No. of points (M)	GH_1			GH_2			GH_3			GH_4		
	Heuristic	Opt.	%Error	Heuristic	Opt.	%Error	Heuristic	Opt.	%Error	Heuristic	Opt.	%Error
10	8	8	0.0	10	10	0.0	10	10	0.0	10	10	0.0
20	13	13	0.0	20	19	5.26	20	20	0.0	20	20	0.0
30	10	10	0.0	29	29	0.0	30	30	0.0	30	30	0.0
40	16	16	0.0	38	38	0.0	40	40	0.0	40	40	0.0
50	21	21	0.0	49	49	0.0	50	50	0.0	50	50	0.0
60	18	18	0.0	55	55	0.0	59	59	0.0	60	60	0.0
70	18	18	0.0	64	63	1.58	70	70	0.0	70	70	0.0
80	21	21	0.0	75	75	0.0	80	79	1.26	80	80	0.0
90	26	26	0.0	84	84	0.0	90	90	0.0	90	90	0.0
100	22	22	0.0	96	92	4.35	100	100	0.0	100	100	0.0

arbitrary hypercube (AH) encoding. In *AH* encoding, the hypercubes may have different dimensions and different handle vectors. As an example, the picture point set $S(f, t)$ can be covered by $AH = \{H(1,3;1,1), H(2,2;2,3)\}$.

The *picture covering problem*, of which the picture paging problem is a special case, can now be stated as follows:

(1) We are given a picture f and its associated picture point set $S(f,t)$.

(2) We wish to find a collection of hypercubes, H_i, to cover $S(f,t)$. If we restrict H_i to have similar $(n - m + 1)$-dimensional handle vectors, this reduces to *GH* encoding problem. Otherwise, this the *AH* encoding problem.

(3) The objective of finding an optimal covering is the minimization of total number of hypercubes.

In [CHANG78d, REUSS78b, LIU81], results are reported on picture paging, which is the picture covering problem for 2-D pictures. The primitive decomposition algorithm described in [LIU81] is illustrated in Figure 1. In Figure 1(a), we first find a *substractor*, which is the largest rectangle that can be found containing only background pixels. After the removal of a substractor, the remaining picture can be decomposed into one to three rectangles, as illustrated in Figure 1(b). The most promising decomposition to minimize the total number of pages is then selected among cost-effective candidates [LIU81]. The procedure can be iteratively applied, until the decomposed rectangle can be covered by one page.

To guide the selection of the most promising decomposition, we can use picture information measures. A measure for the evaluation of picture coverings is based upon the minimum number of pixel gray-level changes to convert a picture into one with constant gray-level. Let $h: \{0, 1, \ldots, L - 1\} \rightarrow N$ represents the *histogram* of f, where $h(i)$ is the number of pixels at gray-level i. We define the *pictorial information measure PIM(f)* as follows:

$$PIM(f) = \left(\sum_{i=0}^{L-1} h(i)\right) - \max_i h(i)$$

Some formal properties and theorems about *PIM* can be found in [SILVER82].

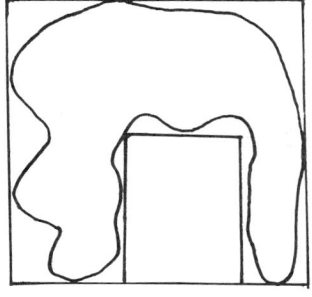 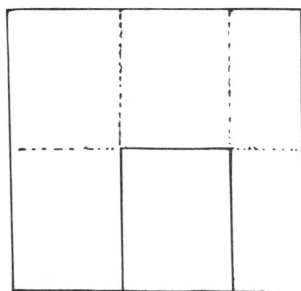

Figure 1a Substractor   **Figure 1b** After removal of substractor

Figure 2 illustrates experimental results of using the decomposition algorithms to decompose a SEASAT image of the Los Angeles area for *HH* encoding. The original image is illustrated in Figure 2(a). It contains 128 × 128 pixels with gray levels between 0 and 255. The image was smoothed using sigma filtering technique developed by Lee [LEE81]. The nonbackground pixels occupy approximately 30% of the image. Using pages of size 10 × 10, the image can be covered by 169 pages. Figure 2(b) illustrates the result using the primitive decomposition algorithm. 135 pages are used, and the total number of nodes in the *HH* encoding tree is 52. Figure 2(c) illustrates the result using the PIM-guided decomposition algorithm. 136 pages are used, and the total number of nodes in the *HH* encoding tree is 51.

Similar experiments were then performed using the PIM-guided algorithm with different threshold values. The 128 × 128 image was first quantized into 64 gray levels. The results are summarized in the following table.

Threshold	No. of nodes	Number of pages
0.10	16	3
0.05	67	55
0.04	66	90
0.03	55	107
0.02	59	116
0.01	52	130
0.00	51	136

It can be seen that as the threshold decreases, more and more pages are considered to be informative. A threshold of 0.10 will select only 3 pages as being informative. If we are given a preset limit on the number of pages, we can also pick a threshold to select the number of pages below that preset page limit.

5 HIERARCHICAL HYPERCUBE ENCODING

The hypercube encoding technqiue can be applied iteratively to create a hierarchy of hypercubes. This technique is called *hierarchical hypercube (HH) encoding*. A hierarchical hypercube encoding is a collection of arbitrary hypercube

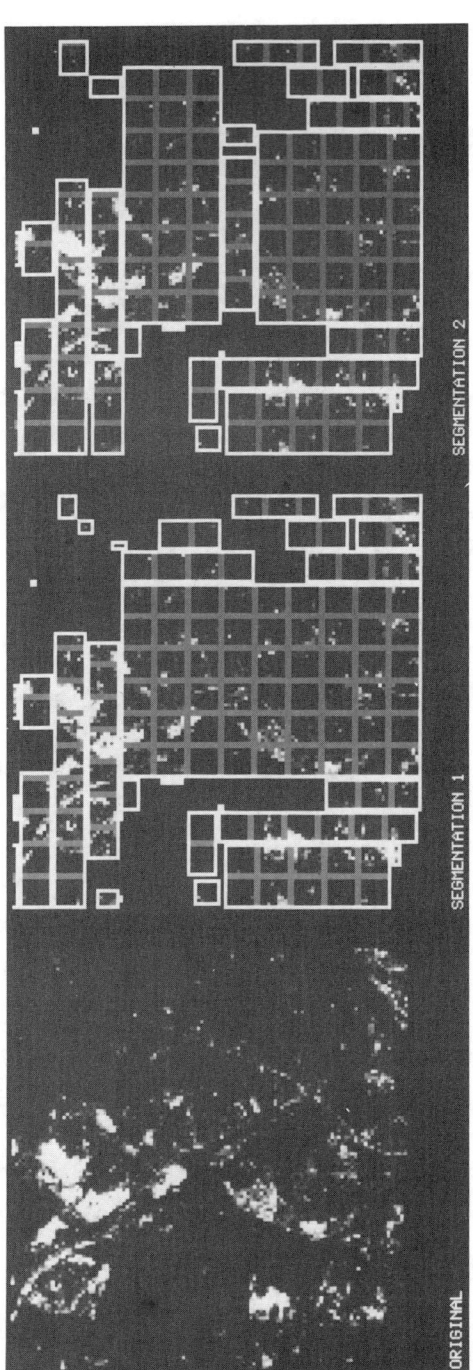

Figure 2a Original Image

Figure 2b Using primitive decomposition algorithm

Figure 2c Using PIM-Guided decomposition algorithm

Pictorial Database 467

encodings AH_1, AH_2, ..., AH_k, satisfying the following conditions:

(1) Each AH_i covers original picture point set S.
(2) AH_1 is a singleton set, i.e., AH_1 consists of one hypercube covering S.
(3) Each AH_{i+1} can be divided into disjoint subsets of hypercubes, such that for each subset, there exists a hypercube in AH_i covering that subset.

As an example, for the picture point set $S(f,t)$, one possible *HH* encoding has three levels: AH_1, AH_2, AH_3, where $AH_1 = \{(1,1;3,3)\}$, $AH_2 = \{H(1,3;1,1), H(2,2;2,3)\}$, and $AH_3 = \{H(1,1;1,1), H(3,3;1,1), H(2,2;2,3)\}$. Another possible *HH* encoding is $AH_1 = \{H(1,1;3,3)\}$, $AH_2 = \{H(1,1;3,3)\}$, $= AH_3 = \{H(1,1;3,3)\}$. Therefore the AH_i's may be identical in *HH* encoding. In Section 7, it will be seen that with the incorporation of micro-pictures, we can then construct picture trees from an *HH* encoding to facilitate pictorial information retrieval.

6 PICTURE OBJECTS, RELATIONAL OBJECTS AND PICTURE RECONSTRUCTION

A *picture object* v_i is represented by v_i ((type, typecode), (x_1, x_{i1}), (x_2, x_{i2}), ..., (x_n, x_{in}), (a_1, a_{i1}), (a_2, a_{i2}), ..., (a_k, a_{ik})), where type is the classification category of the object, $x_1, x_2, \ldots, x_n$ are the spatial coordinates of the object, and $a_1, a_2, \ldots, a_k$ are other attributes (e.g., resolution, average gray level, size orientation, etc). A *picture object set* $V = \{v_1, v_2, \ldots, v_n\}$, is a set of picture objects. As an example, if the picture objects v_i are pixels, then $v_i = ((\text{type,pixel}), (x, x_i), (y, y_i), (c, f(x_i, y_i)))$, where c is pixel value for point (x, y). If the picture objects v_i denote average pixels over a local window area, then $v_i = ((\text{type,pixel}), (x, x_i), (y, y_i), (r, r_i), (c, c_i))$, where r is the size of the local window (the resolution), and c is average pixel value for local area inside the window.

A picture is usually specified by a picture function f, together with a picture object set V and a relational object set R. (V, R) can be regarded as the *logical picture* extracted from the *physical picture* f. This dichotomy is also reflected in the picture encoding.

Each hypercube H_i in a hypercube encoding can be associated with a *micro-picture* I_i, which contains information extracted from the support picture f/H_i. I_i, the micro-picture, consists of two parts: *a logical micro-picture*, IL_i containing picture objects extracted from the support picture and their relations, and a *physical micro-picture* IP_i containing a transformed picture derived from f/H_i. Mathematically, $IL_i = (V_i, R_i)$, where V_i is a subset of V, the original picture object set, and R_i are relational objects constructed on V. In actual implementation, IL_i could be realized as relational tables in a relational database. (See [CHANG78d] where the concepts of logical and physical pictures are discussed, and relational tables are discussed in [CHANGNS79a]). For example, the logical micro-picture may contain information about the average gray level of the support picture f/H_i, the local histogram of f/H_i, and any other relevant information. The physical micro-picture may contain a resolution factor, and a picture function derived from f/H_i at that resolution.

7 PICTURE TREE AND STRUCTURED RETRIEVAL TECHNIQUE

Suppose we are given HH encoding $AH_1, AH_2, \ldots, AH_k$, and the associated micro-picture I_i for each hypercube H_i. This structure is called a *picture tree*, if (1) the original picture f can be reconstructed from the micro-pictures, and (2) the original picture object set V, if specified, is contained in the union of the picture object sets V_i.

As an example, the picture f is

$$\begin{array}{cccc} 1 & 1 & 1 & 2 \\ 1 & 1 & 1 & 2 \\ 2 & 2 & 2 & 2 \\ 2 & 2 & 2 & 3 \end{array}$$

The HH encoding is $AH_1 = \{H(1,4;1,4)\}$, $AH_2 = \{H(1,2;1,2), H(1,2;3,4), H(3,4;1,2), H(3,4;3,4)\}$. For $H_1 = H(1,4;1,4)$, its logical micro-picture IL_1 contains $\{v_1((x,1), (y,4), (r,4), (c,1.68))\}$, indicating the average gray level of entire picture (regarded as a picture object) is 1.68. Its physical micro-picture IP_1 contains $((r,2), (\text{picture}, (2,2.25, 1, 1.5)))$, indicating the resolution is 2 original pixels per blurred pixel, and the subsequent numbers are average gray levels of a blurred support picture. For $H_2 = H(1,2;1,2)$, its logical micro-picture contains $IL_2 = \{v_2((x,1), (y,2), (r,2), (c,2))\}$, indicating the average gray level of the support picture for H_2 is 2. Its physical micro-picture IP_2, contains $((r,1), (\text{picture}, (2\ 2, 2, 2)))$, indicating the resolution is 1 original pixel per blurred pixel, and the subsequent numbers are average gray levels of a blurred support picture—since the resolution is unity, the blurred picture is the support picture itself. Similarly, $H_3 = H(1,2;3,4)$, $IL_3 = \{v_3((x,1), (y,3), (r,2), (c,1))\}$, $IP_3 = ((r,1), (\text{picture}, (1,1,1,1)))$, $H_4 = H(3,4;1,2)$, $IL_4 = \{v_4((x,3), (y,1), (r,2), (c,2.25))\}$, $IP_4 = ((r,1), (\text{picture}, (2,3,2,2)))$, $H_5 = H(3,4;3,4)$, $IL_5 = \{v_5((x,3), (y,3), (r,2), (c,1.5))\}$, $IP_5 = ((r,1), (\text{picture}, (1,2,1,2)))$. From this example, it can be seen that a quad-tree [KLING77] is a special case of a picture tree, where only the leave nodes have associated micro-pictures, and the micro-pictures consist of only restricted physical pictures. In other words, for quad-trees, I_k is empty for all nonterminal nodes H_k.

A *picture query set* QS is a subset of the picture object set V. In other words, QS is the set of picture objects to be retrieved. In the picture query language, the query set QS is usually specified by a picture query Q.

Given a picture query Q and a micro-picture I_i for hypercube H_i, it is assumed that we can compute the following:

> We can derive (TS',Q',S') from (Q,I_i), where TS' is the partial *picture query set* obtainable from (Q,I_i), Q' is the *modified query* specifying the remaining query after the processing of (Q,I_i), and S' is the set of descendent nodes H_j of H_i to be visited next.

We can now describe informally the structured retrieval technique as follows. We start from node H_1 (Q, I_1) and compute (TS',Q',S'). If then query Q cannot be fully answered, we will have nonempty Q'. We then retrieve all hypercubes H_i in AH_2 such that $\{H_i\} \cap S'$ is nonempty. We now use (Q',I') to compute (TS'',Q'',S''), where I' is the micro-picture of a descendent node H_k in S_k. If Q' cannot be fully answered, we will have nonempty Q'', and we should retrieve all hypercubes H_i an AH_3 such that $\{H_i\} \cap S''$ is nonempty. We proceed as above until the lowest level AH_k is reached. It is assumed that any query Q can be answered, if the picture f is given. Since the picture f can be materialized from the micro-pictures in the picture tree, the above procedure can always answer Q, and QS is $Q(TS)$, where TS is the union of the TS's. The above outlined procedure provides a *structured retrieval technique*.

REFERENCES

[CHANGNS79] N.S. Chang and K.S. Fu, "A Relational Database System for Images," TR-EE 79-28, Dept. of Electrical Engineering, Purdue University, May 1979.

[CHANG78a] S.K. Chang, and Y. Wong, "Optimal Histogram Matching by Monotone Gray Level Transformation," Communication of the ACM, Vol. 22, No. 10, ACM, 835-840, October 1978.

[CHANG78b] S.K. Chang, J. Reuss, and B. H. McCormick, "Design Considerations of a Pictorial Database System," International Journal on Policy Analysis and Information Systems, Vol. 1, No. 2, Knowledge System Laboratory, UICC, 49-70, January 1978.

[KLING77] A. Klinger, M.L. Rhode, and V.T. To, "Accessing Image Data," International Journal on Policy Analysis and Information Systems, Vol. 1, No. 2, Knowledge System Laboratory, UICC, 171-189, January 1978.

[KLING79] A. Klinger, "Analysis, Storage, and Retrieval of Elevation Data with Application to Improve Penetration," U.S. Army Corps. of Engineers, Engineer Topological Laboratories, Fort Belvoir, Virginia, 22060, March 1979.

[LEE81] J.S. Lee, "A Simple Speckle Smoothing Algorithm for Synthetic Aperture Radar Images," CGIP, 1981.

[LIU81] S.H. Liu and S.K. Chang, "Picture Covering by 2-D AH Encoding," Proceedings of IEEE Workshop on Computer Architecture for Pattern Analysis and Image Database Management, Hot Springs, Virginia, November 11-13, 1981.

[McKEO77] D.M. McKeown Jr. and D.J. Reddy, "A Hierarchical Symbolic Representation for Image Database," Proceedings of IEEE Workshop on Picture Data Description and Management, IEEE Computer Society, 40-44, April 1977.

[MERRI73] R.D. Merrill, "Representation of Contours and Regions for Efficient Computer Search," Communications of the ACM, Vol. 16, No. 2, ACM, 69-82, February 1973.

[OMOLA79] J. Omolayole and A. Klinger, "A Hierarchical Data Structure Scheme for Storing Pictures," Technical Report, Computer Science Dept., UCLA, 1979.

[REUSS78] J.L. Reuss and S.K. Chang, "Picture Paging for Efficient Image Processing," Proceedings of IEEE Computer Society Conference on Pattern Recognition and Image Processing, IEEE Computer Society, 69-74, May 1978.

[REUSS80] J.L. Reuss, S.K. Chang, B.H. McCormick, "Picture Paging for Efficient Image Processing," in *Pictorial Information Systems*, (S.K. Chang and K.S. Fu, eds.), Spring-Verlag, 1980, 228-256.

[SINGH79] K.K. Singh, S.K. Chang and C.C. Yang, "A Heuristic Method for Generalized Hypercube Encoding," Proceedings of COMPSAC 79, November 6-8, 1979, Chicago, 531-534.

[SILVER82] H. Silver and S.K. Chang, "Picture Information Measures," in *Progress in Pattern Recognition*, (A. Rosenfeld and Kanal, eds.), Vol. 2, 1982.

[SHAPI79a] L.G. Shapiro and R.M. Haralick, "A Spatial Data Structure," Technical Report #CS 79005-R, Dept. of Computer Science, Virginia Polytechnic Institute and State University, p. 35, August 1979.

[TANIM76] S.L. Tanimoto, "An Iconic/Symbolic Data Structuring Scheme," in *Pattern Recognition and Artificial Intelligence*, Academic Press, 452-471, 1976.

[YANG78] C.C. Yang and S.K. Chang, "Encoding Techniques for Efficient Retrieval from Pictorial Databases," Proceedings of IEEE Computer Society Conference on Pattern Recognition and Image Processing, IEEE Computer Society, 120-125, June 1978.

Part VI: Architectures for Signal Processing

Impact of VLSI on Modern Signal Processing*

S. Y. Kung

*Department of Electrical Engineering
University of Southern California
Los Angeles, CA*

1 INTRODUCTION

VLSI microelectronics technology and the emerging computer-aided design methodologies are precipitating a new revolution in signal processing. The area of VLSI signal processing is bound to become a major focus of attention in governmental, industrial, as well as university research activity. To provide a bridge from signal processing theory and algorithms to VLSI processor architectures and implementation, it is critical to have a fundamental understanding of the basic computational requirements of modern signal processing and of the technology constraints of VLSI. This will require a cross-fertilization of the fields of computer software/hardware and signal processing engineering. The relationships between VLSI and modern signal processing can be said to be very intimate. First we assert that the modern signal processing has a need to fully utilize VLSI circuits. The ever-increasing demands for performance, sophistication and real-time signal processing strongly indicate the need for tremendous computation capability, in terms of both volume and speed. The availability of low cost, high density, fast VLSI devices promises the practicality of cost-effective, high speed, parallel processing of large volumes of data [1]. This makes feasible ultra high throughput-rates and presages major technological breakthroughs in real-time signal processing applications.

On the other hand, it is quite obvious that the full potential of VLSI can be realized only when its application domains are discriminatingly identified.

---

*Research supported in part by the Office of Naval Research under contract No. N00014-81-K-0191; by the National Science Foundation under grant No. ECS-80-17081 and by the Defense Advanced Research Project Agency under contract No. MDA903-79-C-0680.

For this purpose, it may be noted that traditional computer architecture designs are no longer suitable for the design of highly concurrent VLSI computing processors. As a major technological constraint, communication in VLSI systems has to be highly restricted, as communication is expensive in terms of area, power and time consumption [1]. Fortunately, most signal processing algorithms have attractive properties of regularity, recursiveness and locality in data-dependence, and thus are very amenable to VLSI implementation [2]. Therefore, the potential of today's fast growing VLSI technology will probably be best exploited by its extensive application to modern signal processing; that is to say, signal processing will be very important for VLSI systems research.

In the past, major work has been done on mapping various kinds of signal processing applications into specific LSI or VLSI architectures. The insight and experience gained from this have already greatly enhanced the understanding of VLSI's impact on signal processing. Unfortunately, such contributions have been very scattered owing to the lack of interaction between different disciplines. To meet the challenge of matching VLSI and modern signal processing, it is critical to have a fundamental and cross-disciplinary understanding of the basic computational requirements and the technology constraints. This will involve a very broad range of research issues, which will be divided into three categories in our discussion:

(1) impact of VLSI on signal processing methods: theory and algorithm;
(2) impact of VLSI on signal processing systems;
(3) integrated design of signal processing systems.

2 IMPACT OF VLSI ON SIGNAL PROCESSING METHODS

Driven by steep increase in the complexity of computations, processing speed requirements and the volume of data handled in various signal processing applications, modern signal processing research and development is undergoing a major revolution. The availability of low cost, high density, fast VLSI devices has opened a new avenue for implementing these increasingly sophisticated algorithms and systems. In other words, the future (of modern signal processing) hinges upon the almost unlimited computation potential of VLSI.

High Performance Signal Processing Techniques

In terms of advanced signal processing research development, the processing techniques can be divided into two categories: the conventional transform based techniques and the modern (nonlinear) spectrum analysis methods [5].

The invention of the FFT provided the first major impetus towards the advancement of signal processing. This technique has drastically reduced computing time for signal and image processing problems. This important advantage has provided the FFT based methods a strong computational footing and sustained their dominance in current signal processing research and industrial developments [4].

From another perspective, FFT based methods, being batch-processing technique, can not claim the recursiveness and self-adaptivity offered by some other methods. For example, Kalman filtering method has possessed such properties and demonstrated much usefulness in many real-time processing problems. Moreover, for high resolution spectrum estimation problems the transform based methods in general cannot yield a satisfactory resolution capability [5]. In contrast, with a little higher computation complexity, modern (nonlinear) spectrum analysis methods often offer much better performance [3], [5], as demonstrated in the next example.

Example of Array Sensor Beam Forming

One of the main functions of an underwater passive sonar system is determining the number of sources present in the medium, and their characteristic parameters, [6-8] from the signals received by N sensors distributed randomly or uniformly depending on the application. The problem is to determine the direction, and intensity of the signal sources and their number from the sensor signals. (cf. Figure 1.)

In passive sonar beam forming problems, resolution is an extremely important factor. As a result, in sonar signal processing community, adaptive, high resolution spectral beam forming (and analysis) algorithms and their implementations have been an active research area.

We shall briefly describe the representative methods that have attracted the most attention. Here we denote

S = Correlation matrix of sensor signals

$a(\theta)$ = phasing vector determined by the geometrical distribution

of the sensors and the angle of target, θ.

The conventional beam-forming technique, fairly well established in signal-processing literature, use the following formulation [6], [9].

$$P_{BF} = a^*(\theta) \ S \ a(\theta) \qquad (1)$$

The Capon maximum likelihood method (MLM) was originally developed in the context of seiemic and sonar array processing [10], [11]. The formulation for computing log likelihood function is [9]

$$P_{ML} = \frac{1}{a^*(\theta) \ S^{-1} a(\theta)}. \qquad (2)$$

For stationary signal processing, the maximum entropy method (MEM) based on an "entropy" concept introduced by Burg [12] is actually equivalent to all-zero autoregressive spectral analysis, linear prediction, and innovations processing. The principal of MEM can be applied to random sensor array processing, with the following formulation [9]

$$P_{ME} = \frac{1}{a^*(\theta) \ C \ C^* \ a(\theta)} \qquad (3)$$

where C is the first column of S^{-1}.

The most promising class of methods are those based on eigenvalue-eigenvector decomposition of the correlation matrix S, as proposed by Owsley [13], Bienvenu [7-8], Schmidt [9], and Johnson and DeGraaf [14]. They are all similar but differ somewhat in the actual use of eigenvectors. As an exam-

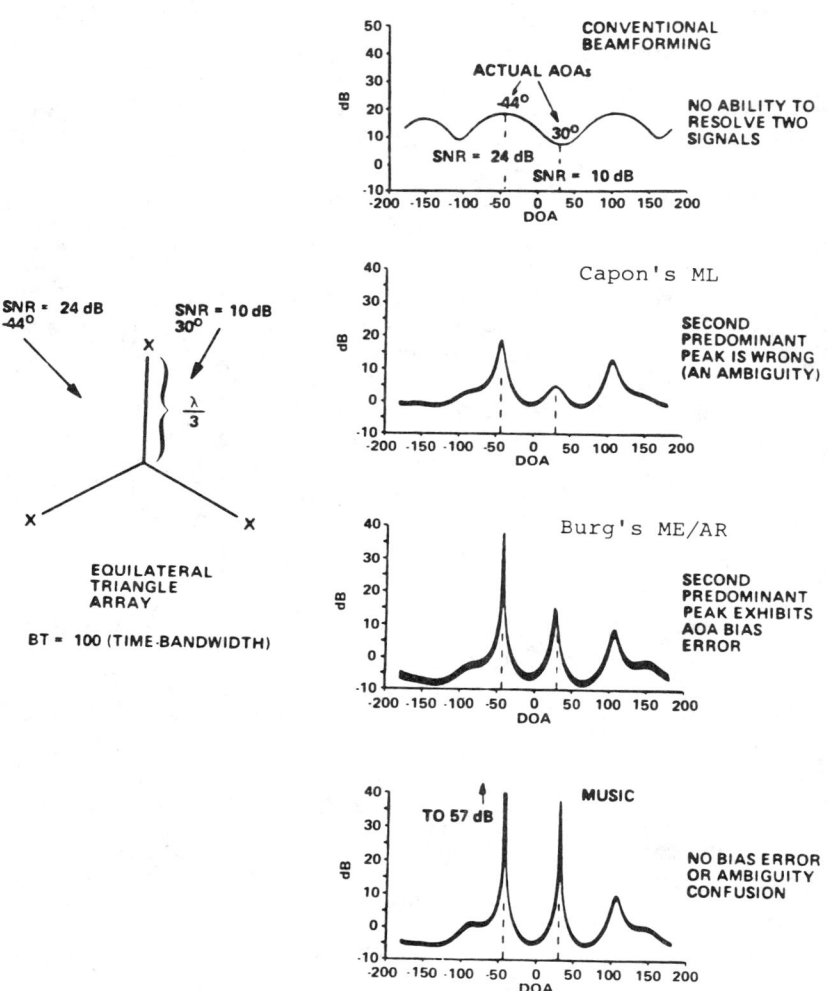

Fig. 1 — Comparisons of the conventional and modern beamforming methods (originally from [9], courtesy from R. Schmidt)

ple, the MUSIC (multiple signal classification) method by Schmidt [9] uses the following formulation:

$$P_{\text{MUSIC}} = \frac{1}{a^*(\theta)\ E_N\ E_N^*\ a(\theta)},\qquad(4)$$

Where E_N is a matrix formed from the eigenvectors associated with the minimum $(N - M)$ eigenvalues of the correlation matrix S, where M is the number of targets. In order to make connection to the classical spectral estimation methods, we make the assumption that all the sensors are uniformly spaced in linear array. There are two noteworthy observations:

(1) In a stationary environment the correlation function will take a Toeplitz form [15], i.e.,

$$S = \begin{bmatrix} R_0 & & R_{N-1} \\ \cdot & & \\ \cdot & & \\ \cdot & & \\ R_{N-1} & & R_0 \end{bmatrix}.$$

This Toeplitz structure can be exploited for fast algorithms and a pipelined processor architecture.

(2) The phasing vector $a(\theta)$ will become

$$a(\theta) = [1, e^{j\omega}, e^{j2\omega}, \ldots, e^{j(N-1)\omega}],$$

where $\omega = 2\pi(d/\lambda)\sin\theta$, where d is the spacing and λ is the wavelength. Note that this is in fact a Fourier coefficient vector, and the Equation (1) will reduce to the classical transform based method for spectrum estimation as shown below:

$$P_{BF} = a^*(\theta) \, S \, a(\theta)$$
$$= \sum_{n=-(N-1)}^{N-1} W(n) \, R(n) \, \exp(e^{jn\omega}),$$

which is exactly the Fourier transform of the correlation function $R(n)$ with Bartlett (triangular) window, $W(n)$.

Figure 1 [9] demonstrates clearly the significant performance differences between classical beam-forming and other modern methods. It is shown that the conventional (transform-type) beam forming technique, though very popular and established in the literature, suffers from severe bias and lack of resolution. The MLM and MEM yield better resolutions while exhibiting some bias. (In this case, the MLM also mistakes one of the predominant peaks). Based on an important assumption, regarding the high coherence (spatial) of the signal sources, these methods can achieve much better performance than other adaptive array processing methods mentioned above. Figure 1 clearly indicates that the MUSIC method out-performs all the other methods. Note that in order to implement high resolution methods, new types of signal processing algorithms and architecture have to be used. These include eigenvalue decomposition, correlation matrix computation, matrix inversion, etc. However, the advent of VLSI and its abundant computing capacities has enabled the use of tremendous parallelism in modern spectrum analysis techniques. This computation facility will result in a new balance between the traditional and modern methods. Therefore, VLSI is in a sense presenting a most vigorous challenge to the transform-based techniques, and should provide a second major impetus towards the advancement of signal processing methods since the invention of FFT.

Massive Parallelism

While a major improvement in VLSI device speed is foreseen, it is in no way comparable to the rate of increase of throughput rate required by modern real-time signal processing. In order to achieve such increases in throughput rate, the only effective solution appears to be highly concurrent processing. Since massive parallelism will have a major impact on the modern signal processing techniques, the traditional performance criterion will have to undergo a major modification.

Parallelism is often achieved by decomposing a problem into independent subproblems or into pipelined subtasks. The constraints which limit the degree of parallelism achievable often varies significantly among different techniques. Therefore, parallelism often represents a major factor in selecting among competing techniques. In several instances, VLSI has made several conventionally inefficient or impossible techniques desirable or even rather attractive. Conversely, several "fast" techniques may start losing ground under this technological impact.

In evaluating the degree of inherent parallelism those methods with frequency or spatial decoupling will offer a greater advantage as they are naturally decomposible into independent subproblems. For example, DFT but not FFT computation can easily focus on the desired frequency ranges. For spectral line estimation problems, a recently developed adaptive notch filtering technique [19] seems to adapt well with the frequency parallelism, i.e., multiple notch filters may be used for multiple frequency ranges. In contrast, the other time-domain methods, such as adaptive least-squares estimation method, can not exploit this advantage of parallelism.

Historically, much of the work on parallel algorithms has concentrated on numerical linear algebra [16-18]. Consequently, the class of methods which are reducible to basic matrix operations are becoming preferred candidates [20], [21].

Parallel Algorithms

There has been a number of efficient parallel algorithms for solving linear equations, triangular and orthogonal matrix decomposition (i.e., for matrix inversion and least-square solution), eigenvalue and singular value decomposition, and linear recurrences. An immediate consequence of the parallel processing capability for matrix operations is that the modern high-resolution methods such as MEM, eigenvalue decomposition methods can be performed at higher speeds.

Although there is a rather long history of the study of parallel processing; however, the revolution of VLSI device and computing technology has aroused new research interests. The preference on regularity, and locality will have a major impact in deriving parallel algorithms, (cf. Section II.a). In our work, the two most critical issues—parallel computing algorithm and VLSI architectural constraint—are considered

1. To structure the algorithm to achieve the maximum parallelism and, therefore, the maximum throughput-rate
2. To cope with the communication constraint so as to compromise least in processing throughput-rate.

Example: A Highly Concurrent Toeplitz System Solver

Toeplitz systems arise in numerous, wide-spread applications ranging from speech, image, to radar, sonar, and geophysics, signal processing. (in our earlier example of sensor array processing, when a stationary situation is considered, solving Equations (2) and (3), will have to call for Toeplitz system solvers.) Based on the above considerations, we have developed a highly concurrent Toeplitz system solver, featuring maximum parallelism which is able to solve a Toeplitz system in $O(N)$ processing time in an array processor, as opposed to $O(N**3)$ for general (sequential) Gauss elimination procedure [22] or $O(N**2)$ for (sequential) Levinson algorithm [23].

In order to achieve full parallelism, we have to fully exploit the Toeplitz structure. For this purpose, we have proposed a new, pipelined version of the Levinson algorithm which allows the "reflection coefficients" to be computed in a pipelined fashion. Mathematically, the scheme finds its root in a classical paper of Schur [24]; and algorithmically, it is closely related to the work in [25], [26]. This avoids the need of the inner product operations, in the original Levinson algorithm, and the total computing time is therefore reduce to $O(N)$.

The configuration of computing structure [15], and a chip layout as a product of joint research project between USC/Hughes [27] are shown in Figure 2.

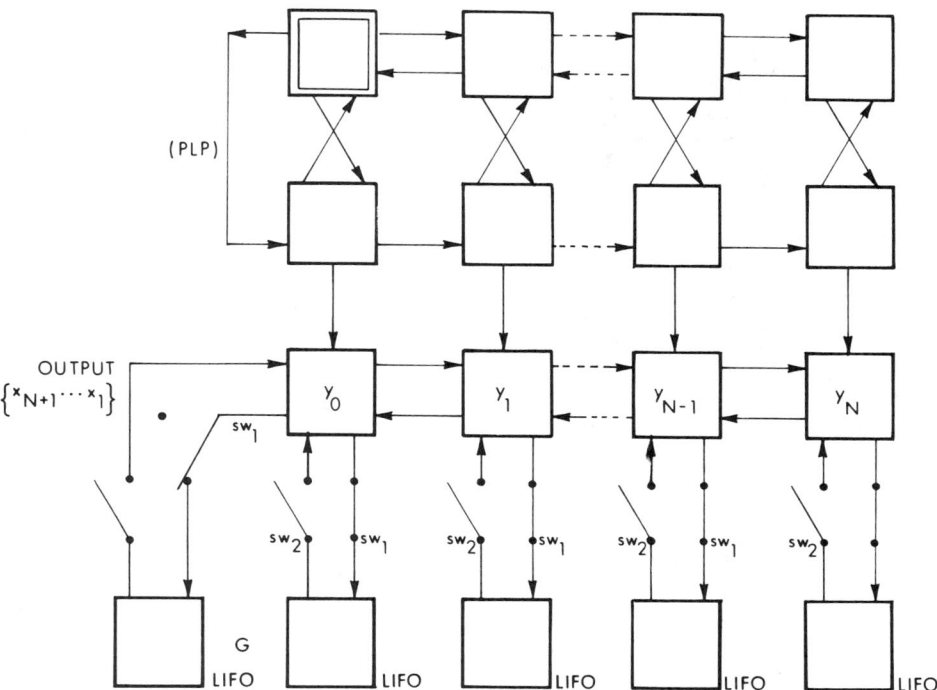

Fig. 2 — Toeplitz system solver: (a) configuration of computing structure, [15]

Fig. 2 (b) VLSI Chip layout, originally from [27]

Challenge of Modern Signal Processing

In the past two decades the transform based methods, has had a major impact on the growth of digital signal processing. The popularity of the FFT based method is due to the fast computation speed, and the many standard software package or commercially available high-speed processors to perform the fast Fourier transform. However, from the viewpoint of advancing signal processing research this overwhelming popularity has somewhat hampered the due attention to other (more analytical) modern methods. The latter kind of research effort, in our opinion, will have a long-term impact to the advanced signal processing technology; and our European colleagues are actually stressing this emphasis and working along these lines [28], [29]. In our projection, the current transform-based (and rather unbalanced) signal processing research will eventually lose its short-term (ready-for-use) advantage, and will have to suffer the disadvantage of lagging behind in the modern methods.

Fortunately, the advent of VLSI offers a new opportunity to review the new signal processing trend. The low-cost, large volume of computations will encourage sophisticated, and high-performance techniques, at the expense of more computations. In order to attract a great popularity for the modern signal processing methods, the availability and accessibility of modern computing hardware, will also become crucial. The new trend should not only inspire signal processing theoreticians to direct in-depth analysis on those modern methods traditionally deemed unfeasible; but also invite signal processing engineers to explore new fast-speed special purpose, parallel processors.

More promisingly, computations for the modern signal processing methods, are often reducible to basic matrix operations which have very high potential for VLSI parallel computing. Thus, there are several very bright aspects for the design of modern signal processing computing systems, which are discussed in the next section.

3 IMPACT OF VLSI ON SIGNAL PROCESSING SYSTEMS

In order to accommodate the ever increasing speeds of processing requirements and volumes of data handled in various modern signal processing applications, the low cost, high density, fast VLSI devices has offered a new avenue implementing the modern signal processing systems. A very critical concern is that the traditional design of parallel computers and languages is deemed unsuitable for VLSI systems [30], since it suffers from heavy supervisory overhead incurred by storage, communication, and scheduling tasks, which severely hamper the real-time signal processing speed. There are several key considerations unique for VLSI design environment, and they are summarized below.

Communication Constraints

Although VLSI provides the capability of implementing a large array of processors on one chip it imposes its own constraints on the system. Among them, the most critical is the restricted (localized) communication, since it costs the most in VLSI chips, in terms of area, time and energy [1]. In general, highly concurrent systems require this locality property in order to reduce interdependence and ensuing waiting delays that result from excessive communication [1]. This locality constraint, discouraging the use of centralized control, leads to the utilization of distributed control and localized data flow on regular and modular array structures. The utilization of a repetitive modular structure also help reduce the high design and test costs in VLSI systems.

Special Purpose VLSI Array Processors

The above restrictions, imposed by VLSI, will render the general purpose array processor very inefficient. It is therefore beneficial to restrict the array processors to a special class of applications, i.e., recursive and local data dependent algorithms, to conform with the constraints imposed by VLSI. On the other hand, this restriction incurs little loss of generality, as a great majority of signal processing algorithms possess these properties.

One typical example is a class of matrix algorithms. It has recently been indicated that a major portion of the computational needs for signal processing and applied mathematical problems can, in fact, be reduced to a basic set of matrix operations and other related algorithms [20], [21]. Therefore, a special purpose parallel machine for processing these typical computational algorithms will be cost effective and attractive in VLSI system design.

For special purpose array processors, the key of designing computing structures is mapping algorithms into computing structures, while keeping a natural topological relationship between the mathematical algorithm and the computing structure. Typical such design examples of the VLSI oriented computed structures are the systolic array processor [31] [1, Chap. 8] and wavefront array processor [2]. They are becoming increasingly popular, due to their simple and regular control/data flows and localized communications.

There are two major classes of parallel array processors: Dedicated Parallel Signal Processor (DPSP) and Programmable Parallel Signal Processor (PPSP). For higher speeds, a DPSP is usually preferred since, the circuits are often optimized for the designated purpose. This is especially so for real-time military applications such as radar and sonar sensor systems. Typical examples are dedicated systolic array processors for correlation or for matrix inversions, or a pipelined Toeplitz system solver for solving high-resolution spectrum estimation problems.

However, in order to offer flexibilities in variables such as applications, filter order, array size, etc. A programmable array processor often attracts a large market interest. Moreover, truly hardwired processors having a separate hardware for each function are usually too costly or too complicated to design, build and test. Therefore, these factors will certainly help justify the concept of PPSP. One such example is a programmable Wavefront Array Processor [2].

Parallel Languages

In order to meet the basic requirement of real-time processing rate and yet to accommodate a reasonable programmability, a new signal processing (parallel) language should be developed in coordination with the hardware structure.

Traditionally, programming language for multiprocessors focus exclusively on the description of parallel data executions with little consideration over the data movements. Therefore, programming an array processor usually involves a heavy burden of scheduling, resource sharing as well as the control of processor interactions. Due to the communication problem in VLSI systems, the issue of data availability and management becomes critical and it has become very desirable to have a new language capable of expressing parallel data movements in a computing network. In other words, an effective parallel programming language should be much more than an ensemble of separate programs: it should also precisely define the coordination and interdependence of the data and the sequencing of the tasks between the processing elements.

Moreover, a truly VLSI oriented parallel language should also take into account the features of regularity, locality, and data-flow nature in a VLSI array processor. A good example is the Matrix Data Flow Language (MDFL) developed at USC, (cf. [2]).

VLSI Signal Processor Architectures

The considerations on the processor architecture design are signal processing oriented Control Unit (CU), Arithmetic Logic Unit (ALU), and Program

Memory (PM), etc. As to the ALU, the selections between fixed/floating point, parallel/serial, or cordic/noncordic arithmetic units depend largely on the type of applications. These algorithms make use of adders, multipliers, dividers, and square-rooters, etc. At one time, the cordic arithmetic units appear very promising, since this can do all the above functions [32], [33]. On the other hand, for the above set of functions, cordic offers slower convergency [34] since it was originally designed for trigonometric functions, which has not found much usefulness in signal processing computations. Therefore, cordic arithmetic units may not be the most efficient choice for signal processing applications. As to CU, a top-down design approach is one based upon an effective classifications of the instruction set of the special-purpose language. For different class of programmability, a reassemble of the modules will lead to a suboptimal circuit. As to PM, it is noted that a signal processing oriented program usually enjoy a systematic "repeat" structure, which, if fully utilized, may lead to a much fast and simpler memory structure [35].

Fault Tolerance Consideration in VLSI System Design

While VLSI offers more affordable chip area, it is also becoming more vulnerable to faulty circuits. In order to achieve a more reliable VLSI computing system, fault tolerance techniques are considered. Basically, this is to trade hardware or time redundancy for yield enhancement. From VLSI array processor perspectives, the regularity and locality properties should be exploited to design an efficient fault tolerant diagnosis scheme. For the reconfigurability consideration, flexibilities on timing and data path have proved very valuable [2].

Design Example: Wavefront Array Processor

The wavefront array processor [2] is conceived as a programmable, data-driven concurrent array processor aiming at solving a majority of matrix algorithms.

The topology of most matrix algorithms can be mapped naturally onto the square, orthogonal $N \times N$ matrix array of processor elements with regular and local interconnections (cf. Figure 3). To create a smooth data movement in a localized communication network, we make use of the computational wavefront concept. A wavefront in the processing array will correspond to a mathematical recursion in the algorithm.

To this end, we shall consider matrix multiplication as being representative. Let

$$A = [a_{ij}], \ B = [b_{ij}], \text{ and } C = A \times B,$$

all be $N \times N$ matrices. The matrix A can be decomposed into columns A_i and matrix B into rows B_j, and therefore,

$$C = A_1 B_1 + A_2 B_2 + \ldots + A_N B_N.$$

The matrix multiplication can then be carried out in N recursions, executing

$$C^{(k)} = C^{(k-1)} + A_k B_k, \text{ with } C^{(0)} = 0,$$

recursively for $k = 1, 2, \ldots, N$.

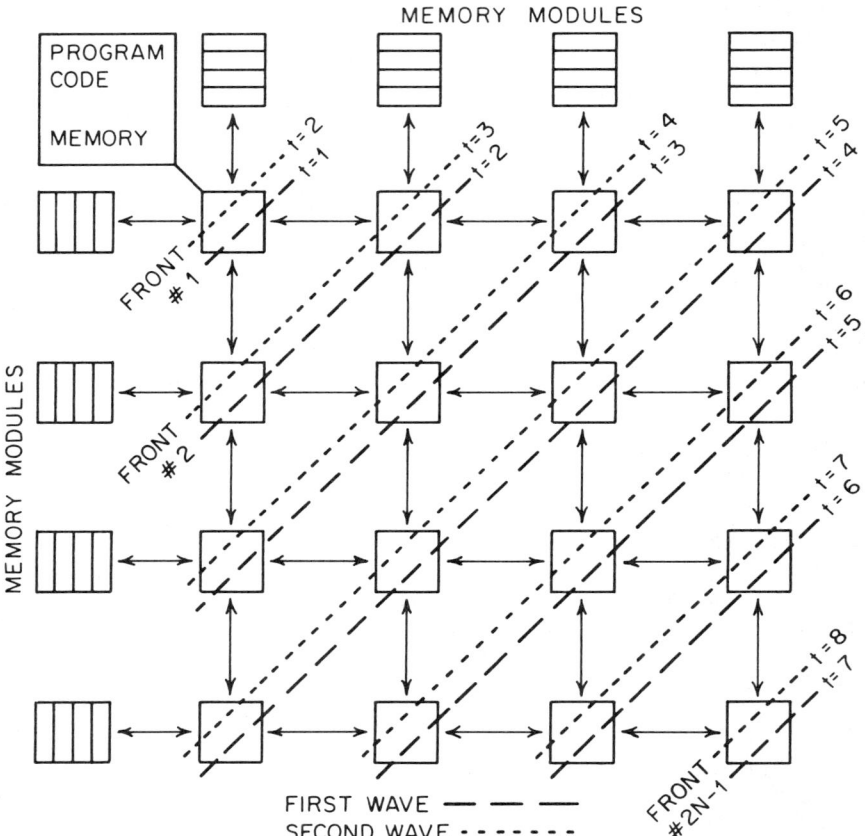

Fig. 3 — The configuration of wavefront array processor (WAP)

Successive pipelining of the wavefronts (as shown in the Figure 3) will accomplish the computation of all recursions. The pipelining is feasible because the wavefronts of two successive recursions will never intersect (Huygen's wavefront principle), as the processors executing the recursions at any given instant will be different, thus avoiding any contention problems. In short, *locality, regularity, recursivity*, and *concurrency* lead to the wavefront phenomenon. Thus, all algorithms which possess these properties will exhibit computational wavefronts.

The wavefront concept provides a firm theoretical foundation for the design of highly parallel array processors and concurrent languages, and it appears to have some distinct advantages.

As to the language aspect, the wavefront notion drastically reduces the complexity in the description of parallel algorithms. The mechanism provided for this description is a special purpose, wavefront-oriented language [2]. Rather than requiring a program for each processor in the array, this language allows the programmer to address an entire front of processors.

Impact of VLSI 485

As to the architecture aspects, the wavefront notion leads to a wavefront-based architecture which preserves Huygen's principle and ensures that wavefronts never intersect. Therefore, a wavefront architecture can provide *asynchronous, data-driven* capability, and consequently, can cope with timing uncertainties, such as local clocking, random delay in communications and fluctuations of computing-times. In short, the notion lends itself to a (asynchronous) data-driven computing structure that conforms well with the constraints of VLSI.

In summary, the integration of the wavefront concept, the wavefront language and the wavefront architecture leads to a programmable computing network, which we will call the WAVEFRONT ARRAY PROCESSOR (WAP). The WAP is, in a sense, an *optimal trade-off* between the globally synchronized and dedicated *systolic array* [1], [30], (that works on a similar set of algorithms), and the general-purpose *data-flow multiprocessors* [36]. It provides a powerful tool for the high speed execution of a large class of algorithms which have widespread applications.

4 INTEGRATED DESIGN OF SIGNAL PROCESSING SYSTEMS

It has now become clear that the VLSI device technology is having major impact on both the signal processing technique and signal processing systems. These impacts naturally implies an imminent challenge to the VLSI signal processing researchers to update the current signal processing methodologies.

The answer to this challenge lies in a cross-disciplinary research encompassing the areas of algorithm analysis, parallel computer design and system applications. We therefore, introduce an integrated research approach aiming at incorporating the vast VLSI computational capability into modern signal processing applications.

Top-down Design

A VLSI integrated system design includes several design phases:

      Applicational Specification
      Theory and algorithm
      Computing Structure/language
      Processor Architecture
      Circuit Layout/Fabrication
      Insertion of Chips into Systems

A top-down design starts with a full understanding of the problem specifications, signal analyses and (parallel) algorithms, and then map them into a suitable computing structures with high-order programming language and dedicated processor architecture. Once these steps are accomplished, several existing CAD packages may then be utilized to facilitate the circuit layout design to produce standard layout codes ready for chip fabrication. Finally, the chips will be tested and integrated into the signal processing systems. It is quite understandable that there will necessarily interactive (top-down and bottom-up) design activities across different design phases, and several iterations of such design processes are often unavoidable. Nevertheless, this proposed design methodology advocates holistically coordinating the algorithm, language, and hardware designs from the very beginning conceptual stages. For example, we have stressed a natural topological mapping between the mathematical algorithms and the VLSI computing structures. Moreover, in linking the language and processor architecture designs, the language construct will definitely affect the processor design, so as to optimize the performance. Finally, to meet the demand of modern signal processing applications, a new ppsp (programmable parallel signal processor), taking into account the overall circuit, speed, and accuracy tradeoff, has become very desirable and important.

Development of VLSI Design Language

For hardware fabrication, register transfer or computer description languages have been developed as convenient models at levels of abstraction in digital design. These hardware description languages need to be extended for system level VLSI design. Naturally, VLSI systems should be described by multilevel, hierarchical models. By taking advantage of the modular and regular structure in VLSI systems and identify those modules functionally rather than in terms of logic/circuitry details, high-level descriptions of VLSI systems should be possible.

In summary, for the CAD research in VLSI system design, a critical task is to develop a suitable mapping between different levels of system descriptions, which will allow translating a (parallel) algorithmic level programming language, e.g., MDFL [2], into their corresponding hardware description codes. A successful research in this area will have revolutionary impact on the design methodology of customized VLSI signal processing chips, which will be the future trend of VLSI signal processor developments.

A System Design Example: Passive Sonar Signal Processing System

As we have mentioned, one of the major problems in VLSI design is the integration of VLSI array processor chips into applicational systems.

This task is in general very involved and we have to limit ourselves to an example on passive sonar systems: Figure 4 shows a possible system configuration for an adaptive passive sonar array signal processing. Briefly, there are two stages of sonar signal processing, (i) preprocessing, i.e., adaptive beamforming for the purpose of improving spatial resolution and rejecting interference noise, and (ii) postprocessing, i.e., detection, estimation, and classification. For example, the classification problem, which is to identify the type of a target ship, will often require a very high resolution capability. Therefore, we suggest the modern beamforming or spectral estimation methods, such as maximum entropy method or eigenvalue decomposition method (e.g., MUSIC [9]), be adopted for adaptive beamforming and classification problems.

In order to offer a real-time processing, the computing hardware for this part should consist of several WAP's, each with $N \times N$ processor array, say $N = 24$. The functions the WAP's will have to perform include matrix correlation, matrix inversion and eigenvalue decompositions, etc. On the other hand, we suggest the tasks of detection and rough estimation be performed via the

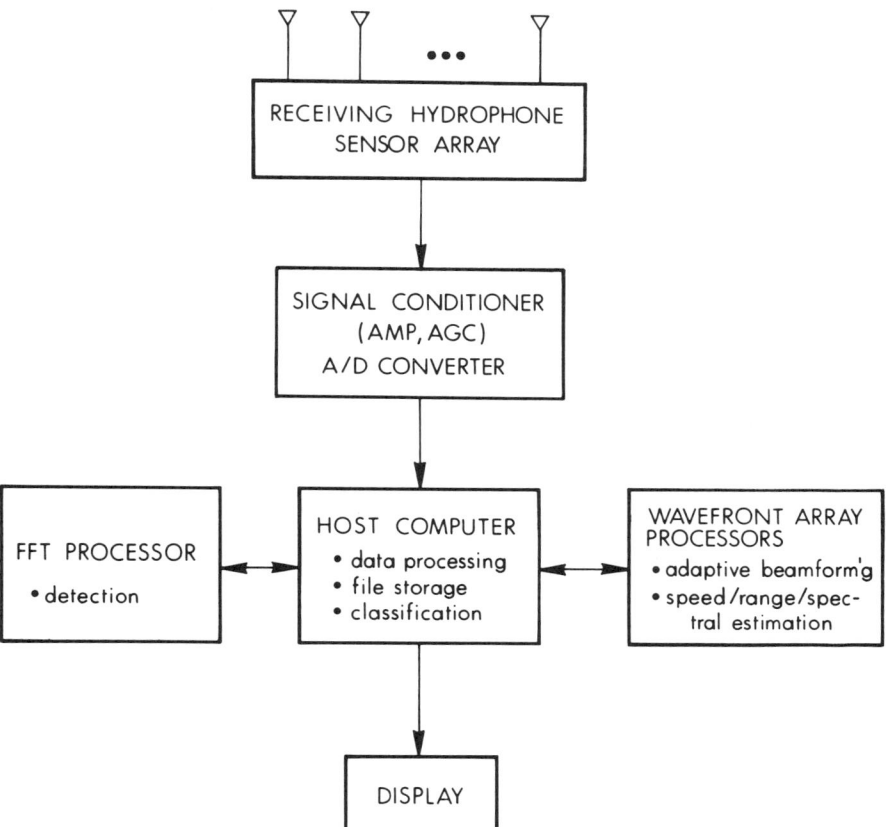

Fig. 4 — Flow-diagram for a passive sonar system

classical transform based methods by FFT processor, as it saves a good portion of computing time.

There are still much to be learned in the future for the integrated system design. Our objective here is just to demonstrate the feasibility of utilizing the wavefront array processors to furnish the major portion of parallel computations needed in modern signal processing applications.

5 CONCLUSION

The advent of VLSI technology provides an opportunity for setting a new signal processing trend. a cross-disciplinary, and integrated, (top-down) design methodology appear to be the most effective for the development of future VLSI signal processing systems. This paper presents some perspectives of the author, and is intended for inviting other viewpoints and focusing the attentions to VLSI signal processing research and development.

ACKNOWLEDGMENT

The author wishes to thank Y. H. Hu, R. Gal-Ezer, D. V. Bhaskar Rao, K. S. Arun, G. Sharma, and C. K. Lo of the University of Southern California, for their research contributions and valuable comments.

REFERENCES

[1] C. Mead and L. Conway, "Introduction to VLSI Systems," Addison Wesley, 1980.

[2] S.Y. Kung, K.S. Arun, R.J. Gal-Ezer, D.V. Bhaskar Rao, "Wavefront Array Processor: Language, Architecture, and Applications," *IEEE Transaction on Computers, Special Issue on Parallel and Distributed Processing*, Vol. 31, No. 11, pp. 1054-1066, Nov. 1982.

[3] S.Y. Kung, D.V.B. Rao, K.S. Arun, "Spectral Estimation: From Conventional Methods to High Resolution Modeling Methods," in S.Y. Kung, H.J. Whitehouse, T. Kailath, Eds., "VLSI and Modern Signal Processing," Prentice-Hall, Inc., Englewood Cliffs, N.J., 1983.

[4] A.V. Oppenheim and R.W. Schafer, "Digital Signal Processing," Prentice-Hall, Inc., Englewood Cliffs, New Jersey, 1975.

[5] H. Cox, "Resolving Power and Sensitivity to Mismatch of Optimum Array Processors," *J. Acoustics Soc. Amer.*, Vol. 54, No. 3, pp. 771-785, 1973.

[6] A.B. Baggeroer, "Sonar Signal Processing," in Applications of Digital Signal Processing, A.V. Oppenheim, Ed. Englewood Cliffs, Prentice-Hall, 331-437, 1978.

[7] G. Bienvenu and L. Kopp, "Adaptivity to Background Noise Spatial Coherence for High Resolution Passive Methods," in *Proc. IEEE ICASSP*, Denver, CO, pp. 307-310, 1980.

[8] G. Bienvenu and L. Kopp, "Source Power Estimation Method Associated With High Resolution Bearing Estimation," in *Proc. IEEE ICASSP*, Atlanta, GA, pp. 153-156, 1981.

[9] R. Schmidt, "A Signal Subspace Approach to Multiple Emitter Location and Spectral Estimation," Ph.D. Dissertation, Dept. of Electrical Engineering, Stanford University, Nov. 1981.

[10] J. Capon, "High-Resolution Frequency-Wavenumber Spectrum Analysis," *Proc. IEEE*, Vol. 57, pp. 1408-1418, Aug. 1969.

[11] R.T. Lacoss, "Data Adaptive Spectral Analysis Method," *Geophysics*, Vol. 36, pp. 661-675, Aug. 1971.

[12] J.P. Burg, "Maximum Entropy Spectral Analysis," Ph.D. Dissertation, Stanford University, Stanford, California, 1975.

[13] N.L. Owsley, "Modal Decomposition of Data Adaptive Spectral Estimates," presented at the *Yale University Workshop on Applications of Adaptive System Theory*, New Haven, CT, 1981.

[14] D.H. Johnson, S.R. Graaf, "Improving the Resolution of Bearing in Passive Sonar Arrays by Eigenvalue Analysis," *IEEE Trans. on Acoustics, Speech, and Signal Processing* Vol. ASSP-30, No. 4, Aug. 1982.

[15] S.Y. Kung and Y.H. Hu, "A highly concurrent algorithm and pipelined architecture for solving Toeplitz Systems," *IEEE Trans. on ASSP*, Vol. 31, No. 1, pp. 66-75, Feb. 1983.

[16] A. Sameh, "Numerical Parallel Algorithm—A Survey," High Speed Computer and Organization, Academic Press, pp. 207-228, 1977.

[17] D. Heller, "A Survey of Parallel Algorithms in Numerical Linear Algebra," *SIAM Review*, Vol. 20, No. 4, pp. 740-777, Oct. 1978.

[18] H.T. Kung, "The Structure of Parallel Algorithms," Advances in Computers, Vol. 19, pp. 70-111, Academic Press, 1980.

[19] S.Y. Kung and D.V. Bhaskar Rao, "Analysis and Implementations of the Adaptive Notch Filter for Frequency Estimation," *ICASSP '82* Paris, May 1982.

[20] S.Y. Kung, "VLSI Array Processor for Signal Processing," *Conference on Advanced Research in Integrated Circuits*, MIT, Cambridge, Mass., Jan. 28-30, 1980.

[21] J. M. Speiser and H.J. Whitehouse, "Architectures for Real Time Matrix Operations," *Proc., GOMAC*, Nov. 1980.

[22] G.W. Stewart, "Introduction to Matrix Computations," Academic Press, 1973.

[23] T. Kailath, "A View of Three Decades of Linear Filtering Theory," *IEEE Trans. Inform. Theory*, Vol. 1T20, No. 2, pp. 145-181, March 1974.

[24] I. Schur, "Uber Potenzreihen die in Innern des Einheitkreises beschrankt sind," *J. fur Math*, Vol. 147, pp. 110-148, 1917.

[25] E.H. Bareiss, "Numerical Solution of Linear Equations with Toeplitz and Vector Toeplitz Matrices," *Numer. Math.*, 3, pp. 404-424, 1969.

[26] M. Morf, "Fast Algorithms for Multivariable Systems," Ph.D. Dissertation, Stanford University, Stanford, CA, 1974.

[27] J.G. Nash, S. Hansen, and G.R. Nudd, "VLSI Processor Array for Matrix Operations and Linear Systems Solution," Internal Report, Hughes Research Laboratories, Malibu, California 90270, March 1982.

[28] V.F. Pisarenko, "The Retrieval of Harmonics from a Covariance Function," *Geophysics, J.R. Astron., Soc.* 33, pp. 346-366.

[29] H.F. Mermoz, "Spatial Processing Beyond Adaptive Beamforming," *J. Acoustics, Soc. Am.*, Vol. 70, No. 1, pp. 74-79, July 1981.

[30] L. S. Haynes, R.L. Lau, D.P. Siewiorek, and D.W. Mizell, "A Survey of Highly Parallel Computing," *IEEE Computer*, Vol. 15, No. 1, pp. 9-26, January 1982.

[31] H.T. Kung, "Let's Design Algorithms for VLSI Systems," *Proc. CALTECH Conf.* on VLSI, pp. 70-90, Jan. 1979.

[32] J.S. Walther, "A Unified Algorithm for Elementary Functions," *Spring Joint Computer Conference*, 1971.

[33] H.M. Ahmed, "Signal Processing Algorithms and Architectures," Technical Report No. M735-21, Information System Laboratory, Stanford University.

[34] G. Sharma, R. Gal-Ezer, and S.Y. Kung, "A Note on Arithmetic Units for Signal Processing," to be submitted for publication.

[35] S.Y. Kung and R. Gal-Ezer, "Hardware Architectures of the Wavefront Array Processor," *to appear in International Computer Symposium*, Taichung, Taiwan, December 1982.

[36] J.B. Dennis, "Data Flow Supercomputers," *IEEE Computer*, pp. 48-56, Nov. 1980.

A VLSI SAR Processor for Ocean Surveillance*

Benjamin Friedlander

Systems Control Technology, Inc.
Palo Alto, CA

1 INTRODUCTION

Imaging of the ocean by Synthetic Aperture Radar (SAR) is a relatively recent development. Observations of ocean waves with airborne SAR were apparently first reported by Brown et al. [1] and Larson et al. [2], in 1976. Since then, numerous other ocean features have been observed with airborne SAR sensors [1-4]. Aircraft observations are somewhat limited in spatial and temporal coverage. Spaceborne SAR offers long-term wide-area coverage of the ocean surface. In 1978 (between June and October) SEASAT provided, for the first time, spaceborne imaging of the ocean, see e.g. [5-6]. Each SEASAT SAR data path covers an area 100 kilometers wide and about 4000 km long with a resolution of 25 m.

Spaceborne SAR provides an all-weather high resolution imaging capability that can be used for a variety of navy missions. Remote sensing of the ocean can be used to study surface wave attributes (sea state, wind direction, and speed) sea ice conditions (location and tracking of icebergs, state of strategic channels, bays, and ports) or cloud and precipitation distribution. In the area of ship surveillance, SAR's can be used to detect, locate, and track vessels at sea or in port. The high resolution capability of the SAR opens the way for ship classification and identification.

SAR imaging of the ocean involves many different issues such as: modeling of the interaction of coherent radar signals with the ocean surface; processing of the radar returns to obtain an image, extraction of the desired information from the image and its transmission/distribution to potential users. In this

---

*This work was supported by the Office of Naval Research under Contract No. N00014-81-C-0300.

paper we discuss mainly issues related to the radar signal processing. Some of the other issues are briefly mentioned next.

The physics of radar wave scattering from the ocean surface are generally known as related to radar scatterometry. However, in radar imaging two new factors have to be accounted for (i) the high resolution capability of the radar and (ii) the use of coherent Doppler information to generate the image. The impact and importance of these two factors on the formation of ocean surface imagery is not fully understood at this time [7-13].

The moving ocean surface interacts with the radar signals in a complicated way. These signals are modulated by variations of the local coherent backscatter cross-section of the surface. Three sources of cross-section modulation seem to play an important role: change of the local tilt angle, variation of the surface roughness, and the waves' orbital velocity. The effect of these phenomena on SAR imagery have been studied in [7-13]. All of this work deals with conventional SAR processing. An interesting open question is whether there exist alternative processing schemes that are better suited to ocean imaging.

Digital processing of SAR data to form an image involves very large amounts of computation: 20-200 million operations (multiplies and adds) per second are typical requirements. Because of this, SAR processing is done mostly off-line, on data recorded during flight. For example, the unprocessed SEASAT video signals were transmitted to ground stations, to be processed later. The possibility of performing part or all of the processing in real-time, on-board the satellite, has been studied intensively in recent years [14-17]. On-board processing will greatly simplify the problems of handling the vast amounts of data produced by a spaceborne radar and the distribution of timely information to users. The advent of VLSI technology is expected to make on-board processing feasible in the near future. This technology will provide the means for a more widespread use of SAR systems not only in spaceborne applications but also in tactical aircraft and homing missiles.

In this paper we consider the feasibility of a VLSI SAR processor based on current silicon technology. The area/size requirements of such a processor are assessed by means of a prototype design (Sections 3 and 4). We start with a brief discussion of the signal processing involved in forming a SAR image.

2 SAR PROCESSING ARCHITECTURES

An extensive literature exists on various processing techniques for generating an image from the radar returns of a SAR: see [18-20] for tutorial type presentations and [21-29] for more specialized topics. The processing can usually be broken into two phases: range processing and azimuth processing. Most coherent radars use some form of modulation or coding of the transmitted waveform to improve resolution [18]. Appropriate processing is performed in the receiving end to demodulate or decode the radar returns. We will refer to this processing as range compression. In the second phase, Doppler processing is applied to achieve high resolution in the azimuthal direction (azimuth compression).

Splitting the processing into two parts results in considerable computational savings. Range compression reduces significantly the data rate for the subsequent azimuth compression. Range processing is usually performed by analog, rather than digital, circuitry which is better equipped to handle the high data rates involved. In some situation (especially in spaceborne applications) an ideal SAR system would require combined range/azimuth processing [24]. However, most practical SAR systems avoid this type of processing because of its increased complexity.

In this paper we consider only azimuth compression. Range compression, prefiltering (to reduce data rates as much as possible, c.f. [27]) and analog to digital conversion are assumed to have been performed separately. Two widely different azimuth processing architectures are briefly described and compared.

2.1 Correlator Architecture

Azimuth processing in a SAR system can be interpreted as correlating the radar return with a reference signal. The reference signal is the expected return from a given point on the ground (a complex signal whose phase is a quadratic function of time, i.e., linear FM). Returns from other points will be essentially uncorrelated with this signal. Thus, the output of each correlator will provide information about the amplitude of the return from one point on the ground. A complete image can be reconstructed by correlating the radar return with reference signals corresponding to all points in the imaged area.

One possible implementation of the correlator processor is depicted in Figure 1. The sampled returns from each radar pulse are multiplied by the proper reference function and circulated in a memory unit. This produces one line of the image corresponding to one azimuthal direction (a range line). Data collected over different time intervals produce different range lines. A collection of these lines produces a complete image. See [30] for a more detailed description.

The correlator architecture has a highly parallel pipelined structure, requiring only local communication of data. From a signal processing standpoint this structure is very flexible: by proper design of the reference function it is possible to focus individual cells of the image, provide velocity compensation, change resolution and perform zooming and panning.

The total calculation rate (C complex multiplies and adds per second) involved in the correlator processor can be shown to be [30]:

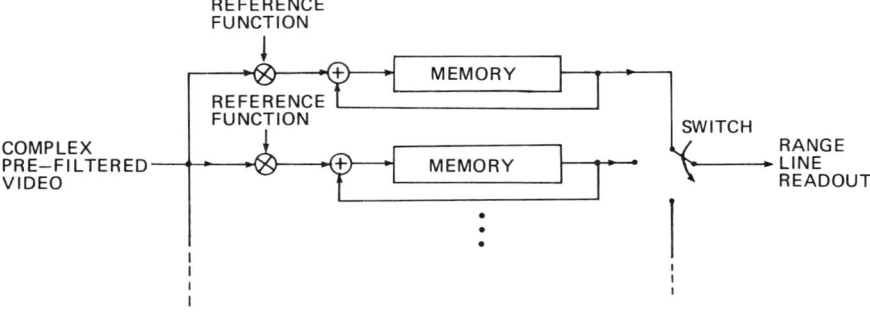

Figure 1 Correlator architecture for SAR processing

$$C = f_s N_r N_F = K_{os} K_s N_F^2 N_r / T \tag{1}$$

where

f_s = sampling rate
N_r = number of range samples
N_F = number of azimuth channels
T = integration time
K_{os} = oversampling factor
K_s = synthetic array weighting constant

The correlator memory requirements (M bits) are given by

$$M = 2k \, N_F N_r \tag{2}$$

where k is the number of bits required to express each noncomplex word.

2.2 FFT Architecture

Another interpretation of azimuth compression is in terms of Doppler processing. Points on the ground at different azimuthal directions introduce different Doppler shifts. A narrow band-pass filter will isolate the returns originating from a particular direction, determined by the center frequency of the filter. To look in many different directions, a bank of band-pass filters is required. These filters can be implemented by Fast Fourier transformation of the video signal at the output of a given range gate. This filtering operation is preceded by a quadratic phase correction, to focus the image. The FFT processor is depicted in Figure 2. Some of the details, such as range-walk correction, have been omitted (see [30] for a more complete description).

The FFT processor has a sequential-parallel structure. Data need to be collected and stored for the entire integration period, before processing can

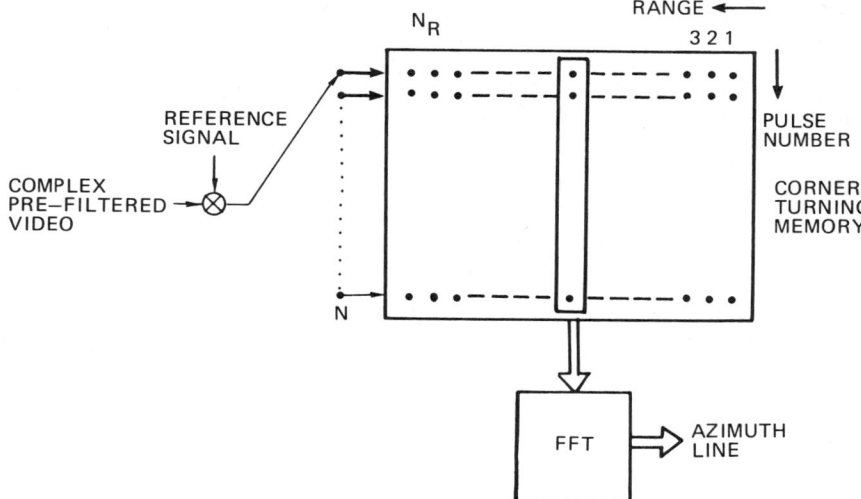

Figure 2 An FFT architecture for SAR processing

commence. Once a complete record is stored, FFT processing can be done in parallel on different azimuth lines, or groups of lines. Compared to the correlator architecture, we note that more memory is needed here to store new data and to perform the FFT. Also, more data communication is required. From a signal processing standpoint the FFT architecture is less flexible: it is not possible to perform focusing of individual cells; all cells within an azimuth line are treated in the same manner.

The total calculation rate of the FFT processor is given by

$$C = NN_r(1 + \log_2 N)/T \qquad (3)$$

where N is the length of the FFT transform ($N = f_s T$ or the nearest power of two).

The memory requirements are given by

$$M = 4kNN_r, \qquad (4)$$

assuming two memories of the type depicted in Figure 2 (one for storing the data that is being processed and one for storing new data). This does not include the memory requirements for performing the FFTs.

The FFT processor requires a calculation rate smaller by a factor of $N/\log_2 N$ compared to the correlator processor. For this reason, most digital SAR processors use some version of the FFT architecture described above.

3 A DESIGN EXAMPLE

In order to develop a clear understanding of these two implementations, it is necessary to consider a specific example. We have chosen a typical spaceborne SAR system, whose parameters are similar to the system used in the SEASAT experiment [5]. The geometry of the prototype system is depicted in Figure 3; the key SAR parameters are summarized in Table 1 (see [30] for more details). We consider the case where data are processed for 25 m resolution. This system is capable of a maximum azimuth resolution of 8.1 m. The former case requires less memory and less computations. Table 2 summarizes computational and memory requirements for the two processing architectures described in Section 2. Note that the correlator based processor requires 21 (143) times the computation rate of the FFT processor. Thus, when performing the computations on a general purpose computer (especially when an array processor is available) the FFT processor has an obvious advantage. However, when considering VLSI implementation of the SAR processor, the situation is quite different, as will be discussed in the next section.

4 THE VLSI SAR

To effectively use the complexity available even now on integrated circuits, it is necessary to begin by choosing an algorithm which maps nicely into silicon. Ten years ago a good algorithm was one that lead to a design that minimized the number of gates. Recently, a good design was one that minimized the number of IC packages (since production cost was a very strong function of package count). Today, with 40,000 transistor chips routinely being designed, and with 100,000 (Intel 432) to 400,000 (HP Focus) transistor chips coming

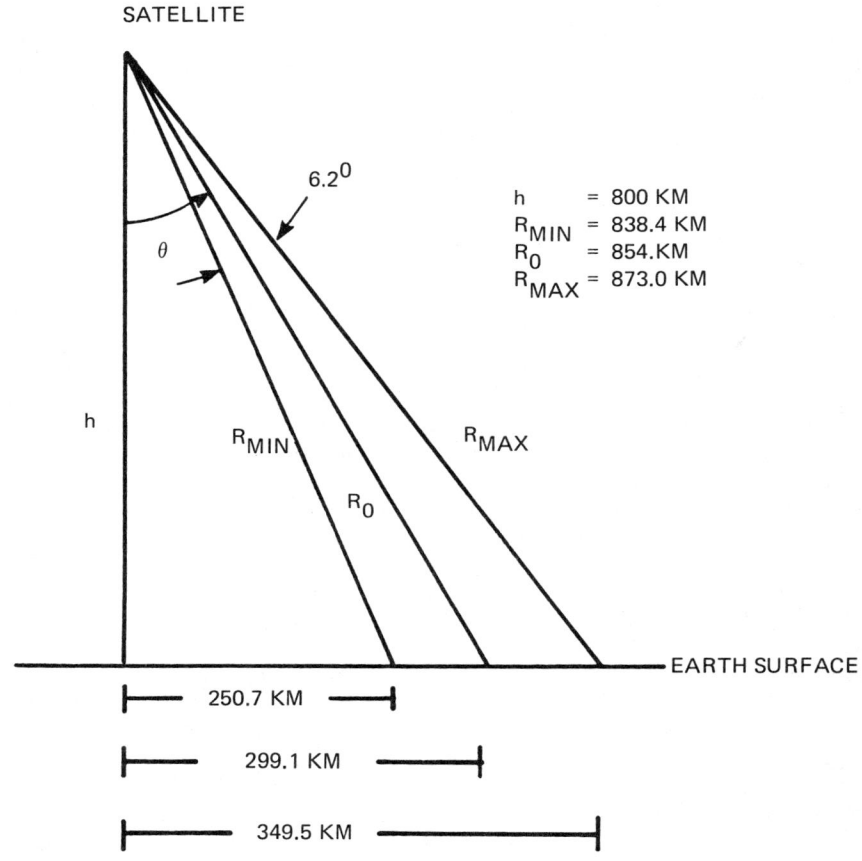

Figure 3 The satellite-earth geometry

into production, minimizing the package count ceases to be meaningful.

As an example, IBM's latest mainframe, the 3081, contains 700,000 gates. At three transistors per gate, this would be 21 of our 100,000 transistor chips. Unfortunately, even ignoring the technical problems, the cost of designing the 21 custom chips would probably be measured in thousands of man years.

Today, the key to exploiting the potential of VLSI is to choose an algorithm that will minimize the design cost. This usually means minimizing the on-chip interconnect. A typical industry rule of thumb is that each wire requires 3 man-days to design, layout and check (and poor designs can have thousands of wires). Good VLSI designs typically have highly replicated, very regular structures. In addition, the designer's emphasis is primarily on those sections of the chip that most strongly effect the speed, area or reliability of the design. (The arithmetic section may be the key to the speed of the design, whereas the interrupt logic plays a small role in both speed and area.)

Table 1 Summary of SAR Parameters

Satellite Speed	$V = 7400$ m/sec
Sea Level Altitude	$h = 800$ Km
Angle of Look	$\theta = 20.5°$ from nadir
Carrier Frequency	$f = 1274.8$ MHz; $\lambda = 23.5$ cm
PRF	$f_r = 1500$
Pulse Bandwidth	$BW = 19.05$ MHz
Pulse Width	$PW = 33.9$ μs
Antenna BW	$ABW = 1°$ AZ, $6.2°$ EL
Swath Width	$SW = 100$ Km
Resolution	$r_c = r_s = 25$ m (8.1 m)
Integration Time	$T = 0.65$ sec (2.01 sec)
Doppler Speed	$\Delta f_D = 550$ Hz
No. Azimuth Lines	$N_r = 3952$ (12228)
No. Azimuth Lines	$N_F = 193$ (1844)
Sampling Frequency	$f_s = 490$ Hz (for $r_c = 25$ m)
Precision of Storage	$k = 12$ (14) bits
Oversampling Factor	$K_{os} = 1.36$
Array Weighting	$K_s = 1.2$
Number of Azimuth Samples	$N = 318$ (3000)

Table 2 Computation and Memory Requirements

	Correlator	FFT
Computational Requirements (MOPS)	373 (33800)	17.4 (236)
Multiplier Precision (Bits)	6×6	12×8 (14×10)
Adder Precision (Bits)	12 (14)	12 (14)
Memory (MBits)	18.3 (631) Serial Addressing	60.3 (2054) Random Access
Memory Precision (Bits)	12 (14)	12 (14)

These guidelines are at odds with the traditions of the IC industry, but then our objective is not to produce a very large number of SAR processors. Our objective is to design a real-time SAR processor quickly, easily and reliably.

In the sequel we will evaluate the two SAR algorithms presented in the previous sections in terms of their potential for VLSI implementation. The only assumption that is critical is that we want to develop a fully-custom, VLSI design. We will target for an available nMOS technology, using well-understood, easily designed digital circuits.

Having selected the algorithm, we will then take a preliminary look at a potential VLSI system. We will anticipate using wafer-scale integration, which will allow us to design for today's technology, but with over a million transistors per "chip." Because it will be nearly impossible to fabricate perfect chips of this size, we will have to build redundancy into our design right from the start. As a secondary benefit, this added redundancy should also improve the system reliability.

4.1 Area Requirements

We begin by making a gross estimate of the silicon area required to implement each algorithm. We will focus only on the dominant computation — the correlation or the FFT — and initially ignore the interconnect cost and the minor area needs of the focusing computation, etc. We consider a straightforward, but compact, memory system and one of the simpler architectures available to the IC designer: serial adders and multipliers. It will become apparent that choosing a higher-performance architecture (shift and add, fully parallel, etc.) would not alter the algorithm selection.

To estimate the area requirement, consider the summary of computational resources in Table 2. From it we can determine the number of functional units (adders and multipliers) needed for a variety of different subsystem architectures (parallel, serial, shift, and add, etc.). First note that a complex multiply requires 4 real multiplications and two additions. A complex add requires 2 real additions. The number of multipliers and adders is given by

$$\text{Number of multipliers} = \frac{4C_M P_M}{R_M B_M} \quad (5a)$$

$$\text{Number of adders} = \frac{2(C_M P_M + C_A P_A)}{R_A B_A} \quad (5b)$$

where

C_M, C_A = computation rate: number of multiplies or adds per second (required by the algorithm).

P_M, P_A = precision: the number of bits in the multiplier or accumulator (required by the algorithm).

R_M, R_A = processing rate: the number of multiplies or adds per second, available from the silicon subsystem, performed on B_M or B_A bits in parallel.

Table 3 summarizes the relevant parameters for several types of subsystems

Table 3 Computation Rates and Area Requirements for Various Subsystems

Subsystem	R Processing Rate (cycles/sec./unit)	B Number of Bits (bits/cycle)	Area (λ^2)
Serial Adder	10^7	1	11,000
Serial Multiplier	10^7	1	236 (100 n + 164) $n = P_M$
Parallel Adder	10^7	P_m, P_A	7200 n $n = P_M, P_A$
Shift and Add Multiplier	$10^7/P_M$	P_M	7900 n $n = P_M$

[31]. The parameter λ is the minimum feature size of the fabrication technology. For today's nMOS technology, λ is typically 2 microns, equating to a line-width of 4 microns. Using the numbers in Tables 2 and 3 and Equation (5) we can compute the required number of multipliers and adders, and their total area. As an example, Table 4 summarizes these requirements for the correlator architecture using serial subsystems.

The multipliers and adders are only half of the picture; we also require memory. Using a straightforward circuit, each bit of storage will require an area of 400 λ^2. Using the numbers in Table 2 we obtain the memory area as shown in Table 5. This leads to the crucial realization that in this silicon technology the memory size completely dominates the total processor size. The processing area takes, in each case, less than 3% of the memory area. Also included in Table 5 is the area allocation for the FFT. From Table 2 we note that the FFT algorithm requires far fewer computations, but at the cost of additional memory. Therefore, the processor area is even less significant than in the case of correlator, and the FFT algorithm is more costly, in proportion to its increased need for memory.

This tradeoff is not uncommon when evaluating algorithms for silicon implementation. As in this case, each processor (multiplier/adder) is used many times during each integration step; speed is traded for area. No such tradeoff is possible for the memory: each bit of memory can be used to store only a single bit of information during one integration period.

Table 4 Number of Multipliers and Adders and Area Requirements

	Numbers of Serial Multipliers	Number of Serial Adders	Area of Multipliers [$10^6 \lambda^2$]	Area of Adders [$10^6 \lambda^2$]	Total [$10^6 \lambda^2$]
Correlator 25 m Resolution	912	1368	165	15	180
Correlator 8.1 m Resolution	80000	140000	15000	1500	16500

Table 5 Total Area Requirements for Serial Adder/Multiplier Implementation

	Memory $[10^6\lambda^2]$	Processor $[10^6\lambda^2]$	Total $[10^6\lambda^2]$
Correlator (25 m resolution)	7,200	180	~7,400
Correlator (8.1 m resolution)	250,000	16,500	~250,000
FFT (25 m resolution)	24,000	—	~24,000
FFT (8.1 m resolution)	800,000	—	~800,000

4.2 Architecture Tradeoffs

Both algorithms lend themselves to replication, an important characteristic of a VLSI architecture. However, the FFT algorithm has a significantly greater complexity in both the flow of data and the control structure. Data is not merely broadcast to all nodes, as with the correlator, but each data element is being mixed with each other data element. Thus, the FFT typically requires random access memory with corner turning capability (data is written by column and read by rows). In contrast, the correlator require a simple, serially addressed memory. In summary, at this level of analysis, the FFT has no structural advantages and the correlator is decidedly simpler.

These observations lead to the interesting and somewhat unexpected conclusion that the correlator architecture is perferable to the FFT architecture when VLSI implementation is considered. The traditional complexity measures of computer science, such as computations rate, are not adequate for the evaluation of VLSI systems. Measures of the structural complexity of the algorithm are much more important.

4.3 The VLSI Processor Structure

The complexity of a VLSI-SAR processor presents a challenging design and manufacturing problem. To achieve reasonable yield we are considering a novel chip structure that incorporates some recent theoretical developments. The complete processing system is expected to fit on a sandwich of about 10 wafers, each having a diameter of about 4″. Each wafer will contain about 150 hexagonal basic units, each with its own processor and memory, as depicted in Figure 4. A communication network is required to broadcast information to the various units on each wafer. The complete processing system will contain on the order of ten million transistors.

The basic memory unit has been designed fabricated and has passed initial tests. The chip layout is depicted in Figure 5. It contains the memory cells, serial addressing, error correction (for yield improvement) and sense amplifiers. This design involves about 10,000 transistors. The design of the communication framework and the processor portion of the basic unit are partially completed.

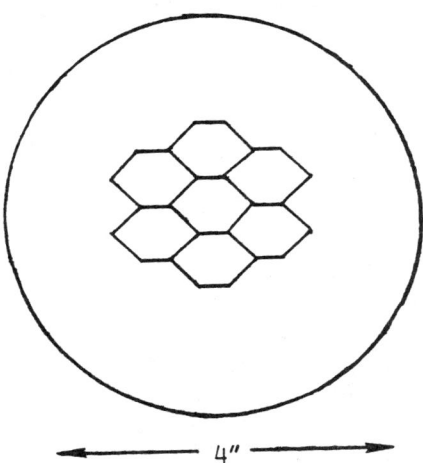

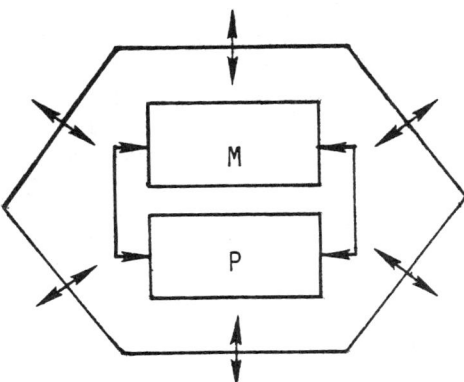

Figure 4 The overall chip layout

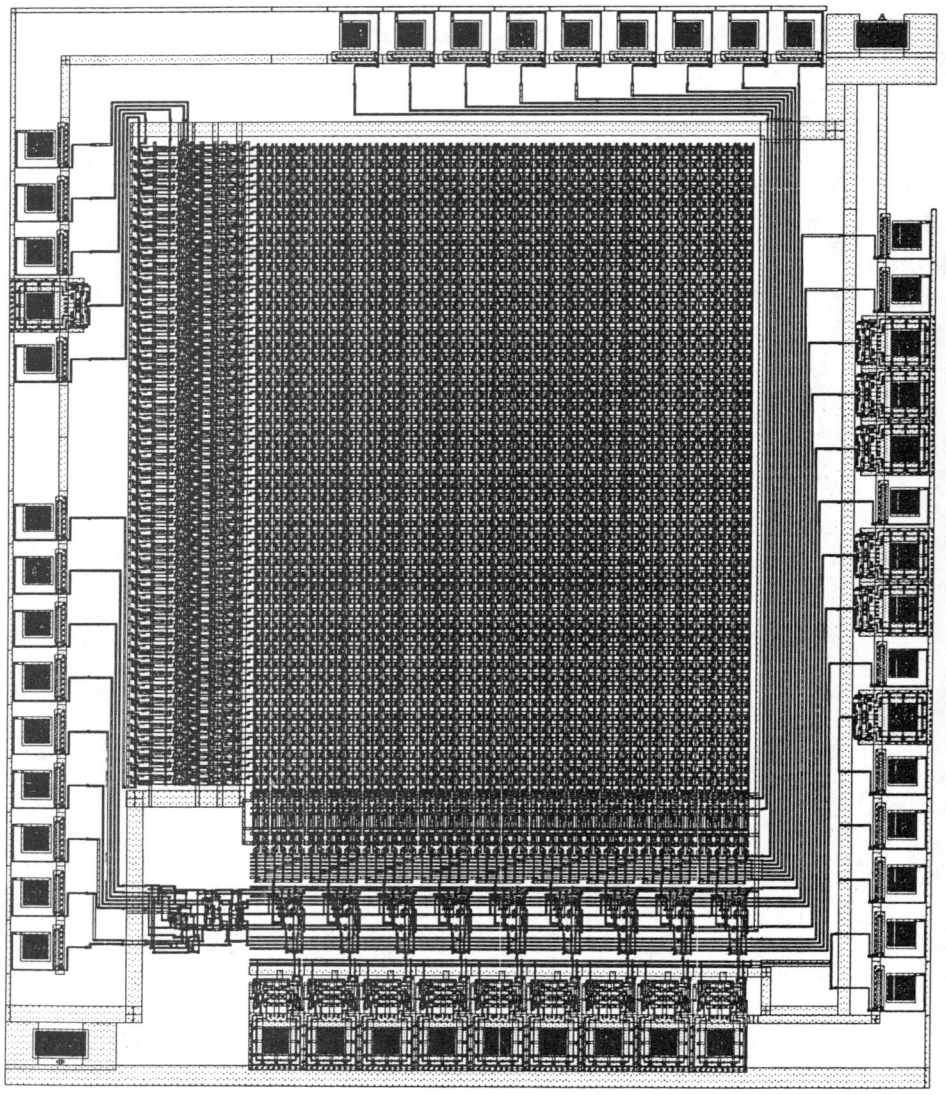

Figure 5 The memory chip

5 CONCLUSIONS

The high-resolution, all-weather capability of SAR makes it an excellent sensor for a variety of remote sensing and surveillance applications. The potential of SAR imaging of the ocean has been explored only to a very limited extent. The development of flexible on-board processing capability will remove one of the main obstacles to the widespread use of SAR systems. As mentioned before, many questions related to the interaction of a coherent radar transmission with the ocean surface and the extraction of useful information from the returns remain open at this time. While theoretical work can answer some of these questions, the experience gained by regular observations of SAR images of the ocean will provide invaluable insight and focus for future research.

ACKNOWLEDGMENTS

The work reported in this paper represents a joint effort by several people: Professor J. Newkirk performed the VLSI design and Dr. Y. Barniv was responsible for the radar system design.

REFERENCES

1. W.E. Brown, C. Elachi, and T.W. Thompson, "Radar Imaging of Ocean Surface Patterns," *Geophys. Res.,* Vol. 81, No. 15, pp. 2657-2667, 1976.

2. T.R. Larson, L.I. Moskowitz, and J.W. Wright, "A Note on SAR Imagery of the Ocean," *IEEE Trans. Ant. and Prop.,* Vol. AP-24, pp. 393-394, May 1976.

3. C. Elachi, "Radar Imaging of the Ocean Surface," *Boundary-Layer Meteorology,* Vol. 13, pp. 165-179, 1978.

4. C. Elachi, "Spaceborne Imaging Radar: Geologic and Oceanographic Applications," *Science,* Vol. 209, pp. 1073-1082, 5 September 1980.

5. R.L. Jordan, "The SEASAT: A Synthetic Aperture Radar System," *IEEE J. Oceanic Eng.,* Vol. OE-5, No. 2, pp. 154-164, April 1980.

6. R.L. Jordan and B.L. Honeycutt, "SEASAT: A Synthetic Aperture Radar Performance," Int. Conf. Communications, Boston, MA, June 10-13, 1979.

7. C. Elachi and W.E. Brown, Jr., "Models of Radar Imaging of the Ocean Surface Waves," *IEEE Trans. Ant. and Prop.,* Vol. AP-25, No. 1, pp. 84-95, January 1977.

8. W.R. Alpers and C.L. Rufenbach, "The Effect of Orbital Motions on Synthetic Aperture Radar Imagery of Ocean Waves," *IEEE Trans. Ant. and Prop.,* Vol. AP-27, No. 5, pp. 685-690, September 1979.

9. C.T. Swift and L.R. Wilson, "Synthetic Aperture Radar Imaging of Moving Ocean Waves," *IEEE Trans. Ant. and Prop.*, Vol. AP-27, No. 6, pp. 725-729, November 1979.

10. W.R. Alpers, D.B. Ross, and C.L. Rufenbach, "On the Detectability of Ocean Surface Waves by Real and Synthetic Aperture Radar," *J. Geoph. Research*, Vol. 86, No. C7, pp. 6481-6498, July 20, 1981.

11. R.K. Raney, "Wave Orbital Velocity, Fade, and SAR Response to Azimuth Waves," *IEEE Trans. Ocean Eng.*, October 1981.

12. R.A. Schuchman, et al., "Synthetic Aperture Radar Ocean Waves Studies," Final Report, Environmental Research Institute of Michigan, September 1978.

13. C.L. Rufenbach and W.R. Alpers, "Imaging Ocean Waves by Synthetic Aperature Radar with Long integration Times," *IEEE Trans. Ant. Prop.*, Vol. AP-29, No. 3, pp. 422-428, May 1981.

14. D.J. Bonfield and J.R.E. Thomas, "Synthetic Aperture Radar Real Time Processing," *Proc. IEEE*, Vol. 127, pt. F, No. 2, pp. 155-162, April 1980.

15. D.J. Bonfield and J.R.E. Thomas, "On-board Signal Processing Technology and Systems," in SAR Processing Architecture and Image Compression Algorithms, ESTEC Contract 2896/76/HP, 1978.

16. W.E. Arens, "The Application of Charge Coupled Device Technology, To Produce Imagery From Synthetic Aperture Radar Data," Proc. of the Conference on AIAA System Design Driven by Sensors, Pasadena, CA.

17. W.E. Arens, "CCD Architecture for Spacecraft SAR Image Processing," JPL Report 77-1392.

18. A.W. Rihaczek, *Principles of High Resolution Radar*, McGraw-Hill, 1969.

19. K. Tomiyasu, "Tutorial Review of Synthetic-Aperture Radar (SAR) with Applications to Imaging of the Ocean Surface," *Proc. IEEE*, Vol. 66, No. 5, pp. 563-583, May 1978.

20. J.J. Kovaly, *Synthetic Aperture Radar*, Artech House, 1976.

21. S.A. Hovanessian, *Synthetic Array and Imaging Radars*, Artech House, 1980.

22. E. Brookner, *Radar Technology*, Artech House, 1977.

23. R.O. Harger, *Synthetic Aperture Radar Systems,* Academic Press, 1970.

24. W.J. van de Lindt, "Digital Technique for Generating Synthetic Aperture Radar Images," *IBM J. Res. Develop.,* pp. 415-432, September 1977.

25. I.G. Cumming and J.R. Bennett, "Digital Processing of SEASAT SAR data," Proc. IEEE Intl' Conf. Acoustics, Speech and Signal Processing, Washington, D.C., April 2-4, 1979.

26. K. Hasselman, "A Simple Algorithm for the Direct Extraction of the Two-Dimensional Surface Image Spectrum from the Return Signal of a Synthetic Aperture Radar," *Int. J. Remote Sensing,* Vol. 1, No. 3, pp. 219-240, 1980.

27. J.C. Kirk, Jr., "A Discussion of Digital Processing In Synthetic Aperture Radar," *IEEE Trans. Aero. Sys.,* Vol. AES-11, No. 3, pp. 326-337, May 1975.

28. R.K. Raney, "Synthetic Aperture Imaging Radar and Moving Targets," *IEEE Trans. Aero. Elec. Sys.,* Vol. AES-7, No. 3, pp. 499-505, May 1971.

29. J.C. Kirk, Jr., "Motion Compensation for Synthetic Aperture Radar," *IEEE Trans. Aero. Elec. Sys.,* Vol. AES-11, No. 3, pp. 338-348, May 1975.

30. Y. Barniv and B. Friedlander, "An Example of Real Time Processing of Spaceborne SAR Images," SCT Technical Memorandum 5448-300, January 1982.

31. J. Newkirk and R. Mathews, "An nMOS Cell Library," Addison-Wesley, 1982.

Parallel Algorithms and Computational Structures for Linear Estimation Problems*

Gerard G. L. Meyer and Howard L. Weinert

Electrical Engineering and Computer Science Department
The Johns Hopkins University
Baltimore, MD

1 INTRODUCTION

In today's environment, real-time signal processing presents significant challenges to the system designer. Modern acquisition techniques produce high volume data sets that must be processed not only very quickly, but also reliably. The computational speed requirements can be achieved only through the use of either faster processing algorithms, or computational components with improved performance, or parallelism, or a combination of the above. As far as the first option is concerned, a large amount of effort has been expended on developing fast algorithms for many types of signal processing problems, and the likelihood of further major improvements in this regard is small. Similarly, the trend in hardware advancement is flattening out, and barring a technological breakthrough, for example in optical processing, the probability of significant improvements in the performance of available computing devices is also small. Therefore, the only realistic means for achieving high throughput is the use of parallelism.

Most signal processing algorithms are intended to be used on a single computational device, and therefore one needs specially designed segmented algorithms in order to use a network of interconnected computers. The segmentation of the software and the modularization of the hardware are not independent. In fact, software segmentation and hardware modularization must be considered simultaneously to achieve efficient use of the hardware. The design of the interconnected network is a typical engineering problem in the

---
*This paper was supported by the Office of Naval Research under contract N00014-81-K-0813.

sense that it involves a number of tradeoffs among speed-up τ, number of processors N_p, number of interconnection links N_t and maximum number of input/output links per processor N_i. The network speed-up, that is the computational performance of the network compared to that of a single processor, depends on N_p, N_t and N_i. The number of processors is limited by cost, size, weight, power and cooling constraints. The number of interconnection links is bounded by the capabilities of the computational devices used in the network, and by the relative complexity of the links versus the computational parts of the devices. Since the time required to effect communication between processors is a non-negligible part of the total computational time, the number of input and output transactions must be kept to a minimum.

In terms of reliability, since we are operating in real time we need the capability of efficient and timely fault detection, location, recovery and repair. The very nature of on-line processing makes it imperative that the fault detection and location schemes be capable of operating on permanent faults, transient faults, data dependent faults and environment dependent faults. Once these faults are detected and located, we need schemes for recovery and repair. As far as recovery is concerned, we want to immediately minimize the effect of the faults so that they do not propagate to the non-faulty hardware and render the signal processing network inoperative, and we want to lose as little data as possible. The repair schemes may use switching techniques to replace the faulty hardware, or to reconfigure the computing network so that the processing tasks still may be performed although in a possibly degraded fashion. Note that the addition of a switching network for fault repair purposes in itself introduces a whole class of possible faults that may have a dramatic effect on the performance of the computational network.

The first part of the paper is devoted to a discussion of real time fault detection requirements. Then the following section considers fault detection and parallel processing of statistical signals using the concept of stochastic redundancy. Finally, parallel algorithms and reliable network designs are presented for a linear regression problem.

2 REAL TIME FAULT DETECTION

To achieve real time fault detection, a concurrent fault detection scheme is needed: one that takes place as the processing network is operating. Currently, three basic techniques exist for carrying out concurrent fault detection: one is total hardware redundancy [AVI76], [GRA76], the second is partial hardware redundancy and the third is functional redundancy [SOG69], [CLA78]. In all three cases, the detection scheme involves the generation of an auxiliary boolean sequence $\{b(i)\}$, followed by a decision based on that sequence. The elements of $\{b(i)\}$ are generated so that if $b(i) = 1$, the system is faulty, and if $b(i) = 0$, the system may or may not be faulty. The decision scheme uses the elements of the sequence $\{b(i)\}$ up to the present time to determine the status of the system.

In the case of total hardware redundancy (see Figure 1), the entire signal processor P_1 is duplicated as P_2. At time t_i outputs $w_1(i)$ and $w_2(i)$ of P_1 and P_2, respectively, are compared to produce the quantity $b(i)$. The variable $b(i)$ is set equal to 0 if $||w_1(i) - w_2(i)|| \leq \epsilon$, and is set equal to 1 otherwise. The

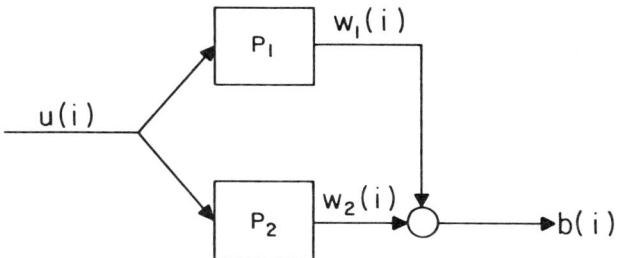

Figure 1 Total hardware redundancy

advantages of this scheme are (i) it is easy to implement because P_2 is identical to P_1 and thus does not require additional system design, and (ii) it is easy to decode the word $(w_1(i), w_2(i))$ by using a simple threshold device. The drawback is that we need complete hardware duplication and this may violate some of the existing system constraints.

In the case of partial hardware redundancy, we do not duplicate the entire signal processor, but only part of it. Although this approach decreases the complexity of the hardware, it does increase design difficulty: the computational device P_2 is not identical to P_1 and must be separately designed. This approach also decreases the fault coverage capability of the fault detection scheme since the device P_1 is not completely duplicated.

With functional redundancy, simulation is used in lieu of hardware duplication. The most important aspects of the operations of P_1 are simulated in the device P_2. This technique is usually reserved for analog and hybrid systems.

The above considerations lead to a general concurrent fault detection model in which, given a sequence $\{z(i)\}$ that is determined by the outputs of the computational elements of the signal processor, one needs to find a map $f(\cdot)$ such that $f(z(i))$ is easily computed, $f(z(i)) \leq \epsilon$ means that the outputs may be correct and $f(z(i)) > \epsilon$ means that the outputs are incorrect. In general, there is no inherently natural map $f(\cdot)$, thus it is necessary to add hardware to artificially create such a map. This produces what we call an $f(\cdot)$-redundant design. As long as no assumptions are made on the nature of the input data, one can obtain an $f(\cdot)$-redundant design only by using total hardware redundancy, partial hardware redundancy, or functional redundancy. In the next section, we show that if the data is stochastic in nature, we can exploit this property to obtain a special type of redundant design, namely a stochastic $f(\cdot)$-redundant design, that may be used for the purpose of fault detection.

3 STATISTICAL SIGNAL PROCESSING AND STOCHASTIC REDUNDANCY

For our purpose, we view statistical signal processing as the transformation of an input sequence of random variables $\{u(i)\}$ into an output sequence of random variables $\{v(i)\}$ posessing desirable properties. This transformation is implemented on a three-part network: a generating network, a computing network and a receiving network (see Figure 2). The generating network breaks

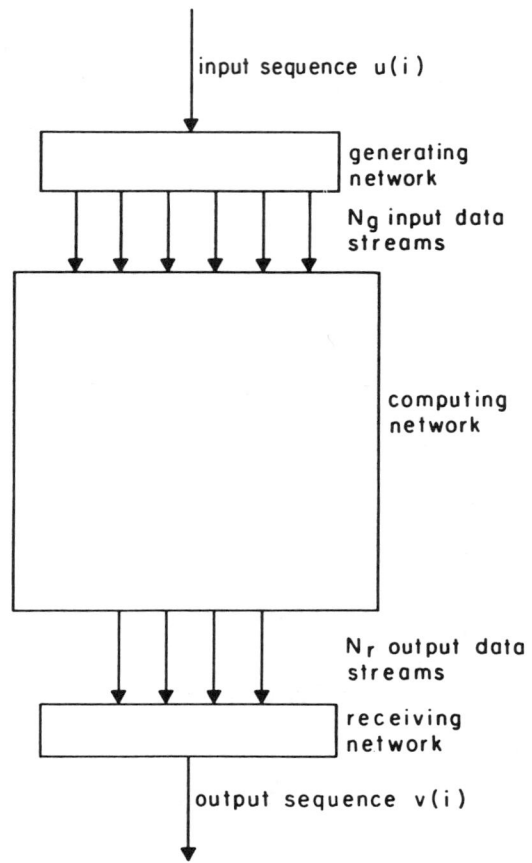

Figure 2 The transformation three-part network

up the input sequence $\{u(i)\}$ into N_g subsequences or input data streams $\{w_{0,0}(i)\}, \{w_{0,1}(i)\}, \ldots, \{w_{0,N_g-1}(i)\}$. The computing network processes the input data streams in parallel and produces N_r output data streams. The receiving network combines the output data streams to produce the desired output sequence $\{v(i)\}$.

In order to achieve our fault detection objectives, the generating network should break up the input sequence $\{u(i)\}$ so as to maximize the average canonical correlation [AND58], [BRI75]. Preliminary investigations show that under mild assumptions this goal may be achieved by the following natural scheme. For $k = 0, 1, 2, \ldots, N_g - 1$, let

$$w_{0,k}(jN_g + k) = u(jN_g + k), \ j = 0, 1, 2, \ldots.$$

The computing network is designed so that the initial redundancy (measured by the canonical correlations) among the input data streams is preserved as the data propagate through the network, and so that the necessary computations are distributed among the computing elements in a balanced fashion. In this way a stochastic $f(\cdot)$-redundant design can be implemented without additional hardware, and the required processing speed can be achieved. A stochastic $f(\cdot)$-redundant design is characterized by the following properties. If $w_{k,l}(i)$ and $w_{m,n}(i)$ are the respective outputs of any two computing elements $P_{k,l}$ and $P_{m,n}$, then there is an easily computed map $f(\cdot)$ such that, in the absence of faults, $f(w_{k,l}(i), w_{m,n}(i))$ has zero mean and a variance that goes to 0 as i goes to infinity. As a result, the status of the computing network can be determined by comparing $f(w_{k,l}(i), w_{m,n}(i))$ to a threshold for various values of $k,l,m,$ and n.

It is important to note that the need for real time fault detection puts non-trivial constraints on the way the data can be processed in parallel. Previous parallel processing studies [MCR74], [MOR79] ignore the issue of network reliability.

In the next section, we illustrate our ideas by considering a linear regression problem.

4 LINEAR REGRESSION

In order to illustrate the concepts discussed in the previous sections, and to demonstrate the trade-offs involved in a stochastic $f(.)$-redundant design, attention is focussed on the one-parameter linear regression problem [DRA81]. This problem was chosen because it possesses many of the features of more general signal processing problems, and yet is simple enough to propose various network designs without burdensome complications.

The problem of interest is the following. Let x be an unknown parameter and let $\{n(i)\}$ be a sequence of uncorrelated zero-mean random variables with common variance σ^2. Suppose we have measurements

$$y(i) = x + n(i), \quad i = 0, 1, 2, \ldots,$$

and we want to compute recursively $\bar{x}(j)$, the minimum variance unbiased linear estimate of x based on $y(0), y(1), \ldots, y(j)$, for each $j = 0, 1, 2, \ldots$. It is well known that the optimal estimator is nothing more than the sample mean of the available measurements, namely

$$\bar{x}(j) = \frac{1}{j+1} \sum_{i=0}^{j} y(i).$$

For this problem, the stochastic $f(\cdot)$-redundancy properties may be implemented by requiring that the output of every element in the computing network be a minimum variance unbiased linear estimate of x. This will insure that the difference of the outputs of any two computing elements will have a zero mean and an asymptotically zero variance in the absence of faults. In other words, we can take $f(w_{k,l}(i), w_{m,n}(i)) = |w_{k,l}(i) - w_{m,n}(i)|$.

Two stochastic $f(\cdot)$-redundant designs and their associated parallel algorithms for this regression problem are discussed below. The first (Figure 3) uses a computing network with two levels, two input data streams and two output data streams. The second (Figure 4) uses a computing network with three

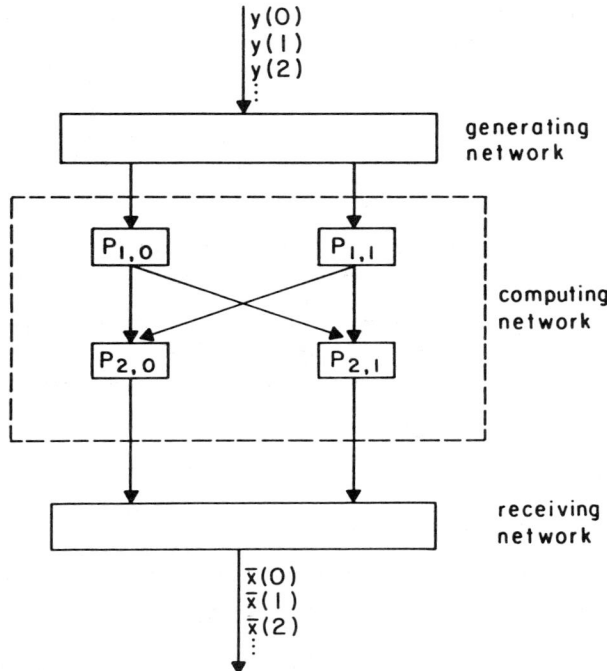

Figure 3 A computing network with two levels

levels, four input data streams and four output data streams. Generalizations to larger computing networks are straightforward.

5 TWO LEVEL-TWO DATA STREAM NETWORK

We start our presentation by considering the simplest design case, namely the two level-two data stream network (see Figure 3). In this case, given the input sequence $y(i)$, $i = 0, 1, 2, \ldots$, the generating network produces the sequences $w_{0,0}(j), j = 0, 2, 4, \ldots$, and $w_{0,1}(j), j = 1, 3, 5, \ldots$ where

$$w_{0,0}(j) = y(j), \; j = 0, 2, 4, \ldots$$

and

$$w_{0,1}(j) = y(j), \; j = 1, 3, 5, \ldots$$

Processor $P_{1,0}$ recursively computes the mean of $\{w_{0,0}(i)\}$ according to

$$w_{1,0}(0) = w_{0,0}(0),$$

and

$$w_{1,0}(2j) = w_{1,0}(2j-2) + \frac{1}{j+1}(w_{0,0}(2j) - w_{1,0}(2j-2)), j = 1, 2, \ldots.$$

Processor $P_{1,1}$ recursively computes the mean of $\{w_{0,1}(i)\}$, according to

Linear Estimation Problems

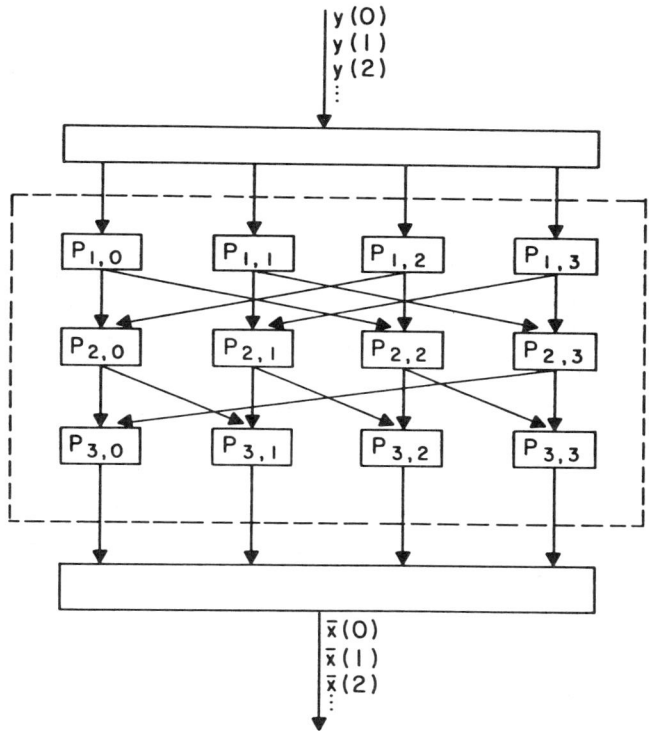

Figure 4 Three-level-four data stream network

$$w_{1,1}(1) = w_{0,1}(1)$$

and

$$w_{1,1}(2j+1) = w_{1,1}(2j-1) + \frac{1}{j+1}(w_{0,1}(2j+1) - w_{1,1}(2j-1)), \, j=1,2,\ldots.$$

Note that $w_{1,0}(2i)$ and $w_{1,1}(2i+1)$ are the minimum variance unbiased linear estimates of x based on the "even" data $\{y(0), y(2), \ldots, y(2i)\}$ and the "odd" data $\{y(1), y(3), \ldots, y(2i+1)\}$, respectively.

The outputs of processors $P_{1,0}$ and $P_{1,1}$ are both sent to two other processors $P_{2,0}$ and $P_{2,1}$. Processor $P_{2,0}$ computes the optimal estimate of x at "even" times based on all present and past data via

$$w_{2,0}(0) = w_{1,0}(0),$$

and

$$w_{2,0}(2j) = \frac{1}{2j+1}((j+1)w_{1,0}(2j) + jw_{1,1}(2j-1)), \, j=1,2,\ldots.$$

and processor $P_{2,1}$ computes the optimal estimate at "odd" times based on all present and past data via

$$w_{2,1}(2j+1) = \frac{1}{2}(w_{1,0}(2j) + w_{1,1}(2j+1)), \; j = 0,1,2,\ldots.$$

Finally, the receiving network generates the estimator sequence $\{\bar{x}(i)\}$ via

$$\bar{x}(j) = w_{2,0}(j), \; j = 0,2,4,\ldots,$$

and

$$\bar{x}(j) = w_{2,1}(j), \; j = 1,3,5,\ldots.$$

We thus have a parallel network of 4 processors. The first two compute the best estimate of x based on the "even" and "odd" data, respectively. The other two combine these estimates to produce the overall best estimates at "even" and "odd" times. This particular decomposition of the problem allows us to compute $\bar{x}$ approximately twice as fast as we could with a single processor (or, in other words, we can handle about twice the data rate). In this case, the numbers N_p, N_t and N_i are equal to 4, 4 and 3, respectively.

6 THREE LEVEL-FOUR DATA STREAM NETWORK

The next case of interest is the three level-four data stream network (see Figure 4). In this case, given the input sequence $y(i)$, $i = 0,1,2,\ldots$, the generating network produces the sequences $w_{0,k}(4j+k), j = 0,1,2,\ldots$, via

$$w_{0,k}(4j+k) = y(4j+k), \; j = 0,1,2,\ldots.$$

The first level processor $P_{1,k}$ recursively computes the mean of $\{w_{0,k}(4i+k)\}$ according to

$$w_{1,k}(k) = w_{0,k}(k),$$

and

$$w_{1,k}(4j+k) = w_{1,k}(4(j-1)+k) + \frac{1}{j+1}(w_{0,k}(4j+k) - w_{1,k}(4(j-1)+k)),$$

for $j = 1,2,\ldots$ and $k = 0,1,2,3$. Note that $w_{1,k}(4i+k)$ is the optimal estimate of x based on $\{y(k), y(4+k), \ldots, y(4i+k)\}$.

Processor $P_{2,0}$ uses the outputs of $P_{1,0}$ and $P_{1,2}$ to compute the optimal estimate of x at times $0, 4, 8, \ldots$ based on the present and past "even" data $\{y(0),y(2),y(4),\ldots\}$, via

$$w_{2,0}(0) = w_{1,0}(0)$$

and

$$w_{2,0}(4j) = \frac{1}{2j+1}((j+1)w_{1,0}(4j) + jw_{1,2}(4j-2))$$

for $j = 1,2,\ldots$.

Processor $P_{2,1}$ uses the outputs of $P_{1,1}$ and $P_{1,3}$ to compute the optimal estimate of x at times $1, 5, 9, \ldots$ based on the present and past "odd" data $\{y(1), y(3), y(5), \ldots\}$, via

$$w_{2,1}(1) = w_{1,1}(1)$$

and
$$w_{2,1}(4j+1) = \frac{1}{2j+1}((j+1)w_{1,1}(4j+1) + jw_{1,3}(4j-1))$$

for $j = 1, 2, \ldots$.

Processor $P_{2,2}$ uses the outputs of $P_{1,2}$ and $P_{1,0}$ to compute the optimal estimate of x at times 2, 6, 10, ... based on the present and past "even" data $\{y(0), y(2), y(4), \ldots\}$, via

$$w_{2,2}(4j+2) = \frac{1}{2}(w_{1,2}(4j+2) + w_{1,0}(4j))$$

for $j = 0, 1, 2, \ldots$.

Processor $P_{2,3}$ uses the outputs of $P_{1,3}$ and $P_{1,1}$ to compute the optimal estimate of x at times 3, 7, 11, ... based on the present and past "odd" data $\{y(1), y(3), y(5), \ldots\}$, via

$$w_{2,3}(4j+3) = \frac{1}{2}(w_{1,3}(4j+3) + w_{1,1}(4j+1))$$

for $j = 0, 1, 2, \ldots$.

Processor $P_{3,0}$ uses the outputs of $P_{2,0}$ and $P_{2,3}$ to compute the optimal estimate of x at times 0, 4, 8, ... based on all present and past data via

$$w_{3,0}(0) = w_{2,0}(0)$$

and

$$w_{3,0}(4j) = \frac{1}{4j+1}((2j+1)w_{2,0}(4j) + 2jw_{2,3}(4j-1))$$

for $j = 1, 2, \ldots$.

Processor $P_{3,1}$ uses the outputs of $P_{2,1}$ and $P_{2,0}$ to compute the optimal estimate of x at times 1, 5, 9, ... based on all present and past data via

$$w_{3,1}(4j+1) = \frac{1}{2}(w_{2,1}(4j+1) + w_{2,0}(4j))$$

for $j = 0, 1, 2, \ldots$.

Processor $P_{3,2}$ uses the outputs of $P_{2,2}$ and $P_{2,1}$ to compute the optimal estimate of x at times 2, 6, 10, ... based on all present and past data via

$$w_{3,2}(4j+2) = \frac{1}{4j+3}((2j+2)w_{2,2}(4j+2) + (2j+1)w_{2,1}(4j+1))$$

for $j = 0, 1, 2, \ldots$.

Processor $P_{3,3}$ uses the outputs of $P_{2,3}$ and $P_{2,2}$ to compute the optimal estimate of x at times 3, 7, 11, ... based on all present and past data via

$$w_{3,3}(4j+3) = \frac{1}{2}(w_{2,3}(4j+3) + w_{2,2}(4j+2))$$

for $j = 0, 1, 2, \ldots$.

Finally, the receiving network produces the estimator sequence $\{\bar{x}(i)\}$ via

$$\bar{x}(4j+k) = w_{3,k}(4j+k), \quad j = 0, 1, 2, \ldots.$$

In this case, $\tau = 4$, $N_p = 12$, $N_t = 16$, and $N_i = 4$.

REFERENCES

AND58 Anderson, T.W., *Introduction to Multivariate Statistical Analysis,* Wiley, New York, 1958.

AVI76 Avizienis, A., Fault Tolerant Systems, *IEEE Trans. Computers,* Vol. C-25, No. 12, December 1976, pp. 1304-1312.

BRI75 Brillinger, D.R., *Time Series: Data Analysis and Theory,* Holt, Rinehart and Winston, New York, 1975.

CLA78 Clark, R.N., Instrument Fault Detection, *IEEE Trans. Aerospace and Electronic Systems,* Vol. AES-14, No. 3. May 1978, pp. 456-465.

DRA81 Draper, N. and Smith, H., *Applied Regression Analysis,* Second Edition, Wiley, New York, 1981.

GRA76 Gray, G.F., and Meyer, J.F., Algebraic Properties of Functions Affecting Optimum Fault Tolerant Realizations, *IEEE Trans. Computers,* Vol. C-25, No. 11, November 1976, pp. 1078-1088.

MCR74 McReynolds, S.R., Parallel Filtering and Smoothing Algorithms, *IEEE Trans. Automatic Control,* Vol. AC-19, 1974, pp., 556-561.

MOR79 Morf, M., Dobbins, J.R., Friedlander, B. and Kailath, T., Squareroot Algorithms for Parallel Processing in Optimal Estimation, *Automatica,* Vol. 15, 1979, pp. 299-306.

SOG69 Sogomonyan, E.S., Failure Diagnosis in Discrete Modular Objects, *Autom. Remote Control,* No. 10, October 1969, pp. 1688-1696.

Performance of Multihops per Bit Binary Frequency-Shift Keying Frequency Hopping (BFSK/FH) Spread Spectrum System in the Partial Band Jamming Environment*

J. S. Lee and Y. K. Hong

J.S. Lee Associates
Arlington, VA

1 INTRODUCTION

The spread spectrum (SS) techniques were originally developed to bring about communication privacy and protection against jamming threat. Recently, however, much effort is being expended to develop Low Probability of Intercept (LPI) Communication systems for both tactical and strategic use. The main objective of LPI Communication System is to achieve a high degree of covertness for the message traffic. An effective method of achieving covert communication is to employ a class of spread spectrum (SS) techniques known as frequency hopping (FH) techniques.

In frequency hopping spread spectrum (FHSS) systems the carrier is switched (or hopped) to a new frequency, and the hopping pattern is pseudorandomly controlled. Thus, the FHSS systems employ extremely wide bandwidth, much greater than that actually required for transmitting the message signal waveform.

To improve the degree of covertness the system designers employ multihops per bit strategy, say L-hops per bit. In this scheme the transmission of one bit is broken into L indpendent transmissions, each of duration $1/L$ of the bit period. The degree of covertness increases in general (with a limit) as L increases. The effective jamming strategy against FH systems is a partial-band

---
*This research was supported by the Office of Naval Research under contract N00014-81-C-0534.

jamming. When the jammer employs an optimum strategy for a given communicator's bit energy, a severe constraint is placed upon the choice of L (the number of hops per bit) if the message bit stream is to be recovered with an acceptable error probability. It, therefore, behooves the designer to know the performance measures of the L-hops/bit system under postulated signal-to-jamming ratio. In this presentation we will consider the performance of a binary frequency-shift keying (BFSK) spread spectrum system in both partial-band jamming and thermal noise.

The required statistical analysis pertaining to this problem is so complex that the workers dealing with this type of problem have always "simplified" the analysis by making certain assumptions such as no thermal noise [1], [2], [3]. Milstein et al. [4] considered both jamming and thermal noise for the case of single-hop per bit ($L = 1$). A study on multiple FSK signal reception treated in [5] employed multihops per bit strategy in both thermal noise and multiple access interference. This work is, conceptually, similar to a special case (wideband jamming) of our problem.

We will present a summary of the complete analysis for the performance of L-hops/bit spread spectrum system employing binary frequency-shift keying (BFSK) modulation in both wideband (ineffective) jamming and optimum partial-band (effective) jamming strategies.

2 SYSTEM DESCRIPTION

The modulation and frequency hopping scheme under consideration is shown in Figure 1. The binary FSK modulator selects one of two frequencies f_1 ("space") and f_2 ("mark"), according to the incoming binary data of $R_b = 1/T_b$ bits/sec, where T_b is the bit period. The selected signal frequency is then mixed (hopped) with the hop frequency f_H generated by a frequency synthesizer which is controlled by a pseudorandom (PN) code generator. While the binary input message data changes at a data rate of R_b bits per second, the frequency synthesizer changes the frequency at a hopping rate R_H hops per second or equivalently L-hops per bit period T_b, which is dictated by the clock rate of $R_H = LR_b$.

Figure 2 shows a receiver block diagram for the BFSK/FH signal employing L-hops/bit strategy described above. The received carrier is downconverted (dehopped) by means of a frequency synthesizer controlled by a pseudorandom code generator. The pseudorandom code is assumed to be synchronized and identical to that in the transmitter. The (dehopped) noisy baseband signal $r(t)$ is then processed by the standard square-law combining binary FSK receiver as shown. It is the purpose of this presentation to show the error rate performance of this receiver under both partial band noise jamming and thermal noise assumption.

3 THE ANALYSIS

The binary FSK signal $s(t)$ is assumed to be, over a hop interval $(0, \tau)$,

$$s(t) = \begin{cases} \sqrt{2S} \cos(2\pi f_1 t + \theta_1), & \text{for "space" signal,} \quad (1a) \\ \sqrt{2S} \cos(2\pi f_2 t + \theta_2), & \text{for "mark" signal,} \quad (1b) \end{cases}$$

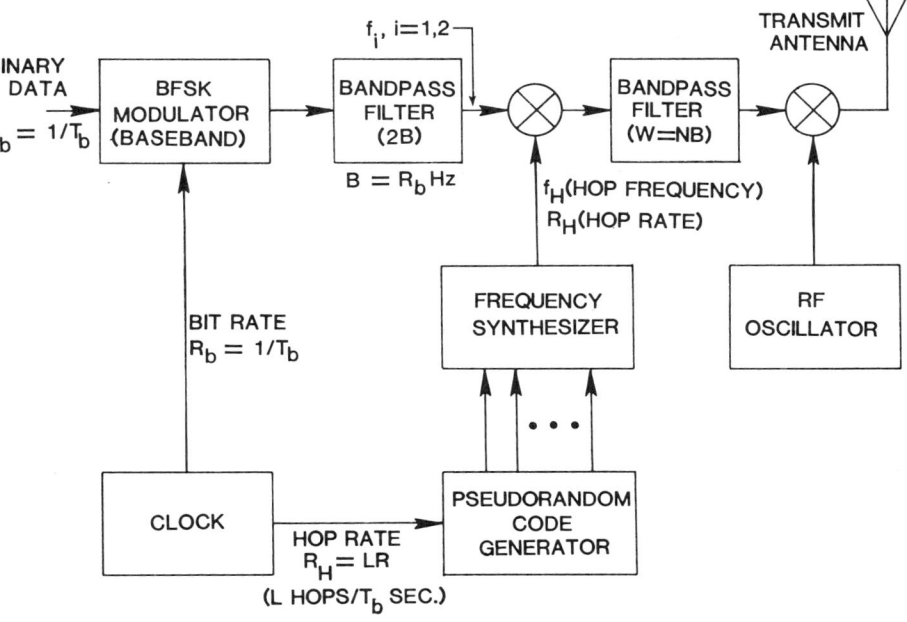

Figure 1 L hops/bit BFSK/FH transmitter

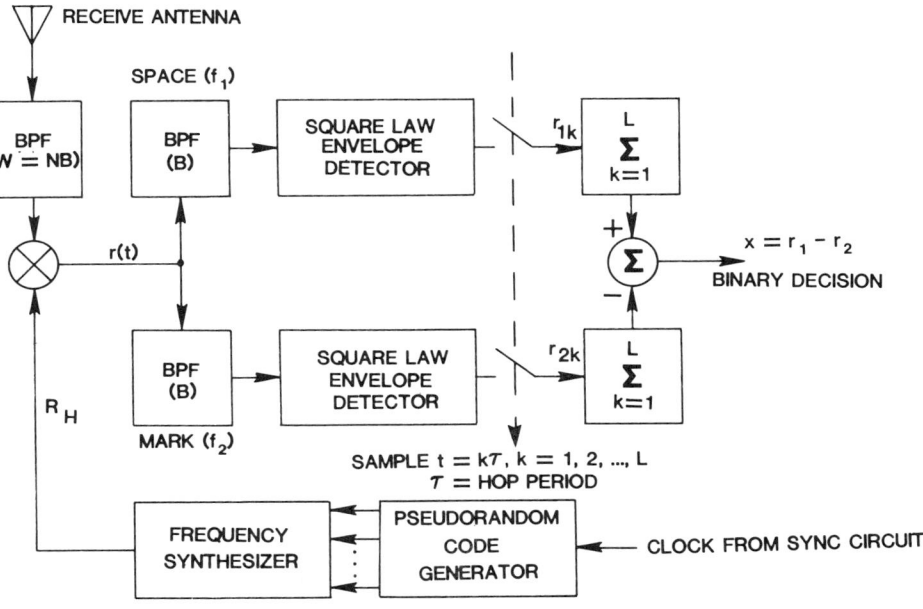

Figure 2 Noncoherent receiver configuration for BFSK/FH signals with L hops/bit strategy

where S is the received (average) signal power, f_1 and f_2 are the "space" and "mark" frequencies, respectively, and θ_1 and θ_2 are independent phases uniformly distributed on $(0, 2\pi)$.

The partial-band noise jammer is assumed to have a total power J, which is uniformly distributed across a fraction γ of the total spread spectrum bandwidth W Hz. Thus, a specific hop is received jamming free with probability $1 - \gamma$ and perturbed by jamming noise of power σ_j^2 with probability γ, where σ_j^2 is given by

$$\sigma_j^2 = \left(\frac{J}{\gamma W}\right) B, \tag{2}$$

where $B = 1/\tau$ is the bandwidth of a frequency cell. We assume that the two adjacent frequency cells of BFSK (modulation band) are jammed simultaneously with probability γ on each hop. The noisy baseband signal $r(t)$, which is recovered after dehopping, may then be represented as

$$r(t) = \begin{cases} s(t) + n(t) + j(t), & \text{with probability } \gamma \\ s(t) + n(t), & \text{with probability } 1 - \gamma, \end{cases} \tag{3}$$

where $s(t)$ is the information-bearing signal given by (1), and $n(t)$ and $j(t)$ are thermal noise and jamming noise, respectively.

Assuming that the thermal noise and jamming noise in any selected cell are Gaussian distributed, we may express the $n_i(t)$ and $j_i(t)$, $i = 1, 2$ at the outputs of the BPF's (or inputs to the square-law envelope detectors), as (Rician decomposition)

$$n_i(t) = n_{ci}(t) \cos 2\pi f_i t + n_{si}(t) \sin 2\pi f_i t; 1, 2, \tag{4a}$$

$$j_i(t) = j_{ci}(t) \cos 2\pi f_i t + j_{si}(t) \sin 2\pi f_i t; i = 1, 2, \tag{4b}$$

where $n_{ci}(t)$, $n_{si}(t)$, $j_{ci}(t)$, and $j_{si}(t)$ at a given time are statistically independent Gaussian random variables with variance (or average power) given by

$$E[n_i^2(t)] = E[n_{ci}^2(t)] = e[n_{si}^2(t)] = \sigma_N^2, \tag{4c}$$

$$E[j_i^2(t)] = E[j_{ci}^2(t)] = E[j_{si}^2(t)] = \sigma_j^2. \tag{4d}$$

General Expression for the Probability of Error

From Figure 2 the decision statistic is given by

$$x = \sum_{k=1}^{L} (r_{1k} - r_{2k}) = \sum_{k=1}^{L} z_k, \tag{5a}$$

where r_{1k} and r_{2k} are samples of the squared envelopes at channel 1 and channel 2, respectively, taken at $t = k\tau; 1, 2, \ldots, L$, where τ and L are the hop interval and the number of hops per bit, respectively, and

$$z_k \triangleq r_{1k} - r_{2k}. \tag{5b}$$

The decision rule based on the statistic x is to choose "space" if $x > 0$ and "mark" if $x < 0$. Thus, assuming the "space" and "mark" symbols are equally likely with probability 1/2 we may write the probability of error

$$P(e) = \tfrac{1}{2} P(e|\text{space}) + \tfrac{1}{2} P(e|\text{mark}) = P(e|\text{space})$$

$$= Pr[x < 0 | \text{space}] = \int_{-\infty}^{0} p_x(\alpha | \text{space}) \, d\alpha, \tag{6}$$

where $p_x(\alpha | \text{space})$ is the probability density function of x given that a "space" is transmitted.

In accordance with (1) through (5), the conditional probability density function may be shown to be

$$p_x(\alpha | \text{space}) = \sum_{l=0}^{L} \binom{L}{l} (1-\gamma)^l \gamma^{L-l} e^{-l\beta_N} e^{-(L-l)\beta_T} \sum_{m=0}^{\infty} \sum_{n=0}^{\infty} \frac{(l\beta_N)^m}{m!} \frac{[(L-l)\beta_T]^n}{n!}$$

$$\begin{cases} \left[\sum_{r=0}^{l+m} A_{1r} \frac{1}{\sigma_N^2} p_{\chi^2}\left(\frac{\alpha}{\sigma_N^2}; 2r\right) + \sum_{r=0}^{L-l+n} A_{3r} \frac{1}{\sigma_T^2} p_{\chi^2}\left(\frac{\alpha}{\sigma_T^2}; 2r\right) \right]; \alpha \geq 0 & (7a) \\ \left[\sum_{r=0}^{l} A_{2r} \frac{1}{\sigma_N^2} p_{\chi^2}\left(\frac{-\alpha}{\sigma_N^2}; 2r\right) + \sum_{r=0}^{L-l} A_{4r} \frac{1}{\sigma_T^2} p_{\chi^2}\left(\frac{-\alpha}{\sigma_T^2}; 2r\right) \right]; \alpha < 0, & (7b) \end{cases}$$

where

$$\beta_N \triangleq S/\sigma_N^2,$$
$$\beta_T \triangleq S/\sigma_T^2 = S/(\sigma_N^2 + \sigma_J^2),$$

with S, σ_N^2, and σ_J^2 being the signal power, thermal noise power, and jamming noise power, respectively, as indicated before, $p_{\chi^2}(u; 2r)$ is the χ^2 density function with $2r$ degrees of freedom and the A_{ir}, ($i = 1, 2, 3, 4$) are some constants involving σ_N^2 and $\sigma_T^2 = \sigma_N^2 + \sigma_J^2$ only. For the error rate expression, only A_{2r} and A_{4r} are required, as will be shown shortly.

We note here that the power parameter S, σ_N^2, and σ_J^2 can be written in terms of energy (or spectral density) parameters as

$$S = E_s/\tau = E_b/(L\tau) \tag{8a}$$

$$\sigma_N^2 = N_0 B, \tag{8b}$$

$$\sigma_J^2 = \frac{1}{\gamma} N_J B; \quad N_J \triangleq J/W, \tag{8c}$$

where E_s is the signal energy per hop, E_b is the bit energy, which is assumed to be equally divided among L hops, N_0 is the one-sided thermal noise spectral density, and $N_J = J/W$ is the effective jamming spectral density.

Substituting (7b) into (6) and using (8), we obtain for the probability of error

$$P(e) = \sum_{l=0}^{L} \binom{L}{l} (1-\gamma)^l \gamma^{L-l} e^{-l\rho_N} e^{-(L-l)\rho_T} \sum_{m=0}^{\infty} \sum_{n=0}^{\infty} \frac{(l\rho_N)^m}{m!} \frac{[(L-l)\rho_T]^n}{n!}$$

$$\cdot \left(\sum_{r=0}^{l} A_{2r} + \sum_{r=0}^{L-l} A_{4r} \right); \quad A_{20} = A_{40} = 0, \tag{9}$$

where

$$\rho_N = E_s/N_0 = E_b/(N_0 L), \tag{10a}$$

$$\rho_T = \frac{E_s}{N_0 + N_J/\gamma} = \frac{1}{1/\rho_N + 1/(\gamma \rho_J)}, \tag{10b}$$

$$\rho_J = E_s/N_J = E_b/(N_J L), \tag{10c}$$

and the A_{2r} and A_{4r} are given by

$$A_{2r} = \left(\frac{1}{2}\right)^{l+m} \left(\frac{1}{1+\delta}\right)^{L-l+n} \left(\frac{1}{1-\delta}\right)^{L-l} \frac{\mu_{2(l-r)}}{(l-r)!}; \; r = 1, 2, \ldots, l, \tag{11a}$$

$$A_{4r} = \left(\frac{\delta}{1+\delta}\right)^{l+m} \left(\frac{\delta}{\delta-1}\right)^{l} \left(\frac{1}{2}\right)^{L-l+n} \frac{\mu_{4(L-l-r)}}{(L-l-r)!}; \; r = 1, 2, \ldots, L-l, \tag{11b}$$

where μ_{ih} (moments), $i = 2, 4$, are related to the κ_{ih} (cumulants), $i = 2, 4$, by

$$\mu_{iq} = \sum_{j=0}^{q-1} \binom{q-1}{j} \mu_{i(q-1-j)}\kappa_{i(j+1)}; \quad i = 2, 4,$$

and

$$\kappa_{2h} = (h-1)! \left[(l+m)\left(\frac{1}{2}\right)^h + (L-l+n)\left(\frac{\delta}{1+\delta}\right)^h + (L-l)\left(\frac{\delta}{\delta-1}\right)^h\right], \tag{12a}$$

$$\kappa_{4h} = (h-1)! \left[(l+m)\left(\frac{1}{1+\delta}\right)^h + l\left(\frac{1}{l-\delta}\right)^h + (L-l+n)\left(\frac{1}{2}\right)^h\right], \tag{12b}$$

where

$$\delta = \frac{N_0 + N_J/\gamma}{N_0} = 1 + \frac{1}{\gamma}\frac{N_J}{N_0} = 1 + \frac{1}{\gamma}\frac{\rho_N}{\rho_J}.$$

The probability of error expression (9) will be computed and plotted with bit energy constraint for a variety of parameter values in conjunction with the optimum jamming strategy to be discussed shortly.

Special Case 1: Single Hop/Bit ($L = 1$)

In this case, the probability of error (9) reduces to

$$P(e) = (1-\gamma)e^{-\rho_N}\sum_{m=0}^{\infty}\frac{1}{m!}\frac{1}{2}\left[\frac{1}{2}\rho_N\right]^m + \gamma e^{-\rho_T}\sum_{n=0}^{\infty}\frac{1}{n!}\frac{1}{2}\left[\frac{1}{2}\rho_T\right]^n$$

$$= \frac{1}{2}(1-\gamma)e^{-\frac{1}{2}\rho_N} + \frac{1}{2}\gamma e^{-\frac{1}{2}\rho_T},$$

where ρ_N and ρ_T are as given in (10). We observe that (13) reduces to the conventional BFSK result for $\gamma = 0$ or 1.

Special Case 2: No Thermal Noise

The analysis under the assumption of "no thermal noise" is of interest. Although this assumption is unrealistic, the result of the analysis provides a lower bound of the probability of error to that of the general case where both thermal noise and jamming noise are present. In our formulation it should be possible to obtain the special case expression by taking the limit of (9) with $\rho_N \to \infty$. The procedure is cumbersome, however, due to the complexity of the error probability expression as a function of ρ_N. Instead, we repeated the derivation by imposing the condition of $\sigma_N^2 = 0$ in the beginning. The result is

$$P(e) = \sum_{l=0}^{L-1} \binom{L}{l} (1-\gamma)^l \gamma^{L-l} e^{-(L-l)\gamma\rho_J} \sum_{n=0}^{\infty} \frac{[(L-l)\gamma\rho_J]^n}{n!}$$
$$\cdot \sum_{r=0}^{L-l} A_{4r} \left[\sum_{k=0}^{r-1} e^{-l\gamma\rho_J} \frac{(l\gamma\rho_J)^k}{k!} \right], \quad (14)$$

where

$$\rho_J = E_s/N_J = E_b/(N_J L)$$

$$A_{4r} = \left(\frac{1}{2}\right)^{L-l+n} \frac{\mu_{4(L-l-r)}}{(L-l-r)!}; r = 1, 2, \ldots, L-l,$$

$$\mu_{4q} = \sum_{i=0}^{q-1} \binom{q-1}{1} \mu_{4(q-1-i)} \kappa_{4(i+1)},$$

$$\kappa_{4h} = (h-1)! (L-l+n) \left(\frac{1}{2}\right)^h.$$

4 NUMERICAL RESULTS

The most effective jamming strategy is to distribute the total jamming power J (choose γ) in such a way as to cause the communicator to have maximum probability of error. There will be an optimum value of γ which will maximize the error probability. The usual method to determine the optimum fraction is to differentiate the error probability expression with respect to γ and set the result equal to zero to find the root. This would be a formidable task due to the complexity of the error probability expression, (9), as a function of γ. Thus, the analytical approach was abandoned. Instead, we carried out the optimizing procedure numerically for every set of E_b/N_0, E_b/N_J, and L of interest using a digital computer. We have selected practical values of E_b/N_0 such as 13.35 dB and 10.94 dB, for which the probability of error becomes 10^{-5} and 10^{-3}, respectively, under a jamming-free condition. For a theoretical interest we computed the no thermal noise case ((14)) also. The number of hops per bit, L, has been taken up to six, beyond which the optimum γ remains virtually unchanged. The (worst case) probability of error for $L > 6$, if needed, may thus simply be calculated, using the optimum γ's for $L = 6$.

To get some feel on the behavior of error probability as a function of γ we first plotted $P(e)$ vs. γ curves in Figure 3 for $E_b/N_0 = 13.35$ dB and $L = 2$ with E_b/N_J as a parameter. As seen, for each curve there exists a value of γ, say optimum γ, that maximizes the probability of error. We observe that the maximum $P(e)$'s for $E_b/N_J = 0$, 5 dB occur at $\gamma = 1$. This implies that when the signal bit energy (E_b) and the jamming density (N_J) are close in magnitude, wideband jamming ($\gamma = 1$) is nearly optimum. As can be seen from the dotted line, however, the optimum γ decreases with increasing E_b/N_J. This is consistent with the observation that when jamming density (or jamming power with fixed system bandwidth) is small relative to the signal energy, the effective jam-

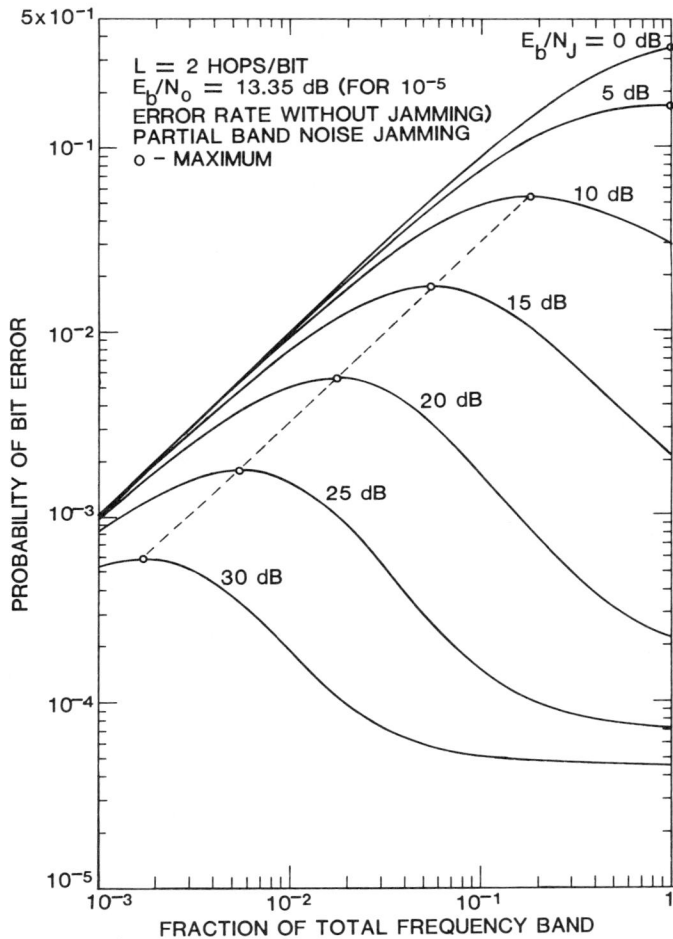

Figure 3 Probability of error vs. fraction of total frequency band jammed for $L = 2$ hops/bit and $E_b/N_0 = 13.35$ dB.

ming strategy would be to distribute the jamming power in a smaller fraction rather than the total system band.

The maximum (worst case) error probabilities corresponding to the optimum γ's are shown in Figure 4. We readily see that each curve approaches asymptotically a certain value as E_b/N_J approaches ∞ ($P(e) \rightarrow 10^{-5}$, 3.8×10^{-5}, 8.8×10^{-5}, 1.7×10^{-4}, 4.3×10^{-4} for $L = 1, 2, 3, 4, 6$, respectively). The reason that all the curves do not approach 10^{-5} is due to the "combining loss." The asymptotic values in Figure 4 coincide with those values of $P(e)$ predicted by combining loss under the thermal noise only case [6]. Figure 5 shows the error rate performance comparisons for the cases of optimum partial-band jamming (optimum γ) and wideband jamming ($\gamma = 1$) for $L = 2$. As a reference,

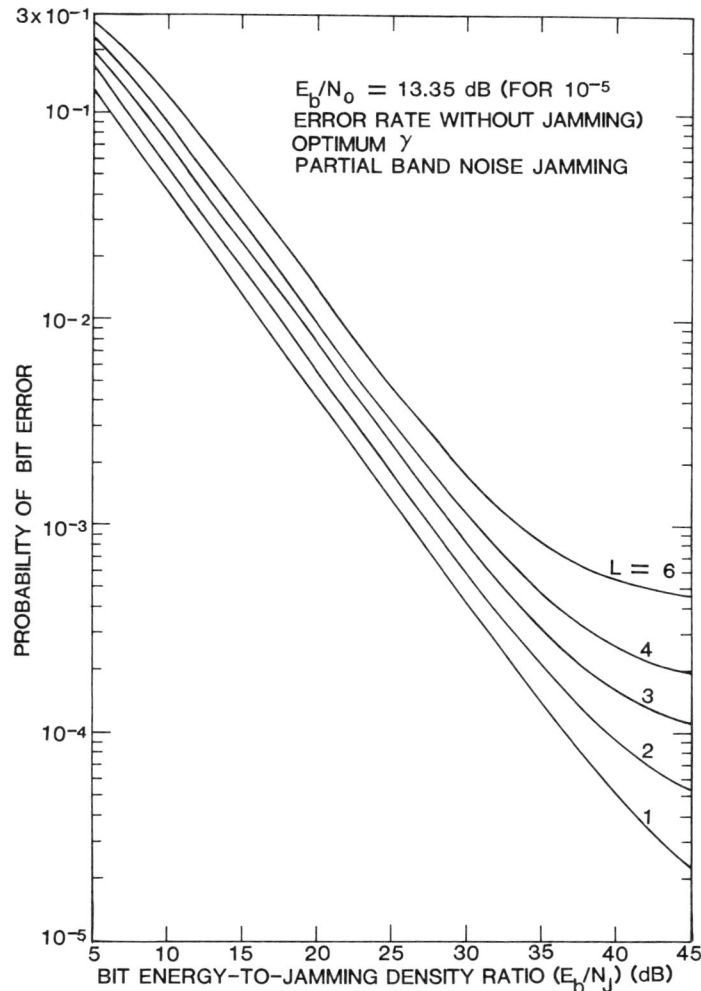

Figure 4 Probability of error vs. E_b/N_J for $E_b/N_0 = 13.35$ **dB**

ideal BFSK performance is plotted with a dotted curve. (For the ideal BFSK curve the abscissa reads E_b/N_0). As seen, the optimum jamming strategy is clearly much more effective (from jammer's point of view) than the wideband jamming strategy.

Finally, performance comparisons for different L's and E_b/N_0's are shown in Figures 6 and 7 for optimum jamming and wideband jamming, respectively. As stated earlier, the difference in the probability of error between $L = 1$ and $L = 3$ cases for the same E_b/N_0's are attributed to the combining loss. The error rate behaviors for optimum jamming and wideband jamming with finite E_b/N_0 are quite analogous to Figure 5. That is, both optimum and wideband jamming curves for the same values of E_b/N_0 and L approach one value of

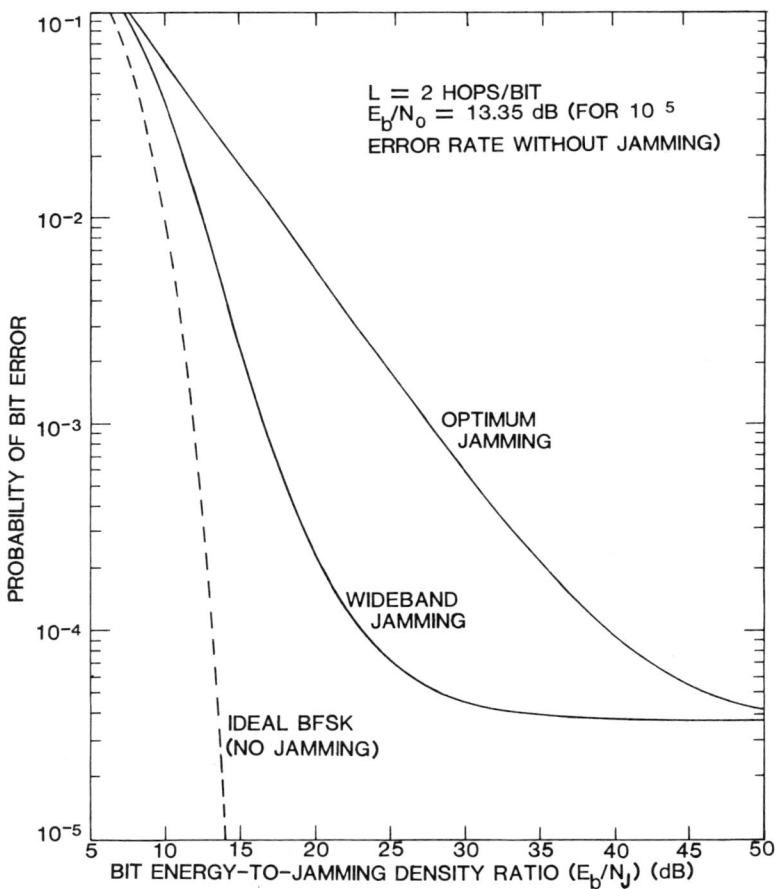

Figure 5 Optimum jamming and wideband jamming performances for BFSK/FH with $L = 2$ hops/bit when E_b/N_0 13.35 dB. (For ideal BFSK curve the abscissa reads E_b/N_0)

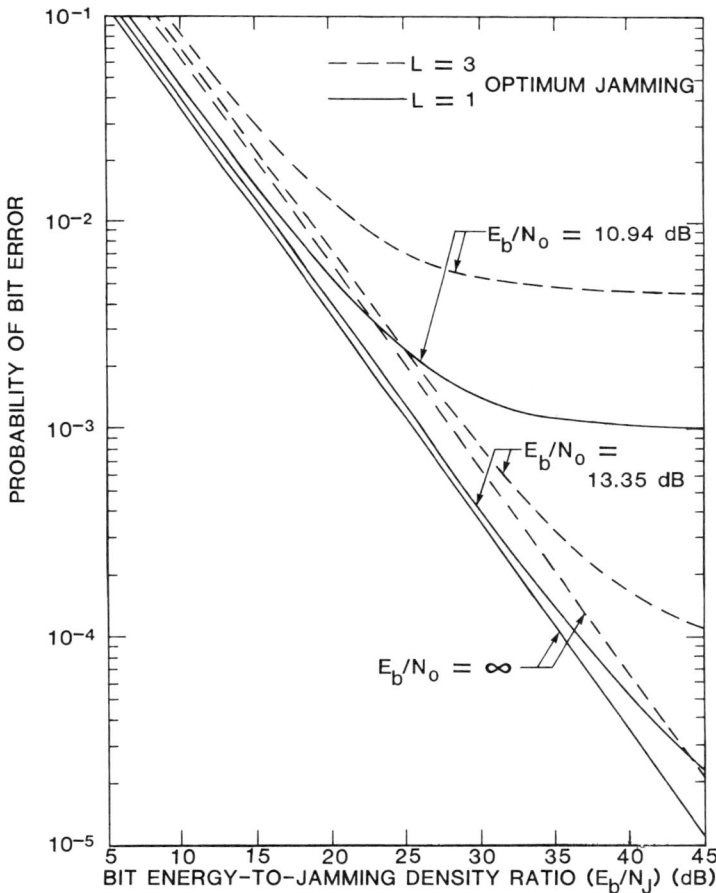

Figure 6 Error performance comparison for $L = 1, 3$ and $E_b/N_0 = 13.35$ dB, 10.94 dB, ∞ with optimim (partial band noise) jamming

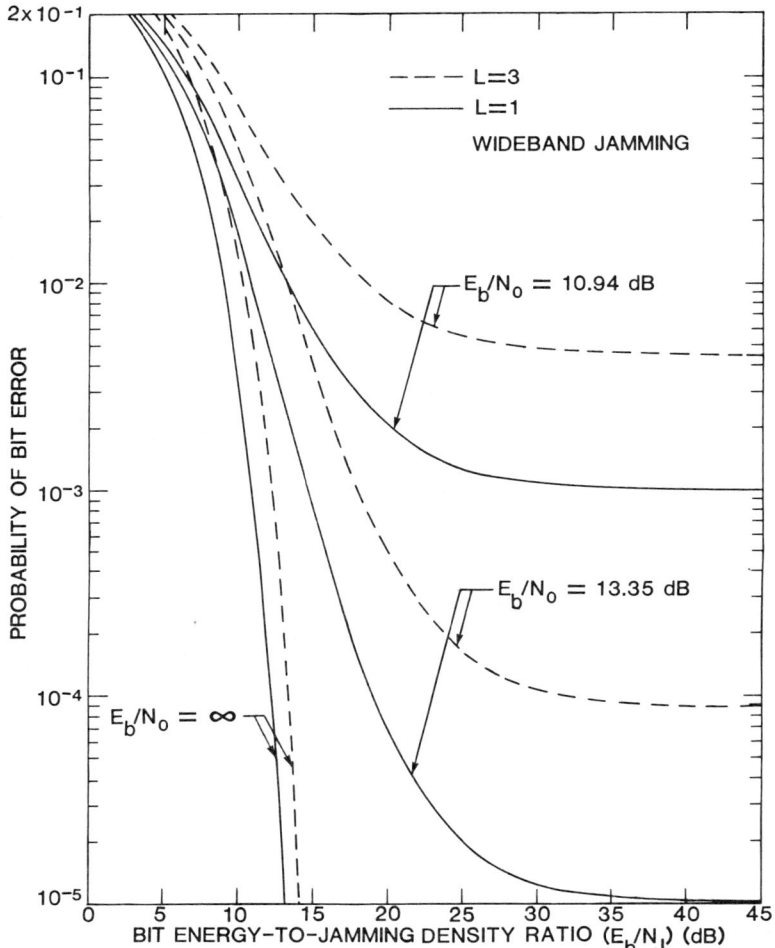

Figure 7 Error performance comparison for $L = 1$, 3 and $E_b/N_0 = 13.35$ dB, ∞ with wideband jamming

$P(e)$ as E_b/N_J increase ($E_b/N_J \geq 45$ dB). In contrast to this, the curves for $E_b/N_0 = \infty$ (no thermal noise) behave quite differently. With wideband jamming (Figure 7) the probability of error decreases quite rapidly (exponentially) with increasing E_b/N_J as predicted by the ideal BFSK performance. However, with optimum jamming (Figure 6) the probability of error decreases linearly as E_b/N_J increases for both $L = 1$ and 3 (also for $L = 2$, 4, and 6 although these curves are not shown here). Also, the error probability becomes larger as L increases for all the range of E_b/N_J shown.

5 CONCLUSION

We have presented a summary of a complete analysis for the derivation of the probability of error for a multihops (L-hops) per bit spread spectrum system employing binary FSK modulation in a partial-band jamming environment. The main contribution of our work was the complete derivation of the error probability expression in the presence of both partial-band jamming and thermal noise. The workers dealing with these types of problems have always simplified the analysis by making certain assumptions, such as neglecting thermal noise, which are not tenable in real situations. We have obtained the solution without the "simplifying assumptions." We presented graphical results for both wideband and optimum jamming strategies. It has been believed that the multihops per bit FH system, which is considered as a form of frequency diversity, provides for an improvement of (worst case) error performance for certain values of L through diversity principles [1]. Our analysis shows that such advantage is not achieved by the conventional square-law combining receiver; the error performance simply becomes worse (from communicator's point of view) with increasing L for all values of bit energy-to-jamming density ratio. The motivation of using this scheme should thus be based on "covertness" rather than antijamming communication purposes.

Regarding the variety of parameters and variables that affect the system performance, the designers of a multihops BFSK/FH system must take various factors into consideration. Our results will be very useful in providing necessary information to the designers of such a system, anticipating its use in an electronic warfare environment. The results are also useful to the dedicated interceptor of such waveforms, which are presumed to be designed to mitigate the threat posed by a potential jammer.

REFERENCES

[1] A.J. Viterbi and J.M. Jacobs, "Advances in Coding and Modulation for Noncoherent Channels Affected by Fading, Partial Band, and Multiple-Access Interference," *Advances in Communication Systems,* Vol. 4, pp. 279-308, Academic Press Inc., New York, 1975.

[2] S.W. Houston, "Modulation Techniques for Communication—Part 1: Tone and Noise Jamming Performance of Spread Spectrum M-ary FSK and 2,4-ary DPSK Waveforms," *Proc. IEEE 1975 National Aerosp. & Electron. Conference,* June 10-12, 1975, pp. 51-58.

[3] B.K. Levitt and J.K. Omura, "Coding Tradeoffs for Improved Performance of FH/MFSK Systems in Partial Band Noise," *NTC Record—1981,* Vol. 2, pp. D9.1.1-D9.1.5.

[4] L.B. Milstein, R.L. Pickholtz, and D.L. Shilling, "Optimization of the Processing Gain of an FSK-FH System," *IEEE Transactions on Communications,* Vol. COM-28, No. 7, pp. 1062-1079, July 1980.

[5] O.C. Yue, "Performance of Frequency-Hopping Multiple-Access Multilevel FSK Systems With Hard-Limited and Linear Combining," *IEEE Transactions on Communications,* Vol. COM-29, No. 11, pp. 1687-1694, November 1981.

[6] S.L. Bernstein, "Error Rates for Square-Law Combining Receivers," Lincoln Laboratory, Technical Note 1971-31, May 14, 1971.

Index

A

absolute lower bound, 287
adaptation, full, 30
adaptive, 30
    partitioning, 162
    thresholding, 349
AH encoding, 464
AIC, 10, 32, 427
Air Force Geophysical
    Laboratory, 409
Akaike's information
    criterion, 10, 427
algebra, Boolean, 399
algorithm
    analysis, 485
    Bartlett, 330
    computationally efficient, 435
    efficient, 435
    fast, 412, 477
    Levinson, 479
    linear predictive, 332
    locally smoothing, 447
    MAP, 391
    ME, 330
    measurement-oriented, 354
    parallel parsing, 381
alias, 141
ALU, 482
ambiguity surface, 365
analysis
    algorithm, 485
    cluster, 379, 399
    frequency Fourier, 243
    frequency wavenumber, 243ff
    image, 375
    large data set, 191
    macro-, 192
    micro-, 192
    nonlinear spectrum, 474ff
    waveform, 380
antenna theory, 245
antijamming, 529
approach
    decision-theoretic, 375
    discriminant, 375

non-backscan, 354
structural, 375
syntatic, 375
target-oriented, 354
approximation, numerical, 284
AR, 24, 169, 436
arbitrary hypercube encoding, 464
architecture
    correlator, 493ff
    FFT, 494ff
    pipelined processor, 477
    VLSI processor, 473
Arctic Submarine Laboratory, 200
ARE, 100
arithmetic logic unit, 482
ARMA, 9, 24, 52
array
    dynamically perturbed, 293
    geometry, 243, 292
    of sensors, 287
    processing, 329
    resolvent, 332
    response, 243
    statically perturbed, 292
    towed, 314
asynchronous, 485
autocorrelation, 67, 218, 422
automata, finite-state, 376
automorphism, 401
autoregression, 4, 7
    bounded-influence, 28
autoregressive, 24, 67ff, 169, 436
    first order, 422
autoregressive-moving average, 24
average mean square error, 169
azimuth, 345
    compression, 493

B

band, narrow frequency, 260
bandwidth, fractional, 294
Barankin bound, 295
Barter Island, 200
Bartlett widow, 477
Bayes
    approach, 261
    classifier, 439
    decision rule, 377

rule, 357
solution, 170
theorem, 3
Bayesian, 168
beam pattern, 330
beamformer, focussed, 290
beamforming, 119, 243, 475
    adaptive, 486
    Bartlett, 329
    classical, 329
    delay and sum, 243
    high resolution spectral, 475
    minimum energy adaptive, 329
    simple, 246
bearing, 244, 345
    error, 293
    source, 243
Beaufort Sea, 200
benchmark, 306
BFSK, 518, 525ff
BFSK/FH, 518
bias error, 298
bijection, 401
bilinear system, 253ff
binary frequency-shift keying, 518
bit, 411ff, 517
    error, 411
    error rate, 411, 415
    most significant, 412
blood pressure, 422
    diastolic, 422
    systolic, 422
Boolean algebra, 399
Boolean conditions, 193
bound, Barankin, 295
boundedness, 180
box plot, serial, 201
Box-Jenkins method, 70
BREAK, 379
breakdown point, 23
broadband wave, 243, 248
Burg method, 70
burst error, 390

C

CART, 192ff
CAT, 10
categorical data type, 193

Index 533

causal linear operation, 114
CDC CYBER 70, 254
CDF, 157, 366
central limit theorem, 215
centralized control, 481
cepstrum, 261
chain, Markov, 386ff, 423
character recognition, 380
Chesapeake Bay, 409
chi-square, 423
circuit, very large scale
  integrated, 473
classification
  Bayesian, 424
  fingerprint, 380
  multiple signal, 476
  ship, 491
  texture image, 436
classifier, Bayes, 439
cluster, 195
clustering
  automatic spatial, 435, 439ff
  hierarchal, 400
  numerically stratified, 400
clutter, 341, 354
  random, 360, 446
code
  arbitrary hypercube, 464
  channel, 416
  folded binary, 412ff
  Hamming, 411, 415
  minimal hypercube, 463
  pseudorandum, 518
  transportable, 181
coefficient
  correlation, 365
  reflection, 479
  weighting, 308
coherent, 330
  component, 117
communication, 386, 481
  constraints, 481
  privacy, 517
computer aided design, 473
Comtal 8000, 453
concurrent array processor, 483
conditional Gaussian, 254
confidence
  bound, 80
  intervals, 226
constrained optimality, 267

contaminant, Laplace, 101
contrast enhancement, 439
control, 386
  unit, 482
convergence property, 181
convexity, 180
convolution, 138, 182
  kernel, 184
cordic, 483
  arithmetic unit, 483
Cornish-Fisher expansion, 94
correlation, 215
  auto-, 67
  canonical, 510
  cepstral, 14
  coefficient, 365
  inverse, 14
covariance, 115, 141
  function, 141
  matrix, 115, 314
  tracker error, 342
coverage problem, 280
Cramer class, 59ff
Cramer-Hida multiplicity
  theory, 111
Cramer-Hida representation, 112
Cramer-Rao inequality, 288
Criterion Autoregressive
  Transfer, 10
cross entropy, 5, 427
CU, 482
cumulant, 53, 522
cumulative sum, 170
curve, star-shaped, 276

D

data
  analysis, 191
  association, 342
  compression, 411, 439
  correlation, 342
  Fisher iris, 430
  Marple impactor, 182
  merchant, 212
  multi-sensor, 385
  ocean acoustic, 211
  pictorial, 461
  remotely sensed, 386

seismic, 218
-driven, 485
-flow multiprocessor, 485
Data General Nova 800, 453
DATA-Gin, 192ff
DCT, 411
DCT, two-dimensional, 412
decomposition
  Rician, 520
  singular value, 67ff
deconvolution, 56
dedicated parallel signal
  processor, 482
deep-water focussing, 262
deflection criterion, 110
delegate sampling, 195
density
  class-conditional, 423
  cross-spectral, 366
  effective jamming spectral, 521
  experimental, 238
  gamma, 83
  Gaussian, 83, 211ff
  generalized Gaussian, 215
  half-normal, 83
  heavy-tailed, 94
  jamming, 523
  Johnson family, 98ff
  kernel, 211
  kernel spectral, 13
  Laplace, 101, 211ff, 226ff
  log spectral, 11
  lognormal, 83, 263
  mixture model, 99
  non-Gaussian probability, 94
  observation prediction, 20
  Pearson, 99
  power spectral, 79ff, 155, 260
  probability, 366, 521
  Rayleigh, 83, 263
  Rice-Nakagami, 263
  search, 266
  spatial clutter, 358
  spectral, 141, 261, 521
  uniform, 83, 149
  Wishart, 367
depth, keel, 200
derivative
  Radon-Nikodym, 109
  quadratic mean, 145, 149
design
  detector/receiver, 343

optimal, 181
random, 147
stratified, 149
DESTREE, 194ff
detection, 93, 142
  Bayes, 109
  data-adaptive, 97, 117
  exponential, 265
  minimax, 109
  Neyman-Pearson, 109
  non-Gaussian, 93
  nonsingular, 109
  probability, 342, 365
  real-time fault, 508
  search, 268
  singular, 109
  target, 273
  threshold, 342, 344ff
detector
  absolute value, 255
  conventional (linear), 161
  locally optimum, 94ff
  nonoptimal, 147
  optimal, 143, 150
  square law envelope, 519
  zero-memory nonlinear, 94
DFT, 79, 125, 154, 249, 366, 478
difference, Doppler, 358
differential, Gateaux, 266ff
direct sensing, 179
discriminant, Fisher's linear, 439
disk
  Poisson random, 275
  random, 275
dispersion, 405
distribution
  asymptotic, 46
  beta, 283
  cumulative, 157, 247
  double exponential, 47
  exponential, 56, 199, 210
  Gaussian, 81, 161, 168, 446
  geometric, 168
  limiting, 41
  longer-tailed, 199
  mixed-beta approximation, 285
  n-dimensional marginal, 135
  near-exponential, 202
  non-Gaussian, 226
  Pareto, 56
  Poisson, 275, 354, 358
  Poisson spatial, 108

Index

posterior, 175
Rayleigh, 163
sculptured exponential, 203
sinusoidal, 218
standard normal, 144
divergence, 4
  filter, 297
  information, 4
  minimum, 3
  wind horizontal, 182ff
domain of attraction, 43
Doppler frequency shift, 258
DPSP, 482
drag-force coefficient, 258
dynamic programming, 386ff

E

EAI Hybrid, 254
ECG, 380, 422
edge preserving, 445
Edgington's normal curve, 224
EEG, 380, 422
efficiency
  asymptotic relative, 100
  parsing, 378, 381
efficient, asymptotically, 33
eigenfunctions, 185
eigenvalue, 68, 111ff, 475, 478
eigenvector, 68, 111ff, 121, 475
EKF (see filter, extended Kalman), 256, 353
elevation, 244
electroencephalogram, 380
ellipsoid, g-sigma, 356
end-effect, 249
entropy, 475
  conditional, 5
  cross, 5
  maximum, 7
environment, reverberation-limited, 107
equalized power pattern, 247
equation
  Abel's integral, 181
  differential, 184
  integral, 181
  Ricatti, 342ff, 353
  stochastic differential, 114
  stochastic Riccati, 344

equations, Yule-Walker, 7, 67ff
error
  burst, 390
  probability, 528
  -correcting parser, 378
estimate
  asymptotic, 55
  bispectral, 55
  M-, 21
  spectrum, 154
estimates, ARMA spectral, 72
estimation, 167
  2-D maximum entropy spectral, 435
  a posteriori, 341
  adaptive state, 309
  Bayes, 173ff, 181
  best linear, 168
  best nonlinear, 168
  data-adaptive, 97
  delay, 313
  generalized, 253
  high resolution spectrum, 482
  iterative Bayes, 436
  kernel, 226
  kernel density, 211
  least squares, 478
  linear, 507ff
  maximum
    likelihood, 249, 424
  minimum variance
    linear, 169
  minimum variance
    nonlinear, 170
  nonlinear, 167
  optimal, 249
  optimal linear, 167
  optimal sequential, 386
  passive sonar delay, 313
  refined density, 196
  source bearing, 329
  spectral, 4
expected value, 246
exponential decay rate, 219
exponentially sculptured
  exponential, 204
exponentially correlated, 304
exterior point, 404
extremal problem, 24
extremal types theorem, 43
extreme value theory, 41

F

fading, 154, 157
false alarm
  probability, 342, 365
  rate, 95, 144, 163, 226
fault tolerance, 483
FBC, 412ff
FDK, 79ff
feature space, 375
feature vector, 375
FEC, 411ff
feedback control, 255
FFT, 13, 76, 86, 155, 215, 246, 260, 474, 477, 480, 487
FH, 517
field
  annular, 274
  Gaussian noise, 226
  Markov, 386ff
  noise, 223
  obstacle, 273
  Poisson, 274
  Poisson random, 273
  stochastic, 387
  trapezoidal, 274
filter, 217, 439
  adaptive notch, 478
  bandpass, 87
  divergence, 297
  extended Kalman, 256, 297
  extended Kalman-Bucy, 353
  finite impulse response, 215
  gradient inverse, 453
  innovations, 9
  Jazwinski's limited memory, 297
  Kalman, 302, 313ff, 386, 475
  Kalman-Bucy, 257, 341, 343, 355
  least square, 445
  low pass, 181
  majority, 390ff
  matched, 103, 119, 346
  median, 445, 450, 453
  Nagao's, 446, 453
  noninvertible, 11
  nonlinear, 253
  nonlinear conditional-Gaussian, 257
  nonrecursive, 69
  optimal, 61
  optimal tracking, 298
  pre-, 405
  sigma, 446, 447ff, 453
  two-dimensional Kalman, 445
  two-step nonlinear (TNF), 256
  unrealizable, 307
  Wiener, 181, 445
filter-cleaner, 33
filtered beam, 245
filtering, 25
fingerprint classification, 380
Fisher matrix, 288ff
Fisher's linear discriminant, 439
fixed width kernel, 226
FLIR, Westinghouse, 408
form, bilinear, 254ff
FORTRAN, 192, 313, 453
forward error correction, 411ff
Fourier transform, 68, 260
  2 dimensional, 119, 121
  2-D fast, 435
  discrete, 79, 125, 154, 249, 366
  fast, 13, 76, 86, 155, 215, 246, 260, 436, 494
  inverse, 367
  spatial, 243ff
FRAM-II, 153ff
frequency, 119
  hopping, 517
  hopping spread spectrum, 517
  mark, 520
  modulated, 154
  spatial, 120, 520
  temporal, 120
function
  characteristic, 367
  confluent hypergeometric, 367
  covariance, 141
  cumulative density, 366
  cumulative distribution, 157
  filter, 261
  forcing, 304
  gamma, 367
  incomplete beta, 367
  joint probability density, 423
  K-, 281
  kernel, 223
  likelihood, 362
  linear discriminant, 424
  log-likelihood, 388, 475
  moment-generating, 205
  penalty, 180

Index

probability density, 366, 521
response, 62
restricted picture, 463
smooth, 179
spectral, 60
spectral density, 141
step, 167
stream, 184
support picture, 463
Walsh, 261
functional, 266
　bounded linear, 179
　concave, 267
　inference, 3
　matrix, 254
　nonlinear data, 181
　point, evaluation, 180
fuzzy language, 377
fuzzy set theory, 377

G

gain, 404
　Kalman, 315
Gaussian elimination, 479
Gaussian martingales, 108
Gaussian random variable, 520
GCV, 180
generalized cross validation, 180ff
generalized search optimization, 266
GH hypercube, 463
gradient inverse weighting, 446
GRAFST2, 209
grammar, 376
　array, 377
　context-free, 380
　context-sensitive, 376
　graph, 377
　pattern, 375
　stochastic, 377
　texture, 380
　transition network, 378
　tree, 377, 378
　web, 376
graph, relational, 375
grating lobes, 245
gray level, 447
gray level set, 462
gross national product, 425
Gulf Stream, 409

H

handle set, 463
handle vector, 463
harmonizable
　strongly, 59
　weakly, 60
heavy-tailedness, 32
Helmholtz theorem, 184
hierarchical hypercube
　encoding, 465
Hilbert space, 179
histogram, 200, 464
Holder's inequality, 147
homogeneity, 191
homogeneous, 197
homomorphism, 403
HORSE, 193
HP 1000, 254
hypercube, 463

I

IBM Watson Research Center, 209
ice
　keel, 199
　subsurface, 201
illposedness, 184
image
　analysis, 375
　compressed, 417
　reconstructed, 416
　satellite, 396
　synthetic aperture radar, 446
imagery, remotely sensed, 435
increment, independent, 134
increments, orthogonal, 61
independent increment, 134
indirect sensing, 179
inequality, 182
　Cramer-Rao, 288
　Holder's, 147
inference, functional, 3
infinitely divisible, 135
influence curve, 23
innovation vector, 355
innovations, 9, 26, 255
input, 298, 405
integral
　Bochner, 62

Dunford-Schwartz, 61
Lebesgue, 142
Morse-Transue, 61
numerical, 284
interactive, 193
inter-arrival times, 168
interference
  broadband, 118
  Gaussian, 118
  impulsive, 232
  narrowband, 153
  plane-wave, 118
  unidirectional, 118
interior point, 404
inversion, 62
ISODATA, 425
isotone, 400
isotropic, 185

J

jackknifing, 32
Jacobian, 265
jammer, 248
  partial-band noise, 520
jamming, 243, 517
  optimum partial-band, 518, 524
  partial-band, 517, 529
  wideband, 518, 524
Jet Propulsion Laboratory, 446
join homomorphism, 403
JPDA, 354ff, 357
JPL, 446

K

Karhunen class, 60
Karush-Kuhn-Tucker condition, 265
kernel, 181, 196, 223
  convolution, 184
  fixed width, 226
key cut, 195
knowledge, a priori, 316
Kolmogorov's Existence
  theorem, 135
Kolmogorov-Smirnov test, 205

Kronecker delta, 67
Kurtosis, 97, 155, 205, 212ff, 230
  frequency domain, 79

L

LANDSAT, 428
language, 376
  fuzzy, 377
  matrix data flow 482
  pattern description, 375, 377
  special-purpose, 483
  stochastic, 377
  VLSI design, 486ff
  wavefront oriented, 484
Laplace-Beltrami operator, 185
large data set analysis, 191
large scale integration, 474
LASA, 381
lead, 199
leptokurtic, 212
likelihood
  conditional, 388
  function, 362
  maximum, 6, 19, 32, 202, 205, 247, 341, 377, 386ff, 424, 438
  maximum imputation, 193
  processor, 98
  ratio, 107, 109, 119, 223
  ratio test, 117
  recursive, 390
LINE, 379
line, bearing/frequency, 358
linear FM, 493
linear predictive processing, 329
linear recurrence, 478
linearly sculptured exponential, 203
link, interconnection, 508
local interconnection, 483
localization, source, 287
localized communication, 481ff
  network, 483
logical micro-picture, 468
long memory, 8
low probability intercept, 517
lower bound vector, 463
low-rank, 117
LSI, 474

M

macroanalysis, 192
MAP, 386
mapping
   isotone, 403
   residual, 400
   residuated, 401
Markov chain, 266
Markov property, 172
martingales, Gaussian, 108
massive parallelism, 478
matched overcontainment, 365
matrix
   Boolean, 406
   confusion, 431ff
   correlation, 475
   covariance, 314, 424
   cross-spectral density, 366
   filter gain, 355
   Fisher, 288ff
   Hermetian, 367
   Hermetian autocorrelation, 366
   inversion, 478
   multiplication, 483
   positive, semidefinite, 356
maximum entropy method, 475
maximum likelihood
   imputation, 193
MDFL, 482
mean, 155
   conditional, 355
   square error, 288
measure
   Levy, 135ff
   of vacancy, 280
   pictorial information, 464
   Poisson, 138
   Poisson random, 136
   product, 266
   time-jump, 135
   visibility, 280ff
measurement
   bearing/frequency, 349
   passive acoustic, 353
   time/Doppler, 349
median, 208
meet homomorphism, 403
MEM, 475, 477
memory
   long, 8
   short, 7

M-estimate, 21, 24
method
   2-D maximum entropy spectral, 435
   Bayes least squares linear, 261
   Box-Jenkins, 70, 261
   Burg, 70
   Capon maximum likelihood, 475
   cross validated spline, 179
   divisive, 402
   eigenvalue decomposition, 486
   flat, 402
   gradient inverse, 450
   hybrid, 205
   local smoothing, 445
   maximum entropy, 486
   modern nonlinear spectrum analysis, 474
   monotone equivariant, 402
   quadrature, 181
   semiflat, 402
   syntatic, 381
metric, Levenshtein, 378
microanalysis, 192
Middleton model, 133ff, 223, 232
minimal cardinality, 403
minimal enclosing hypercube, 463
minimal hypercube encoding, 463
minimum divergence, 3
minimum mean square error, 412
misrecognition, 378
missile guidance, 297
missing value, 193
mixed data type, 193
mixed variables type, 191
mixing, strong, 44
mixture
   additive, 153
   Gaussian-Gaussian, 101
MLM, 475, 477
MMSE, 412
model
   autoregressive texture, 386
   Bayesian, 168, 175
   gamma, 208
   Gaussian, 425
   graph-theoretic, 400
   identification, 4
   Jardine-Sibson, 400
   linear polar state variable, 299
   linearized drag, 300
   Middleton, 133ff, 223, 232

observation, 254
parametric, 429
state-space target, 343
statistical, 429
target, 254, 354
target motion, 298
modeling, parametric, 436
modified query, 468
modular structure, 481
modularization, hardware, 507
modulation, binary FSK, 529
moment, 522
  matching, 205
monotone equivariant, 401
monotonicity, 180
Monte Carlo simulation, 55, 93ff, 415ff
motion, Markovian, 265, 268
moving average, 67ff
MSB, 412
multihop, 517
multipath, 154, 157
multiprocessor, 482
MUSIC, 476, 477, 486
MVLE, 169
MVNE, 170

N

Nanoose (NUWES), 262
Naval Postgraduate School, 209
Naval Research Laboratory, 442
Naval Undersea Center, 200
NCF, 257
nearest neighbor, 196, 358, 379
neighborhood, 403
network
  computing, 509
  generating, 509
  receiving, 509
  reliability, 511
  stream, 514
Neyman-Pearson lemma, 137
noise
  additive, 246
  additive, dependent Gaussian, 107
  ambient, 224, 258
  biological, 229
  class A, 233
  corruption, 386
  field, 223
  Gaussian, 133, 153, 167, 246, 260, 287, 341
  Gaussian White, 168, 314, 446
  glint, 36
  ice, 153
  isolated spot, 453
  Laplace, 101
  multiplicative, 258, 446
  narrow-band, 260
  nearly Gaussian, skewed, 93
  non-Gaussian, 153, 162
  non-Gaussian additive, 301
  power, 370
  radar glint, 20
  reverberation, 137
  spatially correlated, 246
  target radiated, 258
  thermal, 518
  under-ice ambient, 81, 96, 154
  white, 72
  Wiener, 254
noisy time delay, 304
noncordic, 483
non-Gaussian, 27, 51, 93, 117
  environments, 93
nonlinear
  least squares, 205
  prefilter, 300
  processor, 93
  signal processing, 253
nonparametric, 4
  penalty, 4
numerical
  integration, 302
  linear algebra, 478
  weather forecasting, 184
NUWES, 261
Nyquist rate, 142, 218

O

object
  logical picture, 462
  relational, 462
ocean turbulence, 260
on-target, 330
operator, Laplace-Beltrami, 180, 185

optimal
  control, 256
  designs, 145
  detection, 133
  jamming strategy, 522
  resource extraction, 271
  test, 143
order statistics, 203
outlier, 20, 193
  additive, 26
  innovations, 26
overcontainment, 365ff
oversampling factor, 494

P

page, 461
parallelism, 478, 507
parametric, partially, 107
parametric-select, 4
parser
  minimum distance, 378
  modified Earley, 379
partially parametric, 107
partitioning, 161
  adaptive, 162
pattern
  description language, 375
  grammar, 375
  recognition, syntatic, 376
  syntatic, 379
PDA, 343, 353
PDF, 366
pel, 411ff
perfect data association, 360
periodogram, 19, 70ff, 261
phase velocity, 245
phasing vector, 475
physical micro-picture, 467
physically realizable, 62ff
picture
  blurred support, 468
  logical, 467
  object, 467
  object set, 467
  physical, 462, 467
  query set, 468
  tree, 468
pixel, 380, 387, 405, 421, 427, 439, 445ff, 464

value set, 462
plane wave
  equal-energy, 330
  single propagating, 330
  three dimensional, 244
platykurtic, 212
point
  breakdown, 23
  leverage, 24
  set, 462
Point Barrow, 200
polarity, 402
polynomial
  characteristic, 63ff
  trigonometric, 182
polynyas, 199
positivity, 180ff
power, noise, 370
PPSP, 482
prediction tree, 193
predictive theory, 8
prefilter, 405
prefiltering, 493
preliminary screening, 192
preprocessing, 12, 486
prewhitening, 246
primitive, 378
probabilistic data
  association, 343, 353
  joint, 354ff
probability
  detection, 41, 144, 342, 365
  false alarm, 41, 342, 365
  Markov transition, 438
  maximum a posteriori, 386
  of error, 523
  prior, 424
  transition, 390, 423, 427
  worst case error, 524
problem
  ambiguity, 294
  coverage, 280
  data association, 355
  detection, 330
  direct sensing, 180
  resolution, 330
procedure, nonsupervised
  learning, 381
process
  ARCH, 27
  broadband acoustic, 212
  broadband, non-Gaussian, 212

class-conditional, 425
discrete time, 169
first order, Markov, 429
Gaussian, 134
Gaussian random, 293, 247
Gauss-Markov, 146
general linear, 26
infinitely divisible, 135
innovation, 256
input time-correlated
    Gaussian, 298
Karhunen, 59
label, 422
Markov, 298
mean-square continuous, 109
non-Gaussian, 211
non-Gaussian continuous, 260
non-Gaussian linear, 51
non-Gaussian stochastic, 108
observation, 169
Poisson point, 133ff, 142
renewal, 168
semi-Markov, 298
shot noise, 134
signal, 168
Singer correlated Gaussian, 299
stationary stochastic, 141
stochastic, 422
vector stochastic, 113
weakly stationary, 60, 169
Wiener, 108, 113, 255
processing
    digital image, 435
    Doppler, 492
    image, 386, 461
    on-board, 503
    parallel, 511
    real-time signal, 507
    sequential raster, 387
processor
    real-time SAR, 498
    systolic array, 482
    VLSI-SAR, 500
program, slicer, 406
programmable parallel signal
    processor, 482
programming, dynamic, 386ff
property
    convergence, 181
    Markov, 172
propagation, underwater acoustic, 82
pseudorandom, 517

Q

QUAD, 379
quadratic programming, 182
quantile, 204
quantization, 141
    error, 247
quantized, 412
quantizer
    Gaussian assumption, 412
    Laplacian, 412ff
quartile
    lower, 208
    matching, 205
    upper, 208

R

radiances, 181
radiosonde, 184
Radon-Nikodym derivative, 109
random disk, 275
random sampling designs, 142
range compression, 493
range error, 293
rate of convergence, 150
Rayleigh fading, 83
Rayleigh resolution limit 331
receiver
    hard decision, 415
    operating characteristic, 342
    quadrature, 346
    soft decision, 414ff
    square-law combining, 529
recognition
    character, 380
    decision-theoretic, 378
    speech, 380
recursiveness, 474
redescending psi, 32
redundancy, 510
    hardware, 508
    stochastic, 511
region
    annular, 282
    instability, 346
regression, 145
    bounded-influence, 23
    linear, 511ff
    logistic, 181

Index

regular control/data flow, 482
relational graph, 375
remote sensing, 179, 491
reproducing kernel Hilbert
  space, 111
residuals, non-Gaussian AR, 27
residuated, 400
resistance, 20ff
resolution, spatial, 486
resolution, spectral, 435
reverberation, 133, 224, 258
  bottom, 228
  surface, 227
RKHS, 111
RMS, 344
robot, 385
robust
  least square, 205
  pre-whitening, 29
  smoother-cleaner, 29
  tracker, 311
robustness, 20ff
  decision-theoretic, 24
  efficiency, 20ff
  min-max, 20ff
  qualitative, 20ff
  quantitative, 24
ROC, 342ff
root-mean-square error, 344

S

S, 192
sample, sub-195
sampling, 141
  delegate, 195
  designs, 144
  inverse density, 195
  mid-point, 149
  nonperiodic, 141
  periodic, 141
  random designs, 142
SAR, 446, 454, 491
scatterometry, 492
scheduling, 481
Scott Polar Research Institute, 200
sculptured exponential, 204
SDD, 411
sea ice, 199, 491

search
  detection, 268
  multistate target, 269
  optimal, 265
SEASAT, 442, 465, 491ff
  SAR, 446, 491
SEGMENT, 407
segmentation, 377, 385ff, 438, 507
  digital image, 421
  flat, 406
  image, 386ff, 399, 450
  maximum likelihood, 388, 391
  robust, 388
  threshold, 391
  time series, 421
seismic activity, 260
self-adaptivity, 474
semantic information, 376
semi-invariant, 53
sensor
  array, 243
  displacement, 292ff
  space-based, 385
sequence
  Bernoulli, 168
  Gaussian, 45
  m-dependent stationary, 44
  non-Gaussian, 45
  Pareto, 56
set, partially ordered, 399
shadow, 273
shallow water ranging, 301
short memory, 7
signal
  bandlimited, 248
  coherent plane wave, 243
  estimation, 167
  Gaussian, 133
  jamming, 248
  narrowband, 83, 287
  narrowband Gaussian, 83
  non-Gaussian, 107
  non-Gaussian broadband, 88
  processing, 380, 386
  processing, sonar, 486
  sinusoidal, 82, 287
signaling, binary antipodal, 412
signal-to-noise ratio, 144, 302, 342,
  348, 368ff
  array, 330
simulation, 55, 93ff, 175, 184, 302, 415ff

singular value decomposition, 67ff, 120, 478
skewness, 94, 155, 205, 230
skewness- kurtosis plane, 93
skewness-matching, 207
Slepian's lemma, 45
SLICER, 407
smoother-cleaner, 29, 33
smoothing, 25
snapping shrimp, 224, 233
SNR, 144, 161, 302, 342, 348
soft decision demodulation, 411
solver, Toeplitz system, 479
sonar, 258, 260, 313, 475, 479
sound refraction, 262
source
    discrete interfering, 357
    persistent interfering, 354
    separation, 331
space
    feature, 375
    Hilbert, 179
    measure, 266
    Sobolev, 182
spaceborne SAR, 491
spacing, keel, 200
spatially coherent, 243
speckle, 446
SPECTRA, 380
spectral estimation, 4, 7
spectrum, 19, 141
    analysis, 119
    continuous intensity, 446
    maximum entropy, 436
    spread, 517
speech recognition, 380
spline, 179
    cross validated, 179
    polynomial smoothing, 180
    smoothing on a sphere, 181
    thin plate, 181
splitting level, 401
spread spectrum, 517
star-shaped target curve, 274
state
    observable, 253
    unobservable, 253
stationary, 60
    target problem, 265
stereology, 181
stochastically continuous, 137
structure, tree, 461
Student's t, 31

subsample, 195
substractor, 464
support picture, 463
surface, ambiguity, 365
surveillance, 270, 491
SVD, 67ff, 120, 478
syntax, 375
synthetic aperture
    radar, 446, 454, 491
system
    bilinear, 253ff
    dynamic, 354
    linear, 314
    LPI, 517
    remote space-based, 385
    Toeplitz, 479
systolic array, 485

T

target
    dynamics, 258, 297
    moving, 265
    state dynamics, 341
technique, structured retrieval, 469
technology, nMOS, 499
template, 390
    matching, 377
test
    D'Agostino's, 224
    Fisher, 246
    hypothesis, 362
    Kendall rank correlation, 224
    Kolmogorov-Smirnov, 205, 224
    optimal, 143
    Pearson's skew, 224
    Wilcoxon rank sum, 224
texture feature, 436
theory, predictive, 8
thinout, 405
tile, 461
time redundancy, 483
time series
    Gaussian, 261
    Gaussian, stationary, 5
time-bandwidth
    product, 335, 347, 365
time-delay, 258
time/Doppler differences, 358
time-resolution product, 162
TNF, 256
TOC, 342ff

Index

Toeplitz form, 477
tomography, computerized, 181
towed array, 314
track, 358
tracker
  operating characteristic, 342
  robust, 311
tracking, 258
  adaptive range, 297
  geometry, 300
  multitarget ocean, 341, 353
  rigid body, 259
  target, 297
transformation, nonlinear, 161
transient, 248
transition, 423
translation, 403
  invariant, 403
tree
  minimum spanning, 379
  structure, 461
triangular window, 477
TRP, 162
TRUS, 379
two dimensional DCT, 412
two dimensional FFT, 435

U

underwater acoustic, 223
upper bound vector, 463
U.S.S. GURNARD, 200

V

validation gate, 355, 360
variance, 155
vector, feature, 375
vision, computer, 385ff
VLSI, 473, 491, 496
volume, effective cell, 345
vorticity, 182ff

W

Washburn upper bound, 268
waveform, 380
wavefront
  array processor, 482ff
  curvature, 290
wavenumber, 119
  component, 244
  spectrum, 246
weak Cramer class, 60
weakly continuous, 26
weakly harmonizable, 60
wind field, 185
windowing, data, 13
Woods Hole, 260

Y

Yule-Walker equations, 7, 67